Practical Algebra

Wiley Self-Teaching Guides teach practical skills from accounting to astronomy, management to mathematics. Look for them at your local bookstore.

Other Science and Math Wiley Self-Teaching Guides:

Science

Astronomy: A Self-Teaching Guide, by Dinah L. Moche

Basic Physics: A Self-Teaching Guide, Second Edition, by Karl F. Kuhn

Biology: A Self-Teaching Guide, by Steven D. Garber

Math

All the Math You'll Ever Need: A Self-Teaching Guide, by Steve Slavin

Geometry and Trigonometry for Calculus, by Peter H. Selby

Practical Algebra: A Self-Teaching Guide, Second Edition, by Peter H. Selby and Steve Slavin

Quick Algebra Review: A Self-Teaching Guide, by Peter H. Selby and Steve Slavin

Quick Arithmetic: A Self-Teaching Guide, by Robert A. Carman and Marilyn J. Carman

Quick Business Math: A Self-Teaching Guide, by Steve Slavin

Quick Calculus: A Self-Teaching Guide, Second Edition, by Daniel Kleppner and Norman Ramsey

Statistics: A Self-Teaching Guide, by Donald Koosis

Practical Algebra

A Self-Teaching Guide

Third Edition

Bobson Wong
Larisa Bukalov
Steve Slavin

Published by Jossey-Bass
A Wiley Brand
111 River St
Hoboken, New Jersey 07030
www.josseybass.com

Jossey-Bass books and products are available through most bookstores. To contact Jossey-Bass directly call our Customer Care Department within the U.S. at 800-956-7739, outside the U.S. at 317-572-3986, or fax 317-572-4002.

Wiley publishes in a variety of print and electronic formats and by print-on-demand. Some material included with standard print versions of this book may not be included in e-books or in print-on-demand. If this book refers to media such as a CD or DVD that is not included in the version you purchased, you may download this material at http://booksupport.wiley.com. For more information about Wiley products, visit www.wiley.com.

Library of Congress Cataloging-in-Publication Data is Available:

ISBN 978-1-119-71540-5 (Paperback)
ISBN 978-1-119-71541-2 (ePDF)
ISBN 978-1-119-71543-6 (ePub)

Cover Image and Design: Paul McCarthy

SKY10033205_032222

CONTENTS

ACKNOWLEDGMENTS

Writing a book is hard. Writing a book while teaching full-time during a pandemic is even harder. Fortunately, many people helped make this edition of *Practical Algebra* a reality. Our students' mathematical struggles and joys over the years inspired us to write this book. Conversations with our colleagues at Bayside High School and Math for America helped us develop many of the ideas and techniques we describe. Bayside students Juliana Campopiano and Queena Yue helped us proofread the text. The team at Desmos designed a powerful online graphing tool that we used to create the graphs in this book. The staff at John Wiley & Sons (especially Pete Gaughan, Christine O'Connor, Riley Harding, Julie Kerr, and Mackenzie Thompson) have been especially patient and supportive. Larry Ferlazzo introduced us to publishing math books, opening up countless opportunities. Finally, our spouses and children deserve special mention for tolerating our conversations about this book, peppering us with mathematical questions over the years, and helping to keep our work in perspective.

INTRODUCTION

What is algebra? You may associate it with solving equations such as $2x + 7 = 19$. However, both the history of algebra and the way that it's taught today show that algebra is much more. For thousands of years, people solved algebraic problems without symbols such as x and $+$. By the 9th century, people including the Persian mathematician Muḥammad ibn Mūsā al-Khwārizmī had popularized the idea of using an *algorithm* (a set of well-defined instructions) to determine unknown quantities. In fact, the word *algebra* comes from the Arab word *al-jabr*, meaning "the reduction," from the title of al-Khwārizmī's most famous mathematical text, *Kitāb al-jabr wa al-muqābalah*. Symbolic notation didn't become widespread until European mathematicians such as François Viète and René Descartes developed them in the 16th and 17th centuries. Nowadays, algebra courses include not just equations but also functions (the special rules that define mathematical relationships) and real-world modeling with statistics. In short, today's algebra students must know how to understand word problems, make and interpret graphs, create and solve equations, and draw appropriate conclusions from data.

Not surprisingly, algebra makes many people nervous. Maybe you recall endless drills and elaborate procedures from years ago. Perhaps you're a middle school or high school student who's intimidated by the high level of abstract reasoning that's required. If so, you're not alone. We understand how you feel! For many years, we've taught all levels of high school math, so we have a lot of experience working with diverse learners. This book contains concrete strategies that help our students succeed. We strongly believe that people can get better at math if they have access to the right tools.

We wrote this book as a general introduction to algebra. We assume that you're familiar with basic arithmetic (adding, subtracting, multiplying, and dividing numbers) and fractions. If you're *not* comfortable with these topics, don't worry—we briefly review them in Chapters 1 and 2. Even if you *are* comfortable with them, we suggest that you look through these chapters anyway. We explain why these ideas work and how they're related to the algebraic ideas we discuss later on.

Each chapter in this book is divided into sections, with model examples and tips. At the end of each section, you'll find several exercises to help you practice and apply your skills. These exercises include what we call Questions to Think About (open-ended questions designed to help you think about important concepts) as well as dozens of word problems. Each chapter has a test with multiple-choice and open-ended questions. The solutions to all exercises and chapter tests are located at the end of each chapter.

As you work through this book, you'll see some important ideas about algebra that we emphasize:

- **Algebra is a language.** We believe that many people find algebra intimidating because the words and symbols we use, such as polynomial, a_n, and $f(x)$, literally look like a different language. In addition, we don't just *write* math, we also *read* and *speak* it. In the Reading and Writing Tips, we discuss how to write and pronounce mathematical symbols as well as how to use them in context. We also include a glossary of mathematical terms and symbols in the back of the book.
- **Algebra should make sense.** We believe that algebra should be taught in a way that makes sense. In our experience, part of the reason why so many people suffer from math anxiety is that they see it as a collection of disjointed and confusing tricks. Throughout this book, we use techniques (such as the area model for multiplication) that relate to other mathematical topics, such as geometry and statistics. By making these connections, you can extend what you learned in one situation to another context, which will strengthen your mathematical skills and boost your confidence!
- **Algebra requires pictures.** As we taught during the pandemic, we had to adjust our instruction. We couldn't be with our students in person, so they often had to teach themselves more independently. Incorporating graphs, tables, diagrams, and other images into our teaching helped our students make sense of math. Since this book is a *self-teaching* guide, we've included many visual strategies throughout this book.
- **Algebra requires technology.** Calculators, computers, and other technology aren't just shortcuts for menial computations. They are now required for today's complex modeling tasks. Using technology helps us to see patterns more efficiently. Since each of these tools has vastly different user instructions, we don't include specific instructions for each device. Instead, we include Technology Tips that apply *no matter what device you're using*.
- **Algebra is a human endeavor.** We believe that algebra should not be perceived as a set of rigid rules developed by a select group of people. In fact, as we note throughout this book, many mathematical concepts were developed in different cultures around the world over thousands of years. (We mention some of the more interesting stories in the Did You Know? callouts.) In addition, we recognize that making mistakes is a natural part of doing math. In the Watch Out! callouts, we point out many of the common errors that we've seen students make over the years so that you can avoid them!

We hope that as you work through this book, you'll find that algebra can be less intimidating and more meaningful than you originally thought.

— Bobson Wong and Larisa Bukalov

1 BASIC CONCEPTS

In this chapter, we review some of the concepts that students are typically expected to know before learning algebra. Although we don't have the space to fully develop these concepts, we point out some common mistakes and other important points that you should keep in mind. Even if you think that you know these topics, we recommend that you work through this chapter.

1.1 Addition, Subtraction, Multiplication, Division

Throughout this book, we use visual models to represent mathematical ideas. One important model is a **number line**, a line on which each point represents exactly one number. The numbers always increase from left to right. To show the scale, numbers are marked off at equal intervals. We draw an arrow at the end to indicate that the numbers extend infinitely in that direction.

Positive numbers, which we indicate with a + in front of the number, are numbers greater than 0. **Negative numbers**, which we indicate with a − in front of the number, are numbers less than 0. The word **sign** refers to the property of being positive or negative. The term **signed numbers** refers to numbers and their signs. Numbers that don't have a sign in front of them are understood to be positive.

On a horizontal number line (Figure 1.1), positive numbers lie to the right of 0, and negative numbers lie to the left of 0:

Did You Know?

The idea of positive numbers, negative numbers, and 0 may seem obvious to us now, but they actually developed around the world over thousands of years. By the 3rd century BCE, the Chinese were using counting rods of different colors to represent positive and negative numbers in their calculations. The 7th-century Indian mathematician Brahmagupta described rules in terms of "fortunes" (positive numbers) and "debts" (negative numbers). Ancient societies understood the concept of nothing ("we have *no* water"), but many cultures, such as the Egyptians, Romans, and Greeks, created complex mathematics without 0. The use of 0 didn't fully develop until the 5th century CE in India.

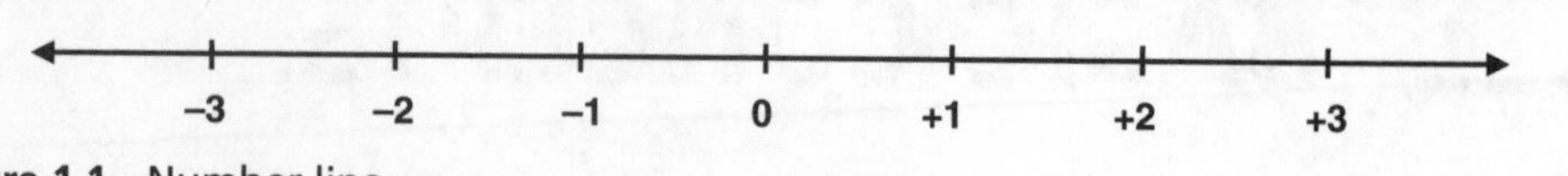

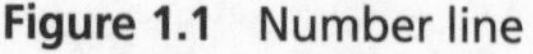
Figure 1.1 Number line

The **absolute value** of a number is its distance from 0 on a number line. Since the absolute value represents distance, it is always positive (unless we're talking about 0, which has an absolute value of 0). We use vertical bars to indicate absolute value. We read $|+2|$ as "the absolute value of positive two." For example, $|+15|$ is equal to 15, $|-15|$ is equal to 15, and $|0|$ is equal to 0. Two numbers that are the same distance from 0 on the number line but have different signs, such as +2 and −2, are **opposites**. Zero is an exception—the opposite of 0 is itself.

In math, we have four basic **operations** (mathematical processes performed on quantities to get a result): addition, subtraction, multiplication, and division. When we combine quantities with operations, we make an **expression**, such as $5 + 3$ and $|+15| - 4$.

Watch Out!

We use the + and − symbols to represent *both* addition and subtraction *and* the sign of a number.

- When + and − represent the sign of a number (which only occurs *before* a number), we read + as "positive" and − as "negative." We *never* put a space between the symbol and the number, so "negative 5" would be written −5, never − 5.
- When + and − represent addition or subtraction (which only occurs *between* two numbers), we read + as "plus" and − as "minus," and we put 1 space before and after the symbol. For example, 4 + 5, which is read as "4 plus 5," means 5 *is added to* 4 to get a sum of 9.

The + and − symbols can represent both operations and signs in the same mathematical sentence. For example, +5 − −3 is read "positive 5 minus negative 3," not "plus 5 minus minus 3." Sometimes, we put parentheses around signed numbers to separate them from the addition or subtraction symbols, so we write +5 − −3 as (+5) − (−3). The parentheses are not pronounced.

You may recall working with number lines in elementary school. In this book, we also use squares to model signed numbers because they enable us to represent far more complicated ideas that we need to work with in algebra. To represent +1, we use a square whose area is +1. To represent −1, we use a square whose area is −1. (Don't worry about what a square with a negative area actually "means"—it's just a model!) A square with area +1 and a square with area −1 have a total area of 0. We call this pair a **zero pair**. We can group zero pairs into rectangles (think of them as "jumbo packs" of +1 or −1 squares) and use them to add signed numbers, as shown in Example 1.1:

Example 1.1 Evaluate −40 + 54.

Solution: When we **evaluate** an expression, we perform mathematical calculations to get a single number.

$-40 + 54$	
$= -40 + 40 + 14$	Split +54 into +40 and +14.
$= (-40 + 40) + 14$	Group the −40 and +40 together to make 40 zero pairs.
$= 14$	The remainder is 14, the final answer.

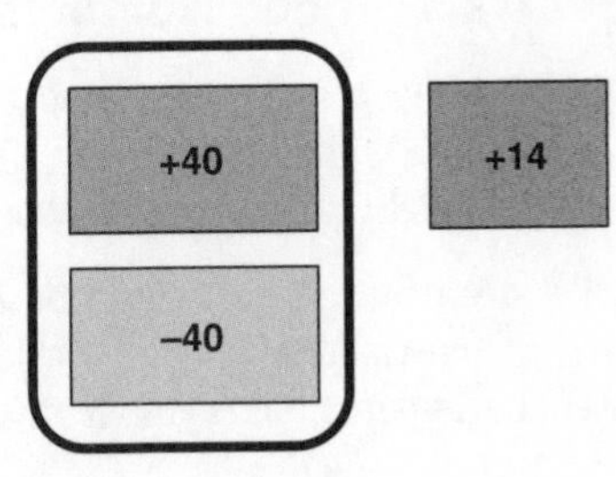

In this example, we use the = symbol (which is called an equal sign and read "equals" or "is equal to"). The equal sign means that the expression on its left has the same value as the expression on its right. A mathematical statement containing an equal sign is called an **equation**. To make your work easier to read, do one part of the calculation at a time and write each step on a different line, starting each line with the equal sign.

Watch Out!

One common mistake when writing several equations on one line is to ignore the meaning of the equal sign. For example, when evaluating 2 + 3 + 4, some students write: 2 + 3 = 5 + 4 = 9. This "run-on" equation implies that 2 + 3, 5 + 4, and 9 are all equal, which isn't what we meant! Instead, write the following:

$2 + 3 + 4$	
$= 5 + 4$	The sum of 2 and 3 is 5.
$= 9$	The sum of 5 and 4 is 9.

How to Add Signed Numbers

1. Determine the number with the larger absolute value.
2. Form zero pairs with the number with the smaller absolute value.
3. The remainder is the final answer, called the **sum**.

Addition and subtraction undo each other. For example, $5 + 3 - 3$ equals 5. More formally, we say that addition and subtraction are **inverse operations**. This means that when we apply inverse operations on a number, the result is the original number. We can think of subtraction in terms of addition.

How to Subtract Signed Numbers

- To subtract a *positive* number, add a negative number with the same absolute value, so $5 - 3 = 5 + (-3)$. The result, called the **difference**, is 2. (This models real-world behavior—adding debt lowers your net worth.)
- To subtract a *negative* number, add a positive number with the same absolute value, so $5 - (-3)$ is the same as $5 + 3$. The difference is 8. (This also models real-world behavior—removing debt raises your net worth.)

Example 1.2 illustrates how these rules work.

Example 1.2 Evaluate $(-30) - (-46)$.

Solution:

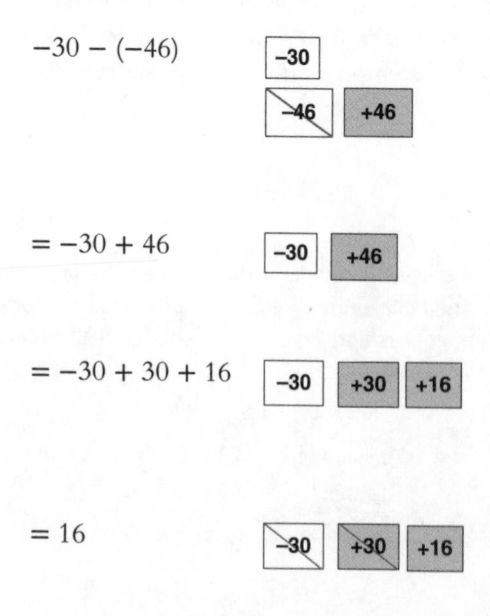

1. From −30, we remove 46 negative unit squares. Since we don't have any more negative unit squares, add 46 zero pairs (46 negative and 46 positive unit squares) and remove the 46 negative unit squares.
2. After removing the 46 negative unit squares, we have 30 negative and 46 positive unit squares.
3. To determine what we have left in step 2, we separate the +46 into 30 positive and 16 positive unit squares (since 46 − 30 equals 16).
4. The 30 negative and 30 positive unit squares make 30 zero pairs, which add up to 0, leaving 16 positive unit squares.

Technology Tip

Many calculators have different buttons for subtraction and negative numbers. Often, the subtraction button is located next to the buttons for addition, multiplication, and division. To change the sign of an entry, they have a button labeled +/- or (–), where the - symbol on the button is shorter than the – symbol. Some calculators will return an error if you try to use the subtraction button to change the sign of a number, so be careful! In contrast, most software applications and mathematical websites don't differentiate between the negative and subtraction symbols, so entering 5 – –3 will result in the correct answer of 8.

When we multiply numbers, we add groups of the same size.

How to Multiply Signed Numbers

- Multiply the absolute values of the **factors** (the numbers being multiplied).
- If we multiply two numbers with *different* signs, the result (called the **product**) is negative.
- If we multiply two numbers with the *same* sign, the result is positive.

We write the multiplication of 3 times 2 using one of these methods:

- with × between the numbers, as in 3×2
- with · between the numbers, as in $3 \cdot 2$
- with parentheses around one or both numbers, as in $(3)(2)$, $3(2)$, or $(3)2$

We recommend not using the × symbol in algebra because it can easily be mistaken for the letter x, which has a special meaning that we discuss in Chapter 3.

Since the area of a rectangle is the product of its length and width, then we can use rectangles to represent multiplication. This idea dates back thousands of years to ancient Mesopotamia, Greece, and the Middle East. Unfortunately, we can't realistically show the difference between positive and negative dimensions with a rectangle, so we label the dimensions with the appropriate signed numbers and use the multiplication rules that we described above to find the correct sign of the product.

Example 1.3 Represent $(-10)(-5)$ using a rectangle and evaluate the result.

Solution: We can represent this as a rectangle whose dimensions are -10 and -5:

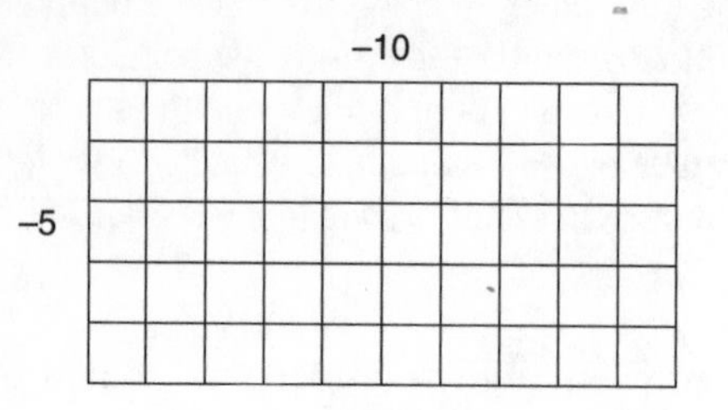

NOTE: We can also think of this as removing 5 groups of -10, which results in a net increase of 50.

Here are some special cases of multiplication:

- **Any number multiplied by 0 equals 0.** For example, 4 groups of 0 is still 0.
- **A number multiplied by 1 equals itself.** We can explain this conceptually by noticing that 1 group of 4 is just that number, so $4(1) = 4$.
- **A number multiplied by -1 equals its opposite.** For example, $4(-1) = -4$ and $-4(-1) = 4$.

When we multiply a number by itself several times, we say that we raise it to a power. For example, we say that $2(2)(2)(2)$ equals 2^4, which we read as "two to the fourth power" or "two to the fourth." In this case, 2 is called the **base** (the number being multiplied) and 4 is the **power** or **exponent** (the number of times the base is being multiplied). The exponent is written above and to the right of the base. The term power refers to both the number 16 (what 2^4 equals) as well as the exponent 4.

Here are some special cases for powers:

- A number raised to the first power is equal to the number, so $2^1 = 2$.
- A number raised to the second power is **squared**, so 4^2 can be read as "four squared," "four to the second power," or "four to the second." (We get this term from the formula for the area of a square, which is the length of its edge multiplied by itself.)
- A number raised to the third power is **cubed**, so 4^3 can be read as "four cubed," "four to the third power," or "four to the third." (We get this term from the formula for the volume of a cube, which is the length of its edge multiplied by itself three times.)

A positive number raised to a positive power is always positive. We can surround the base with parentheses, so $(3)^4$, $(+3)^4$, and 3^4 all represent the same quantity.

When we raise negative numbers to a power, we always surround the base with parentheses, so we write $(-3)(-3)(-3)(-3)$ as $(-3)^4$. If we raise a negative number to powers that are counting numbers, we see an interesting pattern in the signs:

- $(-3)^1 = -3$
- $(-3)^2 = (-3)(-3) = +9$
- $(-3)^3 = (-3)(-3)(-3) = -27$
- $(-3)^4 = (-3)(-3)(-3)(-3) = +81$
- $(-3)^5 = (-3)(-3)(-3)(-3)(-3) = -243$

We summarize this pattern as follows:

- A negative number raised to an odd power is negative.
- A negative number raised to an even power is positive.

Reading and Writing Tip

We have no easy way to express in words the difference between numbers like -3^4 and $(-3)^4$, since both can be pronounced as "negative 3 to the fourth power." We find that people pronounce $(-3)^4$ as "the quantity negative 3 to the fourth power," "parentheses negative 3 to the fourth power," or "negative 3 *(pause)* to the fourth power." This is an example of a situation where mathematical symbols can communicate ideas more clearly and succinctly than words. Pay careful attention to how mathematical symbols are written. In the same way that a missing comma can completely change the meaning of a sentence, missing parentheses can give you a different answer!

When we divide numbers, we separate into groups of equal size. Multiplication and division are inverse operations.

How to Divide Signed Numbers

- Divide the absolute values of the number that we divide (called the **dividend**) and the number that we divide by (called the **divisor**).
- If we divide two numbers with different signs, the result (called the **quotient**) is negative.
- If we divide two numbers with the same sign, the result is positive.

Division is often associated with fractions. A **fraction** is a quantity consisting of one number (called the **numerator**) divided by a nonzero number (called the **denominator**).

We write division using one of these methods:

- Using the ÷ symbol (called the division symbol) between the two numbers, such as $8 \div 2$
- Using the / symbol between the two numbers, all written on the same line, such as 8/2
- Writing one number on top of the other and separating the two with a **fraction bar** (sometimes called a **vinculum**), such as $\frac{8}{2}$

All of these division examples are read as "8 divided by 2." In this book, we prefer using the fraction bar to represent division. It provides the clearest separation of the quantities in division and minimizes the use of parentheses in more complicated mathematical statements.

When we use the ÷ or / symbol, the dividend appears before the symbol and the divisor appears after it. When we use a fraction bar, the dividend appears above it and the divisor appears below it. Figure 1.2 shows the terms associated with division:

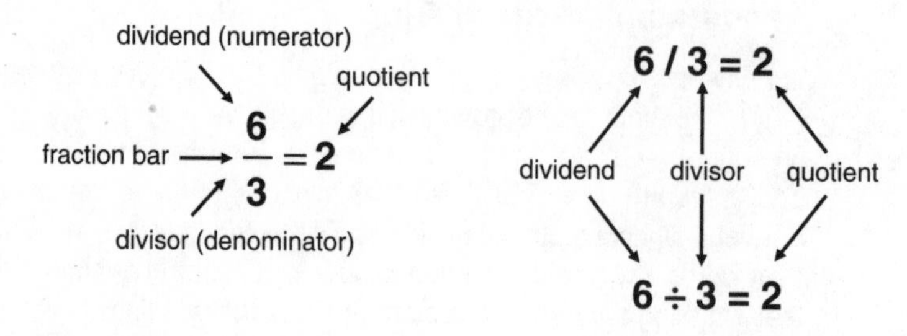

Figure 1.2 Terms associated with division.

Some special cases of division deserve special attention:

- **Any number divided by 1 equals itself.** For example, $\frac{8}{1}$, which means 8 divided into 1 group, equals 8.
- **Any number divided by 0 is meaningless.** Another way of saying this is that a fraction can never have a denominator equal to 0. For example, $\frac{8}{0}$ has no meaning since there is no number that when multiplied by 0 would give a product of 8 (this would have to be true since multiplication and division are inverse operations).
- **Any nonzero number divided by itself equals 1.** For example, $\frac{8}{8} = 1$.
- **Zero divided by a nonzero number equals 0.** The fraction $\frac{0}{8}$ equals 0.
- **The reciprocal of a number is 1 divided by that number.** The reciprocal of 8 is $\frac{1}{8}$.
- **The product of a number and its reciprocal is 1.** For example, $8\left(\frac{1}{8}\right) = 1$.

Example 1.4 Represent $\frac{+6}{-3}$ using a rectangle and evaluate the quotient.

Solution: Using a rectangle, we can think of this as dividing a rectangle that has an area of +6 and a side length of −3. Using the rules for dividing two numbers with different signs, we conclude that the quotient must be negative, so the answer is −2.

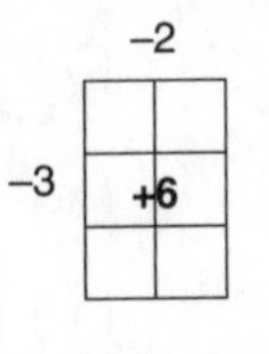

Table 1.1 summarizes the steps for operations with signed numbers:

Table 1.1 Operations with signed numbers.

Operation	Steps	Example
Addition	**1.** Determine the number with the larger absolute value.	$-40+54$ $=-40+40+14$
	2. Form zero pairs with the number with the smaller absolute value.	$=0+14$
	3. The remainder is the final answer.	$=14$
Subtraction	• To subtract a *positive* number, add a negative number with the same absolute value.	$5-3$ $=5+(-3)=2$
	• To subtract a *negative* number, add a positive number with the same absolute value.	$5-(-3)$ $=5+3=8$
Multiplication	• Multiply the absolute values of each number being multiplied.	$10(2)=20$
	• If we multiply two numbers with *different* signs, the result is negative.	$10(-2)=-20$
	• If we multiply two numbers with the *same* sign, the result is positive.	$-10(-2)=20$
Division	• Divide the absolute values of each number being multiplied.	$\frac{20}{2}=10$
	• If we divide two numbers with *different* signs, the result is negative.	$\frac{-20}{2}=-10$
	• If we divide two numbers with the *same* sign, the result is positive.	$\frac{-20}{-2}=10$

One final note: although understanding the rules for operations with signed numbers is important, you can always use technology to help you with these calculations.

Exercises

Write the pronunciation of each expression.

1. $-8-(-12)$
2. $(+1)-(+3)$
3. $(+6)+(-4)$
4. $(+7)(+15)$
5. $(-2)(-32)$
6. $(-6)^4$

Evaluate each expression:

7. $|+7.5|$
8. $|-3|$
9. $|-889|$
10. $(+8)+(+5)$
11. $(+50)+(-10)$
12. $(-30)+(+20)$
13. $(-9)-(+5)$
14. $(-20)-(+30)$
15. $(-100)-(-40)$
16. $(+3)(-5)$
17. $(-3)(-7)$
18. $(-6)(+2)$
19. $(+2)^3$
20. $(+6)^2$
21. $(-7)^2$
22. $-(-1)^2$
23. $\frac{+15}{-3}$
24. $\frac{0}{+9}$
25. $\frac{-25}{-5}$

Questions to Think About

26. What is the difference between the words "plus" and "positive" as they are used in math?
27. What are two examples of real-life quantities that could be modeled by adding negative numbers?
28. What are two examples of real-life quantities that could be modeled by subtracting negative numbers?
29. Is $(-1{,}234{,}567{,}890{,}000{,}000)^{999}$ positive or negative? Explain.

1.2 Order of Operations

For many years, mathematicians didn't have a standardized set of rules for operations. When math education became more widespread in the 19th century, textbooks codified rules for what became known as the **order of operations**, the order in which mathematical operations should be performed. The growing popularity of computers in the last few decades has made the need for a standardized order of operations even more important.

In this book, we use the following convention (Figure 1.3) for order of operations:

1. **GROUPING:** First, evaluate everything surrounded by parentheses, brackets, fraction bars, absolute value symbols, and other grouping symbols, working from the innermost symbols outwards. To group numbers inside parentheses, we use square brackets or another set of parentheses: $2-(3-[5-1])$ or $2-(3-(5-1))$.

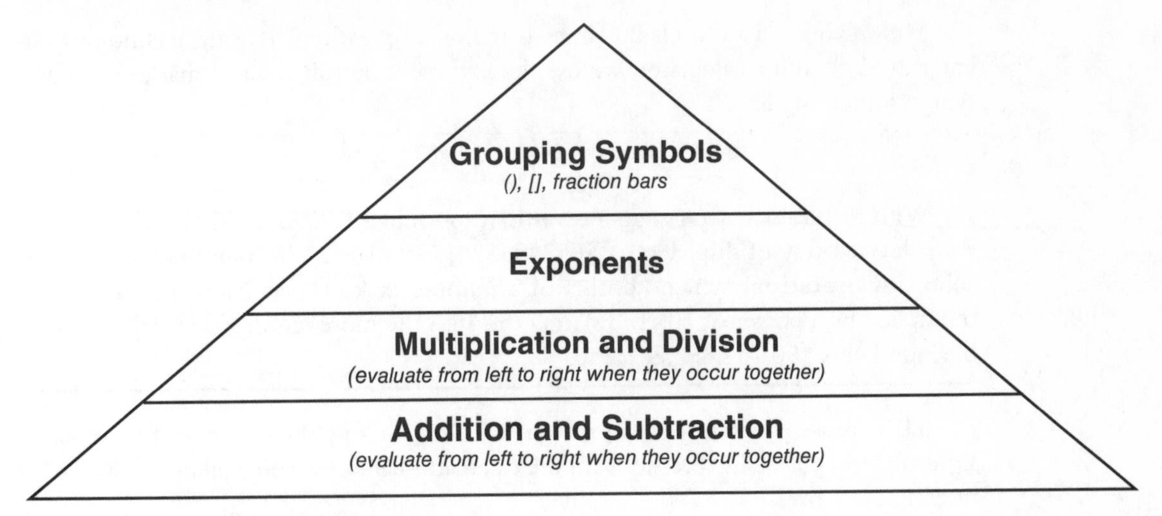

Figure 1.3 Order of operations.

2. **EXPONENTS:** Next, evaluate exponents.
3. **MULTIPLICATION/DIVISION:** Next, when multiplication and division occur together, evaluate them *left to right*.
4. **ADDITION/SUBTRACTION:** Finally, when addition and subtraction occur together, evaluate them *left to right*.

Watch Out!

Some textbooks use the mnemonic PEMDAS (which stands for Parentheses, Exponents, Multiplication, Division, Addition, Subtraction) to remember the order of operations. We recommend you avoid using PEMDAS since it implies that multiplication should be done before division and addition before subtraction. If you prefer using a mnemonic, we suggest PEMA (Parentheses, Exponents, Multiplication, Addition). Unfortunately, PEMA doesn't include division and subtraction, so you'll have to remember which operations are inverse operations (multiplication and division, addition and subtraction) and perform them left to right.

Some problems can be solved more easily using a calculator:

Example 1.5 Evaluate $2(-5)^{20}$.

Solution: The order of operations tells us that we need to evaluate the exponent first (in this case, the twentieth power) *before* the multiplication. The parentheses around -5 indicate that it, not -10 (which is the product of 2 and -5), is the base that is being raised to the twentieth power.

Multiplying -5 by itself 20 times is tedious, so we prefer using technology. To enter $2(-5)^{20}$ into a calculator, we use the exponent button (usually marked ^ or x^y), typing something like:

With technology, we get an answer that looks like 1.9073486328125E14. This is your device's way of displaying $1.9073486328125 \times 10^{14}$. This number is written in **scientific notation**, which consists of a number at least 1 and less than 10 that is multiplied by a power of 10. Translating this into the more familiar standard notation, this number is 190,734,863,281,250.

Just because you can solve a problem with technology doesn't mean that it's easy! Entering complicated expressions on the calculator can be quite challenging, as shown in Example 1.6:

Example 1.6 Evaluate $\frac{12-3+4(3)^2}{33-|-3(1+5)|}$.

Solution: The fraction bar acts as a grouping symbol, separating the numerator from the denominator, so we calculate each separately. In the denominator, we work with grouping symbols from the inside out.

$$\frac{12-3+4(3)^2}{33-|-3(1+5)|}$$

$$=\frac{12-3+4(9)}{33-|-3(1+5)|}$$ Evaluate the power in the numerator (3 is squared).

$$=\frac{12-3+36}{33-|-3(1+5)|}$$ Evaluate 4(9) in the numerator.

$$=\frac{45}{33-|-3(1+5)|}$$ Evaluate addition and subtraction from left to right.

$$=\frac{45}{33-|-3(6)|}$$ Add inside the parentheses in the denominator.

$$=\frac{45}{33-|-18|}$$ Multiply inside the absolute value symbols, which work as grouping symbols.

$$=\frac{45}{33-18}$$ Calculate the absolute value of -18.

$$=\frac{45}{15}$$ Subtract in the denominator.

$$=3$$ Divide 45 by 15.

To enter this problem in the calculator, use the fraction tool if your device has one. This will put the numerator on top of the denominator, separated by a fraction bar. Doing so separates your numerator and denominator more clearly and reduces the likelihood of mistakes. Otherwise, you'd have to enter this problem on one line using many parentheses, which can be very confusing!

Technology Tip

When entering numbers into the calculator, keep the following in mind:

- Use the +/– key, not the addition and subtraction keys, to make a number positive or negative.
- Enter parentheses carefully. For example, $(3)(2+1)^2 = 27$, but $(3(2+1))^2 = 81$.
- Use the exponent key (usually marked ^ or x^y) to enter powers.
- To enter fractions, use the fraction tool if your calculator has one. (See Example 1.6.)

Exercises

Evaluate each expression.

1. $6 - 3 + 2$

2. $1 - 7 + 8$

3. $12 \div 2(3)$

4. $8 \div 2(2+2)$

5. $7 + 5(8)$

6. $9 + (-3)^2$

7. $2(-4)^2$

8. $4(-2)^3$

9. $9 + |1 - 5|^2$

10. $\dfrac{4 + (5 - (4 - 7))}{|-2|}$

11. $\dfrac{2^3 3^2}{5 - 1 + 4}$

12. $\left(\dfrac{5(16+4)}{16 - 2(3)}\right)^2$

Questions to Think About

13. Use the order of operations to explain why $3(4)^2$ is 48 and not 144.

14. Use the order of operations to explain why $5(1)^2$ is 5 and not 25.

15. Use the order of operations to explain why -4^2 is negative and not positive. (HINT: -4^2 can be rewritten as $(-1)(4)^2$.)

1.3 Sets and Properties of Numbers

In math, we work with different groups, or sets, of numbers (Figure 1.4):

- **Counting numbers** are the numbers we use to count: 1, 2, 3, 4, and so on.
- **Whole numbers** are the counting numbers and zero: 0, 1, 2, 3, 4, and so on.

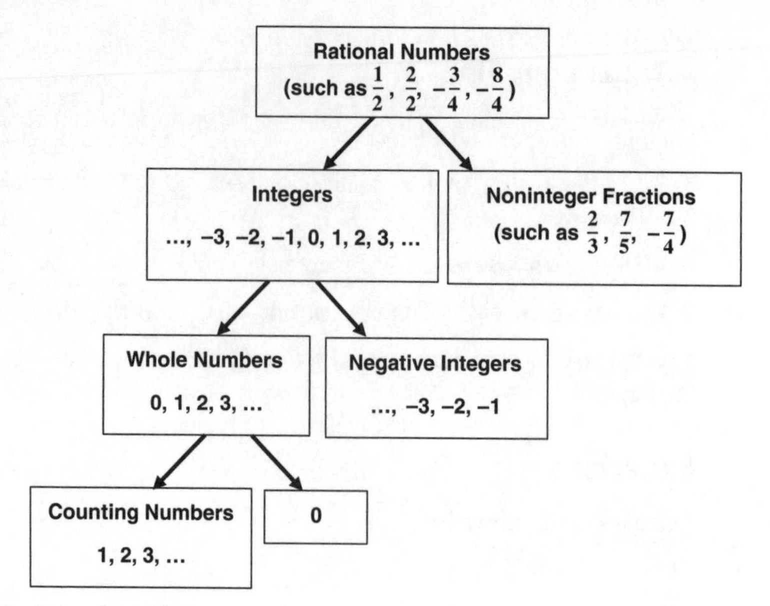

Figure 1.4 Sets of numbers.

- **Integers** are the whole numbers and their opposites: …, −3, −2, −1, 0, 1, 2, 3, …
- **Rational numbers** are numbers that can be expressed as an integer divided by a nonzero integer.

Example 1.7 Is every integer a whole number? Explain.

Solution: No. Whole numbers are the counting numbers and 0 (0, 1, 2, 3, …). Integers are the whole numbers and their opposites (…, −3, −2, −1, 0, 1, 2, 3, …), and negative integers are *not* whole numbers.

Example 1.8 Is every whole number a rational number? Explain.

Solution: The whole numbers are the counting numbers and 0: 0, 1, 2, 3, and so on. The rational numbers are numbers that can be represented as an integer divided by a nonzero integer. Every whole number can be represented as itself divided by 1. Thus, every whole number is a rational number.

Table 1.2 summarizes important properties of numbers, some of which we have already mentioned:

In this table, a, b, and c are **variables**—letters or other symbols that represent quantities that can change in value. We will discuss another important property that relates to addition and multiplication in Chapter 3.

Table 1.2 Properties of Numbers

Property	Description	Symbols	Example
Commutative property of addition	Numbers may be added in any order without changing the result.	$a + b = b + a$	$3 + 4 = 4 + 3$
Commutative property of multiplication	Numbers may be multiplied in any order without changing the result.	$a(b) = b(a)$	$4(3) = 3(4)$
Associative property of addition	Numbers may be grouped in any way for addition without changing the result.	$a + (b + c) = (a + b) + c$	$3 + (4 + 5) = (3 + 4) + 5$
Associative property of multiplication	Numbers may be grouped in any way for multiplication without changing the result.	$a(bc) = (ab)c$	$3(4 \cdot 5) = (3 \cdot 4)5$
Additive inverse property	A number added to its opposite equals 0.	$a + (-a) = 0$	$4 + (-4) = 0$
Additive identity property	A number added to 0 is unchanged.	$a + 0 = a$	$4 + 0 = 4$
Multiplicative inverse property	A number multiplied by its reciprocal equals 1.	$a\left(\frac{1}{a}\right) = 1$	$4\left(\frac{1}{4}\right) = 1$
Multiplicative identity property	A number multiplied by 1 is unchanged.	$a(1) = a$	$4(1) = 4$

Exercises

Determine whether the following statements are true or false. Explain your answer.

1. Every integer is a counting number.
2. Every whole number is an integer.
3. Zero is a whole number but not a counting number.
4. Every counting number is positive.
5. Every whole number is positive.
6. Every integer is either positive or negative.
7. Every rational number is an integer.
8. The number 5 is a rational number.

9. Every whole number is rational.

10. Every rational number can be written as a fraction.

11. Every whole number can be written as a fraction.

12. Every quotient is a rational number.

Questions to Think About

13. Explain why subtracting two counting numbers does not always result in a counting number.

14. How are fractions different from rational numbers?

15. Is subtraction commutative? Explain your answer.

CHAPTER 1 TEST

1. Which number is a counting number?

   (A) 0 (B) −1 (C) 1 (D) $\frac{1}{2}$

2. Which operation is the inverse of multiplication?

   (A) addition (B) subtraction (C) squaring (D) division

3. If a positive number is multiplied by a negative number, the result

   (A) is always positive.
   (B) is always negative.
   (C) can be positive or negative.
   (D) is always 0.

4. According to the order of operations, to calculate $1 + (3 - (6 - 4))^2$, which step must be done first?

   (A) $6 - 4$ (B) 1^2 (C) $1 + 3$ (D) $3 - 6$

5. Which number is equivalent to $(-1)^{15}$?

   (A) +1 (B) −1 (C) +15 (D) −15

6. What does the absolute value of a number represent?

   (A) the directed distance between any two points on a number line
   (B) the sum of two numbers on a number line

(C) the positive distance between any two points on a number line

(D) a number's distance from 0 on a number line

7. Which statement about $-\frac{3}{1}$ is correct?

   (A) It represents a rational number and a fraction, but not an integer.

   (B) It represents a rational number, a fraction, and an integer.

   (C) It represents an integer and a fraction but not a rational number.

   (D) It represents a fraction, but neither a rational number nor an integer.

8. If a negative number is subtracted from a positive number, which statement is correct?

   (A) The result is always positive.

   (B) The result is always negative.

   (C) The result can be positive or 0.

   (D) The result can be positive, negative, or 0.

9. Which of the following is equivalent to adding a negative number?

   (A) subtracting a positive number with the same absolute value

   (B) adding a positive number with the same absolute value

   (C) subtracting a negative number with the same absolute value

   (D) adding a positive number with a different absolute value

10. Is every integer a whole number? Explain.

11. Is 0 rational? Explain.

12. Write the pronunciation of $3^4 + |-8| - (+5)$.

13. Use rectangles to calculate $(-50) - (+40)$.

14. Calculate $(12 - (3 + 1))^2 - 8$.

15. Calculate $\frac{13 - 8 + 1}{2(6 - 7)} + 1$.

CHAPTER 1 SOLUTIONS

1.1. 1. "negative 8 minus negative 12"

2. "positive 1 minus positive 3"

3. "positive 6 plus negative 4"

4. "positive 7 times positive 15"

5. "negative 2 times negative 32"

6. "the quantity negative 6 to the fourth power"

7. 7.5	12. −10	17. +21	22. −1
8. 3	13. −14	18. −12	23. −5
9. 889	14. −50	19. 8	24. 0
10. 13	15. −60	20. 36	25. 5
11. 40	16. −15	21. 49	

26. "Plus" is an operation that shows that two numbers are being added, while "positive" is a characteristic of one number.

27. Answers may vary. Examples include adding debt (which lowers the net worth) or adding ice cubes to a drink (which lowers its temperature).

28. Answers may vary. Examples include cancelling debt (which increases the net worth) or removing ice cubes from a drink (which raises its temperature).

29. Negative. A negative number raised to an odd power is negative.

1.2.

1. 5	4. 16	7. 32	10. 6
2. 2	5. 47	8. −32	11. 9
3. 18	6. 18	9. 25	12. 100

13. The order of operations tells us to calculate powers before multiplication. We calculate 4^2, which equals 16, before multiplying it by 3 to get 48. In contrast, $(3(4))^2 = 12^2 = 144$.

14. The order of operations tells us to calculate powers before multiplication. We calculate 1^2, which equals 1, before multiplying it by 5 to get 5. In contrast, $(5(1))^2 = 5^2 = 25$.

15. The order of operations tells us to calculate powers before multiplication. We calculate 4^2, which equals 16, before multiplying it by -1 to get -16.

1.3.

1. False. Some integers are negative or 0, and every counting number is positive.
2. True. Integers are whole numbers and their opposites.
3. True. Whole numbers are the counting numbers and 0.
4. True. The counting numbers are 1, 2, 3, 4, ... , all of which are greater than 0.
5. False. Zero is a whole number but is neither positive nor negative.
6. False. Zero is an integer but is neither positive nor negative.
7. False. Some rational numbers, such as $\frac{1}{2}$, are not integers ($\frac{1}{2}$ lies between the integers 0 and 1 on the number line).
8. True. The number 5 can be written as the quotient of 5 and 1, or $\frac{5}{1}$.
9. True. Every whole number can be written as the quotient of itself and 1, such as $\frac{5}{1}$.
10. True. A rational number consists of an integer divided by a nonzero integer, while a fraction consists of any quantity (not necessarily an integer) divided by another.
11. True. A whole number can be written as a fraction with a denominator of 1.
12. False. A noninteger fraction divided by an integer, such as $\frac{1}{2} \div 3$, or $\frac{\frac{1}{2}}{3}$, can be written as a fraction but is not rational (since the numerator $\frac{1}{2}$ is not an integer).
13. If the second number is larger than the first, then the result will be negative and will not be a counting number. For example, $2 - 3$ is -1.
14. A rational number consists of an integer divided by a nonzero integer. A fraction consists of any quantity (not necessarily an integer) divided by a nonzero quantity.
15. Subtraction is not commutative. When the order in which numbers are subtracted is reversed, the sign of the difference is also reversed. For example, $5 - 4 = 1$, but $4 - 5 = -1$.

CHAPTER 1 TEST SOLUTIONS

1. (C)
2. (D)
3. (B)
4. (A)
5. (B)
6. (D)
7. (B)
8. (A)
9. (A)
10. No. Some integers are negative, and whole numbers (0, 1, 2, 3, …) cannot be negative.
11. Yes. Zero may be expressed as the quotient of 0 and a nonzero integer: $\frac{0}{1}$.
12. "Three to the fourth power plus the absolute value of negative 8 minus positive 5."
13. −90

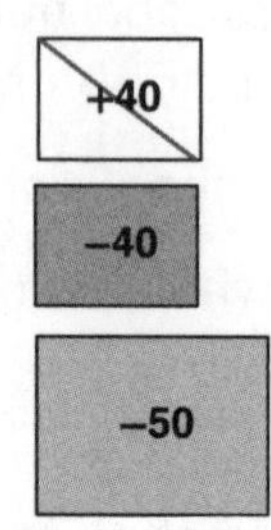

14. 56
15. −2

2 FRACTIONS

Throughout history, many cultures, including the Egyptians, Greeks, Indians, Muslims, Chinese, and Romans, worked with quantities that are not integers. Over time, they developed the idea of fractions. Today, fractions are used throughout math and science. Understanding how fractions behave is critical to being successful in algebra. We find that to succeed in algebra, students don't just need to know *how* to work with fractions but also *why* these rules work.

2.1 Basic Operations

To understand fractions, we start by dividing a quantity into equal parts. For example, if we divide a rectangular pizza into 8 slices, each slice represents $\frac{1}{8}$ (pronounced "one-eighth" or "one over eight") of the rectangle (Figure 2.1):

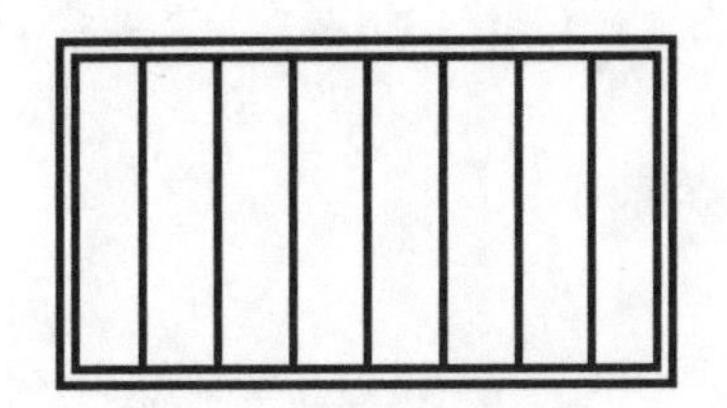

Figure 2.1 Dividing a rectangular pizza into eighths.

In the fraction $\frac{1}{8}$, 1 is the numerator and 8 is the denominator (Figure 2.2):

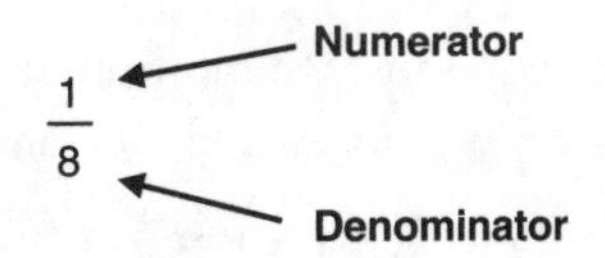

Figure 2.2 Numerator and denominator.

We now review the basic operations with fractions—addition, subtraction, multiplication, and division.

We add or subtract fractions by adding or subtracting their numerators while keeping the denominators unchanged. This rule only works if the fractions have the same denominator (we say that they have **like denominators** or **common**

denominators). For example, if Audrey has $\frac{3}{8}$ of a pizza and Benjamin has $\frac{2}{8}$ of a pizza, then together they have $\frac{3}{8} + \frac{2}{8} = \frac{5}{8}$ of the pizza (Figure 2.3):

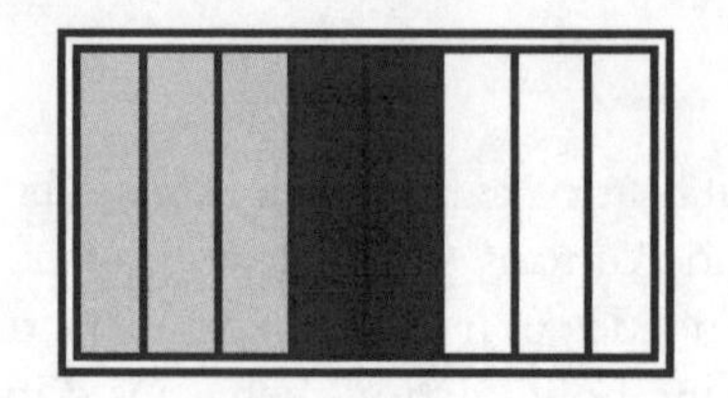

Figure 2.3 Adding fractions.

We summarize this property as follows:

The numerator of the sum or difference of fractions with like denominators is the sum or difference of the numerators. The denominator remains unchanged. In symbols, $\frac{a}{c} + \frac{b}{c} = \frac{a+b}{c}$, where $c \neq 0$.

We discuss adding and subtracting fractions with unlike denominators in Section 2.2.

Example 2.1 Calculate $\frac{3}{7} - \frac{1}{7} + \frac{8}{7}$.

Solution:

$$\frac{3}{7} - \frac{1}{7} + \frac{8}{7}$$

$$= \frac{3-1+8}{7} \qquad \frac{a}{c} + \frac{b}{c} = \frac{a+b}{c}, \text{ where } c \neq 0.$$

$$= \frac{10}{7}$$ Add the numbers in the numerator. (When addition and subtraction are together, evaluate from left to right.)

To multiply fractions, we multiply their numerators and multiply their denominators. For example, if Audrey and Benjamin decide they only want to eat $\frac{3}{4}$ of their $\frac{5}{8}$ of a pizza, then they eat $\frac{3}{4}\left(\frac{5}{8}\right) = \frac{15}{32}$ of the pizza (Figure 2.4):

Figure 2.4 Multiplying fractions.

In other words:

The numerator of a product of fractions equals the product of their numerators. The denominator of a product of fractions equals the product of their denominators. In symbols, $\left(\frac{a}{b}\right)\left(\frac{c}{d}\right) = \frac{ac}{bd}$, where $b \neq 0$ and $d \neq 0$.

Example 2.2 Calculate $\frac{2}{7}\left(\frac{4}{3}\right)$.

Solution:

$\frac{2}{7}\left(\frac{4}{3}\right)$

$= \frac{2(4)}{7(3)}$ — $\left(\frac{a}{b}\right)\left(\frac{c}{d}\right) = \frac{ac}{bd}$, where $b \neq 0$ and $d \neq 0$.

$= \frac{8}{21}$ — Multiply factors together.

Reading and Writing Tip

In Section 1.1, we discussed three different ways to represent multiplication—the × symbol, the ·, and parentheses. When we multiply variables, we often write their letters next to each other with no spaces, so *ab*, *a* · *b*, and (*a*)(*b*) all represent multiplication and are pronounced "*a* times *b*," "the product of *a* and *b*," or "*a*-*b*."

Many times, we need to multiply a fraction by an integer. For example, to multiply 3 by $\frac{1}{8}$, we rewrite 3 as a fraction whose numerator is itself and whose denominator is 1 (recall from Section 1.1 that any number divided by itself equals 1), so $3\left(\frac{1}{8}\right) = \frac{3}{1} \cdot \frac{1}{8} = \frac{3(1)}{1(8)} = \frac{3}{8}$.

Watch Out!

The expression $3\left(\frac{1}{8}\right)$, which is the *product* of 3 and $\frac{1}{8}$ and is read "three times one-eighth," is different from $3\frac{1}{8}$, which is the *sum* of 3 and $\frac{1}{8}$ (equal to $3 + \frac{1}{8}$) and is read "three and one-eighth." The number $3\frac{1}{8}$ is an example of a **mixed number**, which consists of an integer and a **proper fraction** (a fraction whose numerator is less than the denominator).

To divide fractions, we can divide the numerators and divide the denominators to give us, respectively, the numerator and denominator of the quotient. (This shouldn't be surprising since multiplication and division are inverse operations.) For example, $\frac{4}{8} \div \frac{1}{4} = \frac{4}{2}$. However, this rule isn't helpful when the numerators and denominators of the quotient aren't integers, such as $\frac{1}{5} \div \frac{4}{7}$. Fortunately, we have another way to solve division problems. Let's think about this in terms of pizza. Taking 1 part of a pizza that is divided into 2 equal parts is equivalent to taking $\frac{1}{2}$ of the pizza. **In general, for any nonzero number a, dividing by a is equivalent to multiplying by $\frac{1}{a}$,** the reciprocal of a. To find the reciprocal of a fraction, we invert the fraction by switching the positions of the numerator and denominator. Here are some examples:

- The reciprocal of $\frac{2}{3}$ is $\frac{3}{2}$.
- The reciprocal of $-\frac{6}{7}$ is $-\frac{7}{6}$.
- The reciprocal of 5 (which we can rewrite as $\frac{5}{1}$), is $\frac{1}{5}$.
- The reciprocal of $\frac{0}{4}$ does not exist since division by 0 has no meaning.

We can express the rule for division as follows:

To divide two fractions, multiply the first one (the dividend) by the reciprocal of the second one (the divisor). In symbols, $\frac{a}{b} \div \frac{c}{d} = \frac{ad}{bc}$, where $b \neq 0$, $c \neq 0$, and $d \neq 0$.

Example 2.3 Express $\frac{2}{3} \div \frac{5}{7}$ in simplest form.

Solution:

$\frac{2}{3} \div \frac{5}{7}$

$= \frac{2}{3}\left(\frac{7}{5}\right)$ $\quad$ $\frac{a}{b} \div \frac{c}{d} = \frac{ad}{bc}$, where $b \neq 0$, $c \neq 0$, and $d \neq 0$. The reciprocal of $\frac{5}{7}$ is $\frac{7}{5}$.

$= \frac{2(7)}{3(5)}$ $\quad$ $\left(\frac{a}{b}\right)\left(\frac{c}{d}\right) = \frac{ac}{bd}$, where $b \neq 0$ and $d \neq 0$.

$= \frac{14}{15}$ $\quad$ Multiply factors in the numerator and in the denominator.

Exercises

Evaluate each expression.

1. $\frac{5}{8}+\frac{2}{8}$
2. $\frac{18}{13}-\frac{3}{13}$
3. $\frac{45}{73}+\frac{0}{73}$
4. $\frac{15}{28}-\frac{12}{28}+\frac{24}{28}$
5. $\frac{3}{14}-\frac{3}{14}+\frac{3}{14}$
6. $\frac{1}{2}+\frac{2}{2}-\frac{3}{2}+\frac{5}{2}$
7. $\frac{2}{7}\left(\frac{4}{3}\right)$
8. $\frac{5}{11}\left(\frac{1}{9}\right)$
9. $\frac{8}{13}\left(\frac{0}{5}\right)$
10. $\frac{2}{7}\left(-\frac{4}{3}\right)$
11. $-\frac{3}{4}\left(\frac{0}{7}\right)$
12. $-\frac{5}{11}\left(-\frac{9}{4}\right)$
13. $\frac{4}{5}\div\frac{7}{8}$
14. $\frac{1}{5}\div\frac{1}{2}$
15. $\frac{4}{7}\div\frac{21}{8}$

Questions to Think About

16. How is adding fractions similar to adding measurements?
17. Use the properties of multiplying fractions to show that for any nonzero number a, $a\left(\frac{1}{a}\right)=1$.

2.2 Simplifying Fractions

Multiplying the numerator and denominator of a fraction by the same nonzero number results in an **equivalent fraction**, which has the same value as the original fraction. For example, $\frac{3}{4}$ is equivalent to $\frac{6}{8}$.

When a fraction's numerator and denominator have no common factors (recall from Section 1.1 that a factor is one of at least two quantities that multiplied to get a product), we say that the fraction is in **simplest form** or **simplified**. Simplifying fractions allows us to compare them more easily, especially when the numerators and denominators get large. For example, $\frac{3}{4}$ is in simplest form, but $\frac{75}{100}$ is not since 25 is a factor of both 75 and 100. *As a general rule, whenever we get a fraction as a final answer, we express it in simplest form.*

Watch Out!

Some textbooks and websites refer to simplifying fractions as *reducing* fractions since both the numerator and the denominator of a fraction become smaller. We prefer not to use this term since many students mistakenly think that the fraction's value somehow changes.

How to Simplify Fractions

1. Write the prime factorization of the numerator and denominator.
   - **Prime numbers** (2, 3, 5, 7, 11, 13, and so on) have only 2 factors—1 and itself. **Composite numbers** (4, 6, 8, 9, 10, 12, 14, 15, and so on) have more than 2 factors. (The number 1 is considered neither prime nor composite.)
   - **Factoring** is a process in which a number or algebraic expression is written as a product of numbers or expressions. We often factor a quantity by finding one factor and using it as a divisor to find other factors. Thus, factoring is the reverse of multiplication.
   - When talking about factoring, we are usually interested in prime numbers. Typically, we factor a quantity in order to write it as the product of prime factors (what we call its **prime factorization**). To find the prime factorization of a number, divide the number by 2, 3, 5, and other prime numbers until all remaining factors are prime.
2. Identify all factors that the numerator and denominator have in common.
3. Group all remaining factors in the numerator and all remaining factors in the denominator to form the numerator and denominator of the simplified fraction.

Example 2.4 Simplify $\frac{225}{360}$.

Solution: First, we write the prime factorization of 225 and 360. We can draw a **factor tree**, a diagram that shows the process of dividing the number by each prime factor (2, 3, 5, 7, 11, and so on) in order, repeating if possible, until all factors are prime, as shown here:

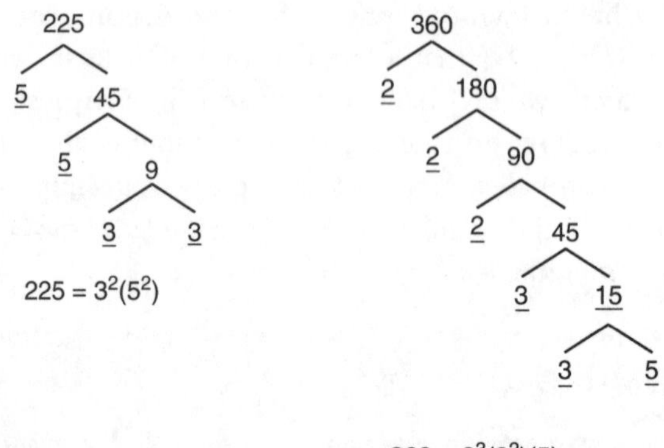

$\frac{225}{360}$	
$= \frac{3(3)(5)(5)}{2(2)(2)(3)(3)(5)}$	Write the prime factorization of the numerator and denominator.
$= \frac{5 \cdot 3(3)(5)}{2(2)(2) \cdot (3)(3)(5)}$	Group common factors in the numerator and denominator: 3(3)(5).
$= \frac{5}{2(2)(2)} \cdot \frac{3(3)(5)}{3(3)(5)}$	Use the property $\left(\frac{a}{b}\right)\left(\frac{c}{d}\right) = \frac{a(b)}{c(d)}$ to separate common factors in the numerator and denominator.
$= \frac{5}{2(2)(2)} \cdot 1$	A nonzero number divided by itself equals 1.
$= \frac{5}{2(2)(2)}$	A number multiplied by 1 equals itself.
$= \frac{5}{8}$	Multiply factors in the numerator and in the denominator.

As you get more experienced with simplifying fractions, you can skip some steps and simply write the following:

$\frac{225}{360}$	
$= \frac{5 \cdot 45}{8 \cdot 45}$	Find the **greatest common factor** (or GCF, the largest number that divides evenly into two or more numbers) of 225 and 360.
$= \frac{5}{8} \cdot \frac{45}{45}$	$\left(\frac{a}{b}\right)\left(\frac{c}{d}\right) = \frac{ac}{bd}$, where $b \neq 0$ and $d \neq 0$.
$= \frac{5}{8}$	A number multiplied by 1 equals itself.

Example 2.5 Express $\frac{14}{35}\left(\frac{5}{16}\right)$ in simplest form.

Solution: Multiplying the numerator and denominator results in big numbers that are difficult to factor. Instead, we can factor the numerator and denominator of each fraction first:

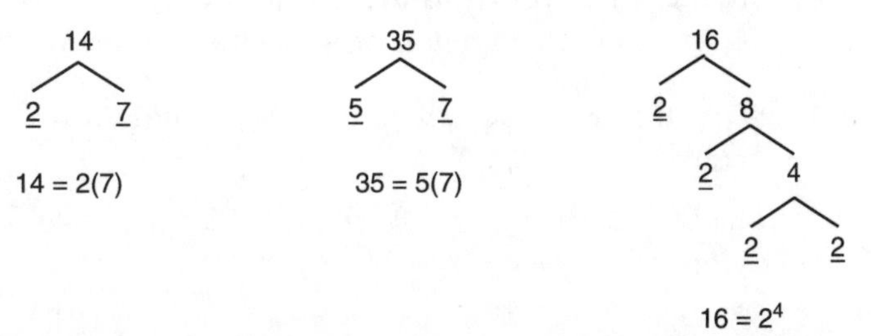

$\frac{14}{35}\left(\frac{5}{16}\right)$

$= \frac{2(7)}{5(7)} \cdot \frac{5}{2(2)(2)(2)}$ Write the prime factorization of each numerator and denominator.

$= \frac{2(5)(7)}{2(2)(2)(2)(5)(7)}$ $\left(\frac{a}{b}\right)\left(\frac{c}{d}\right) = \frac{ac}{bd}$, where $b \neq 0$ and $d \neq 0$.

$= \frac{1 \cdot 2(5)(7)}{2(2)(2) \cdot (2)(5)(7)}$ Group common factors in the numerator and denominator: 2(5)(7).

$= \frac{1}{2(2)(2)} \cdot \frac{2(5)(7)}{2(5)(7)}$ $\left(\frac{a}{b}\right)\left(\frac{c}{d}\right) = \frac{ac}{bd}$, where $b \neq 0$ and $d \neq 0$.

$= \frac{1}{2(2)(2)} \cdot 1$ A nonzero number divided by itself equals 1.

$= \frac{1}{2(2)(2)}$ A number multiplied by 1 equals itself.

$= \frac{1}{8}$ Multiply factors in the numerator and in the denominator.

Watch Out!

When writing equivalent fractions, you may only *multiply* or *divide* by the same number in the numerator and denominator. You may *not* add the same number to or subtract the same number from the numerator and denominator. For example, $\frac{1}{2}$ is equal to $\frac{1(2)}{2(2)}$, which is $\frac{2}{4}$. However, $\frac{1}{2}$ is *not* equal to $\frac{1+1}{2+1}$, which is $\frac{2}{3}$.

Just as we need common units to add or subtract measurements (3 hours + 4 hours = 7 hours, but 3 hours + 4 minutes doesn't equal 7 hours), we need common denominators to add or subtract fractions with unlike denominators. For example (Figure 2.5), to add $\frac{1}{2} + \frac{1}{3}$, we need a common denominator of $2(3) = 6$. When we divide each fraction into 6 equal parts, we see that $\frac{1}{2} + \frac{1}{3} = \frac{3}{6} + \frac{2}{6} = \frac{5}{6}$.

We can always find a common denominator by multiplying the denominators of the fractions. However, doing so can result in large numbers. To keep our numbers smaller and easier to manage, we use the *smallest* possible common denominator. We call this the **least common denominator**, often abbreviated LCD.

To find the least common denominator, do the following:

1. Write the prime factorization of each denominator, using exponents for repeated factors.
2. The LCD is the product of all prime factors, using the highest exponent for each factor.

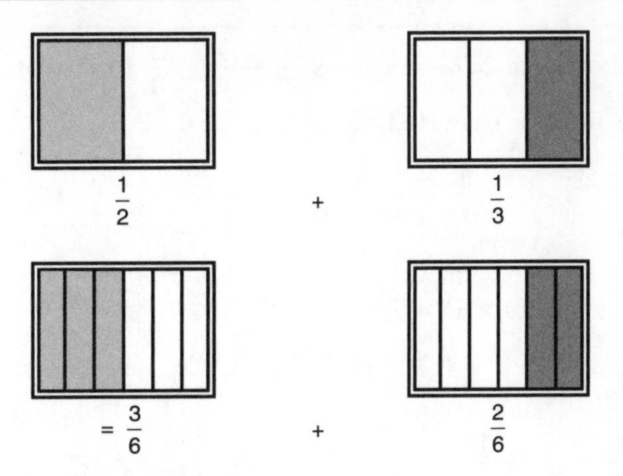

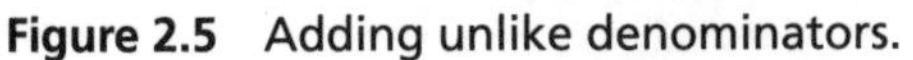

Figure 2.5 Adding unlike denominators.

Example 2.6 Find the least common denominator of $\frac{1}{60}$, $\frac{2}{27}$, and $\frac{5}{18}$.

Solution: First, we draw a factor tree for each denominator.

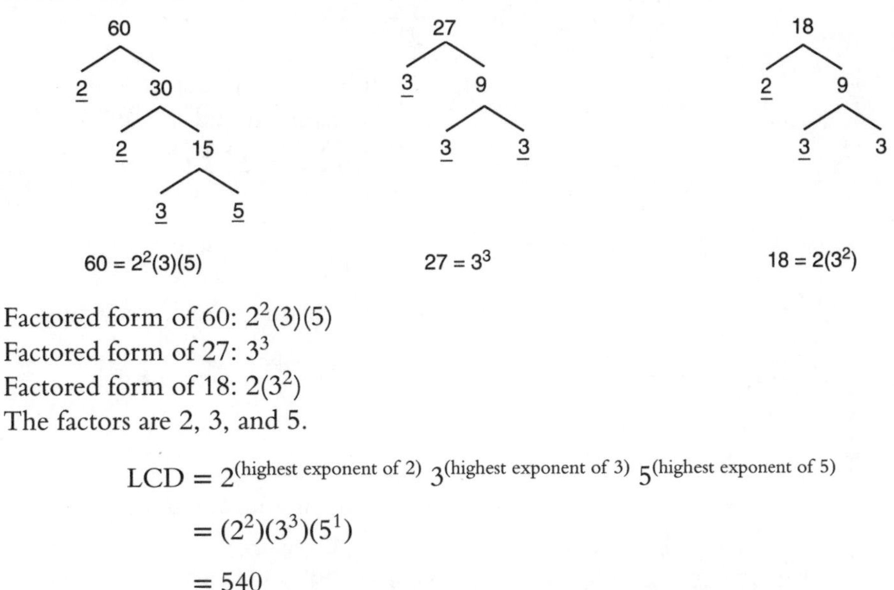

Factored form of 60: $2^2(3)(5)$
Factored form of 27: 3^3
Factored form of 18: $2(3^2)$
The factors are 2, 3, and 5.

$$\begin{aligned}\text{LCD} &= 2^{(\text{highest exponent of }2)}\ 3^{(\text{highest exponent of }3)}\ 5^{(\text{highest exponent of }5)}\\ &= (2^2)(3^3)(5^1)\\ &= 540\end{aligned}$$

To add or subtract fractions with unlike denominators, do the following:

1. Find the LCD of all fractions.
2. Rewrite each fraction as an equivalent fraction with the LCD as the denominator.
3. Add or subtract the numerators and keep the LCD as the denominator.
4. If necessary, simplify the resulting fraction.

Example 2.7 Express $\frac{2}{45} - \frac{3}{35} + \frac{13}{21}$ in simplest form.

Solution: First, we find the LCD of the 3 fractions.

$$45 = 3^2(5)$$
$$35 = 5(7)$$
$$21 = 3(7)$$
$$\text{LCD} = 3^{(\text{highest exponent of 3})}\, 5^{(\text{highest exponent of 5})}\, 7^{(\text{highest exponent of 7})}$$
$$= (3^2)(5)(7) = 315$$

Then:

$\frac{2}{45} - \frac{3}{35} + \frac{13}{21}$ — The LCD is 315.

$= \frac{2}{45}\left(\frac{7}{7}\right) - \frac{3}{35}\left(\frac{9}{9}\right) + \frac{13}{21}\left(\frac{15}{15}\right)$ — Use your calculator to divide the LCD by each denominator to find the additional factor for each numerator and denominator: $\frac{315}{45} = 7$, $\frac{315}{35} = 9$, $\frac{315}{21} = 15$.

OR use the prime factorizations of the LCD and denominators to find the additional factor: $\frac{(3^2)(5)(7)}{(3^2)(5)} = 7$, $\frac{(3^2)(5)(7)}{(5)(7)} = 9$, $\frac{(3^2)(5)(7)}{(3)(7)} = 15$

$= \frac{2(7)}{45(7)} - \frac{3(9)}{35(9)} + \frac{13(15)}{21(15)}$ — $\left(\frac{a}{b}\right)\left(\frac{c}{d}\right) = \frac{ac}{bd}$, where $b \neq 0$ and $d \neq 0$.

$= \frac{14}{315} - \frac{27}{315} + \frac{195}{315}$ — Multiply factors in each numerator and denominator.

$= \frac{14 - 27 + 195}{315}$ — $\frac{a}{c} + \frac{b}{c} = \frac{a+b}{c}$, where $c \neq 0$.

$= \frac{182}{315}$

$= \frac{2(7)(13)}{3(3)(5)(7)}$ — Factor the numerator and factor the denominator.

$= \frac{2(13) \cdot 7}{3(3)(5) \cdot 7}$ — Group common factors in the numerator and denominator: 2(3).

$= \frac{2(13)}{3(3)(5)} \cdot \frac{7}{7}$ — $\left(\frac{a}{b}\right)\left(\frac{c}{d}\right) = \frac{ac}{bd}$, where $b \neq 0$ and $d \neq 0$.

$= \frac{2(13)}{3(3)(5)} \cdot 1$ — A nonzero number divided by itself equals 1.

$= \frac{2(13)}{3(3)(5)}$ — A nonzero number multiplied by 1 equals itself.

$= \frac{26}{45}$ — Multiply factors in the numerator and in the denominator.

Table 2.1 summarizes the methods for adding, subtracting, multiplying, and dividing fractions that are discussed in this chapter.

Table 2.1 Operations with fractions.

Property	In Words	In Symbols	Example
Addition	The numerator of the sum of fractions with like denominators is the sum of the numerators. The denominator remains unchanged. If necessary, write equivalent fractions using the least common denominator (LCD).	$\frac{a}{c}+\frac{b}{c}=\frac{a+b}{c}$, $c \neq 0$	$\frac{3}{8}+\frac{2}{8}=\frac{5}{8}$
Subtraction	The numerator of the difference of fractions with like denominators is the difference of the numerators. The denominator remains unchanged. If necessary, write equivalent fractions using the least common denominator (LCD).	$\frac{a}{c}-\frac{b}{c}=\frac{a-b}{c}$, $c \neq 0$	$\frac{3}{8}-\frac{2}{8}=\frac{1}{8}$
Multiplication	The numerator of a product of fractions equals the product of their numerators. The denominator of a product of fractions equals the product of their denominators.	$\frac{a}{b}\left(\frac{c}{d}\right)=\frac{ac}{bd}$, $b \neq 0, d \neq 0$	$\frac{2}{3}\left(\frac{5}{7}\right)=\frac{10}{21}$
Division	The numerator of a quotient of fractions equals the quotient of their numerators. The denominator of a quotient of fractions equals the quotient of their denominators. To divide fractions, multiply the first fraction by the reciprocal of the divisor.	$\frac{a}{b}\div\frac{c}{d}=\frac{ad}{bc}$ $b \neq 0, c \neq 0, d \neq 0$	$\frac{42}{55}\div\frac{6}{11}=\frac{7}{5}$ $\frac{2}{3}\div\frac{5}{7}=\frac{2(7)}{3(5)}=\frac{14}{15}$

Technology Tip

When possible, use technology to work with numerical fractions:

- To add, subtract, multiply, or divide fractions, enter the expressions into your device. It will express the answer in simplest form.
- To simplify a fraction or convert a decimal into a fraction, use your device's fraction tool.
- To find the LCD, use your device's least common multiple function. On many devices, you can enter lcm (), which means least common multiple, and put the numbers between the parentheses, separated by commas. For example, lcm (4, 6, 8) = 24.

Exercises

Express each fraction in simplest form.

1. $\frac{16}{32}$ **2.** $\frac{45}{24}$ **3.** $\frac{68}{8}$ **4.** $\frac{90}{75}$ **5.** $\frac{144}{350}$ **6.** $-\frac{96}{128}$

Evaluate each expression.

7. $\frac{5}{36}+\frac{6}{18}$

8. $\frac{9}{6}-\frac{4}{9}$

9. $\frac{7}{20}+\frac{1}{30}$

10. $\frac{3}{8}-\frac{7}{24}$

11. $\frac{5}{40}+\frac{11}{32}+\frac{5}{20}$

12. $\frac{1}{54}+\frac{2}{81}-\frac{1}{18}$

13. $\frac{4}{7}\left(\frac{14}{10}\right)$

14. $\frac{10}{2}\left(\frac{5}{12}\right)$

15. $\frac{3}{6}\left(\frac{5}{10}\right)$

16. $\frac{21}{32}\left(\frac{44}{49}\right)$

17. $\frac{81}{22}\left(\frac{11}{27}\right)$

18. $\frac{100}{39}\left(\frac{169}{200}\right)$

19. $\frac{2}{9}\div\frac{50}{18}$

20. $\frac{1}{3}\div\frac{7}{30}$

21. $\frac{9}{25}\div\frac{81}{10}$

22. $\frac{1}{2}+\frac{2}{3}\div\frac{4}{9}$

23. $\frac{4}{7}\div\frac{14}{6}\cdot\frac{2}{5}$

24. $\frac{5}{2}-\frac{1}{4}\div\frac{5}{6}$

Questions to Think About

25. Ariel added $\frac{3}{11}$ and $\frac{5}{11}$ to get $\frac{8}{22}$. Explain in words the mistake that Ariel made.

26. How is the method for multiplying fractions different from the method for adding or subtracting fractions with like denominators?

CHAPTER 2 TEST

1. The expression $\frac{5}{2}+\frac{8}{2}$ equals

   (A) $\frac{13}{2}$ (B) $\frac{13}{4}$ (C) $\frac{40}{4}$ (D) $\frac{40}{2}$

2. Which fraction is in simplest form?

   (A) $\frac{38}{2}$ (B) $\frac{3}{13}$ (C) $-\frac{18}{24}$ (D) $\frac{21}{35}$

3. What is the prime factorization of 240?

   (A) 12(20) (B) 4(3)(4)(5) (C) 2(3)(5) (D) $2^4(3)(5)$

4. What is the reciprocal of -3?

   (A) 0 (B) $\frac{1}{3}$ (C) 3 (D) $-\frac{1}{3}$

5. Which two fractions are equivalent?

   (A) $\frac{8}{12}$ and $\frac{6}{9}$ (B) $\frac{3}{5}$ and $\frac{5}{3}$ (C) $\frac{7}{12}$ and 7 (D) $-\frac{3}{4}$ and $\frac{3}{4}$

6. If $n \neq 0$, which fraction equals n?

   (A) $\frac{n}{1}$ (B) $\frac{n}{n}$ (C) $\frac{1}{n}$ (D) $-\frac{n}{1}$

7. If $w \neq 0$, $x \neq 0$, $y \neq 0$, and $z \neq 0$, which expression always equals $\frac{w}{x} \div \frac{y}{z}$?

   (A) $\frac{w \div y}{xz}$ (B) $\frac{w \div y}{x \div z}$ (C) $\frac{wy}{xz}$ (D) $\frac{xy}{wz}$

8. Which statement about adding fractions with like denominators is always true?

   (A) The numerator of the sum equals the sum of the numerators.

   (B) The denominator of the sum equals the sum of the denominators.

   (C) The numerator of the sum equals the sum of the denominators.

   (D) The denominator of the sum equals the sum of the numerators.

9. If $w \neq 0$, $x \neq 0$, $y \neq 0$, and $z \neq 0$, which expression always equals $\frac{w}{x} - \frac{y}{z}$?

(A) $\frac{w-y}{xz}$ (B) $\frac{w+z}{x+y}$ (C) $\frac{w-y}{x-z}$ (D) $\frac{wz-xy}{xz}$

10. Simplify $\frac{60}{72}$.

11. Determine the least common denominator of $\frac{3}{12}$, $\frac{4}{15}$, and $\frac{5}{9}$.

12. Write the pronunciation of $4\left(\frac{1}{12}\right)$.

13. Express $\frac{25}{10} - \frac{8}{75}$ in simplest form.

14. Express $\frac{1}{3} + \frac{2}{15} \div \frac{4}{9}$ in simplest form.

15. Fill in the blanks in the following explanation of the calculations for $\frac{3}{25}\left(\frac{15}{6}\right)$.

$\frac{3}{25}\left(\frac{15}{6}\right)$

$= \frac{3(3)(5)}{5(5)(2)(3)}$ ____________________

$= \frac{3 \cdot (3)(5)}{5(2) \cdot (3)(5)}$ ____________________

$=$ ________ $\left(\frac{a}{b}\right)\left(\frac{c}{d}\right) = \frac{ac}{bd}$, where $b \neq 0$ and $d \neq 0$.

$= \frac{3}{5(2)} \cdot 1$ ____________________

$= \frac{3}{5(2)}$ ____________________

$=$ ________ Multiply factors in the numerator and in the denominator.

CHAPTER 2 SOLUTIONS

2.1. 1. $\frac{7}{8}$

2. $\frac{15}{13}$

3. $\frac{45}{73}$

4. $\frac{27}{28}$

5. $\frac{3}{14}$

6. $\frac{5}{2}$

7. $\frac{8}{21}$

8. $\frac{5}{99}$

9. 0

10. $-\frac{8}{21}$

11. 0

12. $\frac{45}{44}$

13. $\frac{32}{35}$

14. $\frac{2}{5}$

15. $\frac{32}{147}$

16. The numerators of fractions can be added only when the denominators are the same, just as measurements can be added only when the units are the same.

17. $a\left(\frac{1}{a}\right) = \frac{a}{1}\left(\frac{1}{a}\right) = \frac{a \cdot 1}{1 \cdot a} = \frac{a}{a} = 1.$

2.2. 1. $\frac{1}{2}$

2. $\frac{15}{8}$

3. $\frac{17}{2}$

4. $\frac{6}{5}$

5. $\frac{72}{175}$

6. $-\frac{3}{4}$

7. $\frac{17}{36}$

8. $\frac{19}{18}$

9. $\frac{23}{60}$

10. $\frac{1}{12}$

11. $\frac{23}{32}$

12. $-\frac{1}{81}$

13. $\frac{4}{5}$

14. $\frac{25}{12}$

15. $\frac{1}{4}$

16. $\frac{33}{56}$

17. $\frac{3}{2}$

18. $\frac{13}{6}$

19. $\frac{2}{25}$

20. $\frac{10}{7}$

21. $\frac{2}{45}$

22. 2

23. $\frac{24}{245}$

24. $\frac{11}{5}$

25. Ariel should have left the denominators unchanged instead of adding them.

26. When multiplying fractions, we perform the operation (in this case, multiplication) on the denominators. When adding or subtracting fractions with the same denominator, we add or subtract the numerators but leave the denominators unchanged.

CHAPTER 2 TEST SOLUTIONS

1. (A)
2. (B)
3. (D)
4. (D)
5. (A)
6. (A)
7. (B)
8. (A)
9. (D)
10. $\frac{5}{6}$
11. 180
12. "Four times one-twelfth" or "the product of four and one-twelfth."

13. $\frac{359}{150}$

14. $\frac{19}{30}$

15. $= \frac{3(3)(5)}{5(5)(2)(3)}$ Factor the numerator and factor the denominator.

$= \frac{3 \cdot (3)(5)}{5(2) \cdot (3)(5)}$ Factor common factors from the numerator and denominator.

$= \frac{3}{5(2)} \cdot \frac{3(5)}{3(5)}$ $\left(\frac{a}{b}\right)\left(\frac{c}{d}\right) = \frac{ac}{bd}$, where $b \neq 0$ and $d \neq 0$.

$= \frac{3}{5(2)} \cdot 1$ A nonzero number divided by itself equals 1.

$= \frac{3}{5(2)}$ A number multiplied by 1 equals itself.

$= \frac{3}{10}$ Multiply factors in the numerator and in the denominator.

3 LINEAR EQUATIONS

What number added to 7 equals 12? For thousands of years, humans have asked and answered questions like this in different ways. At first, people relied on words to describe mathematical statements. Over time, people like Diophantus in Egypt (3rd century), Brahmagupta in India (7th century), Abū al-Ḥasan ibn ʿAlī al-Qalaṣādī in Spain (15th century), and François Viète in France (16th and 17th centuries) developed the mathematical symbols that we now use in algebra. In this chapter, we use our modern notation to solve the most basic equations, called linear equations. (In Chapter 7, we'll see why they have this name.)

3.1 Solving One-Step Equations

Nowadays, we express a question like, "What number added to 7 equals 12?" with an equation:

What number	added to	7	equals	12?
x	$+$	7	$=$	12

Here are some important concepts related to equations like this:

- The letter x is a variable, which represents all numbers that make the equation true.
- Our goal is to **solve** the equation, meaning we find the values of the variable that make the equation true. These values are called the **roots** or **solutions** to the equation.
- Equations are solved by writing **equivalent equations**, which have the same solutions as the original. The goal is to write simpler equations by isolating the variable on one **side** of an equation (the expression that appears on one side of the equal sign) and a **constant** (a quantity that has a fixed value, like 7) on the other side.

To solve equations, we follow two basic principles:

- Whatever we do to one side of an equation must be done to the other.
- Use inverse operations to undo the operations being performed on the variables.

We can think of an equation as a scale whose fulcrum is located at the equal sign. For example, if we add a number to one side, we must add the same number to the other side to keep the equation "balanced."

To solve $x + 7 = 12$, we use the additive inverse property, discussed in Section 1.3:

$x + 7 - 7 = 12 - 7$	Subtract 7 from both sides (this "undoes" adding 7 to x).
$x + 0 = 12 - 7$	A number added to its opposite equals 0.
$x = 12 - 7$	A number added to 0 is unchanged.
$x = 5$	Simplify the right side.

Writing each of these steps gets tedious, especially when we solve more complicated equations. Here's how we typically write the work:

$$\begin{array}{r} x + 7 = 12 \\ \underline{\; -7 \;\; -7} \end{array}$$

Write -7 underneath the $+7$ and the 12 since we are subtracting 7 from both.

$$x = 5$$

When we add or subtract the same quantity from both sides, draw a horizontal line underneath it and write the resulting equation under the horizontal line.

After solving an equation, we check our work by **substituting** our solution (meaning that we replace each instance of the variable with the value that we found) into the *original* equation. If a true statement results, then our answer is correct. Here's how we write our check:

Check

$5 + 7 \stackrel{?}{=} 12$	Write a question mark above the equal sign to show we're not yet sure whether the answer is correct.
$12 = 12$	Calculate appropriate calculations on each side until we can show that one side of the equation equals the other. If we don't get a true statement, then we made a mistake somewhere!

The simplest kind of equation is a **one-step equation**, which can be solved by applying one inverse operation to each side.

Example 3.1 Solve for the variable and check the answer: $-4 = p - 4$.

Solution:

$$\begin{array}{l} -4 = p - 4 \\ \underline{+4 \qquad +4} \\ \;\;0 = p \end{array}$$

Add 4 to both sides.

A number added to its opposite equals 0.

When writing the solution, we typically put the variable on the left side and write $p = 0$.

Check:

$-4 \stackrel{?}{=} 0 - 4$	Substitute $p = 0$ into the original equation and see if the resulting statement is true.
$-4 = -4$	

Technology Tip

Although we won't ask you to check your answer for every equation you solve, we recommend you use technology to reduce mistakes.

Here are some important notes on multiplying quantities with variables:

- When multiplying by variables, avoid using the operation symbol ×, which can easily be mistaken for the variable x. Instead, put the number next to the variable or use parentheses, like $8x$ or $8(x)$.
- Use parentheses when multiplying numbers (the · symbol can be confused for a decimal) or when the factors can be misread. For example, when multiplying 2 by $x + 3$, write $2(x + 3)$, not $2 \cdot x + 3$ (which could be misinterpreted as multiplying 2 by x, then adding 3 to the product).
- Parentheses can be omitted around the first factor or around the variables. The product of 8, 2, and x could be written as $8(2)(x)$, $8(2)x$, $(8)(2)x$, or $(8)(2)(x)$.

To solve one-step equations involving multiplication or division, we use the multiplicative inverse property, as shown in Examples 3.2 and 3.3:

Example 3.2 Solve for the variable: $8r = 1$.

Solution: Since r is being multiplied by 8, we perform its inverse, which is division. Don't make the mistake here of dividing both sides by 1 to get $r = 8$. Students often do this because they automatically think that the smaller number in a division problem is always the divisor. In fact, we need to divide both sides by 8 to isolate r:

$8r = 1$	
$\frac{8r}{8} = \frac{1}{8}$	Divide both sides by 8.
$8\left(\frac{1}{8}\right)r = \frac{1}{8}$	Dividing by a number is equivalent to multiplying by its reciprocal.
$r = \frac{1}{8}$	Multiplying and dividing by the same number undo each other.

To save space, we usually condense this work and write the following:

$$8r = 1$$

$$\frac{8r}{8} = \frac{1}{8}$$ Divide both sides by 8.

$$r = \frac{1}{8}$$ A number multiplied by 1 equals itself.

Example 3.3 Solve for the variable: $-\frac{q}{5} = 10$.

Solution:

$$-\frac{q}{5} = 10$$

$$\frac{q}{-5} = 10$$ If two numbers have different signs, their quotient is negative.

$$-5\left(\frac{q}{-5}\right) = -5(10)$$ Multiply both sides by -5.

$$q = -50$$ Multiplying and dividing by the same number undo each other.

Exercises

Solve each equation for the variable. Check each answer.

1. $9 + t = 12$
2. $z - 4 = 5$
3. $18 = 10 + n$
4. $11 = y - 6$
5. $3c = 24$
6. $7 = 49w$
7. $-7u = 21$
8. $-16x = 8$
9. $\frac{b}{2} = 9$
10. $5 = \frac{f}{5}$
11. $\frac{d}{12} = -3$
12. $-10 = \frac{k}{2}$

Questions to Think About

13. Explain why $x + 12 = 14$ and $x = 2$ are equivalent equations.
14. To solve $6x = 3$, Martin did the following:

$$\frac{6x}{3} = \frac{3}{3}$$

$$x = 2$$

Explain the error in his work.

15. What is the left side of the equation $2x = 4$? What is the right side?

3.2 Solving Two-Step Equations

In Section 3.1, we used one inverse operation to solve equations. Equations that require two inverse operations to solve are called **two-step equations**. They are solved by doing either operation first, but often one method is easier than the other, as shown here:

$$\begin{aligned} 3n - 5 &= 7 \\ +5 \ &\ +5 \\ \hline 3n &= 12 \\ \frac{3n}{3} &= \frac{12}{3} \\ n &= 4 \end{aligned}$$

$$\begin{aligned} 3n - 5 &= 7 \\ \frac{3n}{3} - \frac{5}{3} &= \frac{7}{3} \\ n - \frac{5}{3} &= \frac{7}{3} \\ +\frac{5}{3} \ &\ +\frac{5}{3} \\ \hline n &= \frac{12}{3} = 4 \end{aligned}$$

Doing addition or subtraction before multiplication or division often helps us to avoid fractions.

Example 3.4 Solve for the variable: $15 = -r + 4$.

Solution:

$15 = -r + 4$	
$\underline{-4 \qquad -4}$	Subtract 4 from both sides.
$11 = -r$	
$11 = -1r$	A number multiplied by -1 equals its opposite.
$\frac{11}{-1} = \frac{-r}{-1}$	Divide both sides by -1.
$-11 = r$	Final answer, which we write as $r = -11$.

Exercises

Solve each equation for the variable. Check each answer.

1. $2y + 3 = 9$
2. $1 = 6m - 11$
3. $2p - 6 = 4$
4. $7 + 4q = 16$
5. $5p - 8 = 15$
6. $12 + 3n = 20$
7. $6w + 7 = -6$
8. $-5 = 7 - a$
9. $4 - u = -20$
10. $-22 = -x + 12$

3.3 Translating Between Words and Symbols

As we said at the beginning of this chapter, people use both words and symbols to represent mathematical ideas. In this section, we discuss how to translate between them.

Table 3.1 lists some of the most common words and symbols associated with the four operations discussed so far (addition, subtraction, multiplication, and division), equality, and variables.

Table 3.1 Translating words into symbols.

Concept	Words	Symbols
Addition	"**the sum** of 8 **and** 4"	$8 + 4$
	"8 **plus** 4"	
	"4 **added to** 8"	
	"4 **more than** 8"	
	"8 **increased by** 4"	
	"(a number) **exceeds** 8 **by** 4"	
Subtraction	"**the difference** of 8 **and** 4"	$8 - 4$
	"8 **minus** 4"	
	"4 **subtracted from** 8"	
	"4 **less than** 8"	
	"8 **decreased by** 4"	
	"8 **diminished by** 4"	
Multiplication	"**the product** of 8 **and** 4"	$8(4)$
	"8 **times** 4"	$(8)(4)$
	"8 **multiplied by** 4"	$8 \cdot 4$
	"8 **by** 4"	8×4
Division	"**the quotient** of 8 **and** 4"	$\frac{8}{4}$
	"8 **divided by** 4"	$8/4$
	"8 **over** 4"	
	"8 **fourths**"	
Equality	"8 plus 4 **is** 12"	$8 + 4 = 12$
	"8 plus 4 **is equal to** 12"	
	"8 plus 4 **equals** 12"	
	"8 plus 4 **is as much as** 12"	
	"When 4 is added to 8, **the result is** 12"	
Unknown	"**a number**"	x, y, z, or another variable
	"**a quantity**"	

As with any language, the context of words used for mathematics affects their meaning. Here are some especially confusing examples from the table:

- **and:** The word "and" can mean addition ("8 *and* 4 equals 12"), but it can also represent another operation ("the product of 8 *and* 4," "the quotient of 8 *and* 4").
- **exceeds:** If your speed exceeds the speed limit (represented by x) by 10 miles per hour, then your speed is $x + 10$. "A number exceeds 8 by 4" is represented as $x = 8 + 4$, but "8 exceeds a number by 4" is equivalent to "8 is 4 more than a number" and is represented as $8 = x + 4$.
- **less than:** When the phrase "less than" is used for subtraction, we reverse the order in which the numbers appear, so $8 - 4$ means "4 less than 8," not "8 less than 4." In addition, the phrase "is less than" means something completely different when working with inequalities (discussed in Chapter 5).
- **is:** The word "is" can be used for equality ("8 plus 4 *is* 12"), but it can also represent an inequality ("4 *is* less than 8").

Don't read one word at a time and convert each word to a symbol. Pay close attention to phrases and read the text several times if necessary to understand its meaning.

Example 3.5 Translate "5 less than two times a number" into mathematical symbols.

Solution: The phrase "less than" indicates subtraction, but we have to read the entire phrase to know the numbers that are being subtracted. "Two times a number" means 2 multiplied by an unknown number, which we represent with a variable like n. When using the phrase "less than," the numbers being subtracted appear in reverse, so the phrase can be written as $2n - 5$.

We use the phrases "the sum of … and …," "the difference of … and …," "the product of … and …," and "the quotient of … and … " to represent quantities surrounded by parentheses, as shown in the following example:

Example 3.6 Translate $2(3 + 7)$ into words.

Solution: Translating this literally from left to right would result in "2 times 3 plus 7," which could be misinterpreted as $2(3) + 7$, or 13. However, using the order of operations, we calculate the quantity inside the parentheses first, so $2(3 + 7) = 2(10) = 20$, not 13.

To translate this into words, we use the phrase "the sum of … and … " to represent the quantity in parentheses. In this case, $2(3 + 7)$ is written as "two times the sum of 3 and 7."

Exercises

Translate each phrase into mathematical symbols. Use the variable x to represent an unknown number.

1. a number increased by 12

2. the sum of 19 and a number

3. 3 times a number

4. 5 more than a number

5. the quotient of a number and 7

6. twice a number

7. 7 less than a number

8. 9 less than a number

9. 6 less than 4 times a number

10. 8 more than the sum of a number and 12

11. 9 less than the product of 12 and a number

12. 10 less than the quotient of a number and 6

Translate each mathematical expression into words. Use the phrase *a number* to represent a variable.

13. $m + 8$

14. $z - 9$

15. $\frac{p}{3}$

16. $2u + 1$

17. $3y + 17$

18. $4 + 2w$

19. $\frac{d}{30} + 4$

20. $\frac{r}{60} - 10$

21. $\frac{k}{72} + 5$

22. $3(q + 1)$

23. $-4(2y + 3)$

24. $10(2x + 7)$

Questions to Think About

25. Monica says that the word "and" always means addition. Explain the error in her thinking.

26. Joshua says that the word "by" always means division. Explain the error in his thinking.

3.4 Solving Word Problems

Translating words into symbols enables us to solve word problems, in which we find unknown quantities for a given scenario. Once we translate words into mathematical symbols, we can write a mathematical statement with variables and solve for the variables.

Table 3.2 shows a four-step procedure for solving word problems that is based on the problem-solving process outlined by mathematician George Pólya in his book *How to Solve It*.

Table 3.2 How to solve word problems.

1. GIVEN ...

1a. What information am I given?	**1b. What unknown information do I need to find?**
• Highlight key phrases. • Rewrite text with math symbols or words. • Create a table of values or draw a diagram.	• Look at the last sentence. (Include units and rounding.) • Look for words like: find, calculate, express, determine, simplify, create, prove, justify, explain, which?, what?, how many?

2. FIND ...

2a. What does the given information tell me?	**2b. How do I find the unknown information?**
• How is this *similar* to a problem I can solve? • How is this *different* from a problem I can solve?	• Write an equation or inequality. • Use formulas, definitions, or theorems. (Check conditions first!)

3. SOLVE ...

Carry out the plan outlined in step 2 above.

4. CHECK ...

- Did I find what I was supposed to find?
- Does my answer make sense in the context of the problem?

Source: Wong, B. & Bukalov, L. (2020). *The math teacher's toolbox: Hundreds of practical strategies to support your students*. Hoboken, N.J.: Jossey-Bass.

Throughout this book, we will use this four-step method to solve word problems.

Example 3.7 Nine less than a number is 20. Find the number.

Solution:

Step 1: Identify what is given and what we need to find.
We are given that 9 less than a number is 20. We need to find the number. (In practice, you often don't need to write anything down for this step. You may want to highlight or underline important phrases. However, in a short problem like this, every word is important, so that may not be helpful here as it would be in more complicated problems.)

Step 2: *Represent the unknown information.*
Let $n =$ the number.

Nine less than a number	$n - 9$
is 20.	$= 20$

Step 3: *Solve.*

$$\begin{aligned} n - 9 &= 20 \\ +9 \;\; &+9 \\ \hline n &= 29 \end{aligned}$$

Step 4: *Check.*

$$29 - 9 \stackrel{?}{=} 20$$

$$20 = 20$$

The number is 29.

You may have been able to answer this problem by inspection without having to solve an equation. We show the work here as a model for more complicated problems you'll encounter later on.

Watch Out!

When solving a word problem, many people choose x as a variable to represent an unknown quantity. However, you don't have to limit yourself to that letter. Instead, pick a variable that helps you remember what you need to find—d for distance, p for the number of pens, and so on. Some letters are problematic and should be avoided:

- e and i are reserved for special mathematical numbers
- l looks like the number 1
- o looks like the number 0

Other letters should be used carefully because of how they're written:

- a can resemble the number 9 (we suggest writing a with a small hook on the bottom)
- s looks like the number 5
- t can resemble the + symbol (we suggest writing t with a small hook on the bottom)

Example 3.8 Ten exceeds three times a number by 4. Find the number.

Solution:

Step 1: Identify what is given and what we need to find.
We are given that 10 exceeds 3 times a number by 4. Find the number. "Ten exceeds three times a number by 4" is equivalent to "10 is 4 more than 3 times a number."

Step 2: Represent the unknown information.
Let x = the number.

Ten	10
exceeds	=
3 times a number	$3x$
by 4.	$+ 4$

Step 3: Solve.

$$10 = 3x + 4$$
$$\underline{-4 \qquad -4}$$
$$6 = 3x$$
$$\frac{6}{3} = \frac{3x}{3}$$
$$2 = x$$

Step 4: Check.

$$10 \stackrel{?}{=} 3(2) + 4$$
$$10 = 10$$

The number is 2.

Example 3.9 Twice a number increased by 12 equals 26. Find the number.

Solution:

Step 1: Identify what is given and what we need to find.
We are given that twice (or two times) a number increased by 12 equals 26. Find the number.

Step 2: *Represent the unknown information.*
Let $x =$ the number.

Twice a number	$2x$
increased by 12	$+ 12$
equals 26.	$= 26$

Step 3: *Solve.*

$$2x + 12 = 26$$
$$\underline{-12 - 12}$$
$$2x = 14$$
$$\frac{2x}{2} = \frac{14}{2}$$
$$x = 7$$

Step 4: *Check.*

$$2(7) + 12 \stackrel{?}{=} 26$$
$$26 = 26$$

The number is 7.

Exercises

1. A certain number is equal to 35 decreased by 21. What is the number?
2. Find the number such that 15 exceeds it by 5.
3. 13 is equal to a number increased by 7. What is the number?
4. The sum of a number and 11 is 23. What is the number?
5. If a number is decreased by 7, the result is 8. Find the number.
6. Five times a number equals 48. Find the number.
7. The sum of 5 and 8 exceeds a certain number by 2. Find the number.
8. A number increased by 9 exceeds 7 by 10. Find the number.
9. Four times a number, increased by 3, equals 19. Find the number.
10. When four times a number is diminished by 10, the remainder is 26. Find the number.
11. Seven added to 3 times a number is equal to 22. What is the number?
12. Seven exceeds one-half a number by 4. Find the number.

3.5 Equations with Like Terms

Fe has 3 forks, Peter has 4 plates, Fran has 1 fork, and Priya has 6 plates. How many forks and how many plates do they have in all?

We can illustrate this problem with a diagram (Figure 3.1) that shows what each person has:

Figure 3.1 How many forks and how many plates?

Since the question asks how many forks and how many plates they have in all, we don't need to separate items by person. In fact, the persons' names don't matter at all. We only need to group all the forks together and all the plates together (Figure 3.2):

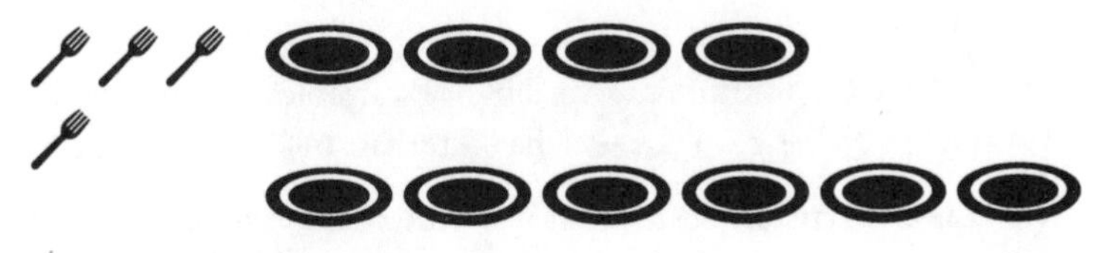

Figure 3.2 Group similar items.

Adding all the numbers together ($3 + 1 + 4 + 6 = 14$) doesn't help us answer the question. We don't have 14 forks, 14 plates, or 14 "fork-plates." We have 4 forks and 10 plates.

We don't need to draw a diagram to solve this problem. We could simply add the numbers of forks and add the numbers of plates:

3 forks + 1 fork = 4 forks

4 plates + 6 plates = 10 plates

In math, we encounter a similar situation when we solve equations like $7n - 3n + 4 = 19$, in which a variable occurs more than once on one side of the equation. Before we go further, we need to introduce some vocabulary to explain what we need to do (Figure 3.3 and Table 3.3):

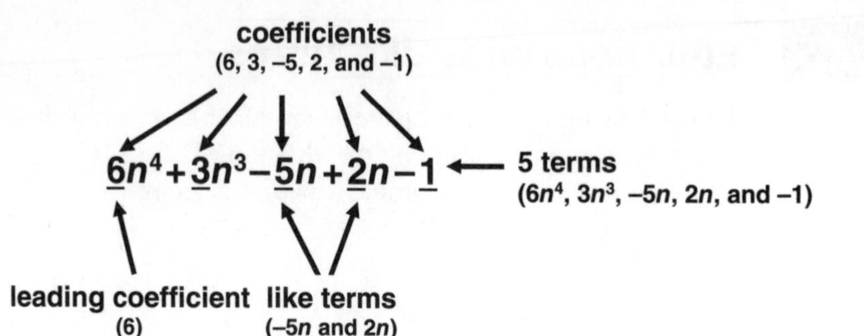

Figure 3.3 Vocabulary for like terms.

Table 3.3 Examples and non-examples of like terms.

Example	Reason	Non-Example	Reason
$8n^2$, $3n^2$	Same variable (n) raised to the same exponent (2)	$8p^2$, $3n^2$	Exponents are the same but the variables (p and n) are different
$-5x$, $4x$	Same variable (x) raised to the same exponent (1)	$3n^2$, $3n$	Variables are not raised to the same exponent
-5, 4	All numbers are like terms	-5, $-5x$	One term (-5) has a variable while the other ($-5x$) doesn't

- A **term** is a number, a variable or variables, or the product of both. The expression $6n^4 + 3n^3 - 5n + 2n - 1$ has 5 terms: $6n^4$, $3n^3$, $-5n$, $2n$, and -1.
- **Like terms** are terms that have the same variables raised to the same exponents. For example, $7n^2$ and $9n^2$ are like terms, and $-5n$ and $2n$ are like terms.
- A **coefficient** is the number multiplied by variables in a term. By convention, we write coefficients in front of the variables. For example, the coefficient of $2n$ is 2. If no number appears before the variable, then the coefficient is 1 (since a number times 1 equals itself).
- The **leading coefficient** is the coefficient of the term in which the variable is raised to the highest exponent. When terms in an expression are ordered so that the variables' exponents are in decreasing order, then the leading coefficient is the coefficient of the first term. In the expression $7n^2 + 2n + 5$, the leading coefficient is 7.

To add or subtract like terms, add or subtract the coefficients and leave the variables unchanged. We call this **combining like terms**.

Here are some examples that involve adding or subtracting like terms:

Example 3.10 Combine like terms: $3x + 4y - 6 + x - 12y - 2y + 18 + 5x$.

Solution: To combine like terms more easily, draw different shapes around each group of like terms (the *x*s, the *y*s, and the numbers). Use the operation in front of each term as the sign of the term.

$$3x + 4y - 6 + x - 12y - 2y + 18 + 5x$$

Then add the coefficients in each group of like terms:

$$3x + x + 5x = 9x$$

(NOTE: The coefficient of the second term is 1, so $3 + 1 + 5 = 9$.)

$$4y - 12y - 2y = -10y$$

$$-6 + 18 = 12$$

Finally, write all the terms in the sum: $9x - 10y + 12$.

Example 3.11 Find 3 consecutive integers whose sum is 57.

Solution:

Step 1: Identify what is given and what we need to find.
Find 3 consecutive integers that add up to 57.
To solve this problem, identify key vocabulary terms. Recall from Section 1.3 that integers are whole numbers and their opposites ($\ldots, -3, -2, -1, 0, 1, 2, 3, \ldots$). **Consecutive integers** are integers that follow each other in order. Since consecutive integers are 1 unit apart, we add 1 to get from one integer to the next. If the first integer is 12, the second is $12 + 1 = 13$, the third is $12 + 1 + 1 = 14$, and so on.

Step 2: Represent the unknown information.
Let $x =$ the first consecutive integer.
Then $x + 1 =$ the second consecutive integer
and $x + 1 + 1 = x + 2 =$ the third consecutive integer.

Three consecutive integers	$x, x + 1, x + 2$
add up to 57.	$x + x + 1 + x + 2 = 57$

Step 3: *Solve.*

$$x + x + 1 + x + 2 = 57$$

$$\begin{aligned} 3x + 3 &= 57 \\ -3 \quad &-3 \\ \hline 3x &= 54 \end{aligned}$$

Combine like terms.
Subtract 3 from both sides.

$$\frac{3x}{3} = \frac{54}{3}$$

Divide both sides by 3.

$$x = 18$$

The first integer is 18, so the other two integers are 19 and 20.

Step 4: *Check.*

$$18 + 19 + 20 \stackrel{?}{=} 57$$

$$57 = 57$$

The three consecutive integers are 18, 19, and 20.

Example 3.12 Find 4 consecutive odd integers that add up to −112.

Solution:

Step 1: *Identify what is given and what we need to find.*
Find 4 consecutive *odd* integers that add up to −112.
Consecutive odd integers are odd integers that follow each other in order. Odd integers differ by 2 (1, 3, 5, 7, ...). Similarly, **consecutive even integers**, which are even integers that follow each other in order, also differ by 2 (2, 4, 6, 8, ...). We represent both consecutive odd and consecutive even integers in the same way: n, $n + 2$, $n + 4$, ...

Step 2: *Represent the unknown information.*
Let x = the first consecutive odd integer.
Then $x + 2$ = the second consecutive odd integer,
$x + 2 + 2 = x + 4$ = the third consecutive odd integer,
and $x + 2 + 2 + 2 = x + 6$ = the fourth consecutive odd integer.

Four consecutive odd integers	$x, x + 2, x + 4, x + 6$
add up to −112.	$x + x + 2 + x + 4 + x + 6 = -112$

Step 3: *Solve.*

$$x + x + 2 + x + 4 + x + 6 = -112$$

$$\begin{aligned} 4x + 12 &= -112 \\ -12 \quad &-12 \\ \hline 4x &= -124 \end{aligned}$$

Combine like terms.
Subtract 12 from both sides.
Divide both sides by 4.

$$\frac{4x}{4} = \frac{-124}{4}$$

$x = -31$

The first integer is -31, so the other integers are $-31 + 2 = -29$, $-31 + 4 = -27$, and $-31 + 6 = -25$.

Step 4: *Check.*

$$-31 + (-29) + (-27) + (-25) \stackrel{?}{=} -112$$

$$-112 = -112$$

The four integers are -31, -29, -27, and -25.

Exercises

Combine like terms in each expression.

1. $(+3a) - (+2a) + (-4a)$

2. $m + 4m - 6m - 10m$

3. $6u^2 + 3u^2 + 4 - 7u - 3u + 1$

4. $2j^2 + 4j^2 + 7 - 5j - 11j + 4$

5. $4ab - 6bc + 3bc - 2ab$

6. $x^2 - y^2 - 3x^2 + 4 - 3y^2$

Solve each equation for the variable. Check your answer.

7. $3x - x = 6$

8. $6z + 2z + 12 = 28$

9. $3 - 8 = 5r - 2r$

10. $4d + 6 - d - 5 = 13$

11. $68 + 5 = 10m + 2m - 2m + 3$

12. $8j + 8 - 7j + 4j + 8 - 11 = 29$

13. $8b - 9 - 5b = 0$

14. $11c - 8 - 2c - 64 = 0$

15. $4k + 2 + 6k + 5 = 12$

16. $12a - 2 - 3a - 8a = 0$

17. $7c + 4 + c - 5c = -5$

18. $2q + q + 2 - q = -7$

19. Ann has 3 bags of apples. Benito has 5 bags of apples. Together, they have 96 apples in all. If each bag holds the same number of apples, how many apples are in each bag?

20. The sum of 4 times a number, 3 times a number, and 40 is 5. What is the number?

21. Find 2 consecutive integers whose sum is 157.

22. Find 3 consecutive integers whose sum is -90.

23. Three consecutive even integers have a sum of 204. What are the 3 integers?

24. Find 4 consecutive odd integers that add up to 328.

Questions to Think About

25. Is $\frac{4}{x}$ a term? Why or why not?

26. Is $19qr$ a term? Why or why not?

27. Determine whether $8xy$ and $4x$ are like terms. Explain your answer.

28. Determine whether $3w^2$ and $16w$ are like terms. Explain your answer.

3.6 Equations with Variables on Both Sides

In all the equations solved so far, the variables have been on one side of the equation. What happens when we have variables on both sides of the equal sign?

We suggest keeping these two ideas in mind when solving these equations:

- **Make the problem similar to problems that you already know how to solve.** In this case, we will add or subtract the same quantities on each side so that all the terms with variables are on one side of the equation and all the terms with numbers are on the other side.
- **Whatever is done to one side of the equation must be done to the other side.** This is a basic rule of solving equations. You can move the variable terms to either side, as shown in Example 3.13:

Example 3.13 Solve for the variable: $7+2y = 5y+28$.

Solution:

METHOD 1: Move the terms with variables to the left side of the equation.

$7 + 2y = 5y + 28$	
$\underline{\quad - 5y - 5y}$	Subtract $5y$ from both sides to move the variables to the left side.
$7 - 3y = 28$	Combine like terms.
$\underline{-7 \qquad -7}$	Subtract 7 from both sides to isolate the term with the variable.
$-3y = 21$	
$\frac{-3y}{-3} = \frac{21}{-3}$	Divide both sides by -3 to isolate the variable
$y = -7$	Final answer.

METHOD 2: Move the terms with variables to the right side of the equation.

$$7 + 2y = 5y + 28$$
$$-2y \quad -2y$$
Subtract $2y$ from both sides to move the variables to the right side.

$$7 = 3y + 28$$
Combine like terms.

$$-28 \quad -28$$
Subtract 28 from both sides to isolate the term with the variable.

$$-21 = 3y$$

$$\frac{-21}{3} = \frac{3y}{3}$$
Divide both sides by 3 to isolate the variable

$$-7 = y$$
Final answer, which can be written as $y = -7$.

As you can see, the side that the variable is on does not affect the final answer, as long as you do the work correctly.

Example 3.14 Nine less than 7 times a number is equal to 19 more than 3 times the number. Find the number.

Solution:

Step 1: Identify what is given and what we need to find.
Nine less than 7 times a number is equal to 19 more than 3 times the number. Find the number.

Step 2: Represent the unknown information.
Let $x =$ the number.

9 less than 7 times a number	$7x - 9$
is equal to	$=$
19 more than 3 times the number.	$19 + 3x$

Step 3: Solve.

$$7x - 9 = 3x + 19$$

$$-3x \quad -3x$$
Subtract $3x$ from both sides.

$$4x - 9 = 19$$
Combine like terms.

$$+9 \quad +9$$
Add 9 to both sides.

$$4x = 28$$

$$\frac{4x}{4} = \frac{28}{4}$$
Divide both sides by 4.

$$x = 7$$
Final answer.

Step 4: *Check.*

$$7(7) - 9 \stackrel{?}{=} 3(7) + 19$$

$$40 = 40$$

The number is 7.

One final note: don't confuse manipulating variables on both sides of the equation with combining like terms. Compare the work in solving the following two examples in Table 3.4:

Table 3.4 Variables on 1 vs. 2 sides of an equation.

Variables on Both Sides	Variables on One Side
$5y = 2y + 21$	$5y + 2y = 21$
$-2y \quad -2y$ Subtract $2y$ from both sides.	$7y = 21$ Combine like terms on the left side.
$3y = 21$	$\frac{7y}{7} = \frac{21}{7}$
$\frac{3y}{3} = \frac{21}{3}$	$y = 3$
$y = 7$	

When variables are on both sides of the equation, change the value of each side (instead of $5y$ on the left side, we now have $3y$). However, when variables are on one side of the equation, combine like terms, which doesn't change the value of that side of the equation ($5y$ added to $2y$ equals $7y$).

We've been able to find solutions for every equation discussed so far. However, some equations are different, as in Example 3.15:

Example 3.15 Solve for the variable: $2x + 4x + 5 = 6x + 4 + 1$.

Solution:

$2x + 4x + 5 = 6x + 4 + 1$

$6x + 5 = 6x + 5$ Combine like terms.

$-6x \quad -6x$ Subtract $6x$ from both sides.

$5 = 5$???

We seem to have "lost" the variable, even though we have done every step correctly! However, if we look at the original equation closely, we see that it is equivalent to $6x + 5 = 6x + 5$, which is true for *any* value of x.

We say here that "x can be any number."

Example 3.16 Solve for the variable: $-6z + 8 = 10 - 6z$.

Solution:

$-6z + 8 = 10 - 6z$	
$\quad - 8 \quad - 8$	Subtract 8 from both sides.
$-6z = 2 - 6z$	Combine like terms.
$+6z = \ + 6z$	Add $6z$ to both sides.
$0 = 2$	???

In this case, we say that the equation has *no solution* since 0 is not equal to 2. If we look at the original equation carefully, we shouldn't be surprised. The equation has $-6z$ on both sides but adds 8 to the left side while adding 10 to the right side – a clear violation of our rule about doing the same thing to both sides of a true equation.

Equations with variables can be divided into three types, depending on their solutions:

- If an equation is true only for certain values of the variable, then the equation is *sometimes* true. We call it a **conditional equation**. All the equations we solved before Example 3.15 are conditional equations.
- If an equation is true for all values of the variable (as in Example 3.15), then the equation is *always* true. We call it an **identity**.
- If an equation has no solutions for any values of the variable (as in Example 3.16), then the equation is *never* true. We call it a **false statement**.

Exercises

Solve each equation for the variable and check the answer, if possible.

1. $2y - 1 = 2 + 3y$
2. $3p + 1 = p + 11$
3. $6q + 4 = 6q - 5$
4. $2w - 10 = 5w + 5$
5. $7p - 2 = 2p + 13$
6. $9y + 3 - 7y + 2 = 2y + 5$
7. $5d + 22 = 2d$
8. $3x = 6 - 4x$
9. $-17u = -17u + 3$
10. $-19x = -8x + 23$
11. $7g + 1 - 2g + 16 = 3g + 29 + 2g - 12$
12. $5r + 21 - 2r = -2 - 13 - r$
13. $9v - 10v + 7 + 3v = 35 + 4v$
14. $1 + 6d + 2d - 13 - d = 6 + 5d - 2$
15. Four times a certain number is equal to 35 decreased by the number. What is the number?

16. If 50 is subtracted from twice a certain number, the difference is equal to the number. What is the number?

17. Four times a certain number, decreased by 5, equals 25 diminished by 6 times the number. What is the number?

Questions to Think About

18. Jon solved the equation $8x - 2x + 18 = 78$ by writing the following:

$$\begin{array}{l} 8x - 2x + 18 = 78 \\ \underline{+2x + 2x} \\ 10x + 18 = 78 \\ \quad\;\; \underline{-\,18 - 18} \\ 10x \qquad\;\; = 60 \\ \dfrac{10x}{10} = \dfrac{60}{10} \\ x = 6 \end{array}$$

Explain the error in his work.

3.7 Equations with Parentheses

Ana has a piece of white chocolate and Bill has a piece of dark chocolate, both of which are shown in Figure 3.4.

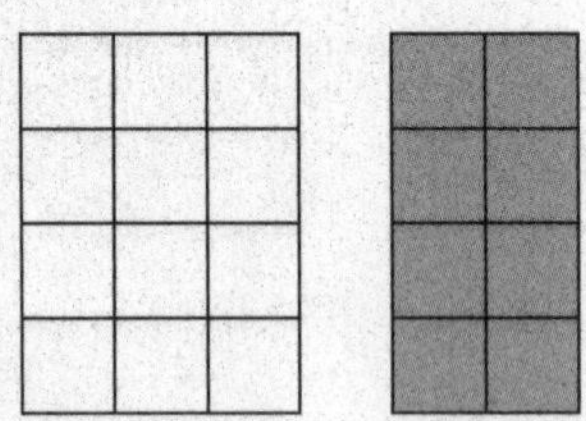

Figure 3.4 How many chocolate squares?

How many squares are in each piece? How many squares do they have in all? We calculate each piece's area by multiplying its length and width. Ana's piece has $3(4) = 12$ squares, while Bill's has $2(4) = 8$ squares. When we combine both pieces, we get Figure 3.5:

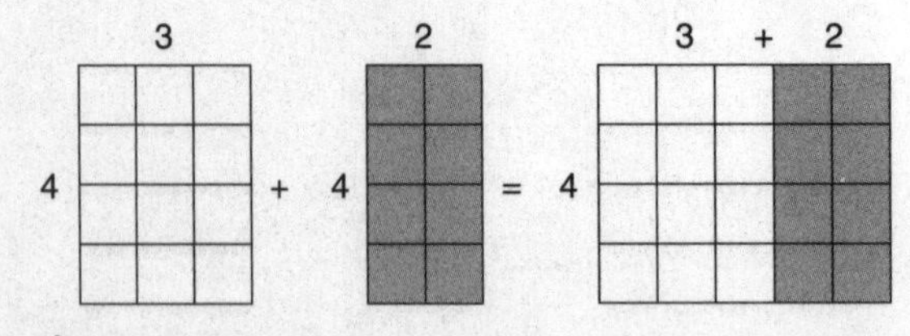

Figure 3.5 Distributive property.

We calculate the total number of squares by adding the total from each piece $(12 + 8 = 20)$ or multiplying the number of squares in the combined rectangle $(4(3 + 2) = 4(5) = 20)$.

We summarize this as follows: $4(3) + 4(2) = 4(3 + 2)$. This illustrates what we call the **Distributive property of multiplication over addition**. We use this name because multiplication "distributes" over addition. Using variables, we write $\boldsymbol{a(b + c) = ab + ac}$, where a, b, and c can be any number.

The distributive property helps explain the procedure for combining like terms:

$3a + 2a$	
$= (3 + 2)a$	Distributive property.
$= 5a$	Order of operations: evaluate parentheses first.

Here are two important ideas about the distributive property and minus signs:

- The distributive property also works over subtraction: $4(5 - 2) = 4(5) - 4(2)$.
- A minus sign $(-)$ that occurs before parentheses is equivalent to a coefficient of -1. (You can think of the minus sign as "flipping" the sign of each term in parentheses.) For example:
  - $8 - (5 - 2) = 8 - 1(5 - 2) = 8 - 1(5) - 1(-2) = 8 - 5 + 2$.
  - $7 - (2x - 5) = 7 - 2x + 5 = 12 - 2x$.

We can solve equations with parentheses in two ways, as shown in Example 3.17:

Example 3.17 Solve for the variable: $20 = 2(9 - 3x)$.

Solution:

METHOD 1: Distributive property.

$20 = 2(9 - 3x)$	
$20 = 2(9) - 2(3x)$	Distributive property.
$20 = 18 - 6x$	Multiply.
$\underline{-18 \quad -18}$	Subtract 18 from both sides.
$2 = -6x$	
$\frac{2}{-6} = \frac{-6x}{-6}$	Divide both sides by -6.
$-\frac{2}{6} = x$	
$-\frac{1(2)}{3(2)} = x$	Group common factors in the numerator and denominator: 2.
$-\frac{1}{3} = x$	Final answer, which we write with the variable on the left side: $x = -\frac{1}{3}$.

METHOD 2: Divide both sides by a common factor.

$20 = 2(9 - 3x)$	
$\frac{20}{2} = \frac{2(9-3x)}{2}$	Divide both sides by 2.
$10 = 9 - 3x$	
$-9 \; -9$	Subtract 9 from both sides.
$1 = -3x$	
$\frac{1}{-3} = \frac{-3x}{-3}$	Divide both sides by −3.
$-\frac{1}{3} = x$	Final answer, which we write with the variable on the left side: $x = -\frac{1}{3}$.

Example 3.18 Solve for the variable: $20 = 8 - 2(9 - 3x)$.

Solution: Students often make the mistake of subtracting $8 - 2$ first. In fact, the order of operations tells us to calculate multiplication first. In this case, we calculate $-2(9 - 3x)$ first and then subtract the result from 8.

$20 = 8 - 2(9 - 3x)$	
$20 = 8 - 2(9) - 2(-3x)$	Distributive property.
$20 = 8 - 18 + 6x$	Multiply.
$20 = -10 + 6x$	Combine like terms.
$+10 \; +10$	Add 10 to both sides.
$30 = 6x$	
$\frac{30}{6} = \frac{6x}{6}$	Divide both sides by 6.
$5 = x$	Final answer, which we write as $x = 5$.

Example 3.19 Find 3 consecutive even integers such that 10 less than 3 times the first is equal to 6 more than 5 times the third.

Solution:

Step 1: Identify what is given and what we need to find.
10 less than 3 times the first of 3 consecutive even integers equals 6 more than 5 times the third.
Find the 3 consecutive integers.

Step 2: Represent the unknown information.
Let $x =$ the first consecutive integer.
Then $x + 2 =$ the second consecutive even integer and
$x + 4 =$ the third consecutive even integer.

10 less than three times the first	$3x - 10$
is equal to	$=$
6 more than 5 times the third.	$6 + 5(x + 4)$

Step 3: *Solve.*

$3x - 10 = 6 + 5(x + 4)$	
$3x - 10 = 6 + 5x + 20$	Distributive property.
$3x - 10 = 5x + 26$	Add.
$-3x \quad -3x$	Subtract $3x$ from both sides.
$-10 = 2x + 26$	Combine like terms.
$-26 \quad -26$	Subtract 26 from both sides.
$-36 = 2x$	
$\frac{-36}{2} = \frac{2x}{2}$	Divide both sides by 2.
$-18 = x$	Since $x = -18$, the integers are -18, $-18 + 2 = -16$, and $-18 + 4 = -14$.

Step 4: *Check.*

$$3(-18) - 10 \stackrel{?}{=} 6 + 5(-14)$$

$$-64 = -64$$

The three integers are -18, -16, and -14.

Example 3.20 A man is 9 times as old as his son. In 3 years, the father will be only 5 times as old as his son. What is the age of each?

Solution:

Step 1: *Identify what is given and what we need to find.*
A father is now 9 times as old as his son. In 3 years, the father will be 5 times as old as his son.
Find the ages of the father and son.

Step 2: *Represent the unknown information.*
Let $a =$ the son's age.
Then $9a =$ the father's age (since the father is 9 times as old as the son).
In 3 years, the son will be 3 years older, so his age will be $a + 3$.
In 3 years, the father will also be 3 years older, so his age will be $9a + 3$.

In 3 years the father	$9a + 3$
will be	$=$
only 5 times as old as his son.	$5(a + 3)$

Step 3: *Solve.*

$9a + 3 = 5(a + 3)$

$9a + 3 = 5a + 15$	Distributive property.
$-5a \quad -5a$	Subtract $5a$ from both sides.
$4a + 3 = 15$	Combine like terms.
$-3 \quad -3$	Subtract 3 from both sides.
$4a = 12$	
$\frac{4a}{4} = \frac{12}{4}$	Divide both sides by 4.
$a = 3$	Final answer.

Step 4: *Check.*

If the son is now 3, then the father is now $9(3) = 27$.

Three years from now, the son will be $3 + 3 = 6$ and the father will be $27 + 3 = 30$.

$$6(5) \stackrel{?}{=} 30$$

$$30 = 30$$

The son is 3 years old and the father is 27 years old.

Exercises

Solve each equation for the variable. Check your answer.

1. $2a - (11 - 2a) = 37$
2. $3r - 4(r + 2) = 5$
3. $2(w - 5) - 3(2w + 1) = 7$
4. $5(b + 4) - 4(b + 3) = 0$
5. $-(5p + 4) = -6p + 3$
6. $4(a - 3) - 6(a + 1) = 0$
7. $5(2z + 3) = 2(4z - 1)$
8. $7(p - 5) = 14 - (p + 1)$
9. $(m - 9) - (m + 7) = 4m$
10. $20 = 8 - 2(9 - 3d)$
11. $7q + 5 - (2q - 15) = 9$
12. $7(5t - 1) - 18t = 12t - (3 - t)$
13. Find 2 consecutive integers such that 3 times the first is 45 more than the second.
14. Find 2 consecutive integers such that 5 times the first exceeds 3 times the second by 35.
15. Find 3 consecutive integers such that 31 less than the first is 18 less than twice the third.

16. Find 3 consecutive integers such that 4 times the first exceeds the sum of the second and twice the third by 19.

17. Find 3 consecutive even integers such that twice the sum of the first and second exceeds the third by 150.

18. Find 4 consecutive odd integers such that the sum of the first and twice the second exceeds the fourth by 68.

19. A man is 28 years older than his son. In 3 years the father will be 5 times as old as the son. Find their present ages.

20. Wen is 8 years older than Ru. In 6 years, 5 times Wen's age will equal 9 times Ru's age. How old is each at present?

21. Belinda is 10 years older than Hector. In 8 years twice Belinda's age will equal 3 times Hector's age. What are their present ages?

22. Two years ago a woman was four times as old as her son. Three years from now the mother will be only 3 times as old as the son. How old is each at present? (Hint: Let a and $4a$ represent their ages 2 years ago.)

23. A man was 30 years of age when his daughter was born. The father's age now exceeds 3 times the daughter's age by 6 years. How old is each at present?

24. Faraz is four times as old as Cyrus. In 10 years he will be only twice as old as Cyrus is then. How old is each?

Questions to Think About

25. Does addition distribute over multiplication? Explain.

26. Helen believes that the distributive property enables her to say that $1 + 4(7 + 2) = 5(7) + 5(2)$. Is she correct? Explain.

27. Write the distributive property of multiplication over subtraction using mathematical symbols.

3.8 Using Tables to Solve Word Problems Involving Values

Some word problems involve items that have a fixed value per unit. Think about what the following situations have in common:

- If nickels have a value of \$0.05 per coin, then the total value of 4 nickels is \$0.05(4) = \$0.20.

- If a person runs at a rate of 5 miles per hour, then the total distance run after 4 hours is $5(4) = 20$ miles.
- If sedans have a capacity of 5 people per car, then the total capacity of 4 sedans is $5(4) = 20$ people.

All three situations follow the same general rule:

(Value per unit)(Number of units) = Total value of units

Another important idea is that if we have units of different kinds, then **the total value of all units is the sum of the total values for each unit**. For example, the total value of the coins shown in Figure 3.6 is the sum of the values of the 3 dimes and 4 nickels: $3(10¢) + 4(5¢) = 30¢ + 20¢ = 50¢$.

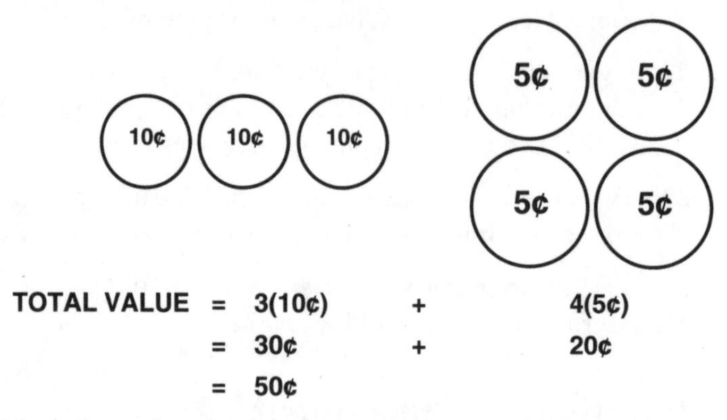

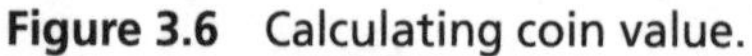

Figure 3.6 Calculating coin value.

For this rule to work, all units must be consistent. For example, if the rate is given in miles per hour, then the time must be in hours (not minutes) and the distance traveled must be in miles.

To solve these problems, we recommend organizing the given information into a table indicating the number of units, the value of 1 unit, and the total value of units, as shown in Example 3.21:

Example 3.21 **The Greenway Taxi company has two types of vehicles in its fleet: large-capacity minivans that hold 6 passengers each and hybrid sedan cars that hold 4 passengers each. Greenway has 10 more cars than minivans. If the total capacity of its fleet is 290 passengers, how many minivans and how many cars does Greenway Taxi have?**

Solution:

Step 1: *Identify what is given and what we need to find.*
Minivans hold 6 passengers each. Cars hold 4 passengers each.
The number of cars is 10 more than the number of minivans.
The cars and minivans hold 290 passengers in all.
Find the number of minivans and number of cars.

Step 2: Represent the unknown information.
Let m = the number of minivans in the company's fleet.
Then $m + 10$ = the number of cars in the fleet.
We can make the following table:

	(Passengers per vehicle)	(Number of vehicles)	= Total passenger capacity
Minivans	6	m	$6m$
Cars	4	$m + 10$	$4(m + 10)$

The total capacity of the minivans and the total capacity of the cars adds up to 290.

$$6m + 4(m + 10) = 290$$

Step 3: Solve.

$$6m + 4(m + 10) = 290$$

$6m + 4m + 40 = 290$ Distributive property.

$10m + 40 = 290$ Combine like terms.

$-40 \quad -40$ Subtract 40 from both sides.

$10m = 250$

$\dfrac{10m}{10} = \dfrac{250}{10}$ Divide both sides by 10.

$m = 25$ Greenway Taxi has 25 minivans and $25 + 10 =$ 35 cars in its fleet.

Step 4: Check.
The passenger capacity of the minivans is $25(6) = 150$. The passenger capacity of the cars is $35(4) = 140$.

$$150 + 140 \stackrel{?}{=} 290$$

$$290 = 290$$

Greenway Taxi has 25 minivans and 35 cars in its fleet.

In problems with bills or coins, the general rule stated above **[(Value per unit)(Number of units) = Total value of units]** becomes:

(Value per bill or coin)(Number of bills or coins) = Total value of bills or coins

---

Example 3.22 Juanita has a bag consisting only of dimes and nickels. The number of nickels is 5 less than twice the number of dimes. The total value of the coins in her bag is \$1.95. Determine the number of dimes and the number of nickels in her bag.

Solution:

Step 1: Identify what is given and what we need to find.
Juanita's bag holds dimes and nickels. The number of nickels is 5 less than twice the number of dimes. The total value of the dimes and nickels is \$1.95. Find the number of dimes and number of nickels in her bag.

Step 2: Represent the unknown information.
Let d = the number of dimes in Juanita's bag.
Then $2d - 5$ = the number of nickels in her bag.
We can avoid mistakes that can arise when working with decimals by converting all values from dollars to cents, as shown in the following table:

	(Value per coin in cents)	(Number of coins)	= Total value of coins in cents
Dimes	10	d	$10d$
Nickels	5	$2d - 5$	$5(2d - 5)$

The total value of the dimes	$10d$
and the total value of the nickels	$+ 5(2d - 5)$
adds up to \$1.95 (= 195 cents)	$= 195$

To express this problem in terms of dollars instead of cents, we would write the equation $0.10d + 0.05(2d - 5) = 1.95$.

Step 3: *Solve.*

$10d + 5(2d - 5) = 195$

$10d + 10d - 25 = 195$ Distributive property.

$20d - 25 = 195$ Combine like terms.

$\underline{\quad + 25 \quad + 25}$ Add 25 to both sides.

$20d = 220$

$\dfrac{20d}{20} = \dfrac{220}{20}$ Divide both sides by 20.

$d = 11$ Juanita had 11 dimes and $2(11) - 5 = 17$ nickels in her bag.

Step 4: *Check.*

$$11(10) + 17(5) \stackrel{?}{=} 195$$

$$195 = 195$$

Juanita had 11 dimes and 17 nickels in her bag.

In problems involving objects that travel at a constant **rate** (defined to be the frequency with which an event happens), the general rule becomes:

(Rate of object)(Time traveled) = Distance traveled

Example 3.23 Two planes leave the same airport at the same time and fly in opposite directions. The faster plane travels 100 miles per hour faster than the slower plane. At the end of 5 hours they are 2,000 miles apart. Find the rate of each plane in miles per hour.

Solution:

Step 1: *Identify what is given and what we need to find.*
Two planes leave the same place at the same time and fly in opposite directions.
The rate of the faster plane is 100 miles per hour faster than the rate of the slower plane.
After 5 hours, the distance between them is 2,000 miles.
Find the rate of each plane in miles per hour.

Step 2: *Represent the unknown information.*
Let r = the rate of the slower plane in miles per hour (mph).
Then $r + 100$ = the rate of the faster plane.
We can make the following table:

	(Rate in mph)	(Time in hrs)	= Distance in miles
Fast Plane	$r + 100$	5	$(r + 100)(5) = 5(r + 100)$
Slow Plane	r	5	$r(5) = 5r$

The distance traveled by the fast plane and the distance traveled by the slow plane adds up to 2,000 miles. $5(r + 100)$ $+ 5r$ $= 2{,}000$

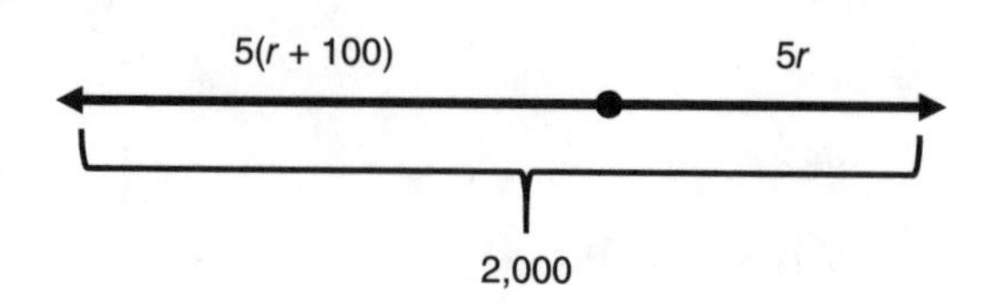

Step 3: *Solve.*

$$5(r + 100) + 5r = 2{,}000$$

$5r + 500 + 5r = 2{,}000$ Distributive property.

$10r + 500 = 2{,}000$ Combine like terms.

$\underline{\quad -500 \quad -500}$ Subtract 500 from both sides.

$$10r = 1{,}500$$

$\dfrac{10r}{10} = \dfrac{1{,}500}{10}$ Divide both sides by 10.

$r = 150$ The rate of the slower plane is 150 miles per hour. The rate of the faster plane is $150 + 100 = 250$ miles per hour.

Step 4: *Check.*

After 5 hours, the slower plane travels (150 mph)(5 hours) = 750 miles and the faster plane travels (250 mph)(5 hours) = 1,250 miles.

$$750 + 1{,}250 \stackrel{?}{=} 2{,}000$$

$$2{,}000 = 2{,}000$$

The slower plane travels at 150 miles per hour and the faster plane travels at 250 miles per hour.

Exercises

1. Elena has \$6.95 in quarters and dimes. She has 4 more quarters than dimes. How many coins of each type does she have?
2. Jacinth sells strawberry-banana shakes at her fruit stand. She bought bananas for \$0.60 per pound and strawberries for \$4.00 per pound. If she bought 10 more pounds of strawberries than bananas and she spent \$109 on bananas and strawberries, how many pounds of strawberries and how many pounds of bananas did she buy?
3. Justin makes a withdrawal from an ATM that dispenses only \$20 and \$5 bills. If the number of \$5 bills in his withdrawal is 5 less than the number of \$20 bills and he withdrew \$400, how many \$20 bills and how many \$5 bills did he receive?
4. After 3 hours two drivers who started from the same location and traveled in opposite directions are 330 miles apart; one is traveling 10 miles per hour faster than the other. Find the rate, in miles per hour, of the slower driver.
5. At speeds of 40 miles per hour and 20 miles per hour, two cars travel in opposite directions for the same amount of time until they are 420 miles apart. How many hours do they travel?
6. Theo and Jagdish, who are 568 miles apart, start driving toward each other in their cars. Theo drives at a rate of 40 miles per hour and Jagdish drives at a rate of 36 miles per hour. By the time they meet, Jagdish has driven one hour more than Theo. How many hours did Jagdish spend driving?
7. Two trains start at the same time from towns 385 miles apart and travel towards each other until they meet in 5 hours. If the rate of one train is 7 mph less than the rate of the other train, what is the rate of each?
8. Mikayla has saved \$170 in \$10, \$5, and \$1 bills. The number of \$1 bills that she saved is 30 more than the number of \$5 bills, which is 3 times the number of \$10 bills that she saved. Determine the number of each type of bill that Mikayla saved.
9. Aiden drove on a cross-country trip to a national park, taking the same route to and from his home. He averaged 50 miles per hour on the way to the park and 45 miles per hour on the way back, taking 2 more hours on the return trip. Determine the number of hours he spent driving to the park.
10. George drove his car from his home to his parents' home in the mountains at an average rate of 35 miles per hour and returned at an average rate of 40 miles per hour. Find his time going and returning if the time returning was one hour less than the time going.
11. Can \$4.50 in change consisting only of dimes and quarters be made such that the number of quarters is 8 more than the number of dimes? Justify your answer.
12. Can \$500 be made with \$20 and \$5 bills if the number of \$5 bills is 10 more than the number of \$20 bills? Justify your answer.

3.9 Transforming Formulas

In Section 3.8, we discussed the relationship between the distance that an object travels (d), the rate at which it travels (r), and the time it travels (t): distance = (rate)(time), or $d = rt$. What if you knew the distance and time and wanted to find the rate?

If Avery traveled 120 miles in 2 hours, Bella traveled 45 miles in 3 hours, and Charlene traveled 160 miles in 4 hours, who traveled the fastest? We calculate each person's rate by substituting into $d = rt$ (Table 3.5).

Table 3.5 Relating rate to distance and time.

Person	Avery	Bella	Charlene
Rate (r)	?	?	?
Time (t)	2 hours	3 hours	4 hours
Distance (d)	120 miles	45 miles	160 miles
Calculation	$d = rt$ $120 = r(2)$ $\frac{120}{2} = \frac{r(2)}{2}$ $r = 60$ miles per hour	$d = rt$ $45 = r(3)$ $\frac{45}{3} = \frac{r(3)}{3}$ $r = 15$ miles per hour	$d = rt$ $160 = r(4)$ $\frac{160}{4} = \frac{r(4)}{4}$ $r = 40$ miles per hour

We see from this table that determining the rate involves similar calculations – dividing the distance by the time. We can create a new equation by using $d = rt$ and solving for the variable r, treating the other variables as quantities:

$d = rt$

$\frac{d}{t} = \frac{rt}{t}$ Divide both sides by t.

$\frac{d}{t} = r$ Final answer, which we write as $r = \frac{d}{t}$.

An equation like $d = rt$ is an example of a **formula**, an identity with two or more variables. Because formulas typically consist mostly or entirely of letters, they are also called **literal equations**. For example, the area of a triangle can be found using the formula $A = \frac{bh}{2}$, where A represents the area, b represents the base of the triangle, and h represents the height.

Many times, we **transform** a formula by rearranging it to solve for another variable. Earlier, we transformed the formula $d = rt$ to solve for r. Although we can't simplify the final answer to a number, we use many of the same strategies to transform a formula that we use to solve equations:

- Eliminate parentheses and fractions.
- Get all terms with the variable we want to solve on one side of the equation and all remaining terms on the other side.

- Combine like terms.
- Use inverse operations to isolate the variable on one side of the equation.

Example 3.24 Solve $F = \frac{mv^2}{2}$ for m.

Solution: We undo everything being done to m by using inverse operations.

$F = \frac{mv^2}{2}$

$2(F) = 2\left(\frac{mv^2}{2}\right)$ Multiply both sides by 2.

$2F = mv^2$ Multiplying and dividing by the same number undo each other.

$\frac{2F}{v^2} = \frac{mv^2}{v^2}$ Divide both sides by v^2.

$\frac{2F}{v^2} = m$ Final answer, which we write as $m = \frac{2F}{v^2}$.

Example 3.25 Solve $ax = bx + c$ for x.

Solution: We get the terms with x on one side of the equation before solving for x.

$ax = bx + c$

$-bx \quad -bx$ Subtract bx from both sides.

$ax - bx = c$ Combine like terms on the right side.

$(a - b)x = c$ Distributive property.

$\frac{(a-b)x}{a-b} = \frac{c}{a-b}$ Divide both sides by $a - b$.

$x = \frac{c}{a-b}$ Final answer.

Exercises

1. Solve $a + 5 = c$ for a.
2. Solve $P = 4s$ for s.
3. Solve $I = prt$ for p.
4. Solve $\frac{PV}{T} = k$ for V.
5. Solve $y = mx + b$ for x.
6. Solve $m = \frac{a+b}{2}$ for a.
7. Solve $A = \frac{(x+y)h}{2}$ for y.
8. Solve $a = f + (n-1)d$ for n.
9. Solve $c(x + y) = d$ for x.
10. Solve $pq = w + pt$ for p.
11. Solve $ax - ry = x + w$ for x.
12. Solve $ax - by = xz + y$ for y.

CHAPTER 3 TEST

1. If $15y = 5$, what is the value of y?

   (A) $\frac{1}{3}$ (B) 10 (C) 5 (D) 3

2. Which expressions are like terms?

   (A) $3n$, $5n^2$, and $-7n$
   (B) $10m$, $-18m$, and $\frac{5}{3}m$
   (C) $4p$, $4q$, and $4a$
   (D) z, z^2, and z^3

3. What is the solution to $\frac{t}{12} + 1 = 0$?

   (A) -12 (B) 12 (C) -1 (D) $-\frac{1}{12}$

4. Which equation is equivalent to $2k + 10 = 14$?

   (A) $k + 10 = 14$
   (B) $-3k - 15 = -21$
   (C) $4k + 12 = 16$
   (D) $3k + 11 = 15$

5. Which expression represents the phrase "12 less than twice a number"?

   (A) $12 - 2n$ (B) $12(2) - n$ (C) $12(2 - n)$ (D) $2n - 12$

6. If $-\frac{y}{6} = 2$, then what is the value of y?

   (A) -3 (B) 3 (C) -12 (D) 12

7. If n is an odd integer, which expressions represent consecutive odd integers?

   (A) $n, n + 1, n + 2$
   (B) $n, n + 1, n + 3$
   (C) $n, n + 2, n + 4$
   (D) $n, n + 2, n + 3$

8. If $y = ax + bw$, then which equation is true?

   (A) $w = \frac{y - ax}{b}$
   (B) $w = \frac{y + ax}{b}$
   (C) $w = y - ax - b$
   (D) $w = ax + by$

9. Which expression illustrates the distributive property?

   (A) $(x + y)z = xz + yz$ (C) $x(yz) = (xy)z$

   (B) $x(yz) = xy + xz$ (D) $x + yz = (x + y)(x + z)$

10. Solve for c in the equation $\frac{12c}{5} = 4$.

11. Is the equation $3w + 2w - 16 = w + 9 - 7w + 12 + 11w$ an identity, a conditional equation, or a false statement? Justify your answer.

12. Write the pronunciation of $3(d + 7)$.

13. Find three consecutive integers such that twice the first added to the second equals 29 less than the third.

14. A father is 3 times as old as his son. In 15 years his age will be 5 years less than twice the son's age. How old is the father now?

15. Starting 280 miles apart, two drivers travel toward each other at rates of 35 mph and 45 mph, respectively, until they meet. What is the travel time of each?

CHAPTER 3 SOLUTIONS

3.1. 1. $t = 3$ 4. $y = 17$ 7. $u = -3$ 10. $f = 25$

2. $z = 9$ 5. $c = 8$ 8. $x = -\frac{1}{2}$ 11. $d = -36$

3. $n = 8$ 6. $w = \frac{1}{7}$ 9. $b = 18$ 12. $k = -20$

13. Equivalent equations have the same solution. If we substitute $x = 2$ into the first equation $(2 + 12 = 14)$ and second equation $(2 = 2)$, we get a true statement.

14. The second line of his work is incorrect. He mistakenly divided 6 by 3 when he should have divided 3 by 6 to get $\frac{6x}{6} = \frac{3}{6}$, or $x = \frac{1}{2}$.

15. The left side is $2x$ and the right side is 4.

3.2. 1. $y = 3$ 3. $p = 5$ 5. $p = \frac{23}{5}$ 7. $w = -\frac{13}{6}$ 9. $u = 24$

2. $m = 2$ 4. $q = \frac{9}{4}$ 6. $n = \frac{8}{3}$ 8. $a = 12$ 10. $x = 34$

3.3.

1. $x + 12$
2. $19 + x$
3. $3x$
4. $x + 5$
5. $\frac{x}{7}$
6. $2x$
7. $x - 7$
8. $x - 9$
9. $4x - 6$
10. $(x + 12) + 8$
11. $12x - 9$
12. $\frac{x}{6} - 10$
13. "8 more than a number" or "the sum of a number and 8"
14. "9 less than a number" or "9 subtracted from a number"
15. "a number divided by 3" or "the quotient of a number and 3"
16. "the sum of twice a number and 1"
17. "the sum of 3 times a number and 17"
18. "the sum of 4 and twice a number" or "4 more than twice a number"
19. "the quotient of a number and 30, added to 4"
20. "10 less than the quotient of a number and 60"
21. "the quotient of a number and 72, added to 5"
22. "3 times the sum of a number and 1"
23. "negative 4 times the sum of twice a number and 3"
24. "10 times the sum of twice a number and 7"
25. The word "and" could be used to represent addition ("2 and 2 make 4"), subtraction ("the difference of 6 and 4"), multiplication ("the product of 4 and 2"), or division ("the quotient of 8 and 4").
26. The word "by" can also be used to describe multiplication, such as "when 3 is multiplied by 4."

3.4.

1. 14
2. 10
3. 6
4. 12
5. 15
6. $\frac{48}{5}$
7. 11
8. 8
9. 4
10. 9
11. 5
12. 6

3.5.

1. $-3a$
2. $-11m$
3. $9u^2 - 10u + 5$
4. $6j^2 - 16j + 11$
5. $2ab - 3bc$
6. $-2x^2 - 4y^2 + 4$
7. $x = 3$
8. $z = 2$
9. $r = -\frac{5}{3}$
10. $d = 4$
11. $m = 7$
12. $j = \frac{24}{5}$
13. $b = 3$
14. $c = 8$
15. $k = \frac{1}{2}$

16. $a = 2$
17. $c = -3$
18. $q = -\frac{9}{2}$
19. 12
20. −5
21. 78 and 79
22. −31, −30, and −29
23. 66, 68, and 70
24. 79, 81, 83, and 85
25. The expression $\frac{4}{x}$ is not a term since it is not a product of a number and a variable. It is a quotient.
26. The expression $19qr$ is a term since it is the product of a number and variables.
27. The terms $8xy$ and $4x$ are not like terms since they don't have the same variables.
28. The terms $3w^2$ and $16w$ are not like terms since the variable w is not raised to the same exponent.

3.6.

1. $y = -3$
2. $p = 5$
3. No solution
4. $w = -5$
5. $p = 3$
6. y can be any number.
7. $d = -\frac{22}{3}$
8. $x = \frac{6}{7}$
9. No solution
10. $x = -\frac{23}{11}$
11. g can be any number.
12. $r = -9$
13. $v = -14$
14. $d = 8$
15. 7
16. 50
17. 3
18. Jon added $2x$ twice to the left side of the equation, violating the principle that whatever is done to one side of the equation must also be done to the other side. He should have combined the like terms $8x$ and $-2x$ to get $6x$.

3.7.

1. $a = 12$
2. $r = -13$
3. $w = -5$
4. $b = -8$
5. $p = 7$
6. $a = -9$
7. $z = -\frac{17}{2}$
8. $p = 6$
9. $m = -4$
10. $d = 5$
11. $q = -\frac{11}{5}$
12. $t = 1$
13. 23 and 24
14. 19 and 20
15. −17, −16, and −15

16. 24, 25, and 26
17. 50, 52, and 54
18. 35, 37, 39, and 41
19. The son is 4 and his father is 32.
20. Ru is 4 and Wen is 12.
21. Hector is 12 and Belinda is 22.
22. The son is 12 and the woman is 42.
23. The daughter is 12 and the father is 42.
24. Cyrus is 5 and Faraz is 20.
25. No. For example, $2 + 3(4) = 2 + 12 = 14$ does not equal $(2 + 3)(2 + 4) = 5(6) = 30$.
26. No. The order of operations states that multiplication must be done before addition, so 4 must be distributed: $1 + 4(7 + 2) = 1 + 4(7) + 4(2) = 1 + 28 + 8 = 37$.
27. If a, b, and c are numbers, then $a(b - c) = ab - ac$.

3.8.

1. 17 dimes and 21 quarters.
2. 15 pounds of bananas and 25 pounds of strawberries.
3. 17 \$20 bills and 12 \$5 bills.
4. The slower driver travels at 50 miles per hour.
5. The cars traveled for 7 hours.
6. Theo drove for 7 hours, so Jagdish drove for 8 hours.
7. The faster train travels at 42 miles per hour, and the slower train travels at 35 miles per hour.
8. Mikayla has 5 \$10 bills, 15 \$5 bills, and 45 \$1 bills.
9. Aiden spent 18 hours driving to the park.
10. George spent 8 hours driving to his parents' home and 7 hours driving back.
11. No, since the equation $10d + 25(d + 8) = 450$ has a solution of $\frac{50}{7}$ dimes, which is not possible.
12. Yes (18 \$20 bills and 28 \$5 bills).

3.9.

1. $a = c - 5$
2. $s = \frac{P}{4}$
3. $p = \frac{I}{rt}$
4. $V = \frac{Tk}{P}$
5. $x = \frac{y - b}{m}$
6. $a = 2m - b$
7. $y = \frac{2A - xh}{h}$ or $y = \frac{2A}{h} - x$
8. $n = \frac{a - f}{d} + 1$ or $n = \frac{a - f + d}{d}$
9. $x = \frac{d}{c} - y$ or $x = \frac{d - cy}{c}$
10. $p = \frac{w}{q - t}$
11. $x = \frac{w + ry}{a - 1}$
12. $y = \frac{ax - xz}{1 + b}$

CHAPTER 3 TEST SOLUTIONS

1. (A)
2. (B)
3. (A)
4. (B)
5. (D)
6. (C)
7. (C)
8. (A)
9. (A)
10. $c = \frac{5}{3}$.
11. Combining like terms on each side gives us $5w - 16 = 5w + 21$, or $-16 = 21$, which is a false statement.
12. "three times the sum of *d* and 7"
13. $-14, -13, -12$
14. The father's age is 30.
15. $\frac{7}{2}$ hours.

4 RATIOS AND PROPORTIONS

As we said in Chapter 2, fractions have been a central part of number systems around the world for thousands of years. In this chapter, we focus on how we use fractions to compare quantities.

4.1 Expressing Ratios in Simplest Form

In Figure 4.1, the distance from point A to point B (AB) is 6 units, and the distance from point B to point C (BC) is 4 units.

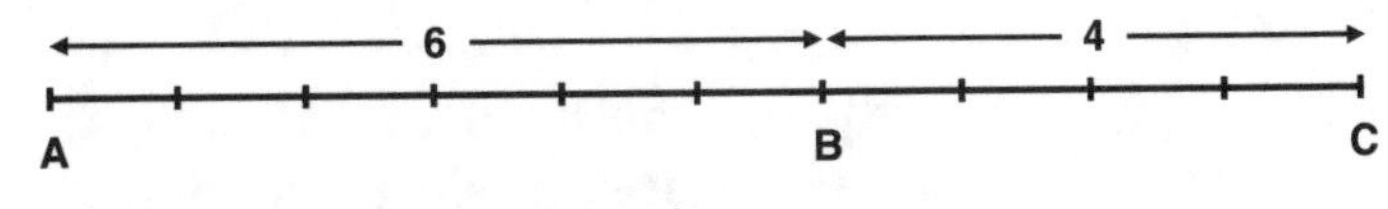

Figure 4.1 Number line showing a 6:4 ratio.

To describe the relationship between AB and BC, we can use a **ratio** – a quantity that indicates how many times one number contains another. We express the number of times that AB contains BC in any of the following ways:

- 6 to 4
- 6:4
- 6/4
- $\frac{6}{4}$

Reading and Writing Tip

Although ratios can be written in different ways, they are pronounced as "[number] *to* [number]." All the ratios listed above are pronounced as "6 to 4." When we write a ratio as a fraction, we don't use the pronunciation used for fractions ("six-fourths" or "six over four").

The ratio of AB to BC should not be confused with the ratio of AB to the entire segment length AC. The ratio of the two parts of the segment to each other (AB:BC) is 6:4, but the ratio of the part AB to the whole (AB:AC) is 6:10. Like fractions, ratios are often expressed in simplest form, so we usually write the ratio AB:BC as 3 to 2, 3:2, 3/2, or $\frac{3}{2}$.

To describe how many times BC contains AB, we write the ratio 2 to 3, 2:3, 2/3, or $\frac{2}{3}$.

If the quantities in a ratio have the same units, we don't have to include these units in the final form of the ratio. For example, the ratio $\frac{30\ miles}{4\ miles}$ is typically written as $\frac{30}{4}$ or $\frac{15}{2}$. When the quantities in a ratio have different units, we include them. In these cases, the ratio $a{:}b$ represents a portion of a that corresponds to one unit of b, so $\frac{30\ miles}{4\ trips}$ is expressed as $\frac{15}{2}$ miles per trip.

Example 4.1 Express 8 cups to 3 cups as a ratio (with units if appropriate) in simplest form.

Solution: Express the ratio in any of the following ways:

- The ratio of cups is 8 to 3.
- The ratio of cups is 8:3.
- The ratio of cups is 8/3.
- The ratio of cups is $\frac{8}{3}$.

Example 4.2 Express 150 books to 60 boxes as a ratio (with units if appropriate) in simplest form.

Solution: First, divide the quantities in the order presented: $\frac{150\ books}{60\ boxes}$. To simplify the fraction, find the greatest common factor of 150 and 60:

$$\frac{150}{60} = \frac{30(5)}{30(2)} = \frac{30}{30}\cdot\frac{5}{2} = \frac{5}{2}$$

(If necessary, go to Section 2.2 to review how to simplify fractions.)
Express the ratio in any of the following ways:

- The ratio of books to boxes is 5 to 2.
- The ratio of books to boxes is 5:2.
- The ratio of books to boxes is 5/2.
- The ratio of books to boxes is $\frac{5}{2}$.
- There are $\frac{5}{2}$ (or 2.5) books per box.
- $\frac{5\ books}{2\ boxes}$

Example 4.3 There are about 1 billion passenger cars and 6 billion mobile phones in use around the world. Express the ratio of passenger cars to mobile phones in simplest form.

Solution: To express the ratio of cars to mobile phones, write the number of cars as the first number in the ratio and the number of mobile phones as the second number. The numbers in this example are very large (1,000,000,000 and 6,000,000,000), but since both numbers are expressed in billions, we can express this relationship using the numbers 1 and 6.

Express the ratio in any of the following ways:

- The ratio of passenger cars to mobile phones is 1 to 6.
- The ratio of passenger cars to mobile phones is 1:6.
- The ratio of passenger cars to mobile phones is $\frac{1}{6}$.
- There is 1 passenger car for every 6 mobile phones.
- 1 passenger car per 6 mobile phones, or $\frac{1 \textit{ passenger car}}{6 \textit{ mobile phones}}$

When we compare more than two quantities, we avoid writing ratios as fractions and use the colon format, as shown in the following example:

Example 4.4 Express 150 books to 20 magazines to 60 boxes as a ratio (with units if appropriate) in simplest form.

Solution: The ratio of books to magazines to boxes is 150:20:60. We simplify this by dividing all numbers by their greatest common factor:

$$150 = 10(15)$$

$$20 = 10(2)$$

$$60 = 10(6)$$

Express the ratio in any of the following ways:

- The ratio of books to magazines to boxes is 15 to 2 to 6.
- The ratio of books to magazines to boxes is 15:2:6.

Section 3.8 included problems involving rates, which are a type of ratio. Recall that a rate expresses the frequency with which an event happens. When we describe rates, we use the word "per" (which means "for each" and indicates division), followed by 1 unit of time. Thus, each of the following:

- the ratio of 55 miles to 1 hour
- traveling 55 miles in 1 hour
- in one hour, 55 miles traveled

is expressed as "55 miles per hour" or 55 $\frac{\textit{miles}}{\textit{hour}}$.

Exercises

Write the pronunciation of each of the following ratios:

1. 4:1
2. $\frac{16}{3}$
3. 17/12
4. 7:12:19
5. 12:2:1
6. 10:20:30:50

Express each of the following as a ratio or rate in simplest form.

7. 3 centimeters to 1 kilometer
8. 2 meters to 1 person
9. 160 kilometers in 4 hours
10. 50 lights to 10 backyards
11. 15 feet in 3 minutes
12. 64 pounds to 32 pounds
13. 36 square centimeters to 3 meters
14. 15 liters to 36 liters
15. 64 centimeters in 18 days
16. 2 kilometers to 2,000 meters to 200,000 centimeters
17. 45 dimes to 12 quarters to 9 pennies
18. 18° to 30° to 15°
19. Every year, the average American produces 700 pounds of material that is recycled and 1,400 pounds of trash. Express the ratio of recycled material to trash as a ratio in simplest form.
20. According to the United Nations Food and Agriculture Organization, about 1,000,000 square kilometers of forest is cleared every 10 years. Express the rate of deforestation in simplest form with appropriate units.
21. To make pie crusts, a pastry chef uses 6 cups of flour, 4 cups of butter, and 2 cups of water. Express the ratio of flour to butter to water used by the chef as a ratio in simplest form.
22. According to the Pew Research Center, the median age of Latinos in the United States is 30 years and the median age of Whites in the United States is 44 years. Express the ratio of the ages of Latinos to Whites as a ratio in simplest form.
23. On weekdays, vehicles in midtown Manhattan in New York City typically travel about 10 miles in 2 hours. Express the rate of vehicular travel in midtown Manhattan with appropriate units.
24. In the United States, there were 47 million registered Democrats, 35 million registered Republicans, and 37 million registered independents and members of other parties. Express the ratio of Republicans to Democrats to independents and members of other parties.

25. According to recommendations from the Centers for Disease Control, a person needs 30 weeks to lose 45 pounds in a healthy manner. Express this weight loss as a rate with appropriate units.

26. A class has 2 teachers and 24 students. Express the ratio of students to teachers in simplest form.

27. A baseball team played 144 games and won 96. If there are no ties, express the number of wins to losses as a ratio in simplest form.

4.2 Using Ratios in the Real World

We often use ratios to find the number of people or items that are part of a known larger quantity. For example, among 30 students responding to a survey on their mood, the ratio of students who responded "happy" to students who responded "sad" was 3:2. If every student responded with one of those two choices, how many students were happy and how many were sad?

The 3:2 ratio means that for every group of 3 happy students, there is a group of 2 sad students (Figure 4.2):

Figure 4.2 5 students in a 3:2 ratio.

Consider Figure 4.3. If we have 2 groups of 3 happy students, then we also have 2 groups of 2 sad students (6 happy + 4 sad = 10 students in all):

Figure 4.3 10 students in a 3:2 ratio.

In Figure 4.4, if we have 3 groups of 3 happy students, then we also have 3 groups of 2 sad students (9 happy and 6 sad = 15 students in all):

Figure 4.4 15 students in a 3:2 ratio.

We summarize our work in Table 4.1:

Table 4.1 Students in a 3:2 ratio.

Number of Groups of 3 Happy Students	Number of Groups of 2 Sad Students	Number of Happy Students	Number of Sad Students	Ratio of Happy Students to Sad Students	Total Number of Students
1	1	1(3) = 3	1(2) = 2	3:2	5
2	2	2(3) = 6	2(2) = 4	6:4	10
3	3	3(3) = 9	3(2) = 6	9:6	15

If we continue this pattern until we reach 30 students in all, we discover that we have 6 groups of 3 = 6(3) = 18 happy students and 6 groups of 2 = 6(2) = 12 sad students (Figure 4.5):

This method works well if we have only 30 students, but what if we were working with larger numbers? For example, if we had 375 students, drawing a picture that would be difficult. Fortunately, we can solve this problem algebraically. From our work above, we see that the number of groups of 3 equals the number of groups of 2, but that number is unknown. We can represent this unknown number of groups with a variable x. If we compare Table 4.1 to Table 4.2, we see a pattern:

There are $3x$ happy students and $2x$ sad students in a 3:2 ratio. Since there are 375 students in all, then:

Number of happy students	+	Number of sad students	= Total number of students
$3x$	+	$2x$	= 375

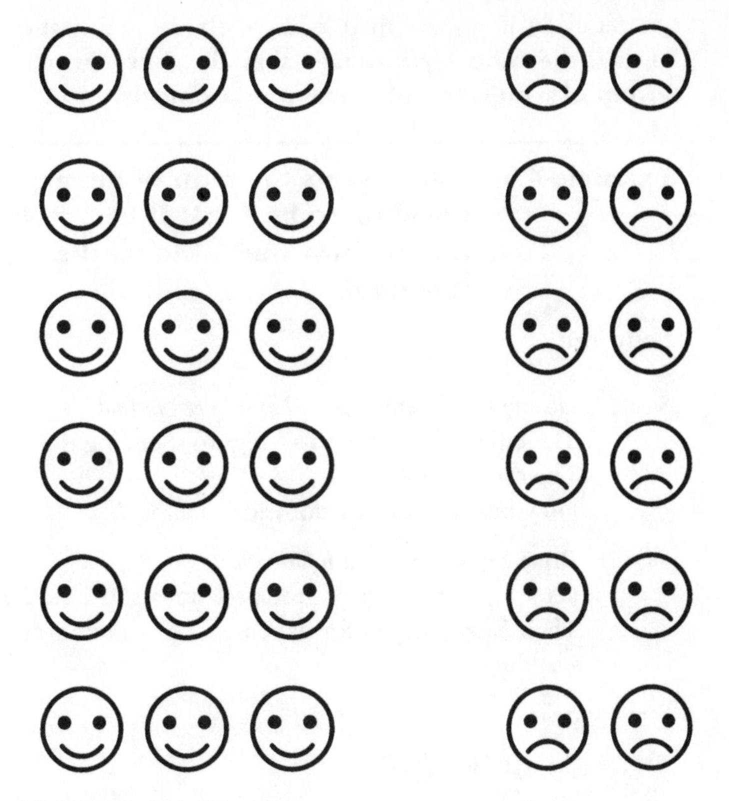

Figure 4.5 30 students in a 3:2 ratio.

Table 4.2 375 students in a 3:2 ratio.

Number of Groups of 3 Happy Students	Number of Groups of 2 Sad Students	Number of Happy Students	Number of Sad Students	Ratio of Happy Students to Sad Students	Total Number of Students
1	1	$1(3) = 3$	$1(2) = 2$	3:2	5
2	2	$2(3) = 6$	$2(2) = 4$	6:4	10
3	3	$3(3) = 9$	$3(2) = 6$	9:6	15
…	…	…	…	…	…
x	x	$x(3) = 3x$	$x(2) = 2x$	$3x{:}2x$	375

Solving this equation for x gives us $5x = 375$, or $x = 75$. This tells us that there are 75 groups of happy students and 75 groups of sad students in a 3:2 ratio. Thus, there are $75(3) = 225$ happy and $75(2) = 150$ sad students.

Table 4.2 shows that we can interpret ratios as follows: **the ratio $x{:}y$ means that for every group of x objects, there is a group of y objects.**

Similarly, if we have three types of objects in the ratio x:y:z, then we say the following: **the ratio x:y:z means that for every group of x objects, there is a group of y objects and a group of z objects.**

Example 4.5 Maria wants the ratio of her expenses to savings to spending to be 5:2:3. If her weekly income after taxes is \$850, how much should she set aside for each category?

Solution:

Step 1: *Identify what is given and what we need to find.*
The ratio of expenses to savings to spending is 5:2:3.
The total is \$850.
How much is set aside for each category?

Step 2: *Represent the unknown information.*
Let x = the number of groups of dollars in a 5:2:3 ratio.
Then $5x$ = amount for expenses, $2x$ = amount for savings, $3x$ = amount for spending.

Step 3: *Solve.*

$$5x + 2x + 3x = 850$$

$$10x = 850 \quad \text{Combine like terms.}$$

$$\frac{10x}{10} = \frac{850}{10} \quad \text{Divide both sides by 10.}$$

$$x = 85$$

She should set aside 85(5) = \$425 per week for expenses, 85(2) = \$170 per week for savings, and 85(3) = \$255 per week for spending.

Step 4: *Check.*

$$425 + 170 + 255 \stackrel{?}{=} 850$$

$$850 = 850$$

Did You Know?

Muslim mathematicians often used their knowledge of ratios and fractions to solve problems involving inheritance, which is an important part of Islamic law. These laws protected women in case their husbands died or divorced them. Example 4.6 is based on a problem that the Persian mathematician al-Khwārizmī wrote in the 9th century.

Example 4.6 **A woman dies, leaving \$400,000 to her husband, son, and three daughters. Her husband receives one-fourth of the amount. The rest is divided among the children so that the son receives twice as much as a daughter. How much does each person receive?**

Solution:

Step 1: Identify what is given and what we need to find.
\$400,000 is divided among a husband, son, and three daughters.
The husband receives one-fourth. The children receive the rest.
The son receives twice as much as each daughter.
How much does each person get?

Step 2: Represent the unknown information.
The husband receives $\frac{1}{4}$(\$400,000)= \$100,000. The remaining \$400,000 − \$100,000 = \$300,000 is divided among the four children. The son receives twice the amount that each daughter receives.
Let x = the amount of money each daughter receives.
Then $2x$ = the amount of money the son receives.

Step 3: Solve.

$x + x + x + 2x = 300,000$ — The 3 daughters and 1 son receive \$300,000 in all.

$5x = 300,000$ — Combine like terms.

$\frac{5x}{5} = \frac{300,000}{5}$ — Divide both sides by 5.

$x = 60,000$ — The amount each daughter receives is \$60,000.

The husband receives \$100,000, the son receives 2(60,000) = \$120,000, and each daughter receives \$60,000.

Step 4: Check.

$$100,000 + 120,000 + 3(60,000) \stackrel{?}{=} 400,000$$

$$400,000 = 400,000$$

Exercises

1. A jar of 144 candies consists of orange and blue Minis in a 3:5 ratio. How many orange Minis and how many blue Minis are in the jar?

2. The ratio of warehouse workers to cashiers at a large retail store with 946 employees is 8:3. How many warehouse workers and how many cashiers are at the store?

3. Travelling in opposite directions at speeds in the ratio of 7:3, a high-speed train and a car are 360 miles apart at the end of 3 hours. Find the rate of each. (Hint: Use $7r$ and $3r$ to represent their respective rates.)

4. People who are food insecure lack reliable access to a sufficient quantity of nutritious, affordable food. In the neighborhood of Elmwood, the ratio of food secure households to food insecure households is 17:3. If Elmwood has 3,200 households, how many food secure households and how many food insecure households live there?

5. The ratio of Whites to non-Whites in the US population is approximately 3:2. If the racial makeup of the 100-member US Senate were representative of the national population, how many senators would be White and how many would be non-White?

6. At a small business, \$288,000 in profits is shared by four partners in the ratio 3:5:7:9. How much money is in each partner's share?

7. To make concrete, a builder mixes cement, sand, and aggregate in a 1:2:3 ratio. How many buckets of each ingredient are required to make 270 buckets of concrete?

8. The ratio of red to yellow to blue in a blend of paint is 4:2:1. If the blend contains only these three colors, how many cups of each color is required to make 14 cups of the blend?

9. According to the US Mint, the ratio of copper to zinc to manganese to nickel in a dollar coin is approximately 89:6:3:2. Each dollar coin weighs about 8 grams. Determine the number of grams of each metal that are present in 500 dollar coins.

4.3 Proportions and Equations with Fractions

A **proportion** is a statement that two ratios are equal. When we write a ratio as a fraction, then we write a proportion as an equation: $\frac{1}{2} = \frac{3}{6}$. When we write a ratio using a colon, then we use two colons or an equal sign to write a proportion: 1:2::3:6 or 1:2 = 3:6, both of which are read as "1 is to 2 as 3 is to 6."

In the proportion $a{:}b = c{:}d$, the quantities a and d are called the **extremes** (since they are farthest apart), and b and c are called the **means** (since they are the middle terms) (Figure 4.6):

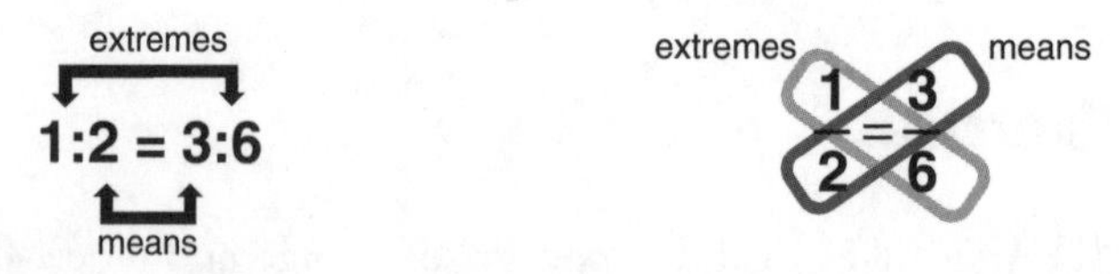

Figure 4.6 Means and extremes in a proportion.

Since ratios can be written as fractions, we use the rules for operations with fractions that we discussed in Chapter 2. Since proportions are equations, we apply the rules for solving equations that we discussed in Chapter 3.

Proportions have the following important property:

In a proportion, the product of the means equals the product of the extremes.

For example, in the proportion $\frac{1}{2} = \frac{3}{6}$ or 1:2::3:6, 2 and 3 are the means and 1 and 6 are the extremes, so $2(3) = 1(6)$.

Watch Out!

Some people refer to this property of proportions as "cross-multiplying." In our experience, people who call this "cross-multiplying" often misapply the term to multiplying fractions, so they mistakenly believe that $\frac{1}{2} \cdot \frac{1}{3} = \frac{1(3)}{2(1)} = \frac{3}{2}$ instead of $\frac{1}{6}$. We recommend you *not* use this term.

Example 4.7 Determine if the ratios $\frac{24}{30}$ and $\frac{12}{15}$ can be used to form a true proportion.

Solution: We could simplify both fractions to see if they are equal. A shorter method is to see if the product of the means equals the product of the extremes in the proportion $\frac{24}{30} = \frac{12}{15}$. Since $24(15) = 360$ and $30(12) = 360$, the product of the means equals the product of the extremes, so the ratios form a true proportion.

We use this fact to help us solve proportions with variables.

Example 4.8 Solve for the variable: $\frac{4}{7} = \frac{x}{11}$.

Solution:

$7x = 4(11)$	In a proportion, the product of the means equals the product of the extremes.
$\frac{7x}{7} = \frac{4(11)}{7}$	Divide both sides by 7.
$x = \frac{4(11)}{7}$	Multiplying and dividing by the same number undo each other.
$x = \frac{44}{7}$	Simplify.

Example 4.9 Solve for the variable: $\frac{4}{5} = \frac{3x-2}{9}$.

Solution: Since the numerator of the second fraction has more than one term, use parentheses and distribute.

$4(9) = 5(3x - 2)$	In a proportion, the product of the means equals the product of the extremes.
$36 = 15x - 10$	Simplify and use the distributive property.
$+10 \quad +10$	Add 10 to both sides.
$46 = 15x$	Combine like terms.
$\frac{46}{15} = \frac{15x}{15}$	Divide both sides by 15.
$\frac{46}{15} = x$	Final answer, which we typically write as $x = \frac{46}{15}$.

Technology Tip

Substituting values into an equation to check your answer can get difficult, especially if the answer is a fraction. For example, to check the answer to Example 4.9, we must determine if $\frac{3\left(\frac{46}{15}\right)-2}{9}$ equals $\frac{4}{5}$. We strongly recommend using your device's fraction function. This will be easier than typing an expression with nested parentheses, like (3(46/15) – 2)/9.

Some equations with fractions may not look like proportions at first glance, but they can be easily rearranged into proportions:

Example 4.10 Solve for the variable: $\frac{9h}{4} + 4 = 11$.

Solution:

$\frac{9h}{4} + 4 = 11$	
$-4 \quad -4$	Subtract 4 from both sides.
$\frac{9h}{4} = 7$	Combine like terms.
$\frac{9h}{4} = \frac{7}{1}$	A number divided by 1 equals itself.
$9h(1) = 7(4)$	In a proportion, the product of the means equals the product of the extremes.
$9h = 28$	Simplify.
$\frac{9h}{9} = \frac{28}{9}$	Divide both sides by 9.
$h = \frac{28}{9}$	Final answer.

When fractions in an equation have the same denominator, combine them and make the equation into a proportion:

Example 4.11 Solve for the variable: $\frac{13u}{3} + \frac{7u}{3} = 4$.

Solution:

$\frac{13u}{3} + \frac{7u}{3} = 4$	
$\frac{20u}{3} = 4$	$\frac{a}{c} + \frac{b}{c} = \frac{a+b}{c}$, where $c \neq 0$.
$\frac{20u}{3} = \frac{4}{1}$	A number divided by 1 equals itself.
$20u(1) = 4(3)$	In a proportion, the product of the means equals the product of the extremes.
$20u = 12$	Simplify.
$\frac{20u}{20} = \frac{12}{20}$	Divide both sides by 20.
$u = \frac{12}{20}$	Multiplying and dividing by the same number undo each other.
$u = \frac{4 \cdot 3}{4 \cdot 5}$	Group common factors in the numerator and denominator.
$u = \frac{4}{4} \cdot \frac{3}{5}$	$\left(\frac{a}{b}\right)\left(\frac{c}{d}\right) = \frac{ac}{bd}$, where $b \neq 0$ and $d \neq 0$.
$u = \frac{3}{5}$	Final answer.

When fractions in an equation have different denominators, multiply both sides of the equation by the least common denominator to eliminate fractions and make the equation easier to solve.

Example 4.12 Solve for the variable: $\frac{k}{2} + \frac{k}{3} = 5$.

Solution:

$\frac{k}{2} + \frac{k}{3} = 5$	
$\frac{k}{2} + \frac{k}{3} = \frac{5}{1}$	A number divided by 1 equals itself.
$\frac{6}{1}\left(\frac{k}{2} + \frac{k}{3}\right) = \frac{6}{1}\left(\frac{5}{1}\right)$	Multiply both sides by the LCD, which is $3(2)(1) = 6 = \frac{6}{1}$.
$\frac{6}{1}\left(\frac{k}{2}\right) + \frac{6}{1}\left(\frac{k}{3}\right) = \frac{6}{1}\left(\frac{5}{1}\right)$	Distributive property.
$\frac{6k}{2} + \frac{6k}{3} = \frac{30}{1}$	$\left(\frac{a}{b}\right)\left(\frac{c}{d}\right) = \frac{ac}{bd}$, where $b \neq 0$ and $d \neq 0$.

$3k + 2k = 30$	Simplify fractions.
$5k = 30$	Add like terms.
$\frac{5k}{5} = \frac{30}{5}$	Divide both sides by 6.
$k = 6$	Final answer.

Some word problems are solved using equations with fractions.

Example 4.13 The formula to convert temperature from the Celsius to Fahrenheit scale is $F = \frac{9}{5}C + 32$, where F represents the temperature in degrees Fahrenheit and C represents the temperature in degrees Celsius. Solve the formula for C.

Solution:

$F = \frac{9}{5}C + 32$	
$\frac{F}{1} = \frac{9}{5}\left(\frac{C}{1}\right) + \frac{32}{1}$	A number divided by 1 equals itself.
$\frac{5}{1}\left(\frac{F}{1}\right) = \frac{5}{1}\left(\frac{9}{5}\left(\frac{C}{1}\right) + \frac{32}{1}\right)$	Multiply both sides by the LCD, which is 5(1) $= 5 = \frac{5}{1}$.
$\frac{5}{1}\left(\frac{F}{1}\right) = \frac{5}{1}\left(\frac{9}{5}\right)\left(\frac{C}{1}\right) + \frac{5}{1}\left(\frac{32}{1}\right)$	Distributive property.
$\frac{5F}{1} = 9C + 160$	$\left(\frac{a}{b}\right)\left(\frac{c}{d}\right) = \frac{ac}{bd}$, where $b \neq 0$ and $d \neq 0$.
$5F = 9C + 160$	Simplify.
$-160 \quad\quad -160$	Subtract 160 from both sides.
$5F - 160 = 9C$	Combine like terms.
$\frac{5F - 160}{9} = \frac{9C}{9}$	Divide both sides by 9.
$\frac{5F - 160}{9} = C$	Final answer, which we write as $C = \frac{5F - 160}{9}$.

Here is a modern translation of a problem from a 3,500-year-old papyrus copied by the Egyptian scribe Ahmes:

Example 4.14 The sum of a number and one-fourth the number is 15. What is the number?

Solution:

Step 1: Identify what is given and what we need to find.

The sum of a number and $\frac{1}{4}$ of the number is 15.

Find the number.

Step 2: *Represent the unknown information.*
Let $n =$ the number.

The sum of a number and	$n +$
one-fourth the number	$\frac{1}{4}n$
is 15.	$= 15$

Step 3: *Solve.*

$$n + \frac{1}{4}n = 15$$

$$\frac{n}{1} + \frac{1}{4}n = \frac{15}{1}$$ A number divided by 1 equals itself.

$$\frac{4}{1}\left(\frac{n}{1} + \frac{1}{4}n\right) = \frac{4}{1}\left(\frac{15}{1}\right)$$ Multiply both sides by the LCD, which is $4 = \frac{4}{1}$.

$$\frac{4}{1}\left(\frac{n}{1}\right) + \frac{4}{1}\left(\frac{1}{4}n\right) = \frac{4}{1}\left(\frac{15}{1}\right)$$ Distributive property.

$$\frac{4}{1}n + \frac{4}{4}n = \frac{60}{1}$$ $\left(\frac{a}{b}\right)\left(\frac{c}{d}\right) = \frac{ac}{bd}$, where $b \neq 0$ and $d \neq 0$.

$$4n + n = 60$$ Simplify fractions.

$$5n = 60$$ Add like terms.

$$\frac{5n}{5} = \frac{60}{5}$$ Divide both sides by 5.

$$n = 12$$ The number is 12.

Step 4: *Check.*

$$12 + \frac{1}{4}(12) \stackrel{?}{=} 15$$

$$15 = 15$$

The number is 12.

We often use proportions to solve problems involving rates.

Example 4.15 A car travels 70 miles in 4 hours. At that rate, how many miles would the car travel in 6 hours?

Solution:

Step 1: *Identify what is given and what we need to find.*
A car travels 70 miles in 4 hours.
Find the distance traveled in 6 hours at the same rate.

Step 2: *Represent the unknown information.*
The car travels at a rate of $\frac{70 \text{ miles}}{4 \text{ hours}}$.
Let d = distance the car travels in 6 hours. The car's rate is also equal to $\frac{d \text{ miles}}{6 \text{ hours}}$.

Step 3: *Solve.*

$$\frac{70}{4} = \frac{d}{6}$$

$70(6) = 4d$ — In a proportion, the product of the means equals the product of the extremes.

$420 = 4d$ — Simplify.

$\frac{420}{4} = \frac{4d}{4}$ — Divide both sides by 4.

$105 = d$ — Final answer. The car travels 105 miles.

Step 4: *Check.*

$$\frac{70}{4} \stackrel{?}{=} \frac{105}{6}$$

$$17.5 = 17.5$$

Problems involving proportions can easily be confused with problems involving ratios, discussed in Section 4.2. Table 4.3 shows three problems that involve ratios or proportions.

The table shows the following:

- In the first problem, we compare the number of red candies to the number of yellow candies.
- In the second problem, we compare the number of red candies in a bag to the number of red candies in several bags.
- In the third problem, we compare the number of red candies to the total number of candies.

To set up problems involving ratios correctly, writing units is important!

Exercises

Write the pronunciation of the following:

1. 9:12::6:8 **2.** 9:18::1:2 **3.** 5:15 = 9:27

Table 4.3 Problems with ratios or proportions.

Problem	Picture	Solution
The ratio of red to yellow candies in a bag is 3:2. Amanda has a bag with **10 yellow candies**. How many red candies does Amanda have?	RRRYY RRRYY RRRYY RRRYY RRRYY	x = number of **red candies** $\frac{3 \text{ red candies}}{2 \text{ yellow candies}} = \frac{x \text{ red candies}}{10 \text{ yellow candies}}$ x = 15 red candies
The ratio of red to yellow candies in a bag is 3:2. Amanda has **10 bags, each with 5 candies**. How many red candies does Amanda have?	RRRYY RRRYY RRRYY RRRYY RRRYY RRRYY RRRYY RRRYY RRRYY RRRYY	x = number of **red candies** $\frac{3 \text{ red candies}}{1 \text{ bag}} = \frac{x \text{ red candies}}{10 \text{ bags}}$ x = 30 red candies
The ratio of red to yellow candies in a bag is 3:2. Amanda has **10 candies in all**. How many red candies does Amanda have?	RRRYY RRRYY	x = number of **groups of candies** in 3:2 ratio $3x$ = number of red candies $2x$ = number of yellow candies Number of red candies + Number of yellow candies = 10 $3x + 2x = 10$ x = 2 groups of candies $3(2)$ = 6 red candies

In each proportion, identify the means and the extremes.

4. $8:24 = 7:21$

5. $11:4 = 33:12$

6. $\frac{100}{3} = \frac{45}{1.35}$

Determine whether each equation is a true proportion. Justify your answer.

7. $\frac{2}{3} = \frac{3}{4}$

8. $\frac{3}{4} = \frac{6}{8}$

9. $\frac{1}{3} = 1 - \frac{2}{3}$

Solve each equation for the variable. Express each answer in simplest form.

10. $\frac{y}{12} = \frac{3}{4}$

11. $\frac{13b}{12} = \frac{52}{2}$

12. $\frac{3}{a} = \frac{6}{4}$

13. $\frac{b-6}{b} = \frac{10}{7}$

14. $\frac{3}{4c} = \frac{9}{4+c}$

15. $\frac{8+2q}{3} = \frac{32}{6}$

16. $\frac{2m+7}{5} = \frac{3m-1}{8}$

17. $\frac{8}{5p-9} = \frac{6}{3p+1}$

18. $\frac{9r+1}{8r+2} = \frac{7}{5}$

19. $\frac{k}{3} + 4 = 8$

20. $\frac{g}{5} - 12 = -15$

21. $19 = \frac{7k}{2} + 9$

22. $\frac{a}{3} + \frac{a}{4} = 7$

23. $\frac{2w}{3} + w = 5$

24. $\frac{p}{3} - p = 2$

25. $\frac{3k}{4} - \frac{2k}{3} = \frac{3}{4}$

26. $\frac{n}{4} + \frac{n}{3} + \frac{n}{2} = 26$

27. $\frac{3d-1}{7} = 2d + 3$

28. If 2 inches of snow fall in 5 hours, how many inches of snow would fall in 16 hours if snow keeps falling at the same rate?

29. One molecule of sulfuric acid contains 2 atoms of hydrogen, 1 atom of sulfur, and four atoms of oxygen. Determine the number of hydrogen atoms in 161 molecules of sulfuric acid.

30. Jenna spends 90 minutes on social media every 2 days. At that rate, how many minutes will she spend on social media in 14 days?

31. The US Geological Survey estimates that 4 drips from a leaky faucet equals 1 milliliter of water. How many milliliters of water will leak from the faucet after 300 drips?

32. The sum of a number and half the number is 16. Find the number. (*Source: Ahmes Papyrus*)

33. Three times a number added to one-third the number equals 1. Find the number. (*Source: Ahmes Papyrus*)

34. The sum of a number and one-seventh of the number is 19. Find the number. (*Source: Ahmes Papyrus*)

35. The sum of a quantity, its two-thirds, its half, and its one-seventh is 388. Determine the quantity.

Questions to Think About

36. Prove that for the proportion $\frac{a}{b} = \frac{c}{d}$, the product of the means equals the product of the extremes. (HINT: Multiply both sides by bd.)

4.4 Converting Units

When discussing rates, we often need to convert between units. Let's say that we need to determine the number of seconds in 30 days. We know that 1 day has 24 hours, 1 hour has 60 minutes, and 1 minute has 60 seconds. We could set up several proportions, one for each unit conversion (each proportion has the same units in its numerators and denominators):

$$\frac{30\ days}{h\ hours} = \frac{1\ day}{24\ hours}, \text{so } h = \frac{30\ days}{1} \cdot \frac{24\ hours}{1\ day} = 720 \text{ hours}$$

$$\frac{720\ hours}{m\ minutes} = \frac{1\ hour}{60\ minutes}, \text{so } m = \frac{720\ hours}{1} \cdot \frac{60\ minutes}{1\ hour} = 43{,}200 \text{ minutes}$$

$$\frac{43{,}200\ minutes}{x\ seconds} = \frac{1\ minute}{60\ seconds}, \text{so } x = \frac{43{,}200\ minutes}{1} \cdot \frac{60\ seconds}{1\ minute} = 2{,}592{,}000 \text{ seconds}$$

Writing all of these proportions is tedious. Fortunately, we can combine these calculations into one line of work as follows:

$$30 \text{ days} = \frac{30\ days}{1} \cdot \frac{24\ hours}{1\ day} \cdot \frac{60\ minutes}{1\ hour} \cdot \frac{60\ seconds}{1\ minute} = 2{,}592{,}000 \text{ seconds}$$

We must write each fraction so that the units divide properly. If the same unit appears in the numerator and denominator of fractions, it will not appear in our final answer.

Sometimes, we need to give approximate answers. In these situations, we round to a given place.

How to Round Numbers

1. Identify the place that you are rounding to. This is the last digit of the number after it's rounded (Table 4.4).
2. Identify the digit to the right of the place you found in step 1 and round up if necessary:
   - If this digit is 5, 6, 7, 8, or 9, add 1 to the digit from step 1.
   - If this digit is 0, 1, 2, 3, or 4, keep the digit from step 1 unchanged.
3. Drop all digits to the right of the digit from step 1.

Table 4.4 Place value.

PLACE	Millions		Hundred-thousands	Ten-thousands	Thousands		Hundreds	Tens	Ones	*Decimal Point*	Tenths	Hundredths	Thousandths	Ten-Thousandths	Hundred-Thousandths
DIGIT	8	,	5	0	3	,	1	6	9	.	4	0	7	2	5

For example:

- 64.352 rounded to the nearest hundredth is 64.35. The 2 in the thousandths place tells us to keep the 5 in the hundredths place unchanged.
- 64.352 rounded to the nearest tenth is 64.4. The 5 in the hundredths place tells us to add 1 to the 3 in the tenths place.

Example 4.16 Convert 9.8 meters per second to the nearest tenth of a mile per hour. (0.62 miles = 1 kilometer, 1 kilometer = 1,000 meters)

Solution: We write each conversion factor as a fraction. First, we look at our conversion factors to map out which units we must convert. To change meters per second to miles per hour, we convert as follows:

meters → kilometers → miles

seconds → minutes → hours

We start by converting meters to kilometers. (We could also start by converting seconds to minutes.) Since meters is in the numerator of a fraction, we write the conversion factor so that meters is in the denominator. This gives us the rate in kilometers per second:

$$\frac{9.8\ \cancel{meters}}{1\ second} \bullet \frac{1\ kilometer}{1,000\ \cancel{meters}}$$

Next, we convert from kilometers to miles, writing the conversion factor so that kilometers is in the denominator. This gives us the rate in miles per second:

$$\frac{9.8\ \cancel{meters}}{1\ second} \bullet \frac{1\ \cancel{kilometer}}{1,000\ \cancel{meters}} \bullet \frac{0.62\ miles}{1\ \cancel{kilometer}}$$

We convert the seconds to minutes. Since seconds appears in the denominator of the first fraction, we write the conversion factor with seconds in the numerator. This gives us the rate in miles per minute:

$$\frac{9.8\ \cancel{meters}}{1\ \cancel{second}} \bullet \frac{1\ \cancel{kilometer}}{1,000\ \cancel{meters}} \bullet \frac{0.62\ miles}{1\ \cancel{kilometer}} \bullet \frac{60\ \cancel{seconds}}{1\ minute}$$

Now we convert the minutes to hours. Since minutes appear in the denominator, we write the conversion factor with minutes in the numerator. This gives us the rate in miles per hour:

$$\frac{9.8\ \cancel{meters}}{1\ \cancel{second}} \bullet \frac{1\ \cancel{kilometer}}{1,000\ \cancel{meters}} \bullet \frac{0.62\ miles}{1\ \cancel{kilometer}} \bullet \frac{60\ \cancel{seconds}}{1\ \cancel{minute}} \bullet \frac{60\ \cancel{minutes}}{1\ hour}$$

Finally, we perform all the calculations using a calculator. This gives us 21.8736 $\approx$ 21.9 miles per hour. (The symbol $\approx$ is pronounced "approximately equal to.")

Example 4.17 When Kendall takes a 5-minute shower, he uses 9 gallons of water. How many gallons of water would he use if he takes a shower that lasts 150 seconds?

Solution:

Step 1: *Identify what is given and what we need to find.*

A shower uses 9 gallons of water in 5 minutes.

A minute has 60 seconds.

How much water is used in a shower that lasts 150 seconds?

Step 2: *Represent the unknown information.*

We recommend including the units in the proportion to make sure that the calculations are correct.

Let w = number of gallons that Kendall uses while taking a shower that lasts 150 seconds.

Kendall's shower uses water at the rate of $\frac{9\ gallons}{5\ minutes}$.

Since the shower time is given in seconds, we convert this rate from gallons per minute to gallons per second, so we must convert 5 minutes to seconds.

Let x = number of seconds in 5 minutes.

Step 3: *Solve.*

First, find the number of seconds in 5 minutes.

$$\frac{1\ minute}{60\ seconds} = \frac{5\ minutes}{x\ seconds}$$

$1(x) = 60(5)$ In a proportion, the product of the means equals the product of the extremes.

$x = 300$ Five minutes is equivalent to 300 seconds.

Next, determine the number of gallons used in 150 seconds.

$$\frac{9 \text{ gallons}}{300 \text{ seconds}} = \frac{w \text{ gallons}}{150 \text{ seconds}}$$

$9(150) = 300w$ — In a proportion, the product of the means equals the product of the extremes.

$1{,}350 = 300w$ — Simplify.

$\frac{1{,}350}{300} = \frac{300w}{300}$ — Divide both sides by 300.

$4.50 = w$ — Final answer.

The shower uses 4.5 gallons in 150 seconds.

Step 4: *Check.*

$$\frac{9 \text{ gallons}}{300 \text{ seconds}} \stackrel{?}{=} \frac{4.5 \text{ gallons}}{150 \text{ seconds}}$$

$$0.03 = 0.03$$

Exercises

1. Convert 55 miles per hour to kilometers per hour. (1 mile = 1.6 kilometers)
2. How many minutes are in 1 year?
3. Convert 500 Saudi riyals to US dollars. (1 US dollar = 3.75 Saudi riyals)
4. Convert 540 grams to pounds. (1 pound = 0.45 kilograms, 1 kilogram = 1,000 grams)
5. How many square meters are in 3 square miles? (1 square mile = 2.6 square kilometers, 1 square kilometer = 1,000,000 square meters)
6. To the nearest hundredth, about how many cups of soda are in a 2-liter bottle? (4 cups = 1 quart, 0.946 liters = 1 quart)
7. A jet plane travels at a speed of up to 900 kilometers per hour. Convert this rate to feet per second. (1 meter = 3.28 feet, 1 kilometer = 1,000 meters)
8. A kitchen faucet has a water flow rate of 3 gallons per minute. Convert this rate to liters per second. (1 gallon = 3.8 liters)
9. A bank offers to exchange US dollars for Chinese yuan at the rate of US$1 to 6.1 Chinese yuan and yuan to Thai bahts at the rate of 1 yuan to 4.5 baht. How many bahts can be exchanged from US$50?

4.5 Percents

One special ratio that is commonly used in the real world is **percent**. Percent indicates a part out of 100, or hundredths. Percent is typically associated with a specific number and is abbreviated with the % symbol (pronounced "percent"). Thus, 3% means 0.03 or $\frac{3}{100}$, as shown in Figure 4.7:

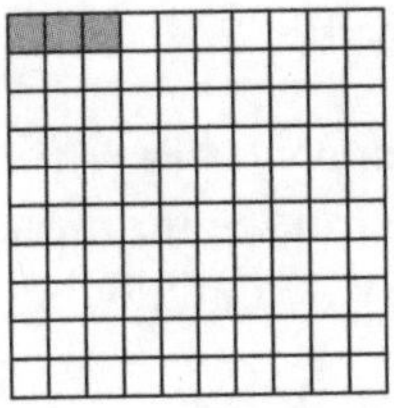

Figure 4.7 3%.

The word **percentage** refers to the general relationship of a part to the whole. For example, we say that "the *percentage* of students who are seniors is 23 *percent*."

To express a fraction as a percentage, we use a proportion in which one of the ratios is the percentage expressed as a fraction with a denominator of 100. All percentage problems are variations of one of the three problems shown in Table 4.5:

Table 4.5 Problems with percentage.

Problem Type	Example	Unknown	Setup
Finding the percent of a number	What is 75% of 40?	Part out of 40	$\frac{x}{40} = \frac{75}{100}$
Finding what percent one is of another	30 is what percent of 40?	Percent	$\frac{30}{40} = \frac{x}{100}$
Finding a number when a percent of it is known	30 is 75% of what number?	Whole of which 30 is a part	$\frac{30}{x} = \frac{75}{100}$

Example 4.18 What is 5% of 240?

Solution:

Let $x =$ the number.

Represent 5% as a fraction: $\frac{5}{100}$.

METHOD 1: Write a proportion.

The ratio of the number to 240 is $\frac{x}{240}$.

$$\frac{x}{240} = \frac{5}{100}$$

$100x = 5(240)$ — In a proportion, the product of the means equals the product of the extremes.

$\frac{100x}{100} = \frac{5(240)}{100}$ — Divide both sides by 100.

$x = \frac{5(240)}{100}$ — A nonzero number divided by itself equals 1.

$x = 12$ — Simplify.

METHOD 2: Multiply by the percentage.

We can see from the next-to-last step for Method 1 that we can solve for x by multiplying 240 by 5%, or 0.05:

$$x = \frac{5}{100}(240) = 0.05(240) = 12$$

Example 4.19 117 is 65% of what number?

Solution:

Let $x =$ the number.

Represent 65% as a fraction: $\frac{65}{100}$.

The ratio of 117 to the number is $\frac{117}{x}$.

$$\frac{117}{x} = \frac{65}{100}$$

$117(100) = 65x$ — In a proportion, the product of the means equals the product of the extremes.

$11{,}700 = 65x$ — Simplify.

$\frac{11{,}700}{65} = \frac{65x}{65}$ — Divide both sides by 65.

$180 = x$ — Final answer, which we typically write as $x = 180$.

The number is 180.

Example 4.20 A US Census Bureau report estimated that 39.5 million Americans live in poverty. If the US population at the time of the report was 321 million, determine to the nearest tenth the percentage of Americans who live in poverty.

Solution:

Step 1: Identify what is given and what we need to find.
39.5 million out of 321 million Americans live in poverty.
How many live in poverty?

Step 2: Represent the unknown information.
Let p = percentage of Americans who live in poverty.
The ratio of Americans who live in poverty to the total population = $\frac{39.5}{321}$.
The percentage of the population that lives in poverty = $\frac{p}{100}$.

Step 3: Solve.

$$\frac{39.5}{321} = \frac{p}{100}$$

$321p = 39.5(100)$ — In a proportion, the product of the means equals the product of the extremes.

$321p = 3{,}950$ — Simplify.

$\frac{321p}{321} = \frac{3{,}950}{321}$ — Divide both sides by 321.

$p \approx 12.305$ — Final answer, which is rounded to 12.3.

About 12.3% of Americans live in poverty.

Step 4: Check.

$$\frac{39.5}{321} \stackrel{?}{=} \frac{12.3}{100}$$

$$0.123 \approx 0.123$$

Example 4.21 A recent survey asked 1,500 registered Black and Hispanic voters if they planned to vote in the next election. (Each voter was labeled as either Black or Hispanic.) The ratio of Black respondents to Hispanic respondents was 2:3. If 60% of Blacks and 50% of Hispanics responded yes, how many respondents said they planned to vote in the next election?

Solution:

Step 1: Identify what is given and what we need to find.
1,500 Blacks and Hispanics were surveyed.
The ratio of Blacks to Hispanics was 2:3.
60% of Blacks answered yes.
50% of Hispanics answered yes.
How many people said yes?

Step 2: *Represent the unknown information.*
Let x = number of groups of Blacks and Hispanics in a 2:3 ratio that answered yes.
Then $2x$ = number of Blacks surveyed and $3x$ = number of Hispanics surveyed.

Step 3: *Solve.*
Number of Black respondents + number of Hispanic respondents = total number of respondents

$$2x + 3x = 1{,}500$$

$$5x = 1{,}500 \quad \text{Combine like terms.}$$

$$\frac{5x}{5} = \frac{1{,}500}{5} \quad \text{Divide both sides by 5.}$$

$$x = 300$$

$2(300) = 600$ Blacks and $3(300) = 900$ Hispanics were surveyed.

Now we find the 60% of 600 Blacks and 50% of 900 Hispanics who answered yes:
Number of Blacks who said yes + Number of Hispanics who said yes

$$= 0.60(600) + 0.50(900)$$

$$= 360 + 450$$

$$= 810$$

810 respondents said yes in the survey.

Step 4: *Check.*

$$\frac{600 \textit{ Blacks}}{900 \textit{ Hispanics}} \stackrel{?}{=} \frac{2}{3}$$

$$\frac{2}{3} = \frac{2}{3}$$

Exercises

1. 34 is what percent of 85?
2. What is 8% of 450?
3. 17 is what percent of 68?
4. 24 is 20% of what number?
5. 7 is what percent of 14?
6. 15 is what percent of 25?
7. 12 is 12% of what number?
8. 68 is 200% of what number?
9. What is 0.1% of $5,000?

10. The 2.2 million high school dropouts in the United States represent 5.8% of the school-age population. To the nearest million, about many people are in the school-age population?

11. A student answered 3 questions on a quiz correctly and 9 questions incorrectly. If the student answered every question, what percentage of the quiz questions did the student answer correctly?

12. A bank account pays $24 of interest on an investment of $400. What was the interest rate paid on the account?

13. In a large high school, 22% of the student population is 14 years old. If 627 students are 14 years old, how many students are in the school?

14. To buy a used car that costs $24,000, Maria wants to put 15% as a down payment. What is the amount of the down payment?

15. A credit card charges interest on any unpaid balance. What is the interest rate charged on a $3,000 balance that generates $479.70 in interest fees?

16. 8% of a number plus 12% of the number is 62. What is the number?

17. A number decreased by 10% of itself equals 405. What is the number?

18. 10% of a number plus 8% of the number, decreased by 6% of the number, equals 42. What is the number?

CHAPTER 4 TEST

1. If a recipe calls for 4 cups of flour and 2 cups of water, what is the ratio of flour to water?

   (A) 2:1 (B) 2:4 (C) 4:1 (D) 1:2

2. If 2:5 = x:6, what is the value of x?

   (A) 3 (B) $\frac{12}{5}$ (C) $\frac{5}{12}$ (D) $\frac{5}{3}$

3. Which statement is a true proportion?

   (A) $\frac{2}{5} = \frac{20}{8}$ (B) $\frac{2}{5} = \frac{7}{10}$ (C) $\frac{2}{5} = \frac{8}{20}$ (D) $\frac{2}{5} = \frac{20}{100}$

4. What is 16% of 720?

   (A) $\frac{200}{9}$ (B) 115.2 (C) 1,843.2 (D) 4,500

5. Which fraction is equivalent to 24%?

   (A) $\frac{6}{25}$ (B) $\frac{24}{25}$ (C) $\frac{100}{24}$ (D) $\frac{1}{24}$

6. The ratio of hardcover books to paperback books in a library is 5:3. All books in the library are either hardcovers or paperbacks. What percentage of the library books are paperbacks?

   (A) 30% (B) 37.5% (C) 50% (D) 62.5%

7. One molecule of water contains two atoms of hydrogen and one atom of oxygen. How many atoms of hydrogen are in 15 molecules of water?

   (A) 10 (B) 15 (C) 30 (D) 45

8. If Sandy runs 200 meters in 27 seconds, what was his total running time in minutes?

   (A) 0.135 min (B) 0.27 min (C) 0.45 min (D) 0.54 min

9. A webcam was sold for $68 after discounts of 10% and 5% off the original price were allowed. What was the original price of the webcam?

   (A) $10.20 (B) $73 (C) $78.20 (D) $80

10. Write the pronunciation of the ratio 7:3:2.

11. 10% of a number increased by 28% of the same number equals 57. What is the number?

12. A sports team won 81.25% of its games over the last three seasons. If it won 91 games in the last three seasons and there were no ties, how many games did it lose during that time?

13. The tallest mountain in the world is Mount Everest, which is 29,035 feet high. If 1 meter = 3.28 feet and 1 kilometer = 1,000 meters, determine Mount Everest's height to the nearest hundredth of a kilometer.

14. The ratio of Black to Hispanic students who attend a large high school is 5:2. If the school has 959 Black and Hispanic students, how many Black students and how many Hispanic students go to the school?

15. A company's annual profit of $150,000 is divided among a senior partner, who earns 40% of the profits, and three junior partners whose share of the profits is split in a 3:2:1 ratio. Determine the amount of each partner's share.

CHAPTER 4 SOLUTIONS

4.1.

1. "4 to 1"
2. "16 to 3"
3. "17 to 12"
4. "7 to 12 to 19"
5. "12 to 2 to 1"
6. "10 to 20 to 30 to 50"
7. 3 centimeters per kilometer
8. 2 meters per person
9. 40 kilometers per hour
10. 5 lights per backyard
11. 5 feet per minute
12. 2:1
13. 12 square centimeters per meter
14. 5:12
15. 32 centimeters per 9 days
16. 1 kilometer to 1,000 meters to 100,000 centimeters
17. 15 dimes to 4 quarters to 3 pennies
18. 6° to 10° to 5°
19. 1 pound recycled material per 2 pounds trash
20. 100,000 square kilometers per year
21. 3:2:1
22. $\frac{15}{22}$
23. 5 miles per hour
24. 35:47:37
25. 1.5 pounds per week
26. 12 students per teacher
27. 2 wins per loss

4.2.

1. 54 orange Minis and 90 blue Minis
2. 688 warehouse workers and 258 cashiers
3. train: 84 miles per hour, car: 36 miles per hour
4. 2,720 food secure households and 480 food insecure households
5. 60 white senators and 40 non-white senators
6. $36,000, $60,000, $84,000, and $108,000
7. 45 buckets of cement, 90 buckets of sand, and 135 buckets of aggregate

8. 8 cups red paint, 4 cups yellow paint, and 2 cups of blue paint
9. 3,560 grams of copper, 240 grams of zinc, 120 grams of manganese, and 80 grams of nickel (500 dollar coins weigh 4,000 grams)

4.3. 1. "9 is to 12 as 6 is to 8."
2. "9 is to 18 as 1 is to 2."
3. "5 is to 15 as 9 is to 27."
4. Means = 24 and 7, extremes = 8 and 21.
5. Means = 4 and 33, extremes = 11 and 12.
6. Means = 3 and 45, extremes = 100 and 1.35.
7. No because $3(3) \neq 2(4)$.
8. Yes because $3(8) = 4(6)$.
9. No because the right side is not a ratio.
10. $y = 9$
11. $b = 24$
12. $a = 2$
13. $b = -14$
14. $c = \frac{4}{11}$
15. $q = 4$
16. $m = -61$
17. $p = \frac{31}{3}$
18. $r = -\frac{9}{11}$
19. $k = 12$
20. $g = -15$
21. $k = \frac{20}{7}$
22. $a = 12$
23. $w = 3$
24. $p = -3$
25. $k = 9$
26. $n = 24$
27. $d = -2$
28. 6.4 inches
29. 322 atoms
30. 630 minutes
31. 75 milliliters
32. $\frac{32}{3}$
33. $\frac{3}{10}$
34. $\frac{133}{8}$
35. 168
36. Multiplying both sides by bd gives us $bd\left(\frac{a}{b}\right) = bd\left(\frac{c}{d}\right)$, or $\left(\frac{b}{b}\right)da = bc\left(\frac{d}{d}\right)$, which equals $da = bc$, or $ad = bc$.

4.4.
1. 88 kilometers per hour
2. 525,600 minutes
3. US$133.33
4. 1.2 pounds
5. 7,800,000 square meters
6. 8.46 cups
7. 820 feet per second
8. 0.19 liters per second
9. 1,372.5 baht

4.5.
1. 40%
2. 36
3. 25%
4. 120
5. 50%
6. 60%
7. 100
8. 34
9. $5
10. 38 million
11. 25%
12. 6%
13. 2,850 students
14. $3,600
15. 15.99%
16. 310
17. 450
18. 350

CHAPTER 4 TEST SOLUTIONS

1. (A)
2. (B)
3. (C)
4. (B)
5. (A)
6. (B)
7. (C)
8. (C)
9. (D)
10. "7 to 3 to 2"
11. 150
12. 21 games
13. 8.85 kilometers.
14. 685 Black and 274 Hispanic students.
15. The senior partner earns $60,000, and the junior partners' shares are $45,000, $30,000, and $15,000.

5 LINEAR INEQUALITIES

So far, we've worked extensively with equations. However, we often deal with situations where we find values of the variable in which one expression is *greater* than or *less* than another. Fortunately, despite some important differences, the methods we use here are similar to the methods we used to solve equations.

5.1 Basic Principles of Solving Inequalities

First, let's start with a basic definition: an **inequality** is a statement that says that one expression is greater than or less than another expression.

Like equations, inequalities must be either true or false—they can't be both at the same time. As with equations, we often want to find the values of the variables that make the inequality true. We call these the **solutions to the inequality**.

In Chapter 1, we used number lines to represent positive and negative numbers. Recall that positive numbers appear to the right of 0, and negative numbers appear to the left of 0. Also, any number that appears to the right of another is the greater number. We use the following inequality symbols:

- $>$ (pronounced "is greater than")
- $\geq$ (pronounced "is greater than or equal to," combines the $>$ and $=$ symbols)
- $<$ (pronounced "is less than")
- $\leq$ (pronounced "is less than or equal to," combines the $<$ and $=$ symbols)

Here are some important points about the direction of inequalities:

- The inequality symbols in statements like $6 > 2$ and $5 > 3$ have the same direction.
- The inequality symbols in statements like $6 > 2$ and $4 < 5$ have the opposite direction.
- If we reverse the order of the numbers, then we reverse the direction of the inequality symbol, so $3 > 1$ is equivalent to $1 < 3$.

Reading and Writing Tip

The > and < symbols are easily confused. Figure 5.1 can help you remember that the inequality symbols "open" in the direction of the larger number:

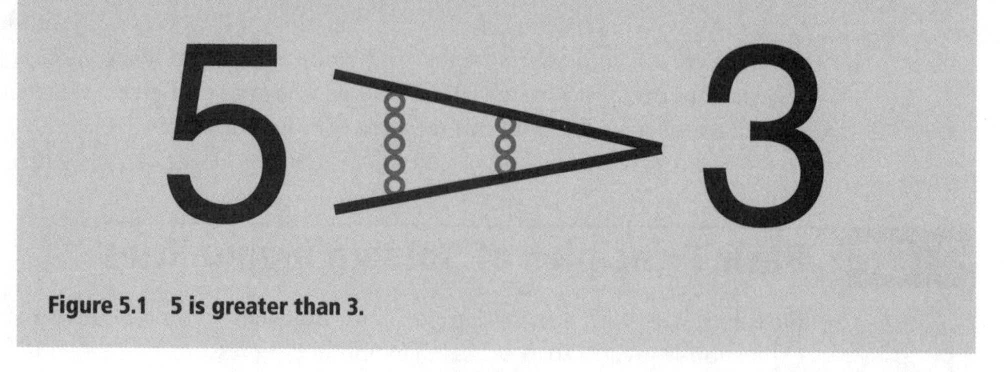

Figure 5.1 5 is greater than 3.

Example 5.1 Determine if the inequality $+1 > -2$ is true. Use a number line to justify your answer.

Solution: We see that -2 is to the left of $+1$ on a number line:

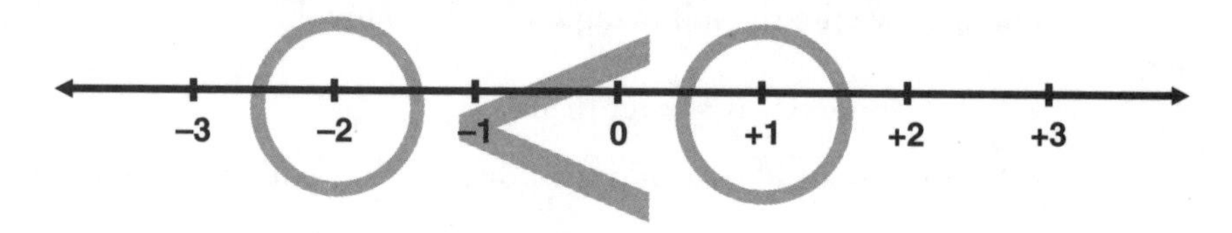

This means that $-2 < +1$, which is equivalent to $+1 > -2$, making our original inequality true.

Exercises

Write the pronunciation of the following inequalities.

1. $5 > 2$

2. $-40 \leq -40$

3. $37 > -15$

4. $-22 < +54$

5. $6 \geq 2$

6. $-15 \leq -8$

Determine if each inequality is true or false. Use a number line to justify your answer.

7. $+5 > +4$

8. $+1 > 0$

9. $-4 < -6$

10. $-1 < +1$

11. $-5 < -1$

12. $0 > 5$

Write a true mathematical statement containing > or < that compares the following:

13. -53 and 4

14. 60 and -10

15. -15 and -24

16. $\frac{7}{3}$ and 2

17. $-\frac{17}{9}$ and $-\frac{18}{11}$

18. $-\frac{1}{3}$ and $-\frac{1}{2}$

5.2 Representing Inequalities

In the real world, we often represent situations that have an upper or lower limit, such as the following:

- The maximum capacity of this room is 10 people.
- The maximum height allowed is 10 inches.
- The account balance is no more than $10.

Although we represent all three with the inequality $x \leq 10$, we see that the possible values of x that make the inequality true differ in each case:

- The capacity of the room can be a whole number from 0 to 10 (0, 1, 2, …, 9, 10).
- The height of an object can be a number greater than 0 but less than or equal to 10 inches, such as 0, 8.5, or $\frac{27}{8}$.
- The account balance can be a number less than or equal to $10, such as −$5, −$0.80, $0, or $9.

The values that make inequalities true depend on the numbers we use for the variables. We define this more formally as follows:

- The **domain** or **replacement set** is the set of all possible values that can be used to replace the variable.
- The **solution** or **solution set** is the set of numbers from the domain that make an inequality true.

In math, a **set** is a collection of objects. We list **elements**, or members, of a set by using curly brackets and listing each element exactly once (the order doesn't matter), separated by commas, such as $\{1, 2, 4, 5\}$. In the examples above, the domains are the set of whole numbers from 0 to 10, the set of all numbers between 0 and 10, and the set of all numbers less than 10, respectively.

We represent the solution set of an inequality on a number line. For example, if the domain is the set of all numbers on the number line, then the inequality $x < 10$ can be graphed as in Figure 5.2:

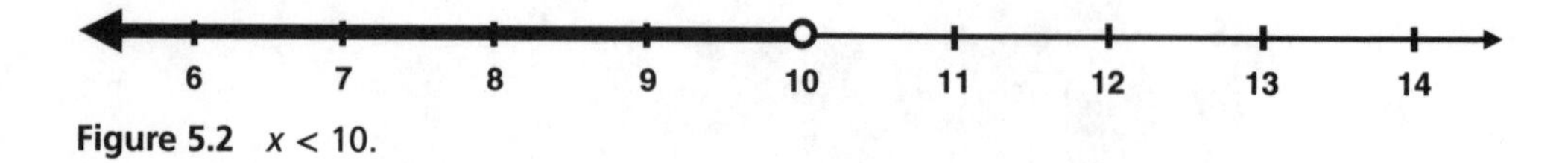

Figure 5.2 $x < 10$.

The graph of an inequality consists of the following:

- An open circle at 10, indicating that 10 is not in the solution set and is the boundary between numbers in the solution set and numbers not in the solution set
- A thick line with an arrow at the end, indicating that the numbers in the solution set continue infinitely in that direction

We graph the equation $x = 10$ on a number line by drawing a point, sometimes called a closed circle, at 10 (Figure 5.3):

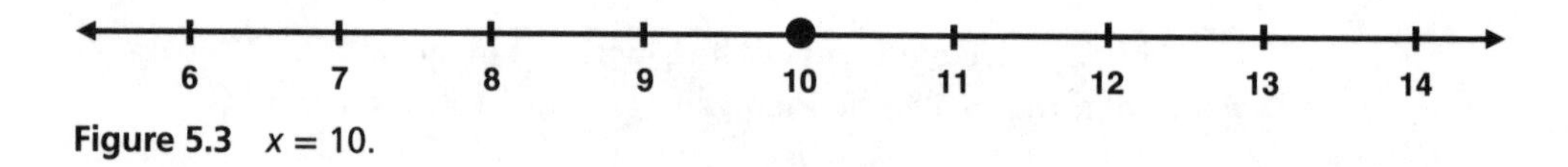

Figure 5.3 $x = 10$.

To show $x \leq 10$, we draw a graph that combines the graphs of $x < 10$ and $x = 10$ (Figure 5.4):

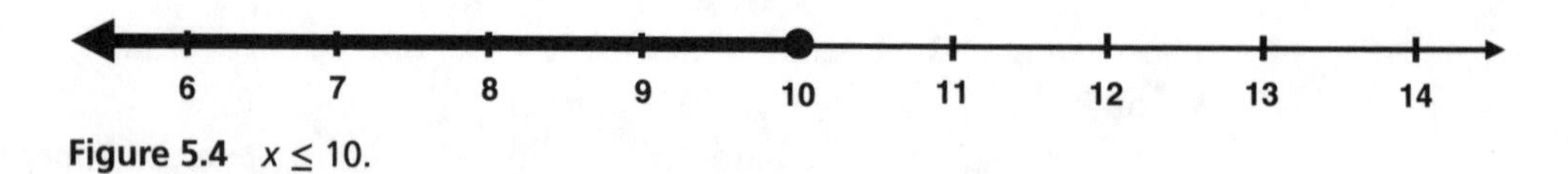

Figure 5.4 $x \leq 10$.

For the inequalities $x > 10$ and $x \geq 10$, we draw similar number lines, but the rays extend to the right instead of to the left. (The table at the end of this section shows examples of graphs illustrating inequalities with $>$ and $\geq$.)

We also represent the solution sets of inequalities more compactly without drawing graphs by using **interval notation**, a shorthand way of describing all numbers between two boundary points. Interval notation contains five parts, as shown in Figure 5.5 showing $x \leq 10$:

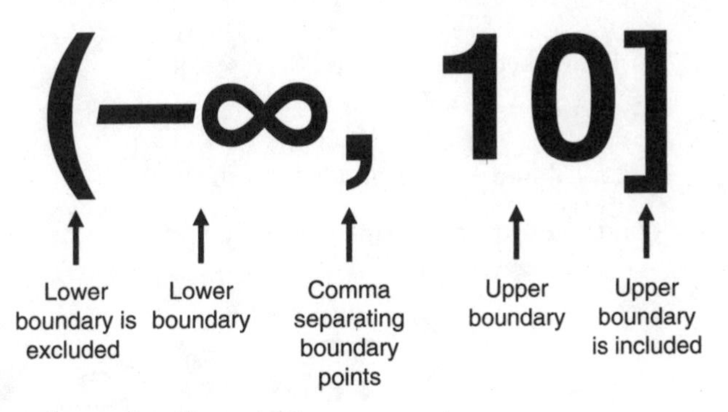

Figure 5.5 Interval notation for $x \leq 10$.

Interval notation contains the following:

- (if the lower boundary is not in the interval or [if it is in the interval
- lower boundary of the interval (if the solution set extends infinitely to the left, we write $-\infty$, which we read as "negative infinity")
- comma separating the boundary points
- upper boundary of the interval (if the solution set extends infinitely to the right, we write ∞, which we read as " infinity")
-) if the upper boundary is not in the interval or] if it is in the interval

Here are some important points about interval notation:

- Since $-\infty$ and ∞ are not numbers, they can't be part of the interval, so we *always* use parentheses when using them in interval notation.
- Interval notation is a compact way of describing all numbers that exist between the two boundary points. Informally, we say that there are no gaps between the boundary points.
- We always put the lower boundary first, then a comma, then the upper boundary.
- If both endpoints are not included in the interval, we call it an **open interval**. We pronounce $(-10, 10)$ as "the open interval from -10 to 10." If both endpoints are included, we call it a **closed interval**. We pronounce $[-10, 10]$ as "the closed interval from -10 to 10."
- If only one endpoint is included in the interval, we pronounce it as "the interval from [lower boundary] to [upper boundary], including [the boundary point in the interval]." We pronounce $(-\infty, 10]$ as "the interval from negative infinity to 10, including 10."

Another way to represent solution sets is **set-builder notation**, which allows us to describe numbers by stating their properties. Set-builder notation contains four parts, shown in Figure 5.6 with its pronunciation:

1. curly brackets (pronounced "the set of")
2. a variable
3. a vertical line | (pronounced "such that")
4. conditions for the numbers, written with mathematical notation

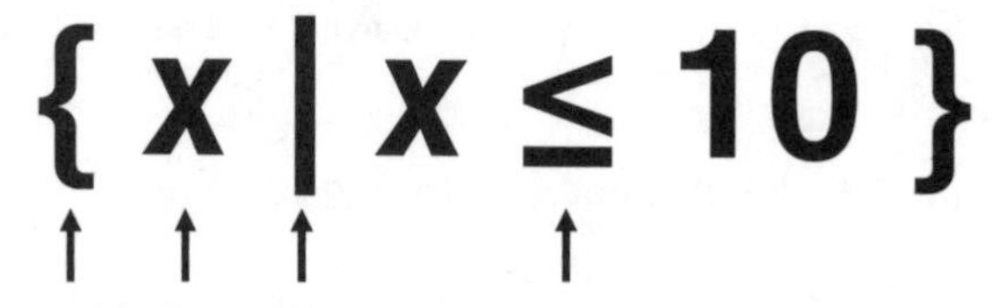

Figure 5.6 Set-builder notation for $x \leq 10$.

Set-builder notation is especially useful for differentiating between solution sets where we restrict possible values, such as "all integers less than 10" and "all whole numbers less than 10." We use the following abbreviations to indicate different sets of numbers:

- **Q** or $\mathbb{Q}$ = the set of rational numbers (TIP: Think **Q** for "quotient.")
- **Z** or $\mathbb{Z}$ = the set of integers
- **W** or $\mathbb{W}$ = the set of whole numbers
- **N**, $\mathbb{N}$, $\mathbf{Z}^+$, or $\mathbb{Z}^+$ = the set of counting numbers (TIP: The counting numbers are also known as natural numbers and can be thought of as positive integers.)

Did You Know?

Why do we use the letter **Z** for the set of integers? European mathematicians used various letters to indicate sets of infinite numbers. In his 1930 book *Grundlagen der Analysis*, German mathematician Edmund Landau denoted the integers with the letter Z (the German word for number is *Zahlen*) written in a special font with a bar over it. Historically, when letters like **Q** or **Z** were used to represent sets in books, they were written in boldface. However, since boldface is difficult to replicate on a classroom blackboard, mathematicians developed the habit of writing in a "blackboard bold" style, in which certain lines (usually the vertical or near-vertical lines) of a letter are doubled, like $\mathbb{Q}$ or $\mathbb{Z}$.

To indicate that a number belongs to one of these sets of numbers, we use the symbol $\in$, which is pronounced "is an element of." For example, we represent the set of whole numbers less than 10 as $\{x \in \mathbf{W} \mid x < 10\}$, which is read as "the set of *x*s that are elements of the whole numbers such that x is less than 10" or "the set of *x*s that are whole numbers such that x is less than 10."

Example 5.2 Express the set of numbers greater than or equal to 25 using a number line, interval notation, and set-builder notation.

Solution: We represent the set of numbers greater than or equal to 25 as the inequality $x \geq 25$.

NUMBER LINE: We draw a closed circle at 25 (to show that 25 is in the set), shade all points to the right, and draw an arrow at the end (to indicate all the numbers greater than 25). The graph appears below.

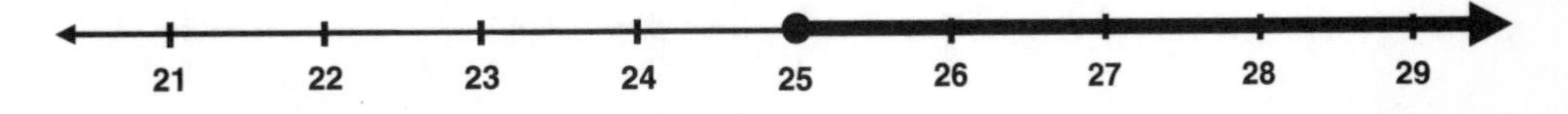

INTERVAL NOTATION: Since the lower boundary is 25 and is included in the interval, then we start with a square bracket symbol. The upper boundary is ∞ and is not in the interval, so we use a parenthesis. The interval notation is $[25, \infty)$, which we read as "the interval from 25 to infinity, including 25."

SET-BUILDER NOTATION: We write $\{x \mid x \geq 25\}$, which we read as "the set of all *x*s such that x is greater than or equal to 25."

Example 5.3 Express $-6 > x$ using a number line, interval notation, and set-builder notation.

Solution: Rewriting this inequality so that the variable is on the left makes determining its direction easier. Here, $-6 > x$ is equivalent to $x < -6$ (if -6 is *greater than* a number, then the number is *less than* -6).

NUMBER LINE: We draw an open circle at -6 (to show that -6 is not in the set), shade all points to the left, and draw an arrow at the end (to indicate all numbers less than -6). The graph appears below.

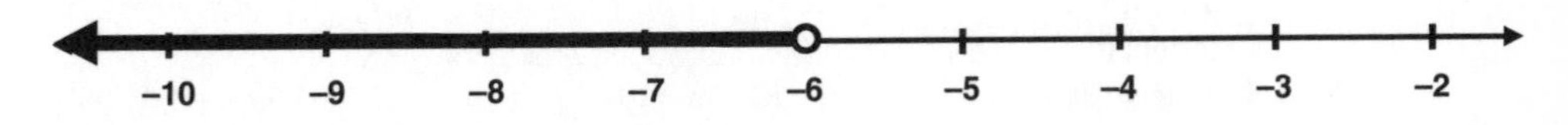

INTERVAL NOTATION: The lower boundary is $-\infty$ and is not included in the interval, so we use a parenthesis. The upper boundary is -6 and is not in the interval, so we use a parenthesis here as well. The interval notation is $(-\infty, -6)$, which we read as "the open interval from negative infinity to -6."

SET-BUILDER NOTATION: We write $\{x \mid -6 > x\}$, which we read as "the set of all *x*s such that -6 is greater than x." We can also write $\{x \mid x < -6\}$.

The inequalities in Examples 5.1, 5.2, and 5.3 have solution sets that are **continuous**, which we can informally define to mean that there are no "gaps" between values. However, some solution sets are **discrete**, meaning that there are "gaps" between values, as shown in Example 5.4:

Example 5.4 Express the set of integers less than 3 using a number line, interval notation, and set-builder notation.

Solution:
GRAPH: We draw closed circles on the points 2, 1, 0, −1, and so on to represent the integers less than 3. We don't connect the points because the numbers between the integers are not part of the set. To indicate that the set extends infinitely to the left, we write an ellipsis on the left side of the number line.

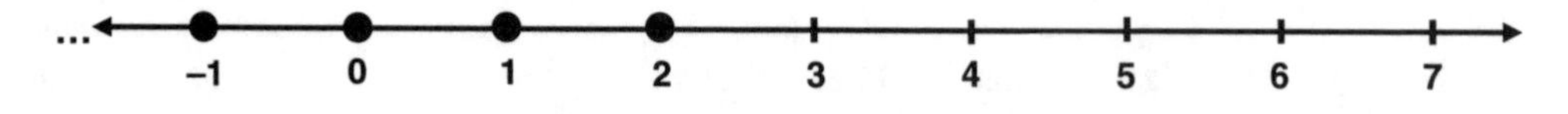

INTERVAL NOTATION: Because there are gaps between points in the set, we can't use interval notation.

SET-BUILDER NOTATION: We write $\{x \in \mathbf{Z} \mid x < 3\}$, which we read as "the set of all xs that are integers such that x is less than 3."

Table 5.1 summarizes how we write inequalities with graphs, interval notation, and set-builder notation:

Exercises

Write the pronunciation of each of the following:

1. $(17, \infty)$

2. $(-\infty, 26)$

3. $(-\infty, 3]$

4. $\{q \mid q > 52\}$

5. $\{d \mid d \leq -78\}$

6. $\{x \mid x \geq 14\}$

7. $\{y \in \mathbf{Z}^+ \mid y \geq 14\}$

8. $\{k \in \mathbf{N} \mid k < 6\}$

9. $\{w \in \mathbf{W} \mid w < 34\}$

Express each of the following sets using a number line, interval notation (if possible), and set-builder notation.

10. $u < 90$

11. $n \leq 82$

12. $a > -3$

13. $x \leq 0$

14. $k < -19$

15. $y \geq 14$

16. $-8 \geq y$

17. $15 < p$

18. $-53 \leq r$

19. $\{2, 3, 4, 5, 6, \ldots\}$

20. the set of integers greater than -3

21. the set of integers less than or equal to 75

Table 5.1 Representing inequalities.

Set	Graph	Interval notation	Set-builder notation
$x < 10$	8 9 10 11 12	$(-\infty, 10)$ "the open interval from negative infinity to 10"	$\{x \mid x < 10\}$ "the set of all xs such that x is less than 10"
$x \leq 10$	8 9 10 11 12	$(-\infty, 10]$ "the interval from negative infinity to 10, including 10"	$\{x \mid x \leq 10\}$ "the set of all xs such that x is less than or equal to 10"
$x > 10$	8 9 10 11 12	$(10, \infty)$ "the open interval from 10 to infinity"	$\{x \mid x > 10\}$ "the set of all xs such that x is greater than 10"
$x \geq 10$	8 9 10 11 12	$[10, \infty)$ "the interval from 10 to infinity, including 10"	$\{x \mid x \geq 10\}$ "the set of all xs such that x is greater than or equal to 10"
Integers greater than 10	8 9 10 11 12 ...	None	$\{x \in \mathbf{Z} \mid x > 10\}$ "the set of all xs that are integers such that x is greater than 10"

Questions to Think About

22. Explain why the interval notation $[-\infty, 18)$ is incorrect.

23. Explain why the set of integers cannot be represented with interval notation.

24. What is the domain of an inequality?

5.3 Solving Linear Inequalities

Solving inequalities has many similarities to solving equations. Adding or subtracting the same number from both sides of an inequality results in a true statement. For example, to solve $x - 4 > 12$, we add 4 to both sides:

$$\begin{array}{r} x - 4 > 12 \\ +4 \quad +4 \\ \hline x > 16 \end{array}$$

Multiplying or dividing both sides of an inequality by the same positive number also results in a true statement. For example:

$$4x > 12$$

$$\frac{4x}{4} > \frac{12}{4}$$

$$x > 3$$

However, multiplying or dividing both sides of an inequality by the same negative number reverses the direction. For example, $+3 > +2$, but $+3(-1) < +2(-1)$ since $-3 < -2$.

Table 5.2 summarizes these properties (the shaded rows indicate cases where the direction of the inequality is reversed):

Table 5.2 Operations with inequalities.

Property	Operation	Effect on Direction of Inequality Symbol	Example	
If $a > b$, then $a + c > b + c$.	**Add** a number	Unchanged	$6 > 4$ $6 + 2 > 4 + 2$ $8 > 6$	$6 > 4$ $6 + (-2) > 4 + (-2)$ $4 > 2$
If $a > b$, then $a - c > b - c$.	**Subtract** a number	Unchanged	$6 > 4$ $6 - 2 > 4 - 2$ $4 > 2$	$6 > 4$ $6 - (-2) > 4 - (-2)$ $8 > 6$
If $a > b$ and $c > 0$, then $a \bullet c > b \bullet c$.	**Multiply** by a **positive** number	Unchanged	$6 > 4$ $6 \bullet 2 > 4 \bullet 2$ $12 > 8$	
If $a > b$ and $c < 0$, then $a \bullet c < b \bullet c$.	**Multiply** by a **negative** number	Reversed	$6 > 4$ $6 \bullet -2 > 4 \bullet -2$ $-12 < -8$	
If $a > b$ and $c > 0$, then $\frac{a}{c} > \frac{b}{c}$.	**Divide** by a **positive** number	Unchanged	$6 > 4$ $\frac{6}{2} > \frac{4}{2}$ $3 > 2$	
If $a > b$ and $c < 0$, then $\frac{a}{c} < \frac{b}{c}$.	**Divide** by a **negative** number	Reversed	$6 > 4$ $\frac{6}{-2} < \frac{4}{-2}$ $-3 < -2$	

To avoid confusion about the direction of the inequality symbol, we recommend writing the solution set's variable on the left side.

Unless otherwise noted, we assume that the domain for inequalities is the set of all numbers on the number line.

Example 5.5 Solve $-3p - 8 < 15$ for the variable. Graph the solution set.

Solution:

$-3p - 8 < 15$	
$\quad +8 \quad +8$	Add 8 to both sides.
$-3p < 23$	Combine like terms.
$\frac{-3p}{-3} > \frac{23}{-3}$	Divide both sides by -3 to isolate p. Dividing both sides of an inequality by a negative number reverses the direction of the inequality symbol.
$p > -\frac{23}{3}$	Simplify.

Check:

$-3(0) - 8 \overset{?}{<} 15$ Substitute a value from the solution set ($x = 0$) into the original inequality.

(NOTE: Since we can pick any value in the solution set to check, we choose a number that we can easily substitute into the original inequality.)

$$-8 < 15$$

To graph the solution set $p > -\frac{23}{3}$, we draw an open circle at $p = -\frac{23}{3}$ (which is $p = -7\frac{2}{3}$, or slightly to the right of $p = -8$), shade all points to the right, and draw an arrow at the end.

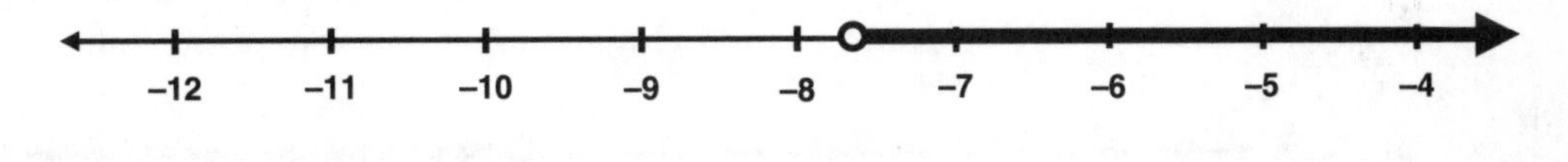

Example 5.6 Solve $3x + 2 < 6x - 4$ for the variable. Graph the solution set.

Solution:

METHOD 1: Keep the variable on the left side of the inequality.

$3x + 2 < 6x - 4$ $-6x - 2 \quad -6x - 2$	Subtract $6x$ and subtract 2 from both sides.
$-3x < -6$	Combine like terms.
$\frac{-3x}{-3} > \frac{-6}{-3}$	Divide both sides by -3. Dividing both sides by a negative number reverses the direction of the inequality symbol.
$x > 2$	Simplify.

METHOD 2: Subtract the term with the smaller coefficient from the term with the larger one.

$3x + 2 < 6x - 4$ $-3x + 4 \quad -3x + 4$	Subtract $3x$ from and add 4 to both sides.
$6 < 3x$	Combine like terms.
$\frac{6}{3} < \frac{3x}{3}$	Divide both sides by 3.
$2 < x$	Simplify.
$x > 2$	Rewrite the inequality with the variable on the left side.

Check:

$6(3) - 4 \overset{?}{>} 3(3) + 2$ Substitute a value from the solution set ($x = 3$) into the original inequality.

$14 > 11$

To graph the solution set $x > 2$, we draw an open circle at $x = 2$, shade all points to the right, and draw an arrow at the end.

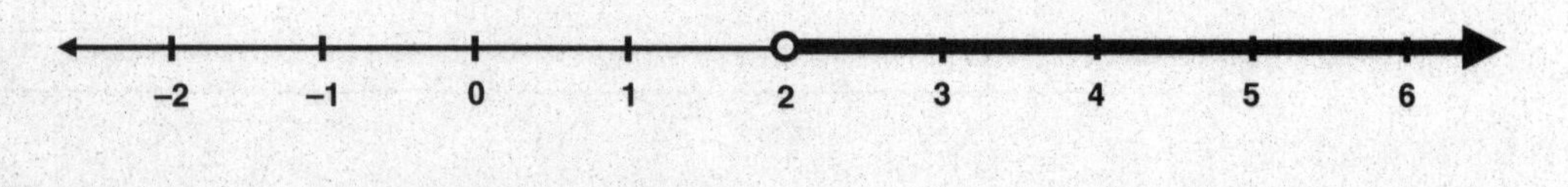

When a domain is specified, we solve the inequality as we did in previous examples, but then we identify numbers from the solution set that are in the domain.

Example 5.7 **Given $\{x \in \mathbf{Z}^+ \mid -9 \le x \le 0\}$, solve**

$$5(x - 7) + 2 > 8x + 3x - 2$$

for the variable. Graph the solution set.

Solution:

$5(x - 7) + 2 > 8x + 3x - 2$	
$5x - 35 + 2 > 8x + 3x - 2$	Use distributive property.
$5x - 33 > 11x - 2$	Combine like terms.
$\underline{-5x + 2 \quad -5x + 2}$	Subtract $5x$ from and add 2 to both sides.
$-31 > 6x$	Combine like terms.
$\frac{-31}{6} > \frac{6x}{6}$	Divide both sides by 6.
$-\frac{31}{6} > x$	Final answer, which we write with the variable on the left: $x < -\frac{31}{6}$.

Since the domain is $\{x \in \mathbf{Z}^+ \mid -9 \le x \le 0\}$, then the set of x-values from which we may choose is $-9, -8, -7, -6, -5, -4, -3, -2, -1$, or 0. The only elements of the domain that are less than $-\frac{31}{6} = -5\frac{1}{6}$ are $-9, -8, -7$, and -6. Thus, the solution is $-9, -8, -7$, or -6.

The graph of the solution set consists of closed circles over $-9, -8, -7$, and -6 on the number line:

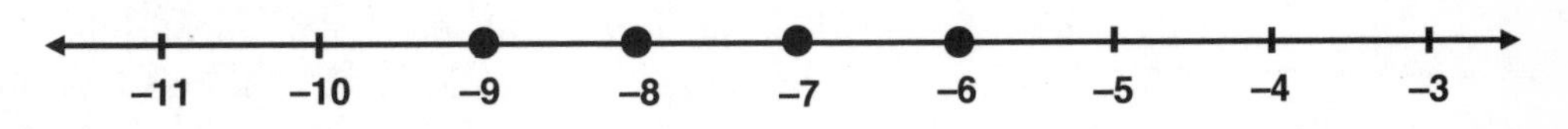

Exercises

Solve each inequality for the variable. Express the solution set using interval notation.

1. $q + 5 > 8$
2. $7 - 2 \le n$
3. $\frac{x}{4} < 8$
4. $5r \ge 30$
5. $8g \le -48$
6. $\frac{x}{18} > -72$
7. $-80j > 400$
8. $-75z \le 34$
9. $-\frac{c}{4} > 24$
10. $\frac{x}{2} - 3 \ge 8$
11. $-8 > 3u - 8$
12. $\frac{v}{12} - 6 \le 3$
13. $7a - 5 \ge 3a + 4$
14. $y + 6 + y < 5y + 9$
15. $5p + 6 < 4 + 3p + 6$

16. $12x - 3 - 8x > 9x + 2$

17. $2y + 1 < 5y - 5 + 9y$

18. $8m - 6m + 3 - m \geq 2m - 7$

19. $5(f - 10) + 6 \geq 111$

20. $-4(d + 6) - 3 < 73$

21. $3x + 4(6 - x) \geq 2$

22. $\frac{7}{8}z + 2 < 8$

23. $\frac{2}{3}p - 17 > 6$

24. $\frac{4}{5}x + \frac{1}{2}x \leq 12$

25. If y is a whole number, solve for y in the inequality $7(y - 1) - 9 < 26$.

26. If q is a counting number, solve for q in the inequality $16 > 3(q + 2) - 5$.

27. Given $\{r \in Z^+ \mid -3 < r < 3\}$, solve for r in the inequality $5r + 2r + 4 \geq 18$.

5.4 Compound Inequalities

So far, we have used an inequality to represent one set of conditions. In the real world, we often deal with situations in which more than one condition must be met. For example:

- Essays must be from 500 to 750 words in length.
- Applicants must have from 3 to 5 years of experience.

We can't represent these situations with the simple inequalities (like $x > 3$ or $x < 750$) discussed so far. In this section, we discuss **compound inequalities**, which combine two or more inequalities.

For example, if a group of fish vary in length from 500 to 750 millimeters, their lengths have a minimum of 500 millimeters ($x \geq 500$, where x = length in millimeters) and a maximum of 750 millimeters ($x \leq 750$). Since *both* inequalities must be true, the solution set is the **intersection**, or overlap, of the solutions of $x \geq 500$ *and* $x \leq 750$ (in other words, the numbers between 500 and 750):

We write this intersection symbolically using any of the following:

- $500 \leq x \leq 750$ (read as "500 is less than or equal to x is less than or equal to 750")
- $\{x \mid x \geq 500 \text{ and } x \leq 750\}$ (read as "the set of xs such that x is greater than or equal to 500 and less than or equal to 750")
- $\{x \mid x \geq 500\} \cap \{x \mid x \leq 750\}$ (read as "the intersection of the set of xs such that x is greater than or equal to 500 and the set of xs such that x is less than or equal to 750")
- [500, 750] (read as "the closed interval from 500 to 750")

To graph $500 \leq x \leq 750$, we sketch the graphs of $x \geq 500$ and $x \leq 750$ above our number line (Figure 5.7). The points where the graphs overlap is the intersection—the interval from 500 to 750, which we shade on the number line. We mark the boundary points on the line with closed circles since they are included in the solution.

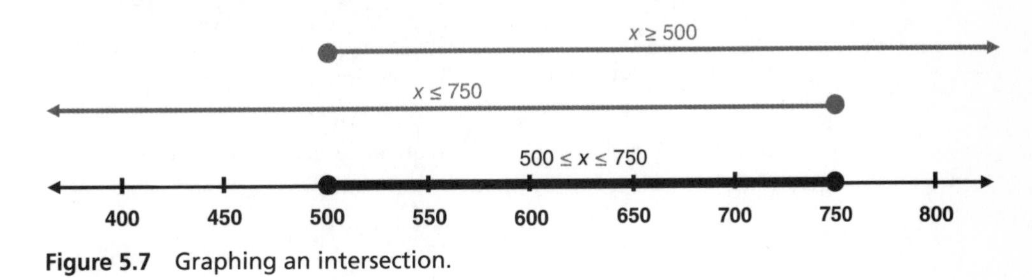

Figure 5.7 Graphing an intersection.

Example 5.8 Graph the solution set of the inequality $3 \geq x > -2$ on a number line and express the solution in interval notation.

Solution: Since quantities increase from left to right on a number line, rewriting the inequality so that the smaller quantity is on the left will make graphing and writing interval notation easier: $-2 < x \leq 3$.

GRAPH: The solution set is all the numbers from −2 to 3, including 3 (but not −2). To graph on a number line, we draw an open circle at −2, then shade all points to 3, then a closed circle at 3.

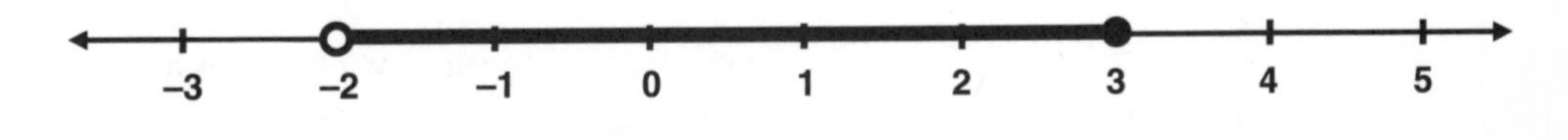

INTERVAL NOTATION: The solution set is $(-2, 3]$.

Example 5.9 Solve the inequality $8 < -2x + 26 < 20$. Graph the solution set on a number line and express it using interval notation.

Solution: We solve each of the two inequalities simultaneously as follows:

$8 < -2x + 26 < \ 20$ $\underline{-26 < \qquad -26 < -26}$	Subtract 26.
$-18 < -2x < -6$	
$\frac{-18}{-2} > \frac{-2x}{-2} > \frac{-6}{-2}$	Divide by −2.
$9 > x > 3$	Final answer, which we write with the smaller number first: $3 < x < 9$.

GRAPH: The solution set is all numbers from 3 to 9. To graph on a number line, draw an open circle at 3, then shade all points to 9, and then draw an open circle at 9.

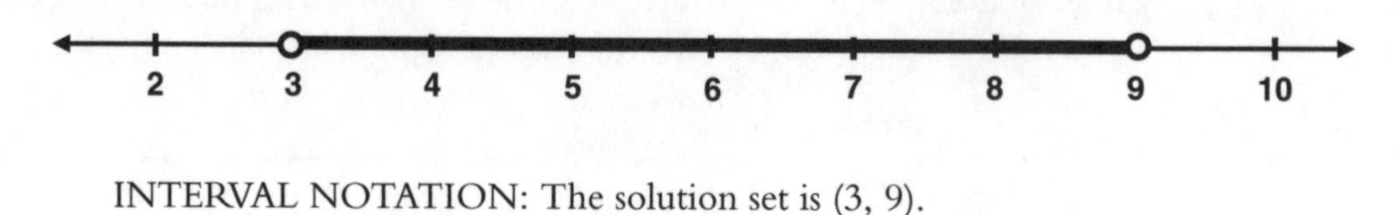

INTERVAL NOTATION: The solution set is (3, 9).

Another type of compound inequality occurs when more than one condition is given but *at least one* has to be met. For example:

- Students whose grades are below 65 or above 80 may not apply.
- People younger than age 5 or older than age 65 are most vulnerable.

For example, the numbers less than 5 or greater than 65 are an example of a **union**, which is the set of numbers that make *at least one* of the inequalities true. In this example, we represent the union of $a < 5$ and $a > 65$ in any of the following ways:

- $\{a \mid a < 5 \text{ or } a > 65\}$, read as "the set of all *a*s such that *a* is less than 5 or greater than 65"
- $\{a \mid a < 5\} \cup \{a \mid a > 65\}$, read as "the union of the set of all *a*s such that *a* is less than 5 and the set of all *a*s such that *a* is greater than 65"
- $(-\infty, 5) \cup (65, \infty)$, read as "the union of the open interval from negative infinity to 5 and the open interval from 65 to infinity"

A union of sets can have numbers that either overlap on a number line or not, as shown in the following examples:

Example 5.10 Graph the solution set of the inequality $x \leq 1$ or $x > 4$ on a number line.

Solution: We graph each inequality above the number line to determine the location of its solution set. We shade the points that belong to one or both solution sets. Since the solution sets do not overlap, we draw both on the number line to represent the union.

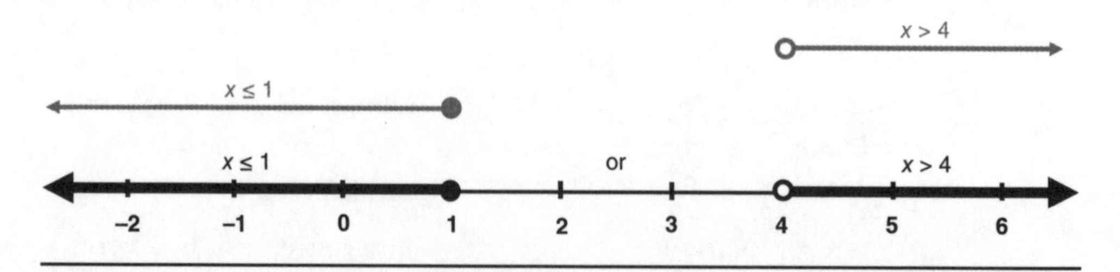

Example 5.11 Graph the solution set of the inequality $x \geq 1$ or $x > 4$ on a number line.

Solution: We graph each inequality above the number line to determine the location of its solution set. To graph the union, we shade the points that belong to one or both solution sets. Since the solution sets overlap, we only graph the more inclusive solution set. (For example, if you need a grade greater than 65 *or* greater than 80 to apply, then any grade above 65 would be acceptable.)

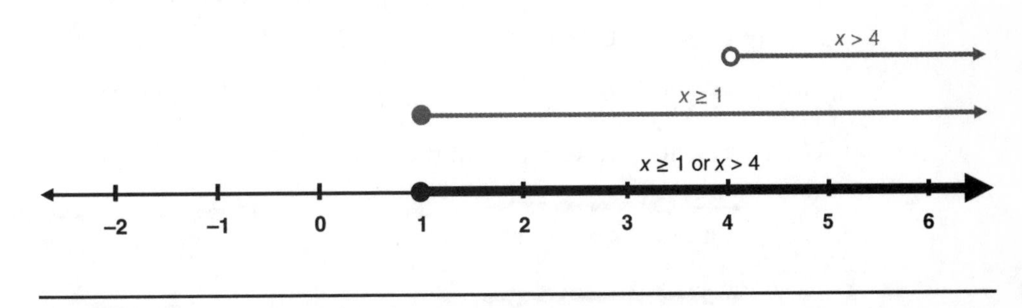

Example 5.12 Solve the inequality $3x + 2 < 8$ or $10 + 8x \leq 10x$. Graph the solution set on a number line and express the solution in interval notation.

Solution: First, we solve each inequality for the variable:

$3x + 2 < 8$	or	$10 + 8x \leq 10x$
$\underline{\quad -2 - 2}$		$\underline{\quad -8x \leq -8x}$
$3x < 6$		$10 \leq 2x$
$\frac{3x}{3} < \frac{6}{3}$		$\frac{10}{2} \leq \frac{2x}{2}$
$x < 2$	or	$5 \leq x$
$x < 2$	or	$x \geq 5$

GRAPH: We graph each inequality above the number line to determine the location of its solution set. To graph the union, we shade the points that belong to one or both solution sets.

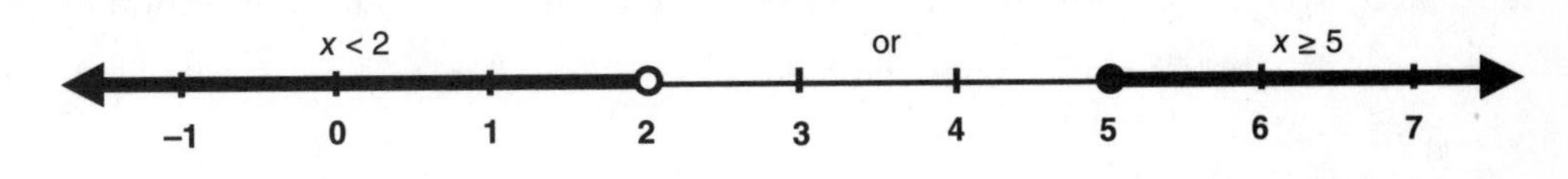

INTERVAL NOTATION: $(-\infty, 2] \cup (5, \infty)$, read as "the union of the interval from negative infinity to 2, including 2, and the open interval from 5 to infinity."

Exercises

Write the pronunciation of each of the following.

1. $\{q \mid q \geq 26 \text{ or } q > 35\}$
2. $45 < x \leq 67$
3. $\{d \mid d > -15 \text{ and } d < 0\}$
4. $\{t \mid t > 12\} \cap \{t \mid t < 15\}$
5. $\{r \mid r \leq -1\} \cup \{r \mid r > 4\}$
6. $(-\infty, 3) \cup [14, \infty)$
7. $\{k \mid k < -6 \text{ or } k > 4\}$
8. $18 \leq p < 20$
9. $\{c \mid c > -2 \text{ and } c < 2\}$
10. $(-\infty, 1] \cap [0, \infty)$
11. $\{b \mid b > -8\} \cup \{b \mid b > 2\}$
12. $(-\infty, 8] \cup [10, \infty)$

Use interval notation to represent the numbers shown in each graph.

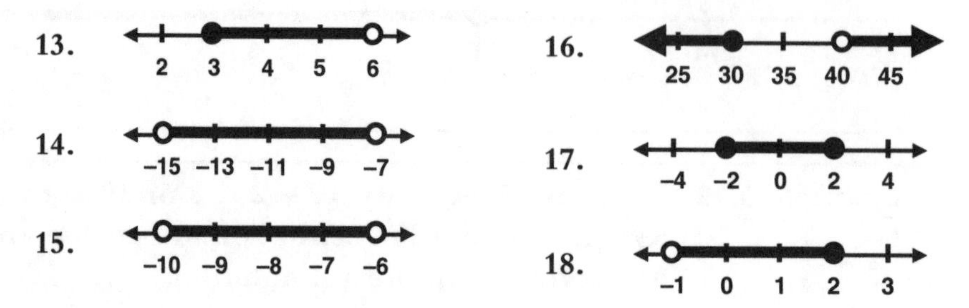

Determine and graph the solution set to each inequality.

19. $4 < x + 3 < 7$
20. $-4 < -2k \leq 4$
21. $2y < -2 \text{ or } y + 7 > 10$
22. $x + 8 < 12 \text{ or } 5x \geq 40$
23. $-3 \leq 2k - 1 < 3$
24. $-40 \leq -4m + 8 \leq 24$
25. $2d + 7 \geq 19 \text{ or } 3d - 22 \leq -10$
26. $4g + 3 < -5 \text{ or } -2g + 3 < 1$
27. $-2a + 3 > 11 \text{ or } 4a > 12$

Questions to Think About

28. Explain the difference between the union and the intersection of two sets.
29. Write a compound inequality with "or" that represents all numbers on the number line.
30. Do the graphs of the intersection of two inequalities always overlap? Explain.

5.5 Word Problems with Inequalities

Inequalities can also be used to represent real-world situations. Table 5.3 lists some common phrases and symbols for expressing inequalities:

Table 5.3 Inequalities in words.

$x > 3$	$x \geq 3$	$x < 3$	$x \leq 3$
is more than 3 greater than 3 exceeds 3 over 3	greater than or equal to 3 a minimum of 3 at least 3 not less than 3 not under 3	less than 3 under 3 up to but not including 3	less than or equal to 3 a maximum of 3 at most 3 not more than 3 not greater than 3 does not exceed 3 not over 3 up to and including 3
$3 \leq x \leq 5$	$3 \leq x < 5$	$x > 3$ or $x < 5$	$x \geq 3$ or $x < 5$
ranges from 3 to 5 between 3 and 5, including 3 and 5	between 3 and 5, including 3	greater than 3 or less than 5 over 3 or under 5	at least 3 or less than 5

Here are some important points about these phrases:

- The phrase "*is more than* 3" ($x > 3$) should not be confused with "*is 5 more than 3*" ($x = 5 + 3$).
- Similarly, the phrase "is less than 3" ($x < 3$) should not be confused with "is 5 less than 3" ($x = 3 - 5$).
- Unfortunately, when used in everyday English, the words "between" or "ranges from" do not specify whether the boundary points of the interval are included. For example, could a book that is *between* 300 and 400 pages long be *exactly* 300 pages long? Adding a phrase like "including … " ("between 300 and 400, including 300") makes the language more precise, but using symbols makes it even clearer.
- The phrase "up to 3" usually implies that 3 is included. Adding the phrase "including 3" or "but not including 3" clarifies the meaning.

Example 5.13 The US Marine Corps will reject women whose height is less than 58 inches. The maximum permissible height for female applicants is 78 inches. Write inequalities representing

(a) heights, in inches, of women who are prohibited from applying to the Marines

(b) heights, in inches, of women who are allowed to apply to the Marines

Solution:

(a) Let p represent the heights, in inches, of women who are prohibited from applying to the Marines. Then the heights less than 58 inches are represented by the inequalities $p < 58$. Since the maximum permissible height is 78 inches, then applicants whose height is greater than 78 inches will be prohibited from applying. This can be represented by $p > 78$. Thus, the heights of women prohibited from applying is represented by the compound inequality $p < 58$ or $p > 78$.

(b) Let a represent the heights, in inches, of women who are allowed to apply to the Marines. These heights are represented by the inequality $58 \leq a \leq 78$.

Many problems with inequalities are solved using methods similar to solving word problems with equations. Use the table at the beginning of this section to help you determine the direction of the inequality symbol that is required to solve these problems.

Example 5.14 To replace its aging technology, a school plans to spend up to $7,000 on new laptops and tablets for its classrooms. The school plans to spend $300 for each laptop and $160 for each tablet. If the school buys 16 laptops, how many tablets can it buy?

Solution:

Step 1: Identify what is given and what we need to find.
Laptops cost $300, and tablets cost $160.
The school spends up to $7,000.
The school buys 16 laptops.
How many tablets can it buy?

Step 2: Represent the unknown information.
Let t = the number of tablets the school can buy.
We can make the following table:

	(Cost per device in dollars)	(Number of devices)	= Total cost of devices
Laptops	300	16	4,800
Tablets	160	t	$160t$

The total cost of the laptops	4,800
and the total cost of the tablets	$+ 160t$
is at most $7,000.	$\leq 7{,}000$

Step 3: Solve.

$$\begin{aligned}
4{,}800 + 160t &\le 7{,}000 \\
\underline{-4{,}800 \qquad\quad} &\underline{\; - 4{,}800} \\
160t &\le 2{,}200 \\
\frac{160t}{160} &= \frac{2{,}200}{160} \\
t &\le 13.75
\end{aligned}$$

Subtract 4,800 from both sides.
Combine like terms.
Divide both sides by 160.
Simplify.

The number of tablets must be *less than* 13.75, and the school cannot buy a part of a tablet. Since it will spend *up to* \$7,000, it can buy *no more than* 13 tablets. (Buying 14 or more tablets would mean spending more than \$7,000.)

Step 4: Check.

$$4{,}800 + 160(13) \overset{?}{\le} 7{,}000$$

$$6{,}880 \le 7{,}000$$

The school can buy no more than 13 tablets.

Example 5.15 Greatway Supermarket plans to generate between \$200,000 and \$300,000 in sales every week. Since the beginning of the week, it has sold \$168,000 worth of products. Write an inequality that represents the additional sales revenue that the supermarket needs to meet its weekly goal.

Solution:

Step 1: Identify what is given and what we need to find.
Sales should be between \$200,000 and \$300,000.
Sales are \$168,000 so far.
How much more in sales are needed?

Step 2: Represent the unknown information.
Let $r =$ the additional sales revenue needed to meet the goal.

Between \$200,000	$200{,}000 \le$
is \$168,000 in sales	168,000
plus additional sales revenue	$+\, r$
and \$300,000.	$\le 300{,}000$

Step 3: Solve.

$$\begin{aligned}
200{,}000 \le 168{,}000 + r &\le 300{,}000 \\
\underline{-\,168{,}000 - 168{,}000 \qquad} &\underline{\; -\,168{,}000} \\
32{,}000 \le r &
\end{aligned}$$

Subtract 168,000.
$\le 132{,}000$

The additional revenue must be between \$32,000 and \$132,000.

Step 4: *Check.*

Select a value in the solution set, such as \$100,000.

$$200{,}000 \overset{?}{\leq} 168{,}000 + 100{,}000 \overset{?}{\leq} 300{,}000$$

$$200{,}000 \leq 268{,}000 \leq 300{,}000$$

Exercises

Represent each phrase with an inequality.

1. A maximum of 8
2. A minimum of 12
3. No more than 60
4. At least 45
5. Not less than 100
6. Not more than 8
7. Less than 32 or greater than 212
8. From 5 to 65
9. Over 12 but under 18
10. To become president of the United States, you must be at least 35 years old. Write an inequality representing the range of allowed ages for presidents.
11. Absolute zero, the lowest temperature that is theoretically possible, is −273.15° Celsius. Represent this temperature restriction as an inequality.
12. The speed limit on interstate highways in the United States is 75 miles per hour. Represent the allowed speeds as an inequality.
13. The SAT is scored on a scale that ranges from 400 to 1,600. Represent this scale range with an inequality.
14. According to NASA, astronauts must be a minimum height of 62 inches and be no more than 75 inches tall. Represent this height range with an inequality.
15. According to the Motion Picture Association, people under the age of 17 may not view a movie with an R rating unless accompanied by an adult. Write an inequality that represents the ages of people allowed to view R-rated movies without being accompanied by an adult.
16. In Ms. Bukalov's 45-minute test review, introductory activities take up to 4 minutes. If she plans to allow students to spend at least three minutes on each problem, how many problems can her class complete?
17. Urban Electricians charges \$75 per visit with a \$60 per hour fee. How many hours can the company spend on a visit if the total fee is at most \$450?
18. For each ride, the Yellow Cab Company charges \$3.50 for the first 0.2 miles and \$0.50 for each additional 0.2 miles. At that rate, how far can a passenger travel in a Yellow Cab if the total fare, excluding tips, is less than \$40?

19. Jaylene wants her target heart rate to be between 100 and 170 beats per minute. Write and solve an inequality that represents how much she needs to raise her resting heart rate, which is currently 63 beats per minute, in order to get it into the target heart rate zone.

20. An electronics store offers a $40 rebate on phones that cost between $150 and $600. Write and solve an inequality that represents the range of costs for the phones after the rebate is applied.

CHAPTER 5 TEST

1. Which of the following inequalities is true?

   (A) $-4 \geq -4$ (B) $0 < -4$ (C) $-4 > 0$ (D) $-4 < -5$

2. If the maximum value of m is 15, which inequality represents the possible values for m?

   (A) $m < 15$ (B) $m > 15$ (C) $m \geq 15$ (D) $m \leq 15$

3. Which number is in the solution set to the inequality $4x > 2x - 1$?

   (A) -2 (B) -1 (C) $-\frac{1}{2}$ (D) 0

4. If v is a negative integer, what is the solution to $2v + 8 > 2$?

   (A) $-3, -2, -1$ (B) $3, -2, -1$ (C) $-2, -1, 0$ (D) $-2, -1$

5. Which inequality is equivalent to $-7x < 49$?

   (A) $x < 7$ (B) $x > 7$ (C) $x > -7$ (D) $x < -7$

6. Which graph represents the union of the sets $\{x \mid x > 9\}$ and $\{x \mid x > 5\}$?

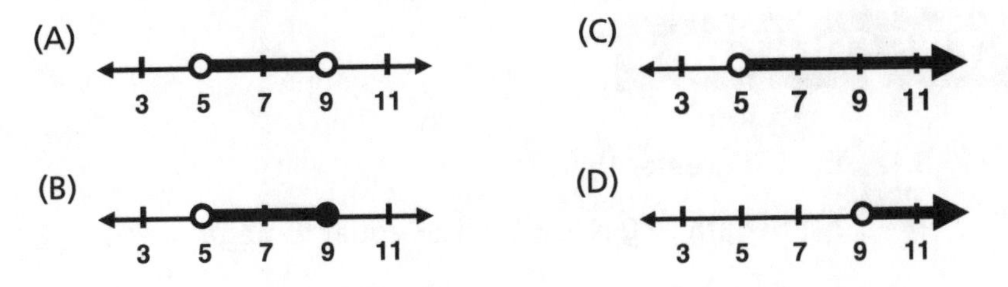

7. What is the solution set to the inequality $5 > x \geq 4$, written in interval notation?

   (A) (4, 5) (B) [4, 5) (C) (4, 5] (D) (5, 4]

8. If $a > b > 0 > c$, then which of the following inequalities is *always* true?

   (A) $ac > bc$ (B) $ac < bc$ (C) $abc > ab$ (D) $abc > 0$

9. Which set represents all numbers on the number line?

   (A) $\{x \mid x \geq 17\} \cup \{x \mid x < 17\}$ (B) $\{x \mid x > 17\} \cap \{x \mid x < 17\}$

   (C) $\{x \mid x > 17\} \cap \{x \mid x \leq 17\}$ (D) $\{x \mid x > 17\} \cup \{x \mid x < 17\}$

10. Write the pronunciation of $[3, 6] \cap (5, 9)$.

11. Solve the inequality $12 > 3k + 8 > 5$ for k.

12. Express the set of integers less than 2 and greater than −3 using set-builder notation.

13. The highest presidential approval rating measured by Gallup was recorded for President George W. Bush shortly after the September 2001 terrorist attacks. Gallup estimated that the percentage of Americans who approved of President Bush was up to 3% above or below 90%. Write a compound inequality that represents this percentage.

14. Solve for p in the inequality $-11(p + 7) - 15 < 15p + 20p$. Graph the solution set.

15. Jahari wants to buy a mobile phone that costs $560. So far, he has saved $120. He earns $172 a week from his part-time job. If he saves one-fourth of his job earnings, how many weeks will he need to work to have enough money to buy the phone?

CHAPTER 5 SOLUTIONS

5.1.

1. "5 is greater than 2"
2. "negative 40 is less than or equal to negative 40"
3. "37 is greater than negative 15"
4. "negative 22 is less than positive 54"
5. "6 is greater than or equal to 2"
6. "negative 15 is less than or equal to negative 8"

7. True because +5 is to the right of +4
8. True because +1 is to the right of 0
9. False because −4 is not to the left of −6
10. True because −1 is to the left of +1
11. True because −5 is to the left of −1
12. False because 0 is not to the right of 5
13. $-53 < 4$
14. $60 > -10$
15. $-15 > -24$
16. $\frac{7}{3} > 2$
17. $-\frac{17}{9} < -\frac{18}{11}$
18. $-\frac{1}{3} > -\frac{1}{2}$

5.2. 1. "the open interval from 17 to infinity"

2. "the open interval from negative infinity to 26"
3. "the interval from negative infinity to 3, including 3"
4. "the set of all *q*s such that *q* is greater than 52"
5. "the set of all *d*s such that *d* is less than or equal to negative 78"
6. "the set of all *x*s such that *x* is greater than or equal to 14"
7. "the set of all *y*s that are positive integers such that *y* is greater than or equal to 14"
8. "the set of all *k*s that are counting numbers such that *k* is less than 6"
9. "the set of all *w*s that are whole numbers such that *w* is less than 34"
10. $(-\infty, 90)$, $\{u \mid u < 90\}$
11. $(-\infty, 82]$, $\{n \mid n \leq 82\}$
12. $(-3, \infty)$, $\{a \mid a > -3\}$
13. $(-\infty, 0]$, $\{x \mid x \leq 0\}$
14. $(-\infty, -19)$, $\{k \mid k < -19\}$
15. $[14, \infty)$, $\{y \mid y \geq 14\}$

16. $(-\infty, -8]$, $\{y \mid y \leq -8\}$

17. $(15, \infty)$, $\{p \mid p > 15\}$

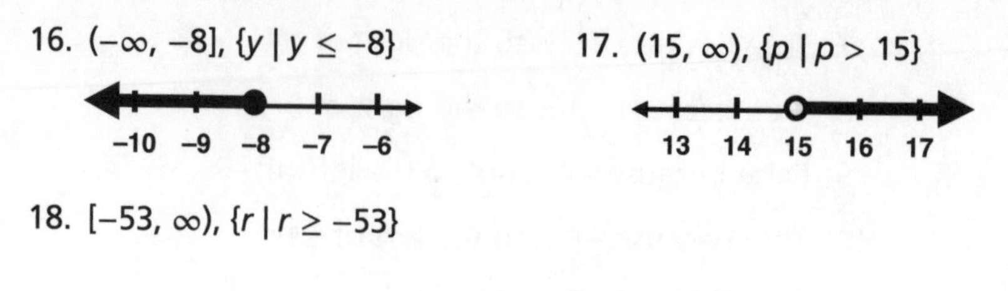

18. $[-53, \infty)$, $\{r \mid r \geq -53\}$

−55 −54 −53 −52 −51

19. no interval notation, $\{x \in \mathbf{Z} \mid x \geq 2\}$ or $\{x \in \mathbf{N} \mid x \geq 2\}$

0 1 2 3 4 ...

20. no interval notation, $\{x \in \mathbf{Z} \mid x > -3\}$

−4 −3 −2 −1 0 ...

21. no interval notation, $\{x \in \mathbf{Z} \mid x \leq 75\}$

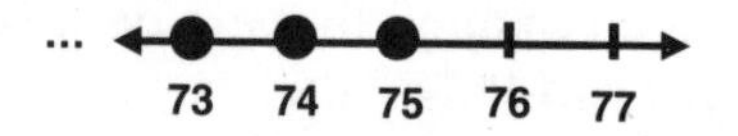

22. Negative infinity is not a number, so it can never be part of an interval. Thus, using a square bracket to include $-\infty$ in the interval is incorrect.

23. Interval notation represents all of the numbers between the endpoints. However, the numbers between the integers are not part of the solution set, and this cannot be represented by interval notation.

24. The domain is the set of all possible values that can be used to replace the variable.

5.3.

1. $q > 3$, $(3, \infty)$
2. $n \geq 5$, $[5, \infty)$
3. $x < 32$, $(-\infty, 32)$
4. $r \geq 6$, $[6, \infty)$
5. $g \leq -6$, $(-\infty, -6]$
6. $x > -1{,}296$, $(-1{,}296, \infty)$
7. $j < -5$, $(-\infty, -5)$
8. $z \geq -\frac{34}{75}$, $\left[-\frac{34}{75}, \infty\right)$

9. $c < -96$, $(-\infty, -96)$
10. $x \geq 22$, $[22, \infty)$
11. $u < 0$, $(-\infty, 0)$
12. $v \leq 108$, $(-\infty, 108]$
13. $a \geq \frac{9}{4}$, $\left[\frac{9}{4}, \infty\right)$
14. $y > -1$, $(-1, \infty)$
15. $p < 2$, $(-\infty, 2)$
16. $x < -1$, $(-\infty, -1)$
17. $y > \frac{1}{2}$, $\left(\frac{1}{2}, \infty\right)$
18. $m \leq 10$, $(-\infty, 10]$
19. $f \geq 31$, $[31, \infty)$
20. $d > -25$, $(-25, \infty)$
21. $x \leq 22$, $(-\infty, 22]$
22. $z < \frac{48}{7}$, $\left(-\infty, \frac{48}{7}\right)$
23. $p > \frac{69}{2}$, $\left(\frac{69}{2}, \infty\right)$
24. $x \leq \frac{120}{13}$, $\left(-\infty, \frac{120}{13}\right]$
25. $\{0, 1, 2, 3, 4, 5\}$
26. $\{1, 2, 3, 4\}$
27. $\{2\}$

5.4.

1. "the set of all *q*s such that *q* is greater than or equal to 26 or greater than 35"
2. "45 is less than *x* is less than or equal to 67"
3. "the set of all *d*s such that *d* is greater than negative 15 and less than 0"
4. "the intersection of the set of all *t*s such that *t* is greater than 12 and the set of all *t*s such that *t* is less than 15"
5. "the union of the set of all *r*s such that *r* is less than or equal to negative 1 and the set of all *r*s such that *r* is greater than 4"
6. "the union of the open interval from negative infinity to 3 and the interval from 14 to infinity, including 14"
7. "the set of all *k*s such that *k* is less than negative 6 or greater than 4"
8. "18 is less than or equal to *p* is less than 20"
9. "the set of all *c*s such that *c* is greater than negative 2 and less than 2 "
10. "the intersection of the interval from negative infinity to 1, including 1, and the interval from 0 to infinity, including 0"
11. "the union of the set of all *b*s such that *b* is greater than negative 8 and the set of all *b*s such that *b* is greater than 2"

12. "the union of the interval from negative infinity to 8, including 8, and the interval from 10 to infinity, including 10"

13. [3, 6)

14. (−15, −7)

15. (−10, −6)

16. $(-\infty, 30] \cup (40, \infty)$

17. [−2, 2]

18. (−1, 2]

19. $1 < x < 4$,

0 1 2 3 4

20. $-2 \le k < 2$,

−2 −1 0 1 2

21. $y < -1$ or $y > 3$,

−3 −1 1 3 5

22. $x < 4$ or $x \ge 8$,

2 4 6 8 10

23. $-1 \le k < 2$,

−1 0 1 2 3

24. $-4 \le m \le 12$,

−4 0 4 8 12

25. $d \ge 6$ or $d \le 4$,

3 4 5 6 7

26. $g < -2$ or $g > 1$,

−2 −1 0 1 2

27. $a < -4$ or $a > 3$,

−4 −2 0 2 4

28. The intersection is the set of numbers that makes all the inequalities true, while the union is the set of numbers that make at least one of the inequalities true.

29. Answers will vary. For example, $x > 5$ or $x \le 5$.

30. No. For example, $x > 6$ and $x < 3$ do not overlap since the inequalities point in different directions on the number line and the solution set doesn't include any number between the endpoints.

5.5.

1. $x \le 8$
2. $x \ge 12$
3. $x \le 60$
4. $x \ge 45$
5. $x \ge 100$
6. $x \le 8$
7. $x < 32$ or $x > 212$
8. $5 \le x \le 65$
9. $12 < x < 18$

10. $a \geq 35$
11. $t \geq -273.15$
12. $s \leq 75$
13. $400 \leq s \leq 1{,}600$
14. $62 \leq h \leq 75$
15. $a \geq 17$
16. no more than 13 problems
17. no more than 6.25 hours
18. no more than 14.8 miles ($3.50 for the first 0.2 miles and $36.50 for the remaining 14.6 miles)
19. $100 \leq 63 + b \leq 170$, $37 \leq b \leq 107$
20. $150 \leq 40 + c \leq 600$, $110 \leq c \leq 560$

CHAPTER 5 TEST SOLUTIONS

1. (A)
2. (D)
3. (D)
4. (D)
5. (C)
6. (C)
7. (B)
8. (B)
9. (A)
10. "the intersection of the closed interval from 3 to 6 and the open interval from 5 to 9"
11. $-1 < k < \frac{4}{3}$
12. $\{x \in \mathbf{Z} \mid -3 < x < 2\}$
13. If r = George W. Bush's presidential approval rating, then $87 \leq r \leq 93$.
14. $p > -2$.

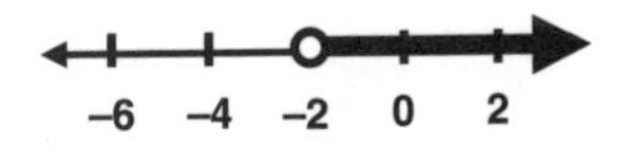

15. at least 11 weeks

6 FUNCTIONS AND GRAPHS WITH TWO VARIABLES

So far, we've worked with equations and inequalities with one variable. We often work with equations with *two* variables. In this chapter, we explain why working with two variables is important in math and discuss different ways to represent mathematical relationships in two variables.

6.1 Functions and Function Notation

Have you ever used an electronic vending machine? You put in money, press a button or series of buttons, and the machine dispenses an item, such as a drink or snack. The item that comes out of the machine depends on what buttons you press. We can represent a vending machine's inputs (the buttons pressed to select items) and outputs (the items dispensed by the machines) with a table, as shown in Table 6.1:

Table 6.1 Functional and non-functional vending machines.

Functional Vending Machines

Machine #1

Button	Item
1	Nachos
2	Pretzels
3	Popcorn
4	Crackers

Machine #2

Button	Item
11	Water
12	Water
13	Water
14	Juice

Non-Functional Vending Machines

Machine #3

Button	Item
A	Juice Water
B	Soda
C	Iced Tea
D	Diet Soda

Machine #4

Button	Item
A1	
A2	
A3	Water
A4	Juice

Here are some important notes about these tables:

- If a vending machine is functional (in other words, if it's working properly), then it will dispense one item after a button is selected. Some items are so popular that they can be chosen using several buttons. For example, in Machine #2, picking 11, 12, or 13 would result in getting a bottle of water.
- Sometimes, vending machines don't work properly. On non-functional vending machines, selecting some numbers will result in *more than one* item being dispensed (in Machine #3, juice and water come out when A is selected) or *nothing* being dispensed (in Machine #4, nothing happens when A1 or A2 is selected).

- Whether the vending machines are functional or non-functional, they relate information in a particular order. In this example, 3 → popcorn relates the number 3 and the word "popcorn," meaning "if you press button 3, you get a bag of popcorn."

We use similar thinking to define mathematical relationships between sets. A **relation** connects elements of one set (called the **domain**) to elements of another set (called the **codomain**). Since relations pair elements in a particular order, we can also define a relation as a set of **ordered pairs**. Each ordered pair has a first value (called an **input**) and a second value (called an **output**).

If the variables are x and y, then the x-values correspond to the domain or input values, and the y-values correspond to the range or output values. We call x the **independent variable** and y the **dependent variable** since the value of y *depends* on the value of x.

Figure 6.1 summarizes the vocabulary that we use with relations.

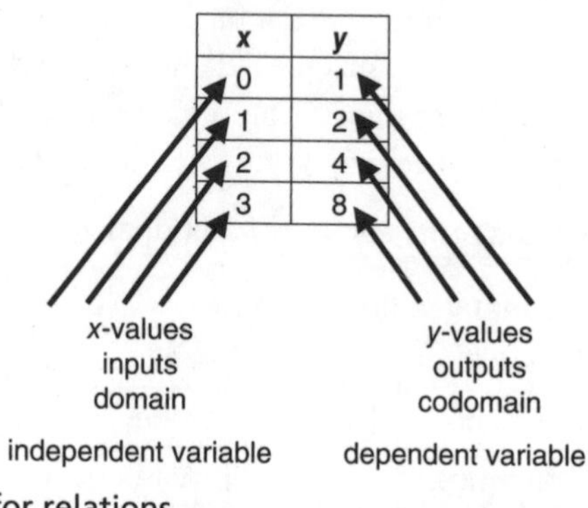

Figure 6.1 Vocabulary for relations.

Since many relations share a useful property, we give them a special name. A **function** is a relation in which *every element of the domain is mapped to exactly one element of the codomain*. (In other words, if an input is mapped to *no* outputs or *more than one* output, then the relation is not a function.) The set of a function's output values is called the **range**.

Functions are important in math, computers, and science because we use them to define relationships between variables precisely. When we define a function, we can compute the output for a given input value.

We represent a function with a table of values, a diagram, or a set of ordered pairs, as shown in Figure 6.2:

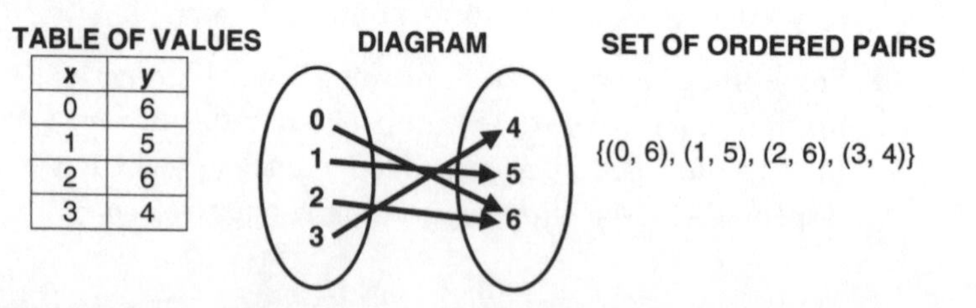

Figure 6.2 Function as a table, diagram, or set of ordered pairs.

Here are some notes about each:

- The table of values can be written vertically or horizontally.
- In the diagram, we write each input value once in the oval on the left and each output value once in the oval on the right. We then draw arrows from left to right to indicate each ordered pair.
- We use parentheses to indicate an ordered pair. The ordered pair (0, 6), read as "the ordered pair 0, 6," indicates that an input of 0 produces an output of 6. This set of ordered pairs is pronounced "the set of ordered pairs 0, 6 (pause), 1, 5 (pause), 2, 6 (pause), and 3, 4."
- For this function, the domain is the set $\{0, 1, 2, 3\}$ and the range is the set $\{4, 5, 6\}$.

Figure 6.3 shows a relation that is not a function. The input 2 produces more than 1 output (4 and 6).

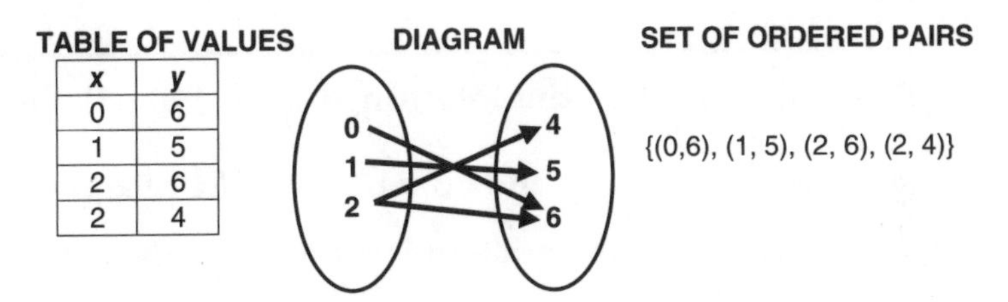

Figure 6.3 Non-function as a table, diagram, or set of ordered pairs.

Watch Out!

The notation for ordered pairs looks like the notation for intervals. How do you know if (3, 4) represents an ordered pair or the interval $\{x \mid 3 < x < 4\}$? Often, you can tell from the context, so (3, 4) in a situation with only one variable probably refers to an interval, while (3, 4) in a situation with two variables probably refers to an ordered pair. To avoid confusion, say or write "the interval (3, 4)" or "the ordered pair (3, 4)." While we use parentheses for an ordered pair, we use curly brackets {} to indicate a set. Don't confuse them! ({3, 4}, {5, 6}) or ((3, 4), (5, 6)) are *not* correct.

We also say "y is a function of x." We write it in symbols as $y = f(x)$, which we read as "y equals f of x." Here, f is a letter that we use to name the function, x is a placeholder that represents the input, and $f(x)$ represents the output of f. We use the function notation $f(x)$ to represent y to emphasize that it is the output when x is the input. To summarize:

y	is a	function of x	can be written as
y	$=$	$f(x)$	which we read as "y equals f of x"

When we substitute a specific value for the input variable, we say that we are **evaluating** the function for that value. For example, if we define f as the set of ordered pairs $\{(0, 6), (1, 5), (2, 6), (3, 4)\}$, then evaluating $f(0)$ means calculating the output value when the input is 0. In this case, the output is 6, so we write $f(0) = 6$ (read as "f of 0 equals 6"). Since $f(0)$ is a number, we can perform calculations with it just as we can with any other number. Thus, $f(0) + 1 = 6 + 1 = 7$.

Reading and Writing Tip

Many mistake function notation like $f(x)$ for multiplication. We use letters, not numbers, to name functions, so 4(2) is read as "4 times 2," *not* "4 of 2." To indicate the product of two variables, we recommend writing the variables next to each other without any spaces or other marks. For example, we write the product of m and x as mx, not $m(x)$, which could be misinterpreted as the function $m(x)$.

Example 6.1 For the relation $\{(4, 5), (4, -8), (6, 9), (7, 3)\}$:

(a) State the domain and codomain of the relation.

(b) Determine if the relation is a function. Explain your answer.

Solution:

(a) The domain is $\{4, 6, 7\}$ and the codomain is $\{-8, 3, 9, 5\}$.

(b) In a function, every input is mapped to exactly one output, but in this relation, the input 4 is mapped to both 5 and -8. Thus, it is not a function.

Watch Out!

In our experience, many students explain that a relation is not a function by saying that "the *x*s repeat." Similarly, they say that a relation is a function by saying that "the *x*s don't repeat." We recommend that you *not* use this language to explain your answer because it's not precise enough. For example, the relation $\{(4, 1), (2, 4)\}$ is a function even though 4, which is an x-value, repeats as a y-value. Instead, use the definition of functions:

- If each input is mapped to *exactly one* output, then the relation is a function.
- As we said before, if an input is mapped to *no* outputs or *more than one* output, then the relation is not a function.

Example 6.2 Use function notation to express the relationship between the highlighted numbers.

t	0	1	3	5
$g(t)$	0	3	8	11

Solution: The independent variable is t, and the **highlighted** input value is $t = 3$. The function is denoted with the variable g, and the output value when $t = 3$ is 8. Thus, we write $g(3) = 8$ (pronounced "g of 3 equals 8").

Example 6.3 Given the function shown in the accompanying table, evaluate $7 - 4f(3)$.

a	−2	3	4	40
$f(a)$	10	2	0	3

Solution: The table for the function f tells us that when the input value is 3, the output value is 2. Symbolically, we write $f(3) = 2$. Then

$7 - 4f(3)$	
$= 7 - 4(2)$	Substitute 2 for $f(3)$.
$= 7 - 8$	By the order of operations, we multiply before subtracting.
$= -1$	Final answer.

Exercises

Write the pronunciation of each equation.

1. $f(3) = 5$

2. $g(0) = 6$

3. $h(-2) = 0$

4. $g(5) = 5 + 7(9)$

5. $8f(3) = 8(9 - 3)$

6. $-f(4) = -7 - 4$

State the domain and codomain of each relation. Then determine if each relation is a function. Explain your answer.

7. $\{(-7, 10), (5, 7), (6, 10), (2, 5)\}$

8. $\{(-1, 9), (0, 9), (3, 9), (4, 9)\}$

9. $\{(8, 7), (2, 0), (-16, -1), (2, 39)\}$

10. $\{(16, 2), (5, 8), (16, 8), (3, 17)\}$

11.

x	25	16	9	4
y	5	4	3	2

13.

s	4	9	0	9
r	2	3	0	−3

12.

t	5	2	8	6
d	10	3	10	4

14.

t	4	8	12	16
a	27	81	96	144

Use function notation to express the relationship between the numbers in boldface.

15.

x	0	1	**2**	3
$f(x)$	10	20	**30**	40

16.

t	0	1	2	**10**
$h(t)$	0	4.9	19.6	**490**

17. Given the function shown in the table, evaluate $8f(4)$.

x	0	2	4	6
$f(x)$	1	7	13	19

18. Given the function shown in the table, evaluate $1 + 3g(-2)$.

x	−4	−2	0	4
$g(x)$	6	2	−2	−6

Questions to Think About

19. Name a real-life example of a relation that is a function. Explain your answer.

20. Name a real-life example of a relation that is not a function. Explain your answer.

21. If the input and output values in each ordered pair of a function is reversed, is the resulting relation always a function? Explain.

6.2 Introduction to Graphing

In Section 6.1, we represented the relationships between two sets of variables with ordered pairs, diagrams, or tables. In this section, we discuss how we represent these relationships visually with a rectangular grid.

You've seen rectangular grids before. We use them to identify locations on department store maps (shoes are at B3), chessboards (queen to g7), and computer spreadsheets (cell B42). In math, we use a rectangular grid, shown in Figure 6.4:

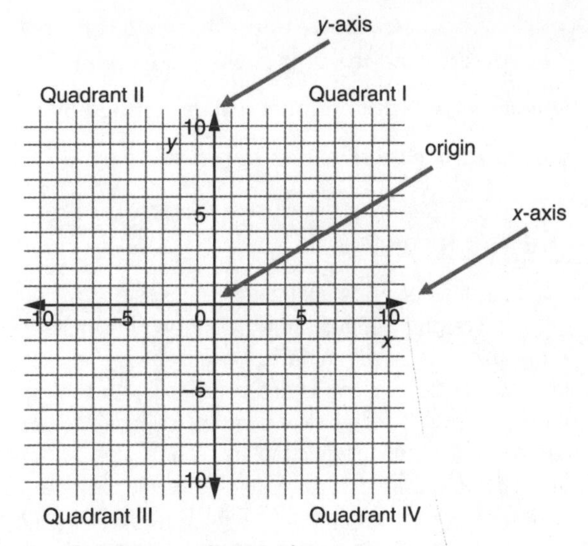

Figure 6.4 Vocabulary for the coordinate plane.

Here are important points about coordinates:

- Horizontal and vertical number lines, called **axes**, indicate position. We often use the variable x to represent horizontal distance and the variable y to represent vertical distance. In these situations, we call the axes the x-axis and y-axis, respectively. We label each axis with the symbol representing the input (such as x for the horizontal axis) or output (such as y or $f(x)$ for the output.
- The point where the axes meet is called the **origin**. At the origin, $x = 0$ and $y = 0$.
- **Coordinates** are the individual numbers used to indicate position when we write ordered pairs. The first number represents the horizontal distance from the origin (called the ***x*-coordinate** when the input variable is x), and the second number represents the vertical distance from the origin (called the ***y*-coordinate** when the output variable is y). As we said in Section 6.1, we use parentheses to indicate ordered pairs, so the point (4, 3) has an x-coordinate of 4 and a y-coordinate of 3. The coordinates of the origin are (0, 0).
- To graph ordered pairs, we start at the origin, count to the left or right the number of units indicated by the x-coordinate, and count up or down the number of units indicated by the y-coordinate. We put a dot at the ordered pair's location.
- On the horizontal axis, values increase from left to right. Negative numbers are to the left of the origin, and positive numbers are to the right. On the vertical axis, values increase from bottom to top. Negative numbers are below the origin, and positive numbers are above.
- The axes divide the coordinate plane into four **quadrants**. We name them using Roman numerals from Quadrant I (top right), going counterclockwise to Quadrant IV (bottom right), but we usually don't label them. (We number them

counterclockwise for reasons that will become clearer when you learn about trigonometry and the unit circle in Algebra II.)

- The entire two-dimensional surface defined by the axes is the **coordinate plane**.
- A **graph** is a representation of ordered pairs on a coordinate plane.

Did You Know?

For thousands of years, mathematicians have used rectangular grids to describe location. In the 3rd century BCE, the Greek scientist Eratosthenes placed grids of overlapping lines over his maps of the Earth's surface. Ancient Hindu and Islamic astronomers developed complex methods for determining longitude. In the 14th century, French philosopher Nicole Oresme proposed using a rectangular coordinate system to describe the relationship between *any* two quantities, not just geographical coordinates. Most of the graphing conventions that we use today come from 18th-century European textbooks that elaborated on the writings of 17th-century French philosopher René Descartes. In his honor, the coordinate plane is sometimes called the **Cartesian plane**.

To show the relationship between multiple points, we plot them on the same coordinate plane, as shown in Example 6.4:

Example 6.4 Graph the relation $\{(6, -1), (3, 0), (0, 1)\}$ on the coordinate plane. State its domain and codomain.

Solution: The point $(6, -1)$ has an x-coordinate of $+6$ (meaning 6 units right from the origin) and a y-coordinate of -1 (meaning 1 unit down from the origin). Similarly, the point $(3, 0)$ is three units right from the origin along the horizontal axis, and $(0, 1)$ is one unit up from the origin along the vertical axis.

We draw a dot to represent each point on the coordinate plane.

The domain is the set of x-values: $\{0, 3, 6\}$. The codomain is the set of y-values: $\{-1, 0, 1\}$.

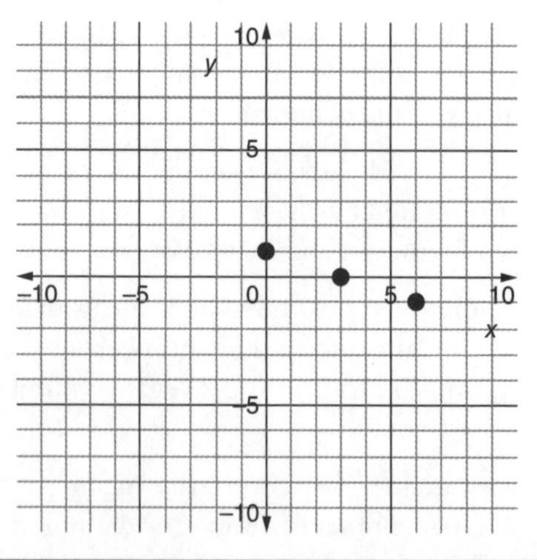

Example 6.5 shows how we can use the graph of a relation to determine if it is a function.

Example 6.5 Is the relation shown in the accompanying graph a function? Explain.

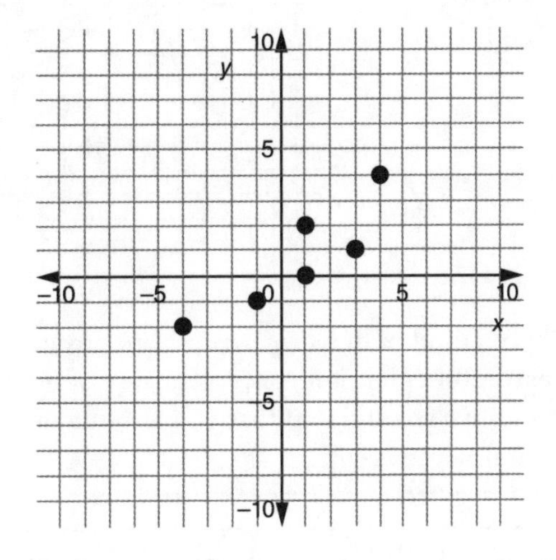

Solution:

METHOD 1: *Convert the points on the graph to ordered pairs.*
The ordered pairs shown in the graph are (−4, −2), (−1, −1), (1, 0), (1, 2), (3, 1), and (4, 4). From the coordinates, we see that the input value 1 has two output values (0, and 2). Since every input in a function can have only one output, then this relation is *not* a function.

METHOD 2: *Use vertical lines drawn from each input value.*
We draw vertical lines on the graph from each input value. If every vertical line passes through only one point on the graph, then each input has only one output, and the relation is a function. If a line passes through more than one point, then it has more than one output value. We see that the vertical line at $x = 1$ (the dashed line in the graph) passes through *two* points, which means that this input has two outputs. Since every input in a function can have only one output, then this relation is *not* a function.

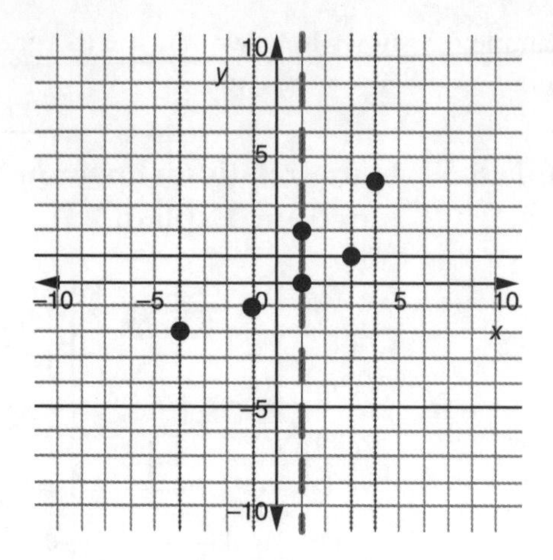

This method of drawing vertical lines from each input value is called the **vertical line test**. Although it is a convenient way of using a graph to determine if a relation is a function, we recommend that you also understand why it works and be prepared to explain it if necessary.

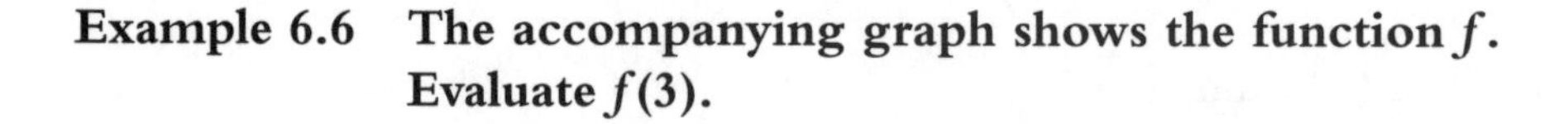

Example 6.6 The accompanying graph shows the function f. Evaluate $f(3)$.

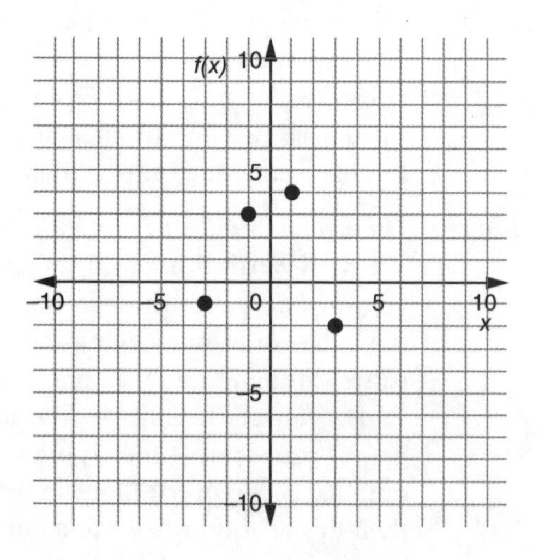

Solution: We need to evaluate $f(3)$ from the graph of f. As we said in Section 6.1, evaluating a function means we calculate the output (y-coordinate) when the input (x-coordinate) is 3. To evaluate $f(3)$, we find the point whose x-coordinate is 3 by drawing a vertical line at $x = 3$ and identifying the point that intersects the line. Then we find its y-coordinate by drawing a horizontal line from the point to see where it intersects the y-axis. The y-coordinate is -2, so $f(3) = -2$.

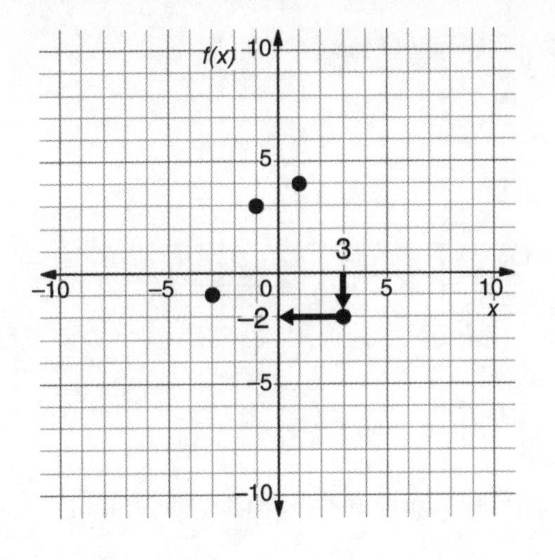

Exercises

Graph each relation on the coordinate plane and state its domain and codomain.

1. $\{(-5, 6), (-4, 2), (-3, 1), (-2, 0)\}$
2. $\{(-1, -6), (7, -1), (2, 6), (2, 4)\}$
3. $\{(5, -2), (-1, -2), (0, 7), (2, -4), (6, 0)\}$
4. $\{(0, 0), (1, 3), (2, 5), (3, 7), (4, 9)\}$

Determine whether each relation is a function. Justify your answer.

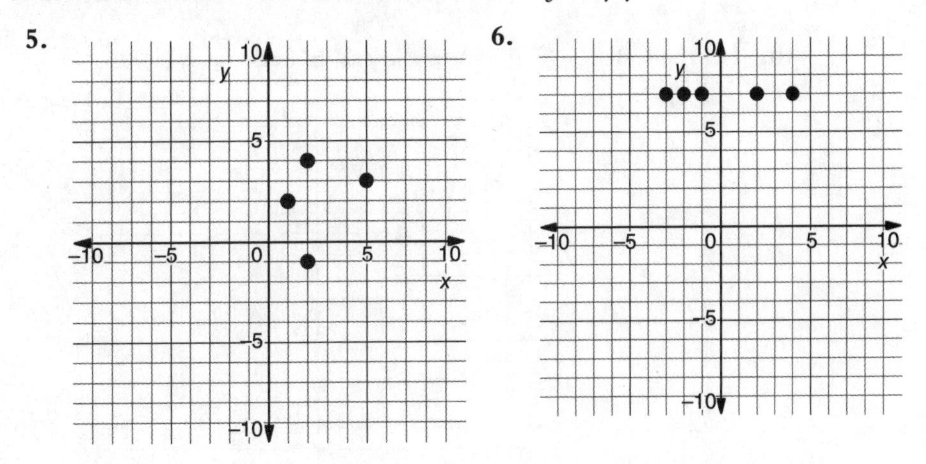

7. **8.**

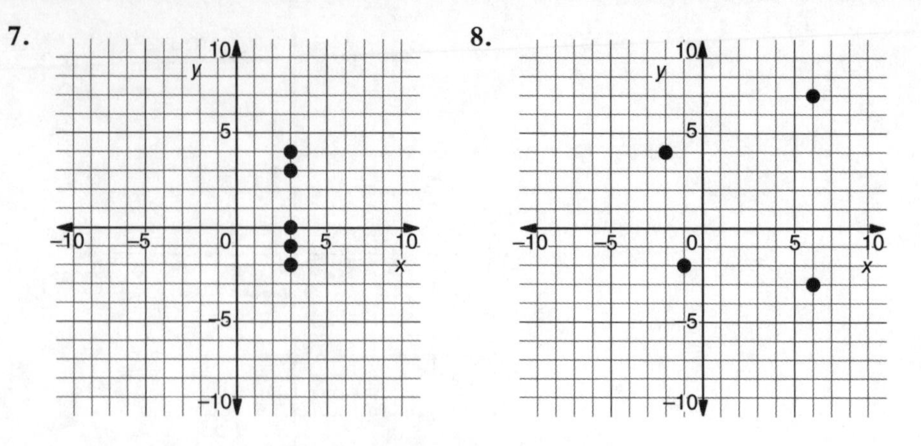

9. For the function f shown in the graph, evaluate $f(6)$.

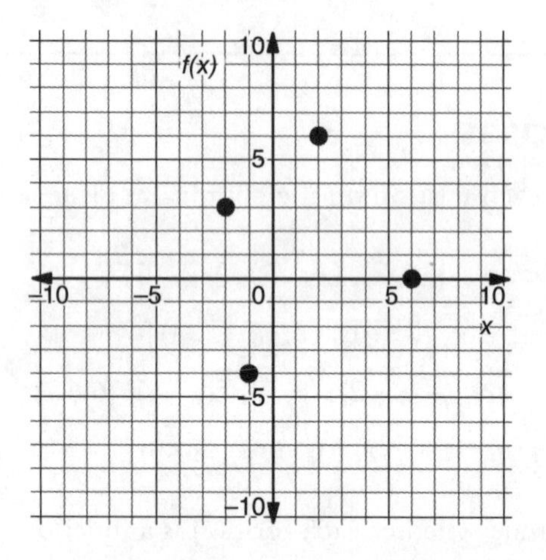

10. For the function f shown in the graph, evaluate $2f(1)$.

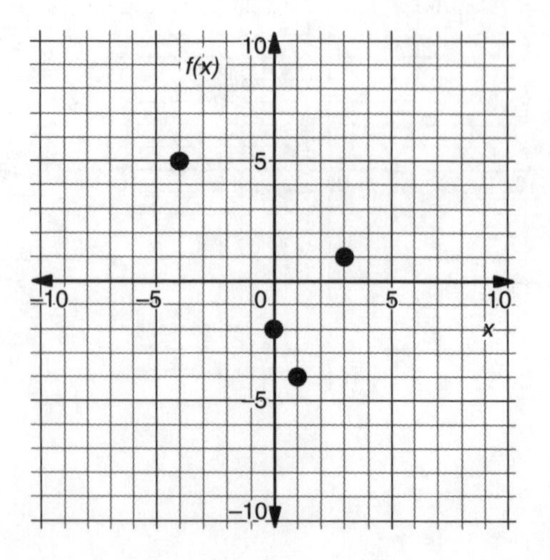

11. For the function f shown in the graph, evaluate $-f(4)$.

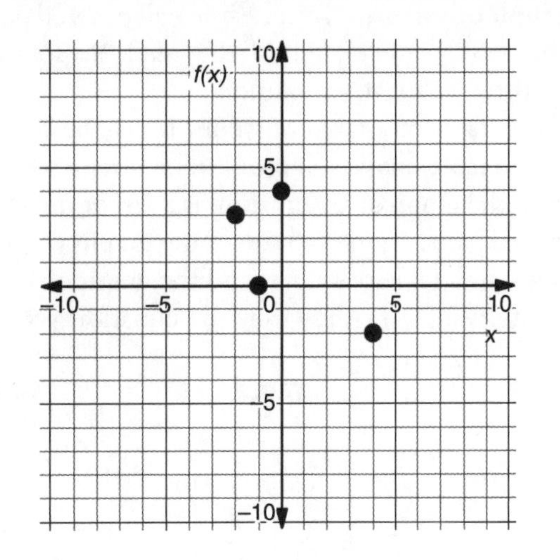

12. For the function f shown in the graph, evaluate $8 - 3f(-2)$.

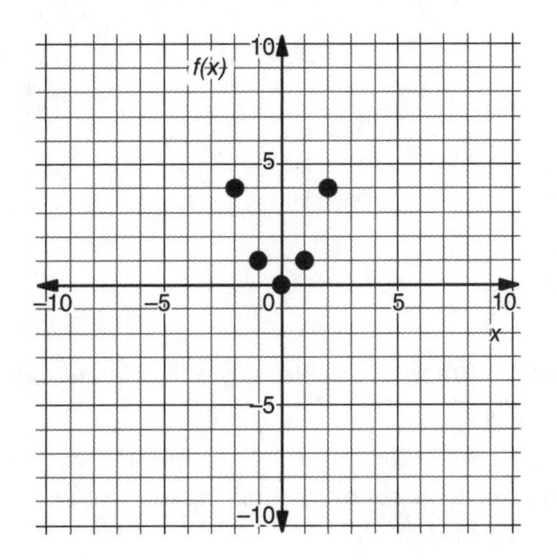

6.3 Characteristics of Graphs

So far, we've worked with relations that have a fixed number of points, using tables of values to represent them. Many times, we also work with relations that can be represented as equations. Equations give us a compact way of representing the relationship between two variables when we have an infinite number of inputs.

To represent an equation on the coordinate plane, we draw a line or curve called the **graph of an equation**. It shows all ordered pairs that are solutions to the equation. In other words, every point on the graph is a solution to the equation, and every point *not* on the graph is *not* a solution.

The graph of a table of values is *not* the same as the graph of an equation. The graph of a table consists of *points* since it represents a fixed number of ordered pairs. In contrast, the graph of an equation is a *line* or *curve* since it represents an infinite number of ordered pairs. Figure 6.5 shows the graphs of a relation whose domain is 1, 2, 3, and 4 (shown by a table of values) and the graph of a relation whose domain is all numbers in the interval [1, 4] (shown by the equation $y = 2x$).

TABLE

x	y
1	2
2	4
3	6
4	8

GRAPH OF TABLE

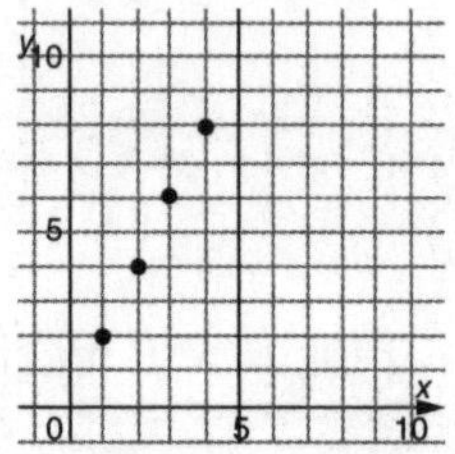

EQUATION

$y = 2x$
Domain: $1 \leq x \leq 4$

GRAPH OF EQUATION

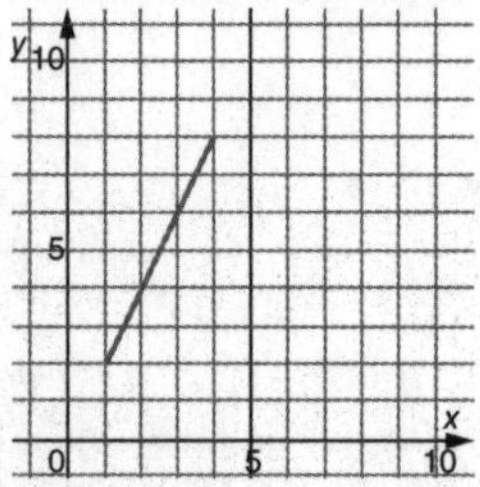

Figure 6.5 Graph of a table vs. graph of an equation.

Throughout this book, we examine the graphs of different types of equations. In this section, we focus on identifying important characteristics of relations based on their graphs.

First, we use the graph of a relation to determine its domain and codomain and determine if it is a function, as shown in Example 6.7:

Example 6.7 The accompanying graph shows a relation.

(a) Determine the domain and codomain of the relation.

(b) Determine if the relation is a function.

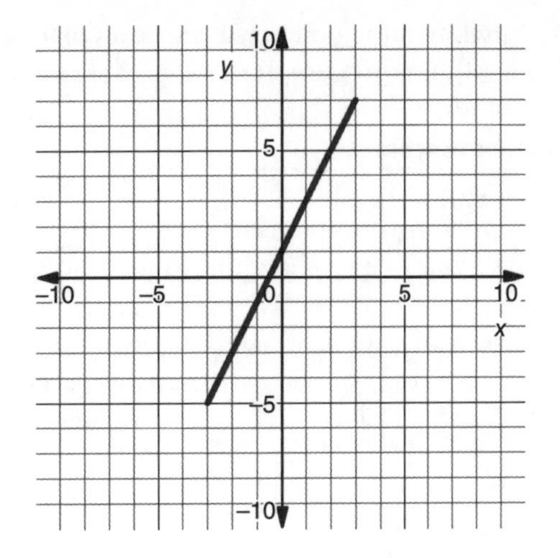

Solution:

(a) To find the domain, we find the minimum and maximum x-values. We imagine a vertical line passing through the x-axis. Wherever the vertical line intersects a point on the graph, that point's x-value is part of the domain. From the graph, we see that the domain is $\{x \mid -3 \leq x \leq 3\}$.

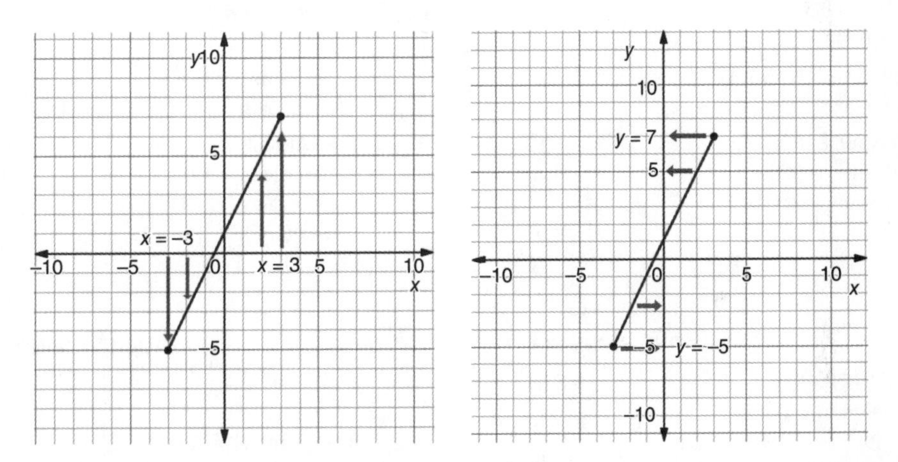

Similarly, to find the codomain, we imagine a horizontal line passing through the y-axis. Wherever the horizontal line intersects a point on the graph, that point's y-value is part of the codomain. From the graph, we see that the range is $\{y \mid -5 \leq y \leq 7\}$.

(b) To determine if the relation is a function, we use the vertical line test, which we discussed in Section 6.2. From our work above to find the domain, we see that a vertical line will not pass through the graph more than once, which means that every input has exactly one output. Thus, the relation is a function.

We can determine other characteristics of functions based on their graphs, as shown in Table 6.2:

Table 6.2 Characteristics of functions.

INCREASING OVER AN INTERVAL

DEFINITION: As x increases over the interval, y increases.

TIP: Graph moves up from left to right.

EXAMPLE: f is increasing over the interval (0, 3).

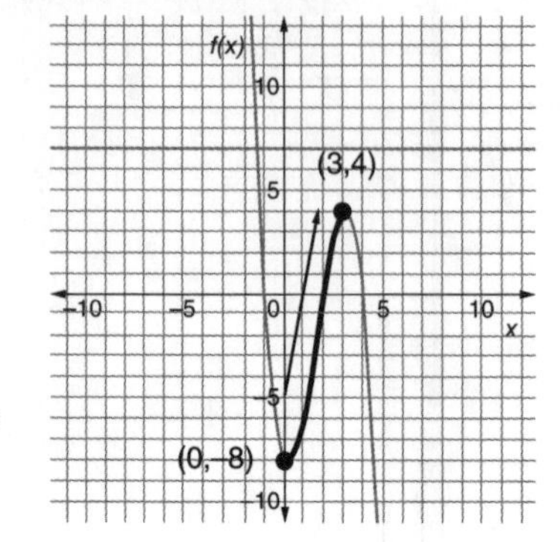

DECREASING OVER AN INTERVAL

DEFINITION: As x increases over the interval, y decreases.

TIP: Graph moves down from left to right.

EXAMPLE: f is decreasing over the interval (0, 3).

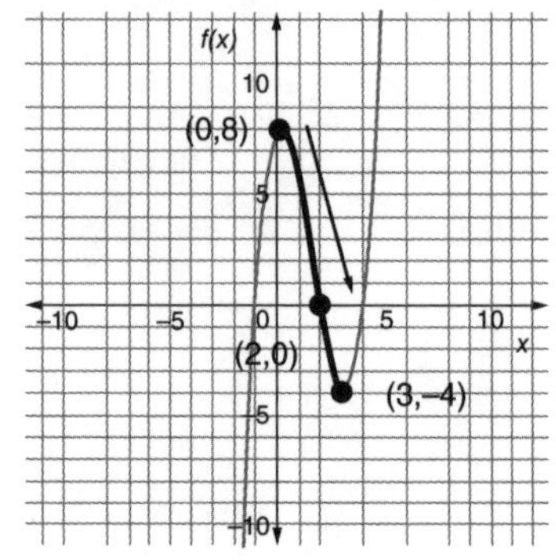

POSITIVE OVER AN INTERVAL

DEFINITION: All y-values in the interval are positive.

TIP: Graph is above the x-axis.

EXAMPLE: f is positive over the intervals (–2, –1) and (3, 5).

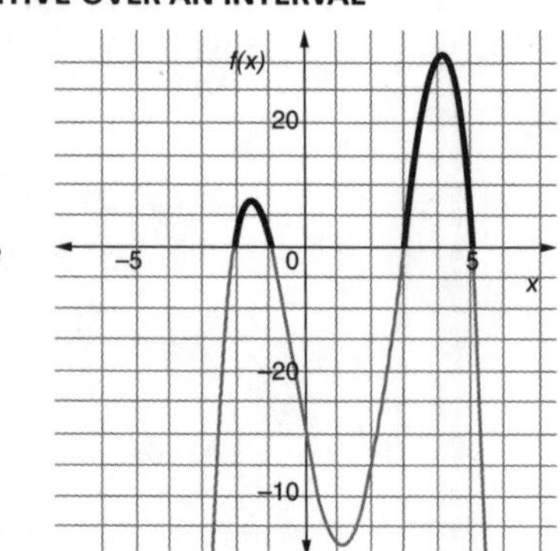

NEGATIVE OVER AN INTERVAL

DEFINITION: All y-values in the interval are negative.

TIP: Graph is below the x-axis.

EXAMPLE: f is negative over the intervals (–2, –1) and (3, 5).

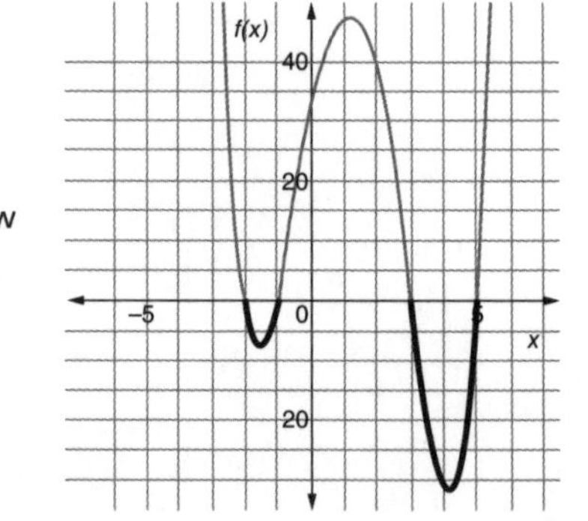

RELATIVE MAXIMUM

DEFINITION: y-value of point in an interval where function changes from increasing to decreasing.

TIP: Highest y-value ("peak") in an interval.

EXAMPLE: f has a relative maximum of 13 at $x = -0.5$.

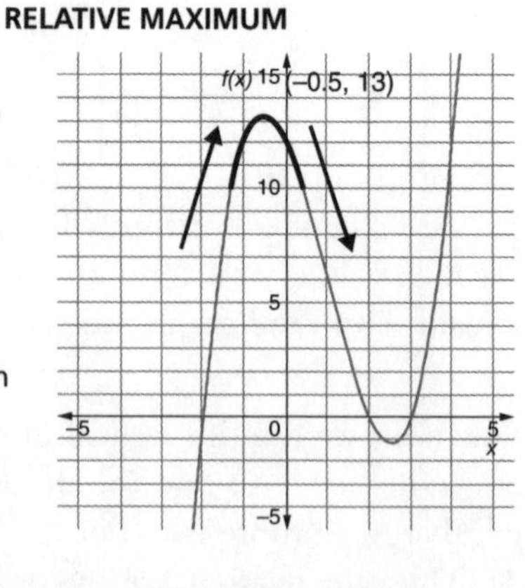

RELATIVE MINIMUM

DEFINITION: y-value of point in an interval where function changes from decreasing to increasing.

TIP: Lowest y-value ("valley") in an interval.

EXAMPLE: f has a relative minimum of –1 at $x = 2.5$.

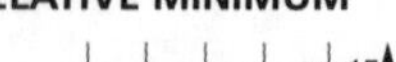

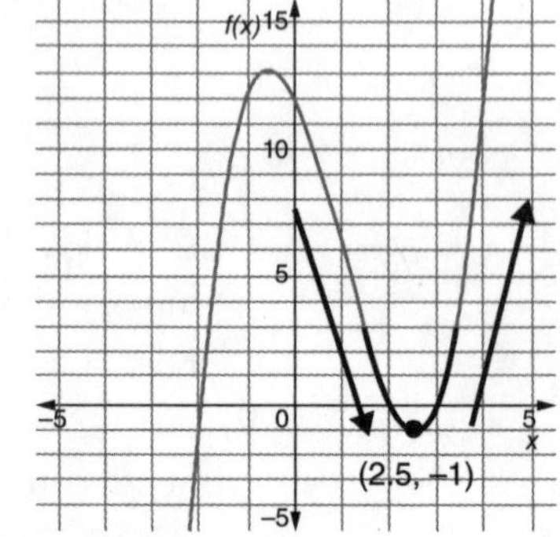

Table 6.2 (*Continued*)

	***X*-INTERCEPT**		***Y*-INTERCEPT**
DEFINITION: point where the graph intersects the *x*-axis (at $y = 0$). A **zero** of a function f is an *x*-value that makes $f(x) = 0$. TIP: Point where graph crosses *x*-axis. EXAMPLE: The *x*-intercepts of f are $(-2, 0)$ and $(4, 0)$. The zeros of $f(x)$ are -2 and 4.		DEFINITION: point where the graph intersects the *y*–axis (at $x = 0$). TIP: Point where graph crosses *y*–axis. EXAMPLE: The *y*–intercept of f is $(0, -8)$.	

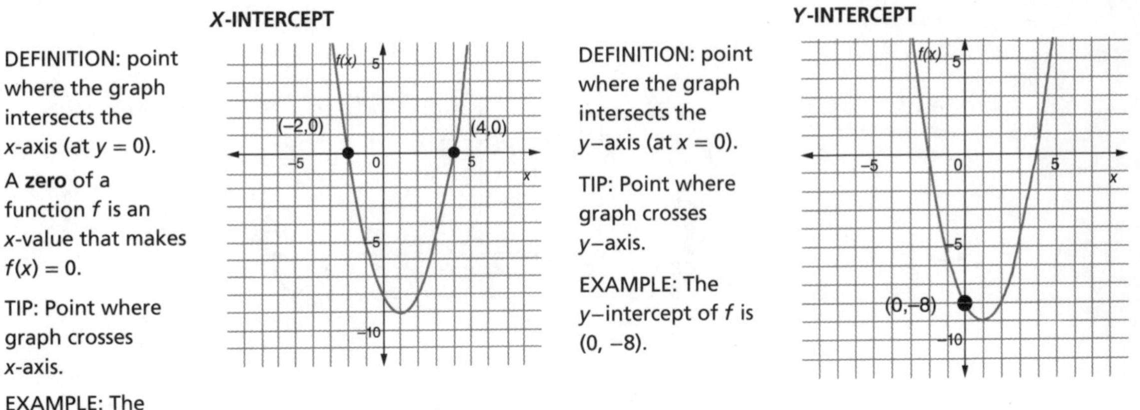

Here are some important notes about these characteristics:

- When we use the terms *increasing*, *decreasing*, *positive*, *negative*, *relative maximum*, and *relative minimum*, we must also refer to a specific interval of a relation. Thus, a relation can be increasing over one interval and decreasing over another.
- The *x*-intercept(s) and *y*-intercept of a function are ordered pairs, but the zeros are *x*-values.
- If a point has a *y*-coordinate that is greater than or equal to the *y*-coordinates of all other points on the graph (in other words, if it is the highest point on the graph), then the *y*-coordinate is called an **absolute maximum** (shortened to **maximum**). If a point has a *y*-coordinate that is less than or equal to the *y*-coordinates of all other points on the graph (in other words, if it is the lowest point on the graph), then the *y*-coordinate is called an **absolute minimum** (shortened to **minimum**).
- We also use these words to describe the behavior of relations that are not functions. The only difference is that we don't use function notation.
- At the *x*-intercepts, the function is neither positive nor negative since $y = 0$ at those points.
- At the relative minimum and maximum, the function is neither increasing nor decreasing.

Example 6.8 The graph of a function is shown below.

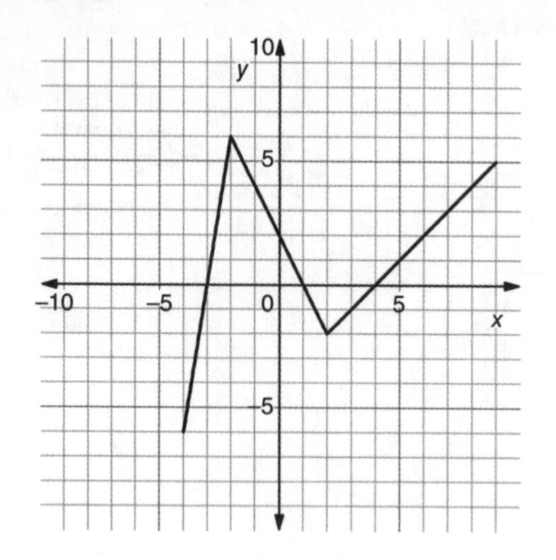

(a) Use interval notation to state the intervals for which the function is positive, negative, increasing, and decreasing.

(b) Determine the intercepts of the function.

(c) State the zeros of the function.

(d) Determine the relative minimum and relative maximum of the function over the interval [−3, 5].

Solution:

(a) The function is positive over the intervals for which the graph is above the x-axis: $\{x \mid -3 < x < 1\}$ and $\{x \mid 4 < x \leq 9\}$. (We write $\leq$ at 9 since the function is positive at $x = 9$ but $<$ for 4 since the function is neither positive nor negative there.) The function is negative over the intervals for which the graph is below the x-axis: $\{x \mid -4 \leq x < -3\}$ and $\{x \mid 1 < x < 4\}$. It is increasing over the intervals for which the graph goes up from left to right: $\{x \mid -4 < x < -2\}$ and $\{x \mid 2 < x < 9\}$. It is decreasing over the interval in which the graph goes down from left to right: $\{x \mid -2 < x < 2\}$.

(b) The x-intercepts of the function (the points where the graph intersects the x-axis) are $(-3, 0)$, $(1, 0)$, and $(4, 0)$. The y-intercept (the point where the graph intersects the y-axis) is $(0, 2)$.

(c) The zeros are the x-coordinates of the x-intercepts: -3, 1, and 4.

(d) The relative maximum is the y-coordinate of the highest point in the interval between $x = -3$ and $x = 5$. Since this point has coordinates $(-2, 6)$, then the relative maximum is 6. The relative minimum is the y-coordinate of the lowest point. This point has coordinates $(2, -2)$, so the relative minimum is -2.

Exercises

For each relation shown below, determine its domain and codomain. Then determine if it is a function.

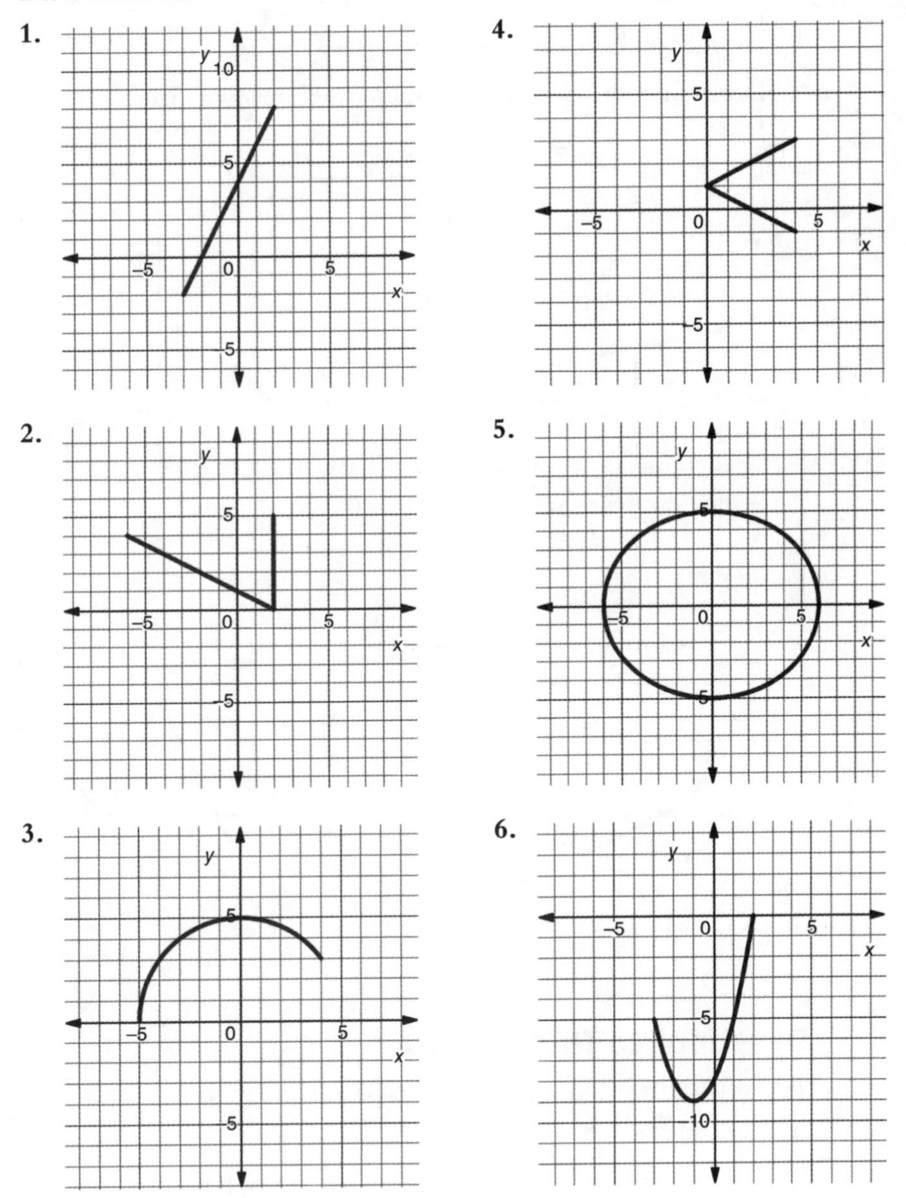

For each function:

(a) State the intervals for which the function is positive, negative, increasing, and decreasing.

(b) Determine the intercepts and the zeros.

(c) Determine the relative minimum and relative maximum over the interval $[-5, 5]$.

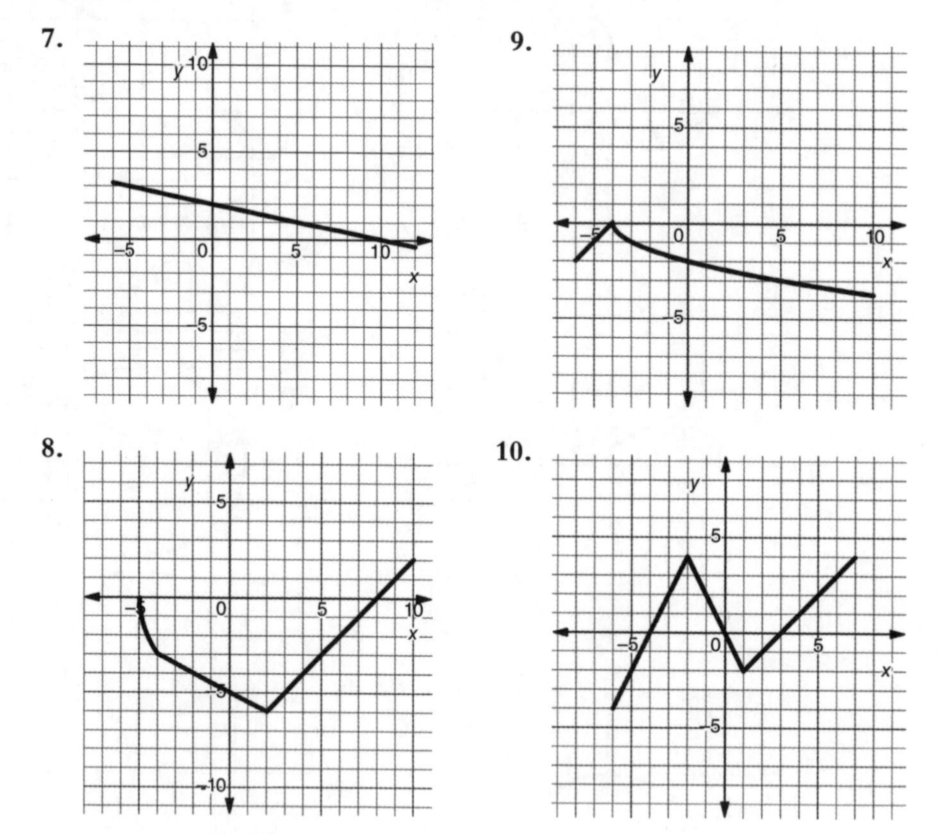

Questions to Think About

11. What is the relationship between the graph of an equation and the solutions to an equation?

12. Explain the difference between a function that is positive over an interval and one that is increasing over an interval.

13. Can a function have more than one y-intercept? Use the definition of a function to explain your answer.

6.4 Evaluating Functions from Equations and Graphs

In Section 6.1, we discussed how we evaluate functions based on a table of values by finding the output value that corresponds to a particular input. In this section, we evaluate functions based on their equations or graphs.

To evaluate functions using their equations, we substitute the input value for x and follow the order of operations (discussed in Section 1.2) to simplify the expression.

Example 6.9 If $f(x) = 3(x + 1)^2 + 4(x + 1) + 7$, evaluate $f(-2)$.

Solution:

$f(-2) = 3(-2 + 1)^2 + 4(-2 + 1) + 7$	Substitute -2 for x in the function equation.
$= 3(-1)^2 + 4(-1) + 7$	Evaluate inside parentheses first.
$= 3(1) + 4(-1) + 7$	Evaluate exponents.
$= 3 - 4 + 7$	Evaluate multiplication.
$= 6$	When addition and subtraction are together, evaluate left to right.

To evaluate functions based on their graphs, we find the point whose x-coordinate is the input value. The point's y-coordinate is the output value.

Example 6.10 The graph of f is shown below. What is $f(3)$?

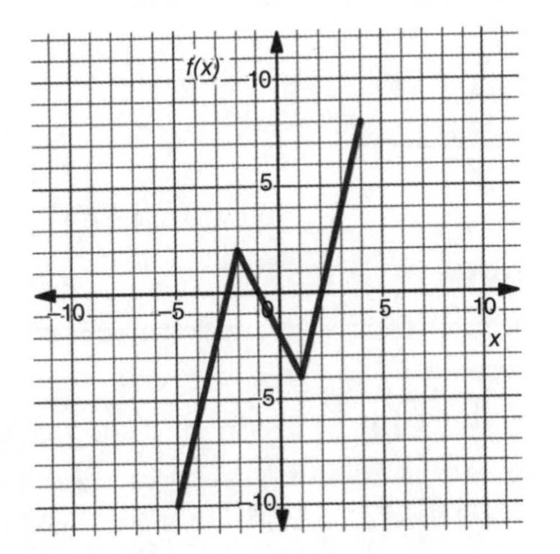

Solution: We identify the point on the graph whose x-coordinate is 3. This point is (3, 4). Since the y-coordinate is 4, then $f(3) = 4$.

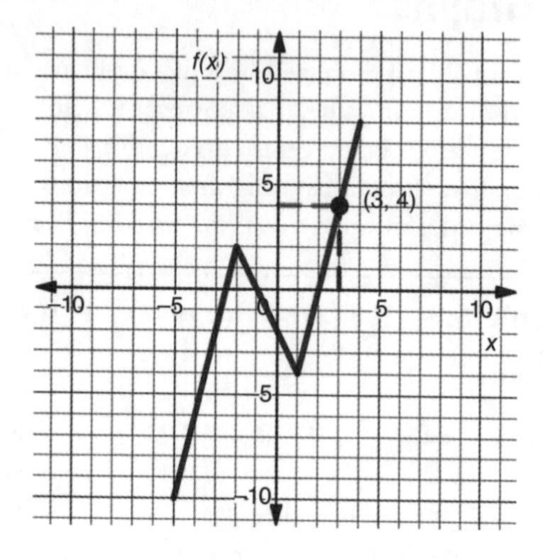

Example 6.11 Given that $f(x) = 2x + 11$ and $g(x) = 5x + 5$, determine the solution to the equation $f(x) = g(x)$.

Solution: Since the wording of this question is tricky, we suggest reading through it in phrases to determine its meaning:

Given that $f(x) = 2x + 11$	$f(x) = 2x + 11$.
and $g(x) = 5x + 5$,	$g(x) = 5x + 5$.
determine the solution to the equation $f(x) = g(x)$.	Find the value of x that makes $2x + 11 = 5x + 5$ true.

$2x + 11 = 5x + 5$	
$-2x - 5 \quad -2x - 5$	Subtract $2x$ and 5 from both sides.
$6 = 3x$	Combine like terms.
$\frac{6}{3} = \frac{3x}{3}$	Divide both sides by 3.
$2 = x$	Final answer, which we write as $x = 2$.

Technology Tip

We can use technology to evaluate functions given their equations or graphs. Enter the function's equation into your device. Then look at the function's table of values (look for the TABLE function) or trace along the function's graph to find the output value for the given input value.

Exercises

1. If $f(x) = 3x - 7$, evaluate $f(2)$.
2. If $g(x) = 10x + 4$, evaluate $g(7)$.
3. If $a(t) = 7t^2 - 5t$, evaluate $a(-4)$.
4. If $k(x) = 4x^2 - 9x + 1$, evaluate $k(3)$.
5. If $q(n) = -5(n + 1)^2 - 10n$, evaluate $q(-1)$.
6. If $h(t) = 14t^2 + 6t + 2$, evaluate $h(0)$.
7. Given this graph of m, what is $m(4)$?

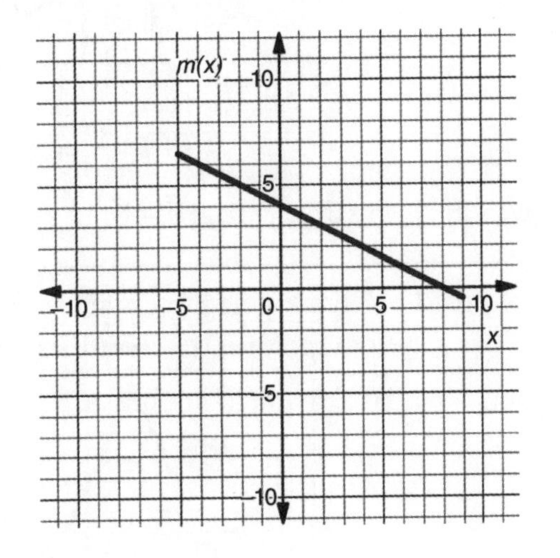

8. Given this graph of g, what is $g(-2)$?

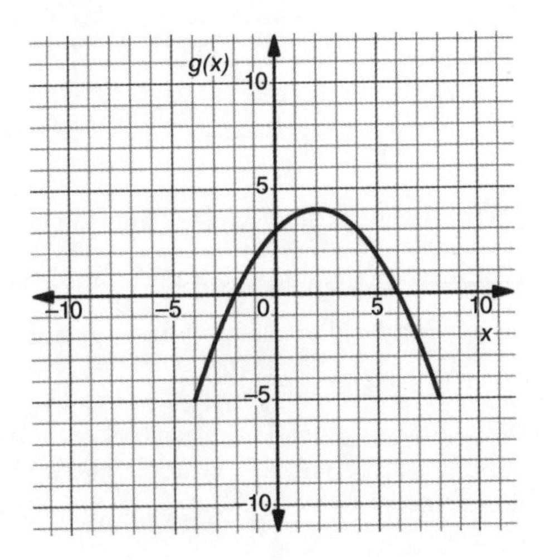

9. Given this graph of j, what is $j(7)$?

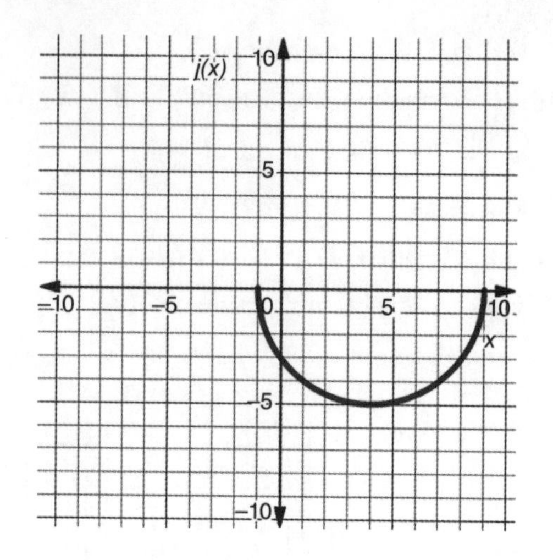

10. Given this graph of p, what is $p(-4)$?

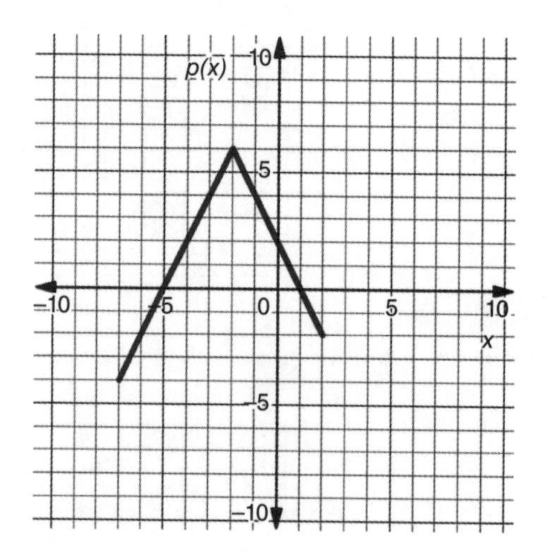

11. If $f(x) = 7x$ and $g(x) = 2x + 5$, find the value of x such that $f(x) = g(x)$.

12. If $p(x) = 3x - 1$ and $q(x) = 9x + 5$, find the value of x such that $p(x) = q(x)$.

13. If $a(x) = 10x + 11$ and $b(x) = 8x - 7$, find the value of x such that $a(x) = b(x)$.

CHAPTER 6 TEST

1. What ordered pair is shown in the accompanying diagram?

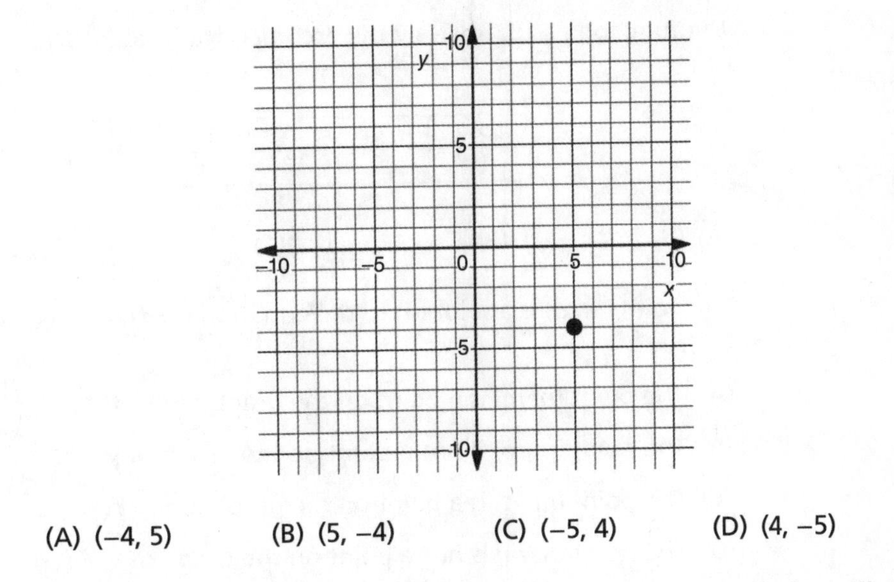

(A) (−4, 5) (B) (5, −4) (C) (−5, 4) (D) (4, −5)

2. If the function f is defined by the set {(−2, 5), (0, 5), (5, 7), (7, −2)}, what is $f(5)$?

(A) −2 (B) 0 and −2 (C) 0 or −2 (D) 7

3. If $f(x) = 2x^2 - 5$, then which expression is equivalent to $f(-3)$?

(A) $2(-3)^2 - 5$ (B) $2(-3)^2(-5)$ (C) $-2(3)^2 - 5$ (D) $-2(3)^2(-5)$

4. Which relation is a function?

(A) {(0, 7), (2, 7), (5, 7), (8, 0)}

(B)

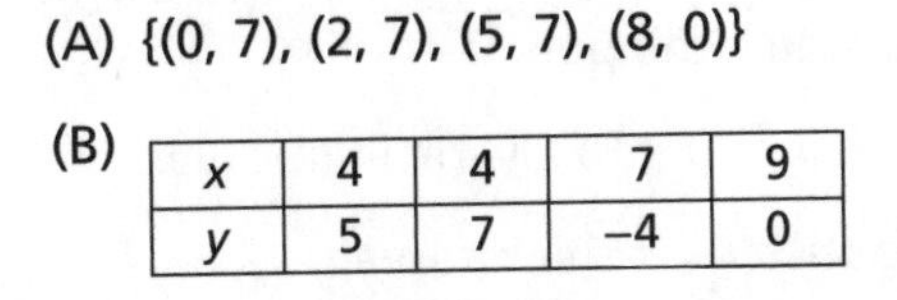

x	4	4	7	9
y	5	7	−4	0

(C)

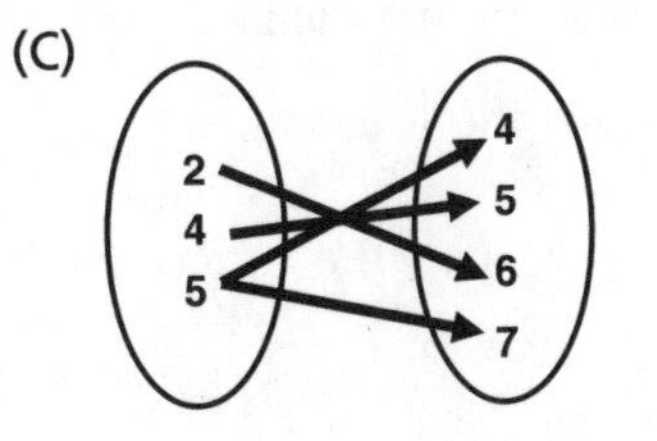

(D)

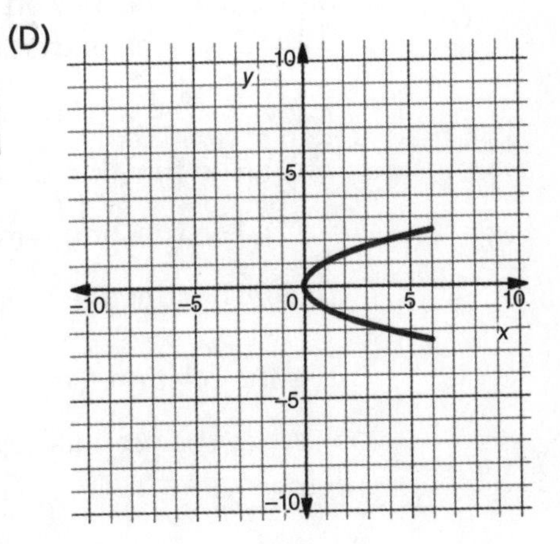

5. In the function {(2, 5), (3, 6), (7, 9), (9, 10)}, what is the domain?

   (A) {5, 6, 9, 10}
   (B) {2, 5}
   (C) {2, 3, 5, 6}
   (D) {2, 3, 7, 9}

6. The function f is defined by the accompanying table. What is the minimum of f?

x	0	5	10	15
$f(x)$	20	12	6	8

   (A) 0
   (B) 6
   (C) 8
   (D) 20

7. The point (m, n) is a solution to the equation $y = f(x)$. Which statement must be true?

   (A) The point (n, m) is a point on the graph of $y = f(x)$.
   (B) The point (m, n) is *not* a point on the graph of $y = f(x)$.
   (C) The point (m, n) is a point on the graph of $y = f(x)$.
   (D) The point (n, m) is *not* a point on the graph of $y = f(x)$.

8. The point $(a, 0)$ is on the graph of $y = f(x)$. Which statement must be true?

   (A) $(0, a)$ is a y-intercept of the graph.
   (B) $(a, 0)$ is a y-intercept of the graph.
   (C) $(a, 0)$ is an x-intercept of the graph.
   (D) $(0, a)$ is an x-intercept of the graph.

9. The point (4, 3) is on the graph of $y = f(x)$. Which statement *cannot* be true for any interval that contains the point?

   (A) (4, 3) is a relative maximum of f over the interval.
   (B) The function is negative over the interval.
   (C) The function is increasing over the interval.
   (D) The function is decreasing over the interval.

10. Is the relation {(14, 5), (14, 10), (7, 8), (9, 0)} a function? Explain.

11. Write the pronunciation of $f(-3) = 6 - 2(-3)$.

12. Use function notation to express the relationship between the highlighted values in the table.

x	2	4	6	8
$q(x)$	−2	−5	−8	−11

13. If $p(x) = 12x - 7$ and $q(x) = 8x - 23$, find the value of x such that $p(x) = q(x)$.

14. State the domain and range, minimum, and maximum of the function whose graph is shown here.

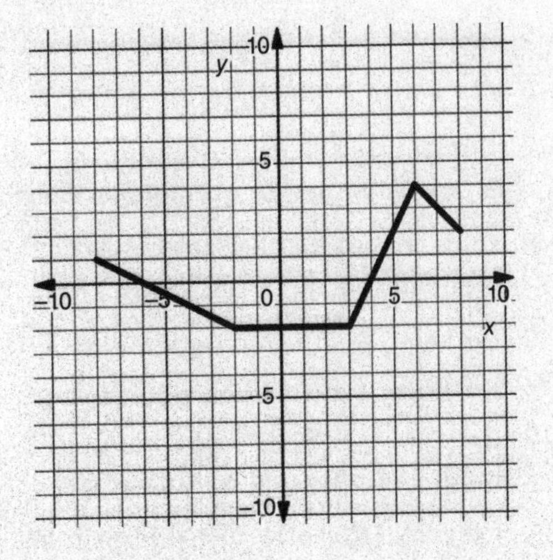

15. The graph of the function f is shown here.

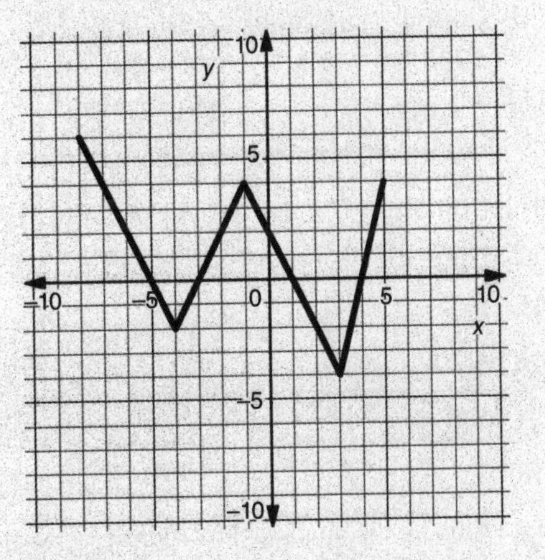

(a) Over what intervals is the function decreasing?

(b) Over what intervals is the function positive?

(c) State the intercepts of the function.

CHAPTER 6 SOLUTIONS

6.1. 1. "f of 3 equals 5."

2. "g of 0 equals 6."

3. "h of negative 2 equals 0."

4. "g of 5 equals 5 plus 7 times 9."

5. "8 times f of 3 equals 8 times the difference of 9 and 3."

6. "Negative f of 4 equals negative 7 minus 4."

7. Domain: {−7, 5, 6, 2}, Codomain: {10, 7, 5}. It is a function since every input is mapped to exactly one output.

8. Domain: {−1, 0, 3, 4}. Codomain: {9}. It is a function since every input is mapped to exactly one output.

9. Domain: {8, 2, −16}. Codomain: {7, 0, −1, 39}. It is not a function since the input 2 is mapped to both 0 and 39.

10. Domain: {16, 5, 3}. Codomain: {2, 8, 17}. It is not a function since the input 16 is mapped to both 2 and 8.

11. Domain: {25, 16, 9, 4}. Codomain: {5, 4, 3, 2}. It is a function since every input is mapped to exactly one output.

12. Domain: {5, 2, 8, 6}. Codomain: {10, 3, 4}. It is a function since every input is mapped to exactly one output.

13. Domain: {4, 9, 0}. Codomain: {2, 3, 0, −3}. It is not a function since the input 9 is mapped to both 3 and −3.

14. Domain: {4, 8, 12, 16}. Codomain: 27, 81, 96, 144}. It is a function since every input is mapped to exactly one output.

15. $f(2) = 30$

16. $h(10) = 490$

17. 104

18. 7

19. Answers may vary. For example, the mapping of people to birthdays is a function since every person has exactly one birthday.

20. Answers may vary. For example, the mapping of hair color to people is not a function since a hair color can be matched to more than one person.

21. Not necessarily. For example, the function {(0, 3), (1, 3)} maps each input to exactly one output. However, when we reverse the input and output values, the resulting relation {(3, 0), (3, 1)} is not a function since the input 3 is mapped to both 0 and 1.

6.2. 1. Domain: {−5, −4, −3, −2}, Codomain: {6, 2, 1, 0}.

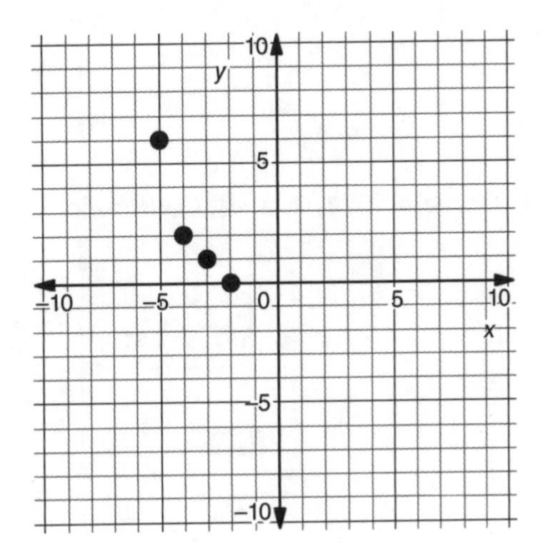

2. Domain: {−1, 7, 2}, Codomain: {−6, −1, 6, 4}.

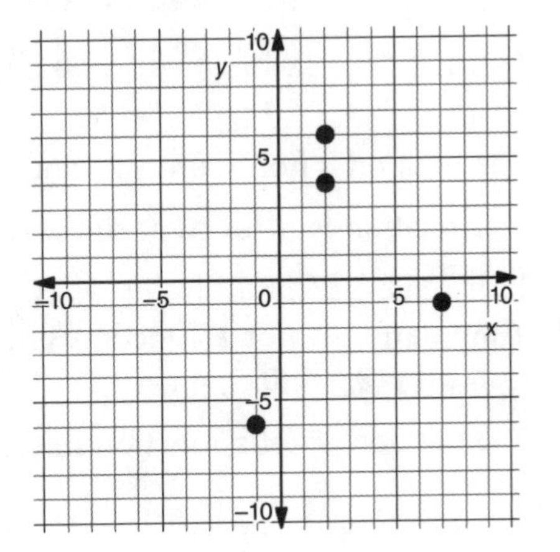

3. Domain: {5, –1, 0, 2, 6}, Codomain: {–2, 7, –4, 0}.

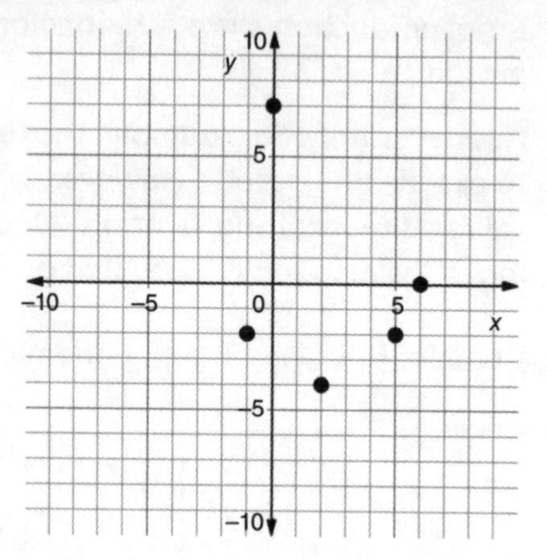

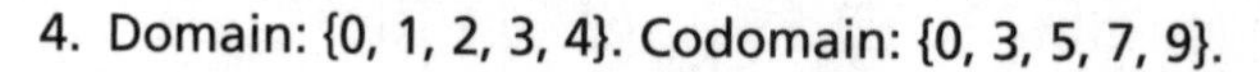
4. Domain: {0, 1, 2, 3, 4}. Codomain: {0, 3, 5, 7, 9}.

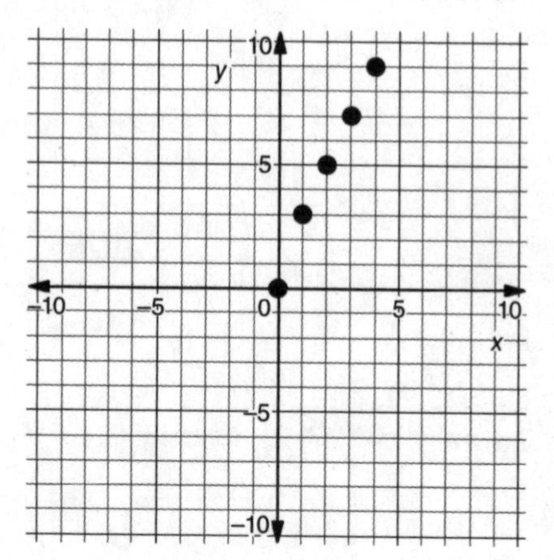

5. The relation is not a function since the input 2 is mapped to 4 and –1.
6. The relation is a function since every input is mapped to exactly one output.
7. The relation is not a function since the input 3 is mapped to several outputs (–2, –1, 0, 3, 4).
8. The relation is not a function since the input 6 is mapped to 7 and –3.
9. 0
10. –8
11. 2
12. –4

6.3.

1. Domain: $\{x \mid -3 \le x \le 2\}$, codomain: $\{y \mid -2 \le y \le 8\}$. The relation is a function since every input is mapped to exactly one output.
2. Domain: $\{x \mid -6 \le x \le 2\}$, codomain: $\{y \mid 0 \le y \le 5\}$. The relation is not a function since the input $x = 2$ is mapped to several outputs.
3. Domain: $\{x \mid -5 \le x \le 4\}$, codomain: $\{y \mid 0 \le y \le 5\}$. The relation is a function since every input is mapped to exactly one output.
4. Domain: $\{x \mid 0 \le x \le 4\}$, codomain: $\{y \mid -1 \le y \le 3\}$. The relation is not a function since the inputs $\{x \mid 0 < x \le 4\}$ are mapped to more than one output.
5. Domain: $\{x \mid -6 \le x \le 6\}$, codomain: $\{y \mid -5 \le y \le 5\}$. The relation is not a function since the inputs $\{x \mid -6 < x < 6\}$ are mapped to more than one output.
6. Domain: $\{x \mid -3 \le x \le 2\}$, codomain: $\{y \mid -9 \le y \le 0\}$. The relation is a function since every input is mapped to exactly one output.
7. Positive: $\{x \mid -6 \le x < 10\}$, negative: $\{x \mid 10 < x \le 12\}$, increasing: none, decreasing: $\{x \mid -6 < x < 12\}$, x-intercept: (10, 0), y-intercept: (0, 2), zero: 10, relative minimum: 1, relative maximum: 3.
8. Positive: $\{x \mid 8 < x \le 10\}$, negative: $\{x \mid -5 < x < 8\}$, increasing: $\{x \mid 2 < x < 10\}$, decreasing: $\{x \mid -5 < x < 2\}$, x-intercepts: (–5, 0) and (8, 0), y-intercept: (0, –5), zeros: –5 and 8, relative minimum: –6, relative maximum: 0.
9. Positive: none, negative: $\{x \mid -6 \le x < -4\}$ and $\{x \mid -4 < x \le 10\}$, increasing: $\{x \mid -6 < x < -4\}$, decreasing: $\{x \mid -4 < x < 10\}$, x-intercept: (–4, 0), y-intercept: (0, –2), zero: –4, relative minimum: –3, relative maximum: 0.
10. Positive: $\{x \mid -4 < x < 0\}$ and $\{x \mid 3 < x \le 7\}$; negative: $\{x \mid -6 \le x < -4\}$ and $\{x \mid 0 < x < 3\}$; increasing: $\{x \mid -6 < x < -2\}$ and $\{x \mid 1 < x < 7\}$; decreasing: $\{x \mid -2 < x < 1\}$; x-intercept: (–4, 0), (0, 0) and (3, 0); zeros: –4, 0, and 3; y-intercept: (0, 0), relative minimum: –2, relative maximum: 4.
11. Every point on the graph is a solution to the equation, and every point not on the graph is not a solution.
12. A function that is positive over an interval has y-values there that are all greater than 0. A function that is increasing over an interval has y-values that are increasing as x increases. These y-values don't have to be positive.
13. A function cannot have more than one y-intercept. At the y-intercept, the input is $x = 0$, so only one output is allowed.

6.4.

1. -1
2. 74
3. 132
4. 10
5. 10
6. 2
7. 2
8. 0
9. -4
10. 2
11. $x = 1$
12. $x = -1$
13. $x = -9$

CHAPTER 6 TEST SOLUTIONS

1. (B)
2. (D)
3. (A)
4. (A)
5. (D)
6. (B)
7. (C)
8. (C)
9. (B)
10. No, it is not a function since the input 14 has two outputs (5 and 10). In a function, each input has exactly one output.
11. "*f* of negative 3 equals 6 minus 2 times negative 3."
12. $q(8) = -11$
13. $x = -4$
14. Domain: $\{x \mid -8 \le x \le 8\}$, range: $\{y \mid -2 \le y \le 4\}$, minimum: -2, maximum: 4.
15. (a) The function is decreasing over the intervals $\{x \mid -8 < x < -4\}$ and $\{x \mid -1 < x < 3\}$.

    (b) The function is positive over the intervals $\{x \mid -8 \le x < -5\}$, $\{x \mid -3 < x < 1\}$, and $\{x \mid 4 < x \le 5\}$.

    (c) The *x*-intercepts are $(-5, 0)$, $(-3, 0)$, $(1, 0)$, and $(4, 0)$. The *y*-intercept is $(0, 2)$.

7 LINEAR FUNCTIONS AND THEIR GRAPHS

In Chapter 6, we introduced the idea of graphing relations in two variables on the coordinate plane. In this chapter, we discuss graphing an important family of functions called linear functions. We use their equations to graph them and discuss how we use them to model real-world situations.

7.1 Introduction to Linear Functions

How would you describe the patterns in Figure 7.1?

#1

x	y
0	1
1	3
2	5
3	7

#2

x	y
0	−2
1	0
2	2
3	4

#3

x	y
0	4
1	1
2	−2
3	−5

Figure 7.1 Describe the patterns in the tables.

Each table has a constant difference between consecutive y-values: +2 in Table #1, +2 in Table #2, and −3 in Table #3 (Figure 7.2):

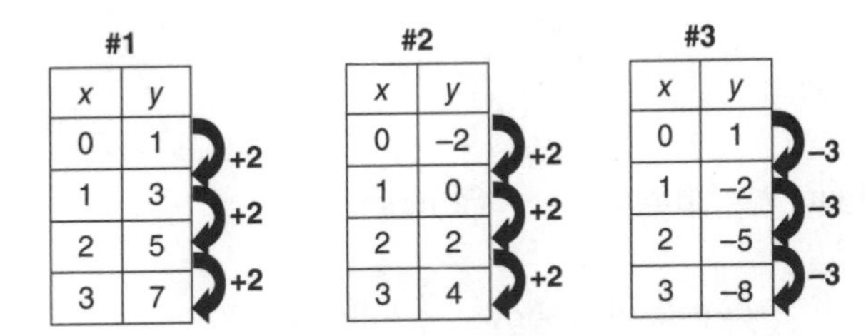

#1

x	y
0	1
1	3
2	5
3	7

#2

x	y
0	−2
1	0
2	2
3	4

#3

x	y
0	1
1	−2
2	−5
3	−8

Figure 7.2 Constant difference between consecutive y-values.

However, simply stating the constant difference doesn't fully describe the patterns. We also need to state the starting points for each table: (0, 1), (0, −2), and (0, 1). We summarize the patterns as follows:

- Table #1: We start at the point (0, 1). As x increases by 1, y increases by 2.
- Table #2: We start at the point (0, −2). As x increases by 1, y increases by 2.
- Table #3: We start at the point (0, 1). As x increases by 1, y decreases by 3.

We also express this relationship more compactly as a formula with variables. A formula enables us to substitute any value of x and immediately find the value of y. The verbal descriptions of the patterns above are usually written as the equations $y = 2x + 1$, $y = 2x - 2$, and $y = -3x + 1$, respectively.

When we substitute many x-values into each equation and find their corresponding y-values, we can graph each equation. Let's compare the tables, equations, and graphs in Figure 7.3:

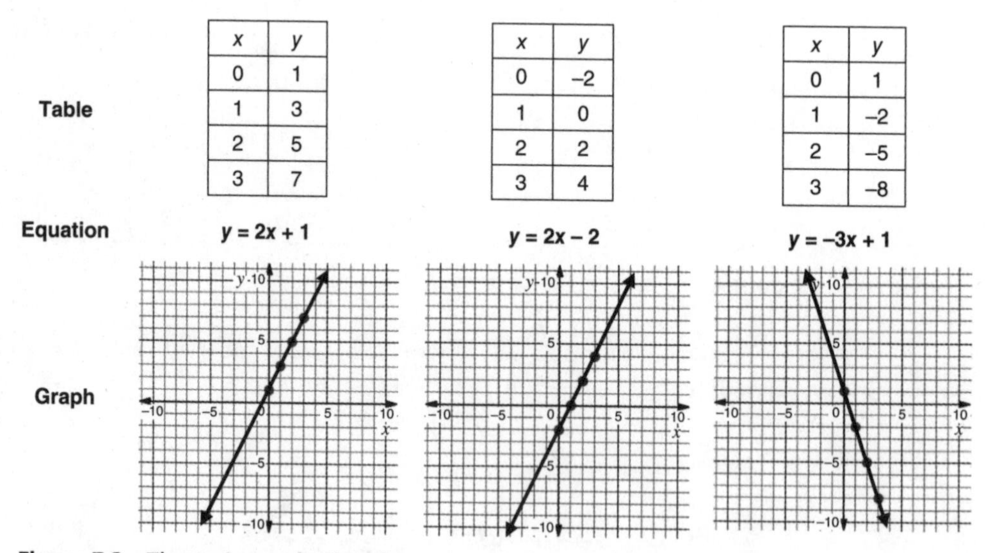

x	y
0	1
1	3
2	5
3	7

x	y
0	–2
1	0
2	2
3	4

x	y
0	1
1	–2
2	–5
3	–8

Figure 7.3 Three views of a function.

These examples show the following:

- Points with a constant difference between y-values lie on a line. That's why functions whose points have a constant difference between y-values are called **linear functions**.
- Some lines are steeper than others. They can tilt upward or downward from left to right.

- Lines can have the same amount of steepness but pass through different points (parallel lines).
- When we describe lines using equations such as $y = 2x + 1$ or $y = -3x + 1$, the first number is related to the line's steepness and the second is related to a point on the line.

We describe the steepness of a line using the word *slope*. The **slope** of a line is the ratio of the amount of vertical change (or change in y) to the amount of horizontal change (or change in x). By convention, mathematicians use the letter m to represent slope. We express slope using the formula $m = \frac{\textit{vertical change}}{\textit{horizontal change}}$.

Slope expresses the rate of change of a linear function. Since a linear function has a constant rate of change, we interpret the slope of a line as follows: **as x increases by __ units, y increases (or decreases) by __ units.**

Here are some examples:

- A slope of -2 means that as x increases by 1 unit, y decreases by 2 units.
- A slope of $\frac{4}{3}$ means that as x increases by 3 units, y increases by 4 units. (We could also say that as x increases by 1 unit, y increases by $\frac{4}{3}$ units.)
- A slope of 0 means that as x increases by 1 unit, y increases by 0 units. (In other words, y does not change, so the line is horizontal.)

To calculate the slope of a line from its graph, do the following:

1. Find any two points on the line. (We recommend choosing points whose coordinates are integers so that you can read them more easily.)
2. Draw a horizontal line from the first point and a vertical line from the second point until they intersect.
3. Measure the vertical change (sometimes called **rise**), which is the vertical distance (positive if moving up, negative if moving down) from the second point to the intersection point.
4. Measure the horizontal change (sometimes called **run**), which is the horizontal distance (positive if moving from left to right, negative if moving from right to left) from the intersection point to the first point.
5. Divide the vertical change by the horizontal change. Simplify the fraction if necessary.

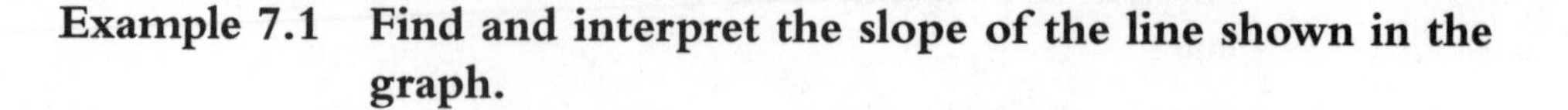

Example 7.1 Find and interpret the slope of the line shown in the graph.

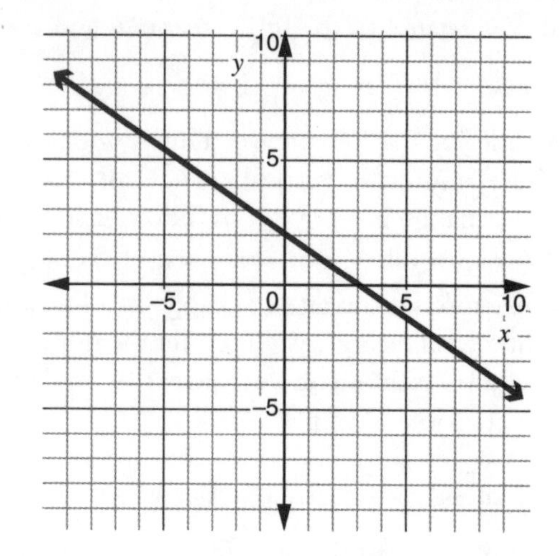

Solution: As shown in the next diagram, we pick two points on the line and measure the horizontal and vertical distance from the first point (−3, 4) to the second point (0, 2). The vertical distance is −2 and the horizontal distance is +3, so the slope = $m = \frac{\text{vertical change}}{\text{horizontal change}} = \frac{-2}{+3} = -\frac{2}{3}$. (Recall from Chapter 1 that when two numbers have different signs, their quotients are negative.)

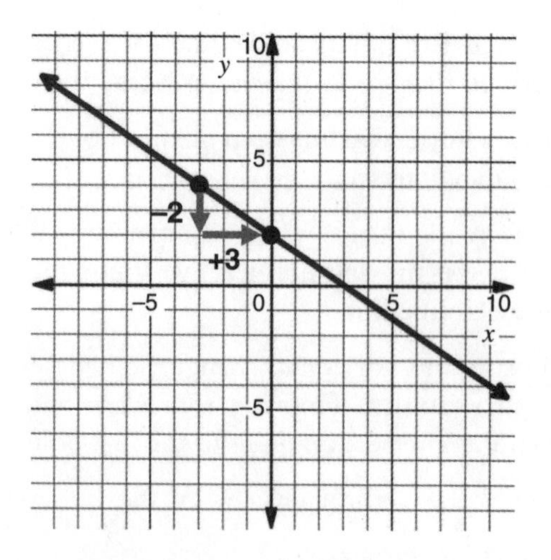

We interpret the slope as follows: As x increases by 3 units, y decreases by 2 units.

Slopes can be grouped into four types, as shown in Table 7.1:

Table 7.1 Types of slope.

Slope	Positive	Negative	Zero	Undefined
Graph				
Description	As x increases, y increases.	As x increases, y decreases.	As x increases, y is unchanged.	x is unchanged.
Notes	$\frac{\text{change in } y}{\text{change in } x} = \frac{+}{+}$ is positive.	$\frac{\text{change in } y}{\text{change in } x} = \frac{-}{+}$ is negative.	$\frac{\text{change in } y}{\text{change in } x} = \frac{0}{x}$ is 0.	$\frac{\text{change in } y}{\text{change in } x} = \frac{y}{0}$ is undefined.

We see from the table that lines that slope upward from left to right have positive slope (think of this as going up a mountain), while lines that slope downward from left to right have negative slope (think of this as going down a mountain).

Watch Out!

Here are some important points to remember when working with slope:

- **Don't confuse positive and negative *slopes* with positive and negative *functions*.** Remember that functions are positive or negative over an interval depending on whether their y-values are greater than or less than 0. A linear function could have a positive slope but be negative over an interval if it is below the x-axis.
- **Don't confuse zero slope with undefined slope.** A line with zero slope is horizontal (instead of going up or down a mountain, you're on a moving platform that runs along a flat surface). Avoid using the term "no slope" since it could be easily confused with "zero slope."

If we know the slope and y-intercept of a line, then we can graph it on the coordinate plane. When we write the equation of a line in the format that we used at the beginning of this section, we can determine the slope and y-intercept. We call this format the **slope-intercept form** of a line. We represent it with variables as $\boldsymbol{y = mx + b}$, where x and y are the variables, m is the slope, and b is the y-coordinate of the y-intercept.

Example 7.2 Graph $2y - 4x = 10$ on the coordinate plane.

Solution: First, we write the equation in slope-intercept form:

$2y - 4x = 10$	
$\quad + 4x \qquad + 4x$	Add $4x$ to both sides.
$2y = 4x + 10$	Combine like terms.
$\frac{2y}{2} = \frac{4x}{2} + \frac{10}{2}$	Divide both sides by 2.
$y = 2x + 5$	Equation in slope-intercept form.

Here, $m = 2$ and $b = 5$, so the slope is 2 and the y-intercept is (0, 5). Starting from the point (0, 5), we draw a line with a slope of 2, or $\frac{+2}{+1}$. Expressing slope as a fraction enables us to identify the vertical change (in the numerator) and the horizontal change (in the denominator). A slope of 2 means that as x increases by 1 unit, y increases by 2 units. We draw at least two points on the line. We recommend drawing a third point to check your work—in this example, we plot the points (0, 5), (1, 7), and (2, 9). We connect the points with a line, putting arrows on either end to indicate that the line extends infinitely in both directions.

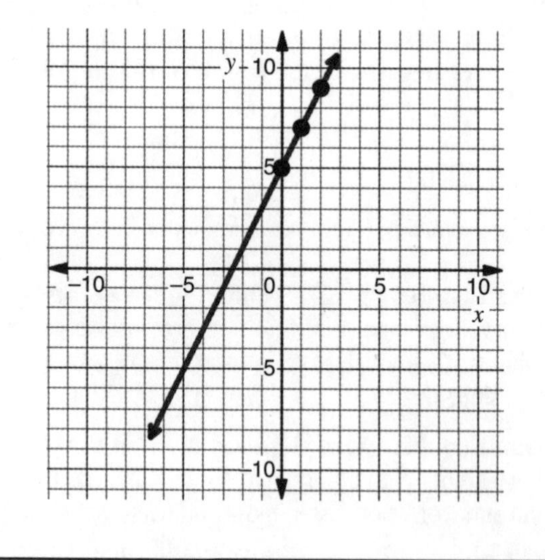

Example 7.3 Graph $y = 4$ on the coordinate plane.

Solution: This line has a slope of 0 and a y-intercept of (0, 4). In slope-intercept form, the equation is $y = 0x + 4$, which we can write more simply as $y = 4$ since 0 times any number is 0. All points on this horizontal line—such as (0, 4), (1, 4), (−2, 4), and (−4, 4)—have a y-coordinate of 4.

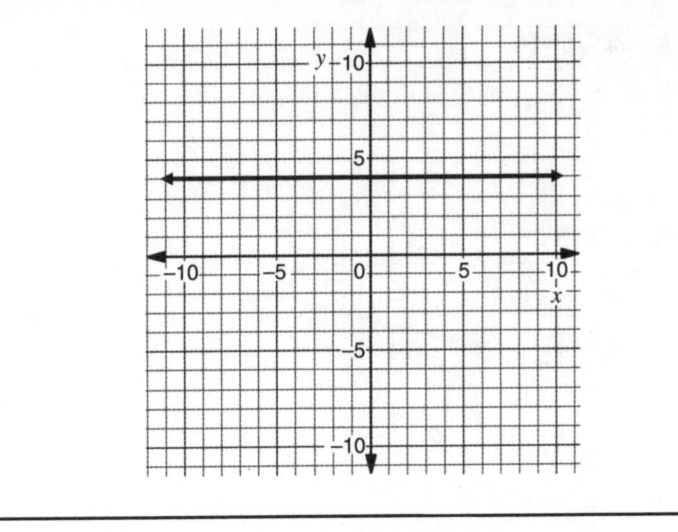

Example 7.4 Graph $x = 3$ on the coordinate plane.

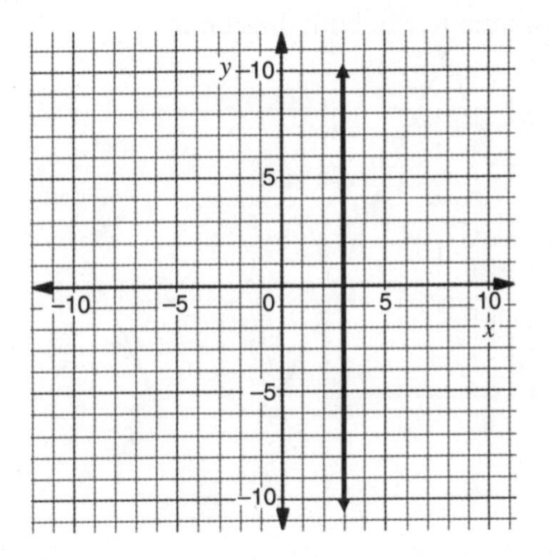

Solution: All points on this line—such as (3, 0), (3, 1), (3, −2), and (3, −4)—have an x-coordinate of 3. When we graph them, we see that they fall on a vertical line.

Technology Tip

Some older calculators won't graph equations unless they are functions that are solved for y. Newer technology will allow you to graph equations like $2y + 3x = 10$ without having to solve for y.

Exercises

Find and interpret the slope shown in each graph.

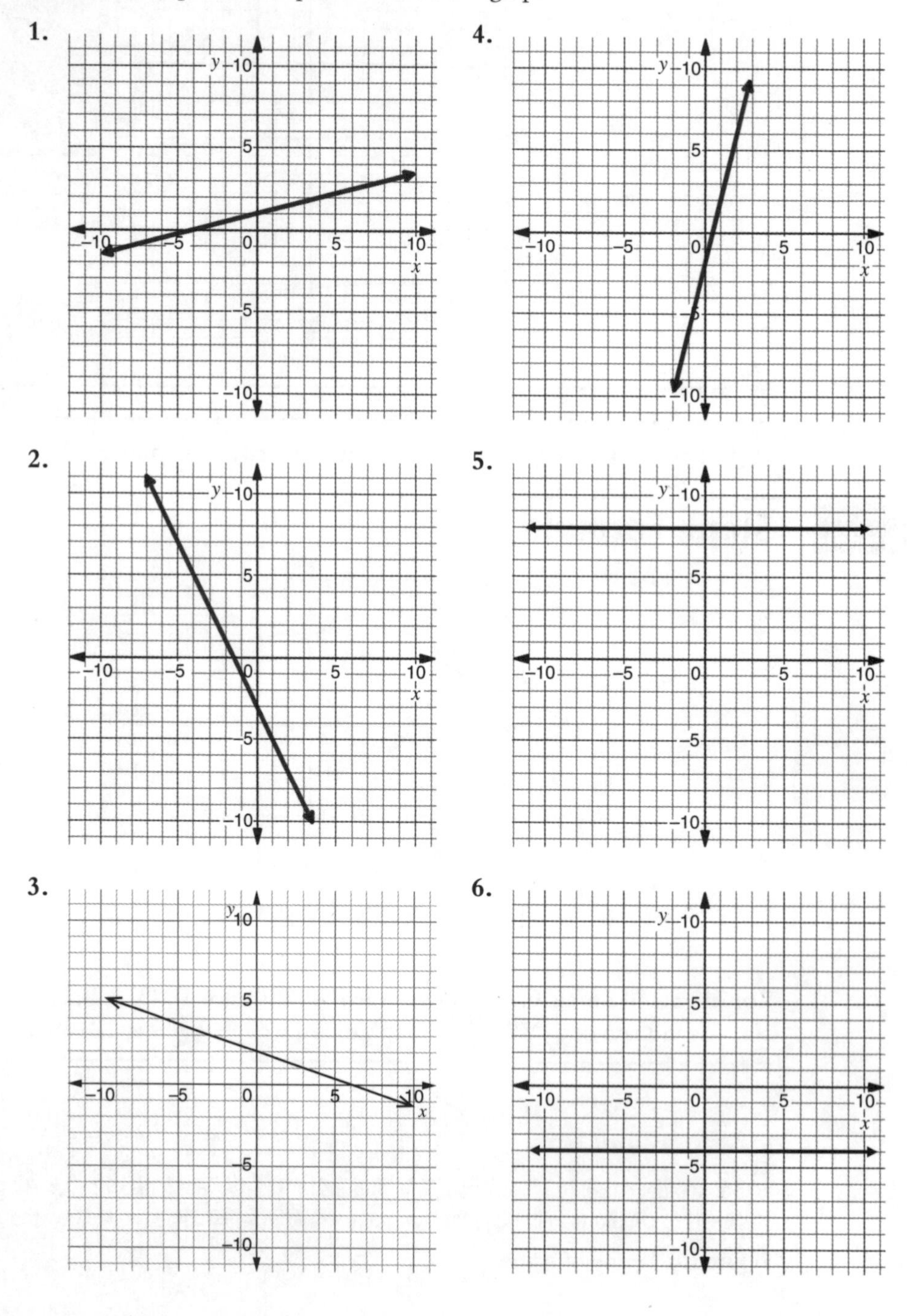

Graph each of the following equations on the coordinate plane.

7. $y = 2x$

8. $y = \frac{1}{3}x + 1$

9. $y = -\frac{4}{5}x + 2$

10. $10y = 5x - 10$

11. $x = 1$

12. $x = 6$

13. $y = -2$

14. $y = -4$

Questions to Think About

15. Explain why the equation of a vertical line cannot be written in slope-intercept form.

16. What is the equation of the x-axis? Explain.

17. Explain why the term "no slope" can be confusing.

7.2 Slope Formula

We can also find the slope of a line without graphing. Since the rate of change stays the same throughout the line, we can find any two points on the line and use their coordinates to determine its slope.

Let's find the slope of the line that passes through the points $A(105, 103)$ and $B(305, 203)$. We see from the coordinates that to move from point A to point B, we add +200 to the x-coordinate (since $305 - 105 = +200$) and +100 to the y-coordinate (since $203 - 103 = +100$). This means that the slope is $m = \frac{\textit{vertical change}}{\textit{horizontal change}} = \frac{203-103}{305-105} = \frac{100}{200} = \frac{1}{2}$ (Figure 7.4).

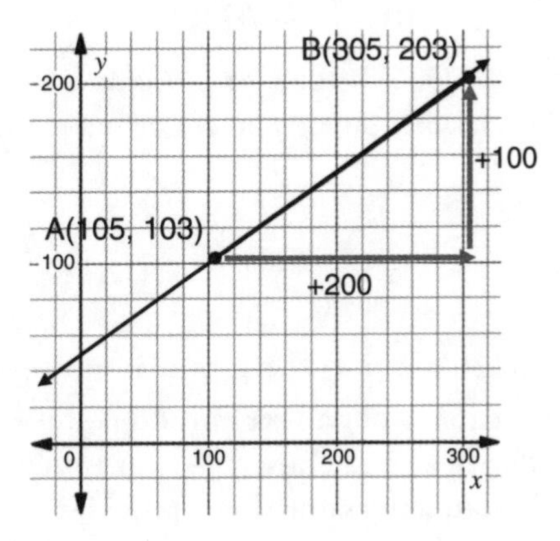

Figure 7.4 Slope of a line.

To find the slope of the line passing through *any* two points, we can subtract their y-coordinates, subtract their x-coordinates, and divide the two numbers. We write this using mathematical symbols as follows:

The slope of the line passing through two points (x_1, y_1) and (x_2, y_2) is $m = \frac{y_2 - y_1}{x_2 - x_1}$.

We use the subscripts to differentiate between the coordinates of the first point and the coordinates of the second point. This formula is read as "m equals the quotient of y-sub-2 minus y-sub-1 and x-sub-2 minus x-sub-1."

Watch Out!

In our experience, students often make mistakes when using the slope formula. To avoid these errors, we recommend doing the following:

- **Don't read or write the subscripts in x_1, y_1, x_2, and y_2 as exponents.** Remember that x_2 is read as "x-sub-two," not "x to the second power," "x squared," or "x-two."
- **Don't mix up the order of the coordinates.** We strongly recommend labeling the coordinates of each point as x_1, y_1, x_2, and y_2 first before doing any calculations. If you subtract $y_2 - y_1$ in the denominator and $x_1 - x_2$ in the numerator, you will not get the correct slope.
- **Don't flip the x- and y-coordinates.** Remember that slope is $\frac{\text{vertical change}}{\text{horizontal change}}$ or $\frac{\text{change in } y}{\text{change in } x}$, not $\frac{\text{change in } x}{\text{change in } y}$.

Example 7.5 Find the slope of the line passing through (5, −3) and (−1, 6).

Solution: Call (5, −3) the first point and (−1, 6) the second point. We recommend writing the variables directly above or below the coordinates, as shown here:

$$\underset{x_1\ \ y_1}{(5, -3)} \qquad \underset{x_2\ \ y_2}{(-1, 6)}$$

Then substitute into the formula: $m = \frac{y_2 - y_1}{x_2 - x_1} = \frac{6-(-3)}{-1-5} = \frac{+9}{-6} = -\frac{3}{2}$. (As with any fraction, we express our answer in simplest form.)

Note that if we had chosen (−1, 6) as the first point and (5, −3) as the second, we would have gotten the same slope, provided that we correctly labeled each coordinate: $x_1 = -1$, $y_1 = 6$, $x_2 = 5$, $y_2 = -3$, so $m = \frac{y_2 - y_1}{x_2 - x_1} = \frac{-3-6}{5-(-1)} = \frac{-9}{+6} = -\frac{3}{2}$.

Example 7.6 Find the slope of the line passing through (8, 4) and (−12, 4).

Solution: Call (8, 4) the first point and (−12, 4) the second point. Label the coordinates of each point.

$$\underset{x_1\ y_1}{(8, 4)} \qquad \underset{x_2\ y_2}{(-12, 4)}$$

Then substitute into the formula: $m = \frac{y_2 - y_1}{x_2 - x_1} = \frac{4-4}{-12-8} = \frac{0}{-20} = 0$.

Note that since the y-coordinates are the same, the points lie on a horizontal line, which has zero slope.

Example 7.7 Find the slope of the line passing through (7, 2) and (7, 5).

Solution: Call (7, 2) the first point and (7, 5) the second point. Label the coordinates of each point:

$$\underset{x_1\ y_1}{(7, 2)} \qquad \underset{x_2\ y_2}{(7, 5)}$$

Then substitute into the formula: $m = \frac{y_2 - y_1}{x_2 - x_1} = \frac{5-2}{7-7} = \frac{3}{0}$. Since division by 0 is undefined, then the slope of the line is undefined.

This is not surprising. Since the x-coordinates are the same, the points lie on a vertical line, which has an undefined slope.

Exercises

For each example, find the slope of the line passing through the two points. Express your answer in simplest form.

1. (4, 6), (7, 8)
2. (8, 4), (2, 5)
3. (7, 11), (10, 15)
4. (0, 2), (5, 8)
5. (−2, 0), (7, −5)
6. (−3, 6), (−5, 7)
7. (−4, 8), (−6, 10)
8. (−7, 10), (−12, 13)
9. (−4, 19), (−3, 19)
10. (7, 17), (−4, 17)
11. (4, 5), (4, −12)
12. (−5, 12), (−5, 6)

Questions to Think About

13. Write the pronunciation of the formula for slope.

14. To find the slope of the line passing through the points (3, 17) and (−1, 4), Marissa used the calculation $m = \frac{17-4}{-1-3}$. Explain the error in her thinking.

15. Use the formula for slope to explain why a horizontal line has zero slope.

7.3 Determining If an Ordered Pair Is a Solution to an Equation

In previous sections, we discussed how to substitute values into an equation and how to graph a line given its equation. In this section, we apply these skills to determine if an ordered pair is a solution to the equation. We solve these problems by using algebra and by graphing.

We summarize the relationship between an ordered pair, an equation in two variables, and the graph of the equation as follows:

If an ordered pair is on the graph of an equation, then its coordinates are a solution to the equation (meaning that when we substitute the coordinates into the equation, the resulting statement is true).

We can determine algebraically or geometrically if an ordered pair is a solution to an equation, as shown in Examples 7.8 and 7.9:

Example 7.8 Is (−3, 1) a solution to the equation $y = \frac{1}{3}x + 2$? Explain.

Solution:

METHOD 1: Substitute $x = -3$ into the equation and see if $y = 1$.

$$1 \stackrel{?}{=} \frac{1}{3}(-3) + 2$$

$$1 \stackrel{?}{=} -1 + 2$$

$$1 = 1$$

Since substituting the ordered pair (−3, 1) results in a true equation, then (−3, 1) is a solution to $y = \frac{1}{3}x + 2$.

METHOD 2: *Graph the equation and see if $(-3, 1)$ lies on the graph.*
The line $y = \frac{1}{3}x + 2$ has a slope of $\frac{1}{3}$ and a y-intercept of $(0, 2)$. When we graph the line, we see that the point $(-3, 1)$ lies on the graph. Thus, it is a solution to the equation.

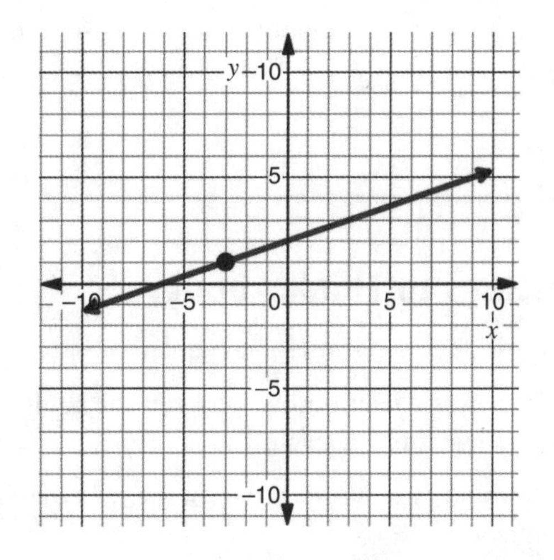

Technology Tip

Use technology to generate a table of values from the equation. If the coordinates of a point are in the table, then the ordered pair is a solution to an equation.

Example 7.9 Does the point (250, 448) lie on the graph of the equation $y = 3x - 301$?

Solution: Since the problem asks us if a point lies on the graph, we might think that drawing a graph would be the best way to solve the problem. However, we'd have to count several hundred units on the coordinate plane.

In this case, solving algebraically is easier. When we substitute $x = 250$ (the x-coordinate) into the equation $y = 3x - 301$, we get $y = 3(250) - 301 = 449$, which is close to but not exactly 448. We conclude that the point (250, 448) does *not* lie on the graph. Note that if we had attempted to graph the equation on paper, we might have mistakenly concluded that the point is on the graph because the points (250, 448) and (250, 449) are so close together. However, they're not exactly the same!

Exercises

Determine if the given ordered pair is a solution to the given equation.

1. $(3, 6)$, $y = 2x$
2. $(2, 11)$, $y = 7x + 3$
3. $(-1, 4)$, $x + 5 = y$
4. $(2, 3)$, $x = 2$
5. $(0, 12)$, $y = 11$
6. $(0, 12)$, $2x + y = 12$
7. $(-17, -25)$, $y = 15x + 230$
8. $(40, 20)$, $3x - 4y = 30$
9. $(6, 15)$, $y = \frac{5}{2}x - 1$

Questions to Think About

10. What does the graph of an equation represent?
11. What is the relationship between the solution set on an equation and the graph of an equation? Explain.
12. Describe two ways to determine whether an ordered pair is a solution to an equation.

7.4 Writing the Equation of a Line

In this section, we discuss different methods for writing the equation of a line based on different information that we are given in the problem.

If we know the slope and y-intercept, then we can write the equation of the line in slope-intercept form.

Example 7.10 Write an equation of the line that has a slope of 4 and a y-intercept of (0, −8).

Solution: The equation in slope-intercept form is $y = mx + b$, where $m =$ slope and $b =$ y-coordinate of the y-intercept. Here, $m = 4$ and $b = -8$, so the equation is $y = 4x - 8$.

Example 7.11 Write an equation of the line that has zero slope and a y-intercept of (0, 12).

Solution: The equation in slope-intercept form is $y = mx + b$. Here, $m = 0$ and $b = 12$, so the equation is $y = 0x + 12$, which we typically write as $y = 12$.

What if we're given the slope and a point on the line that's *not* the y-intercept? For example, imagine that we had to write the equation of a line that has a slope of $\frac{2}{3}$ and passes through the point (4, 3). Let (x, y) be the coordinates of any point on the line (Figure 7.5). The formula for slope is $m = \frac{\textit{vertical change}}{\textit{horizontal change}}$, where the horizontal change is the difference between the x-coordinates $(x - 4)$ and the vertical change is the difference between the y-coordinates $(y - 3)$.

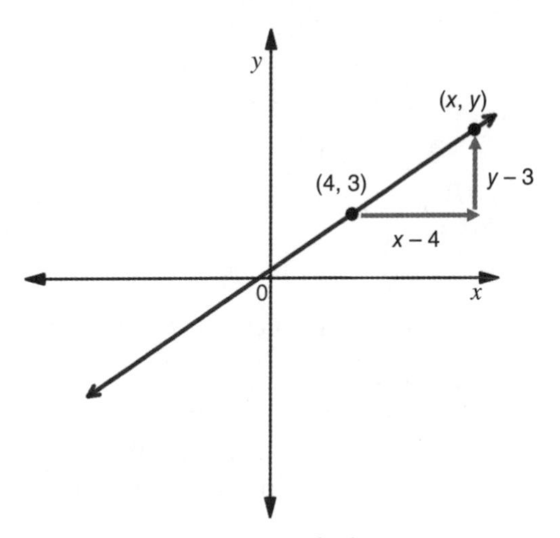

Figure 7.5 Point-slope form of the equation of a line.

We know that $m = \frac{2}{3}$, so substituting into the slope formula gives us $\frac{y-3}{x-4} = \frac{2}{3}$. If we multiply both sides of the equation by $x - 4$, we get $y - 3 = \frac{2}{3}(x - 4)$, which is an equation of the line that has a slope of $\frac{2}{3}$ and passes through the point (4, 3).

This form of the equation of a line is called the **point-slope form** since it requires a *point* on the line and the *slope* of the line. In general, the point-slope form of the equation of a line is $\boldsymbol{y - y_1 = m(x - x_1)}$, where m is the slope and (x_1, y_1) is a point on the line.

The following example shows why the point-slope form is useful.

Example 7.12 Write an equation of the line that has a slope of $-\frac{2}{5}$ and passes through (6, −5).

Solution: We can use either point-slope or slope-intercept form of the equation.

METHOD 1: Use point-slope form.

Here, $m = -\frac{2}{5}$, $x_1 = 6$, and $y_1 = -5$. When we substitute into the formula for point-slope form, we get $y - (-5) = -\frac{2}{5}(x - 6)$, or $y + 5 = -\frac{2}{5}(x - 6)$.

METHOD 2: *Use slope-intercept form.*

The equation in slope-intercept form is $y = mx + b$. Here, $m = -\frac{2}{5}$. Since x and y represent the set of all ordered pairs (x, y) that satisfy the equation, we can substitute the ordered pair $(6, -5)$. We substitute $x = 6$ and $y = -5$ into the slope-intercept form to solve for b:

$$y = mx + b$$

$$-5 = -\frac{2}{5}(6) + b$$

$$-5 = -\frac{12}{5} + b \qquad \text{Multiply.}$$

$$+\frac{12}{5} \quad +\frac{12}{5} \qquad \text{Add } \frac{12}{5} \text{ to both sides.}$$

$$-\frac{13}{5} = b \qquad \text{Answer, which we write as } b = -\frac{13}{5}.$$

The equation of the line in slope-intercept form is $y = -\frac{2}{5}x + \left(-\frac{13}{5}\right)$, or $y = -\frac{2}{5}x - \frac{13}{5}$.

Although these two equations look different, they are actually equivalent. We can rearrange the point-slope form into slope-intercept form by distributing and rearranging terms:

$$y + 5 = -\frac{2}{5}(x - 6)$$

$$y + 5 = -\frac{2}{5}x - \frac{2}{5}(-6) \qquad \text{Distributive property.}$$

$$y + 5 = -\frac{2}{5}x + \frac{12}{5} \qquad \text{Multiply.}$$

$$-5 \qquad\qquad -5 \qquad \text{Subtract 5 from both sides.}$$

$$y = -\frac{2}{5}x - \frac{13}{5} \qquad \text{Final answer.}$$

As you can see, the point-slope form of a line can simplify our work tremendously!

To write the equation of a line in point-slope or slope-intercept form, we need the slope of the line. If we're not given the slope—for example, if we're only given the coordinates of two points on the line—then we use the coordinates to find the slope.

Example 7.13 Write an equation of the line that passes through the points (7, 12) and (−2, −3).

Solution: First, we find the slope of the line. Here, we use the slope formula (you could also graph the points to find the slope). Call (7, 12) the first point and (−2, −3) the second point. Then $m = \frac{y_2 - y_1}{x_2 - x_1} = \frac{-3-12}{-2-7} = \frac{-15}{-9} = +\frac{15}{9} = \frac{5}{3}$.

METHOD 1: Use point-slope form.
We may use either of the two points that pass through the line. If we use the first one, then $m = \frac{5}{3}$, $x_1 = 7$, and $y_1 = 12$. When we substitute into the formula for point-slope form, we get $y - 12 = \frac{5}{3}(x - 7)$. If we had used the second point, we would have gotten $y + 3 = \frac{5}{3}(x + 2)$.

METHOD 2: Use slope-intercept form.
Substitute the coordinates of the ordered pair (7, 12) (we could also substitute the ordered pair (−2, −3)) and $m = \frac{5}{3}$ into the slope-intercept form to solve for b:

$$y = mx + b$$

$$12 = \frac{5}{3}(7) + b$$

$$12 = \frac{35}{3} + b \qquad \text{Multiply.}$$

$$-\frac{35}{3} \quad -\frac{35}{3} \qquad \text{Subtract } \frac{35}{3} \text{ from both sides.}$$

$$\frac{1}{3} = b \qquad \text{Write with the variable on the left side: } b = \frac{1}{3}.$$

The equation of the line in slope-intercept form is $y = \frac{5}{3}x + \frac{1}{3}$. You can confirm that this equation is equivalent to both $y - 12 = \frac{5}{3}(x - 7)$ and $y + 3 = \frac{5}{3}(x + 2)$.

Sometimes, we are only given the graph of the line. To write its equation, we identify two points on the line and then proceed as we did when we used the coordinates of two points on the line.

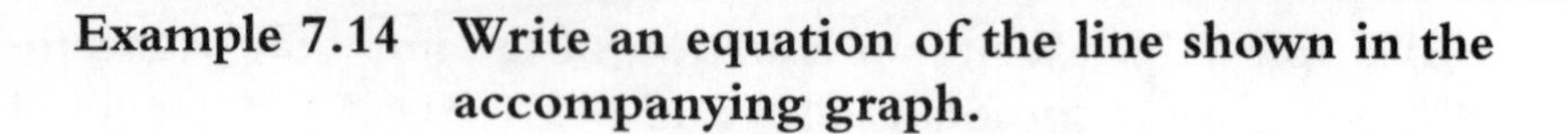

Example 7.14 Write an equation of the line shown in the accompanying graph.

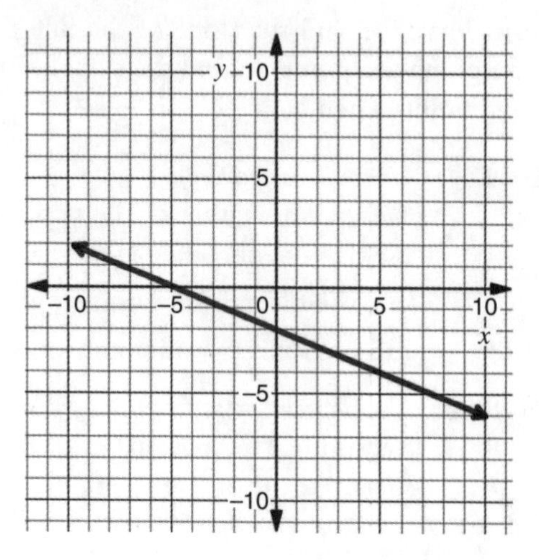

Solution: First, we determine the slope of the line. We identify two points on the line: (0, −2) and (5, −4). We see from the graph that from (0, −2), the line goes down 2 units and to the right 5 units. Thus, the slope of the line is $m = \frac{\textit{vertical change}}{\textit{horizontal change}} = \frac{-2}{+5} = -\frac{2}{5}$. (We can also use the slope formula to get the same result.)

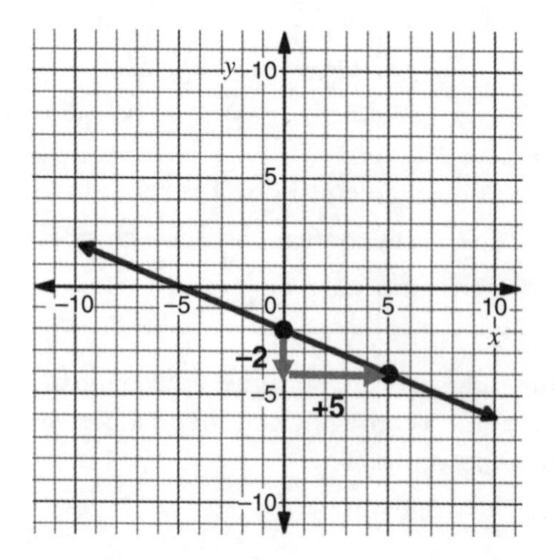

One of the points that we identified is the y-intercept, so we can write the equation of the line in slope-intercept form. The y-intercept of the graph is (0, −2), so its y-coordinate is $b = -2$. Since the slope is $m = -\frac{2}{5}$, then an equation of the line is $y = -\frac{2}{5}x - 2$.

Table 7.2 summarizes the different methods for writing the equation of a line that we discussed in this chapter.

Table 7.2 How to write the equation of a line.

Given	Steps	Example
Slope and y-intercept	Use slope-intercept form: $y = mx + b$	Write an equation of the line whose slope is 5 and y-intercept is (0, 2). $y = 5x + 2$
Slope and point on the line	Use point-slope form: $y - y_1 = m(x - x_1)$	Write an equation of the line that has a slope of 5 and passes through (6, 7). $y - 7 = 5(x - 6)$
2 points on the line	**1.** Find the slope of the line. **2.** Use slope and 1 point to write the equation in point-slope or slope-intercept form.	Write an equation of the line that passes through (3, 5) and (7, 13). $m = \frac{13-5}{7-3} = \frac{8}{4} = 2$, point: (3, 5) $y - 5 = 2(x - 3)$ or $y = 2x - 1$
Graph of the line	**1.** Find 2 points on the line. **2.** Find the slope of the line. **3.** Use slope and 1 point to write the equation in point-slope or slope-intercept form.	Write an equation of the line shown in the graph. 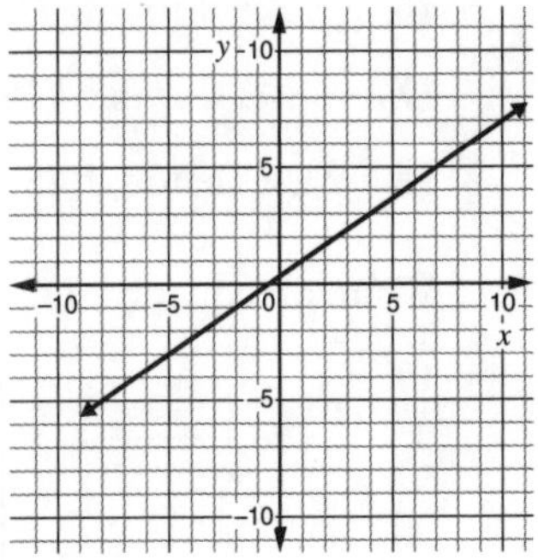 2 points: (10, 7) and (7, 5) $m = \frac{7-5}{10-7} = \frac{2}{3}$, point: (10, 7) $y - 7 = \frac{2}{3}(x - 10)$ or $y = \frac{2}{3}x + \frac{1}{3}$

Exercises

Write an equation of the line that has the following characteristics:

1. slope of -5 and a y-intercept of (0, 3)

2. slope of $\frac{2}{7}$ and y-intercept of (0, 4)

3. slope of $\frac{8}{3}$ and y-intercept of $\left(0, \frac{1}{2}\right)$
4. slope of 0 and y-intercept of (0, 7)
5. undefined slope and x-intercept of (2, 0)
6. undefined slope and x-intercept of (−18, 0)
7. slope of −2 and passes through (−2, 11)
8. slope of $\frac{4}{5}$ and passes through (10, 6)
9. slope of $\frac{1}{8}$ and passes through (8, 2)
10. slope of 5 and passes through (3, −1)
11. slope of $-\frac{3}{7}$ and passes through (4, 7)
12. slope of 0 and passes through (6, 8)
13. passes through (1, 7) and (5, 11)
14. passes through (−2, 7) and (8, −3)
15. passes through (6, 12) and (−2, 8)
16. passes through (0, 5) and (10, 16)
17. passes through (18, 2) and (18, 5)
18. passes through (5, −11) and (7, −11)

Write an equation of the line shown in each graph.

19. 20.

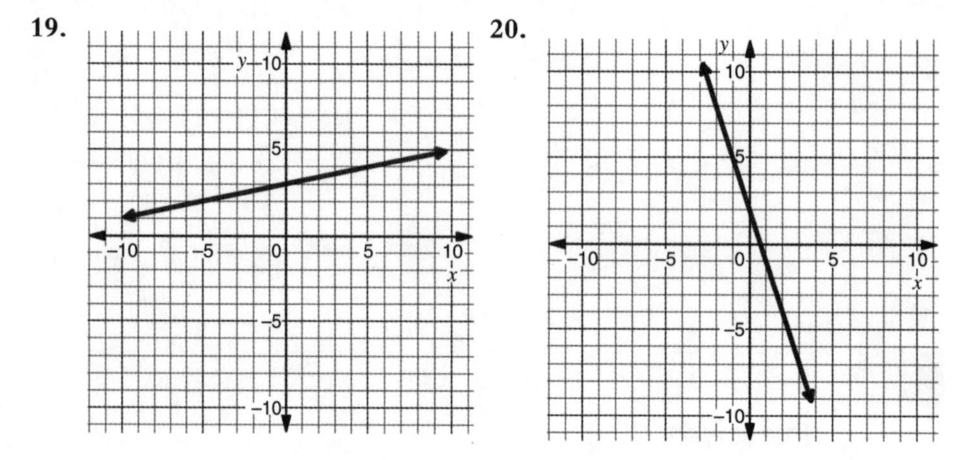

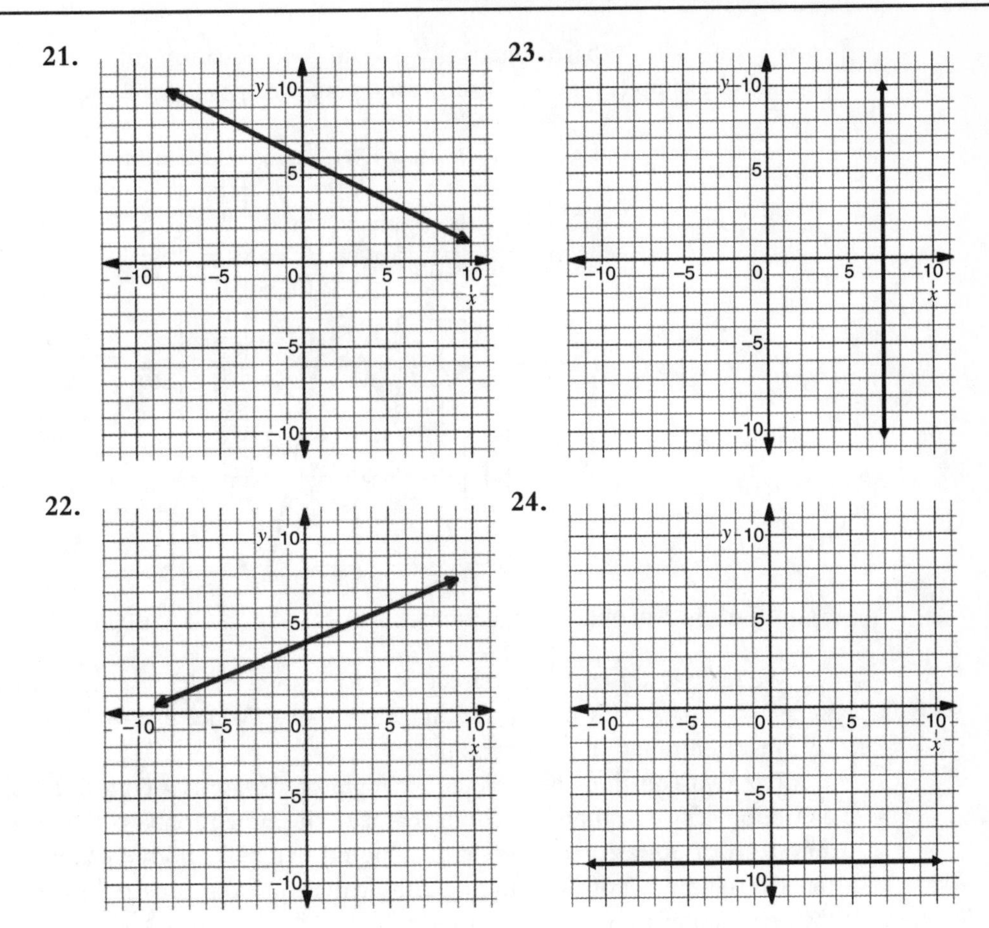

Questions to Think About

25. Can you write an equation of a line given only its slope? Explain.

26. Do the equations $y = 3x + 5$ and $y + 13 = 3(x + 6)$ represent the same line? Explain.

27. Can a vertical line be written in point-slope form? Why or why not?

7.5 Solving Systems of Linear Equations by Graphing

As covered in Section 6.3, the graph of an equation represents all the points that are solutions to the equation. So far, we have graphed only one equation at a time on the coordinate plane. What happens if we graph two equations on the same coordinate plane? For example, Figure 7.6 shows the graphs of $y = 2x + 1$ and $y = -4x + 7$:

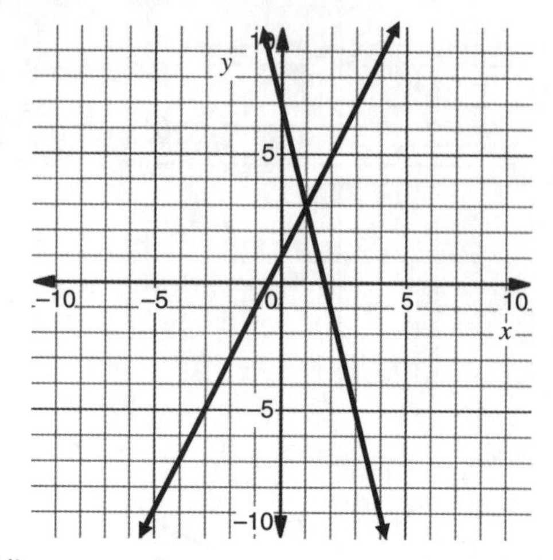

Figure 7.6 System of linear equations.

We see that the graphs intersect at the point (1, 3). It's a solution to $y = 2x + 1$ since $3 = 2(1) + 1$. It's also a solution to $y = -4x + 7$ since $3 = -4(1) + 7$. We say that (1, 3) is a solution to both equations.

The two equations here are an example of a **system of equations**, which is a collection of two or more equations that have the same variables. The **solution set to a system of equations** is the set of all points that make both equations true. We can find the solution set of a system by graphing the equations and finding the points where the graphs intersect.

Watch Out!

When we state the solution set of a system of equations, we must specify the value for each variable. We do so in any one of the following ways:

- an ordered pair, such as (3, 1)
- an equation for each variable, connected by the word "and," such as $x = 3$ and $y = 1$

Don't just say "3 and 1."

When solving a system of equations graphically, remember to do the following:

- Label each line or curve with its equation.
- Draw a dot at each intersection.
- Label each intersection with its coordinates.

Example 7.15 Solve the system of equations graphically:

$$4x + 8 = 32$$
$$3y - 1 = 11$$

Solution: First, we solve each equation for its variable:

$$\begin{aligned} 4x + 8 &= 32 \\ -8 \quad &\ -8 \\ \hline 4x &= 24 \\ \frac{4x}{4} &= \frac{24}{4} \\ x &= 6 \end{aligned} \qquad \begin{aligned} 3y - 1 &= 11 \\ +1 \quad &\ +1 \\ \hline 3y &= 12 \\ \frac{3y}{3} &= \frac{12}{3} \\ y &= 4 \end{aligned}$$

Then we graph $x = 6$ and $y = 4$ on the same coordinate plane. We label each line with its equation and label the intersection point (6, 4) with its coordinates. The intersection point is the only point that has an x-coordinate of 6 *and* a y-coordinate of 4.

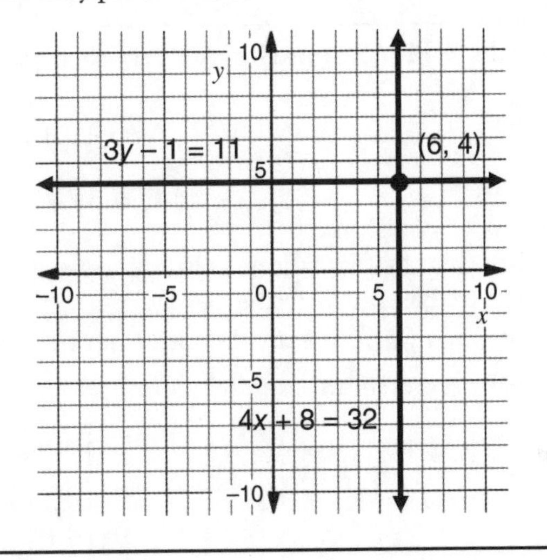

In Section 3.6, we saw that some equations have no solution or infinitely many solutions. This sometimes occurs with systems of equations, as Examples 7.16 and 7.17 show:

Example 7.16 Solve the system of equations graphically:

$$x + y = 4$$
$$y = -x + 6$$

Solution: If we rewrite the first equation in slope-intercept form by subtracting x from both sides, we get $y = -x + 4$, Both equations have a slope of -1 but have different y-intercepts: (0, 4) for the first and (0, 6) for the second.

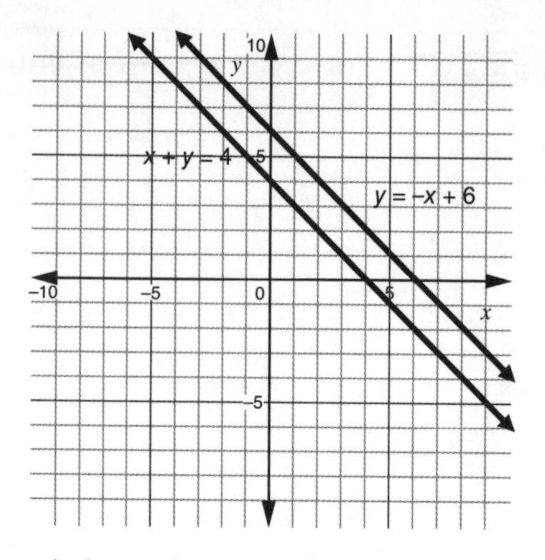

We see from the graph that since the lines have the same slope, they will not meet. The lines have no point of intersection, so the system has no solution.

Example 7.17 Solve the system of equations graphically:

$$q = \frac{3}{4}p + 3$$

$$4q = 3p + 12$$

Solution: The first equation is already in slope-intercept form. (Here, p is the independent variable and q is the dependent variable.) To put the second equation in slope-intercept form, we divide both sides by 4: $\frac{4q}{4} = \frac{3p}{4} + \frac{12}{4}$. This simplifies to $q = \frac{3}{4}p + 3$! We represent this on the coordinate plane by graphing only one line:

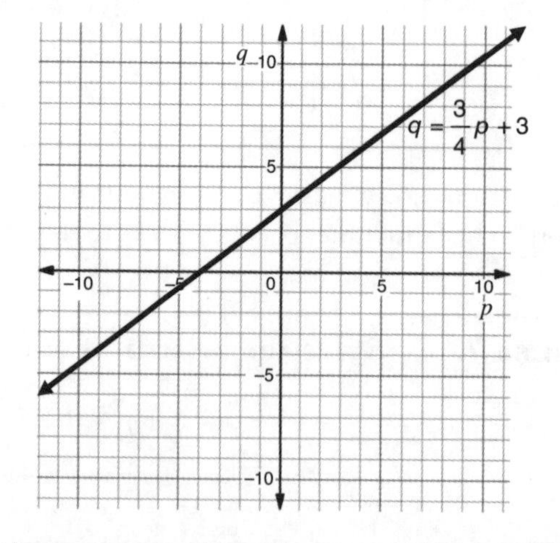

Since both equations are equivalent, then any solution of the first equation will also be a solution of the second. Thus, the system has infinitely many solutions.

Technology Tip

Here are some tips for using technology to solve systems of equations:

- **Use an appropriate window for the graph.** Many functions are not visible in the default window set by graphing utilities ($-10 \leq x \leq 10$, $-10 \leq y \leq 10$). To see important features of the function on your device's screen, you may need to adjust the scale (number of units between tick marks) on the *x*- or *y*-axes or the minimum or maximum values for *x* and *y* (often called the Xmin, Xmax, Ymin, and Ymax values).
- **Find the point of intersection using the intersection tool.** Using technology enables you to find the point where the graphs intersect. Some devices enable you to click on the point to determine its coordinates. If you can see the table of values for each equation, the point of intersection has the same *x*- and *y*-values.
- **If necessary, replace the variables in the equations with *x* and *y* when typing into your device.** Most devices will only graph equations if they are expressed in terms of *x* and *y*. If you need to graph the result on paper, remember to switch the variables back!
- **Use systems of equations to check the solution for an equation with one variable.** For example, to solve the equation $2x + 11 = 5x + 5$ (see Example 6.11), let the left side of the equation be the function $f(x) = 2x + 11$ and the right side be the function $g(x) = 5x + 5$. Finding the *x*-values that make $f(x) = g(x)$ is equivalent to solving the equation $2x + 11 = 5x$. If we graph the equations $y = 2x + 11$ and $y = 5x + 5$ using technology, we see that the point of intersection is (2, 15). The *x*-coordinate, or $x = 2$, is the solution to $2x + 11 = 5x + 5$.

Exercises

Solve each system of equations graphically.

1. $5x + 7 = 42$
$y + 5 = -4$

2. $10x + 1 = 21$
$3y - 2 = 7$

3. $4x + 30 = 22$
$9y - 2 = 25$

4. $y = 2x + 2$
$y = \frac{1}{2}x + 5$

5. $y = \frac{3}{4}x - 1$
$y = -\frac{1}{2}x - 6$

6. $y = -3x + 3$
$y = 2x - 7$

7. $y = 2x + 6$
$y + 3x = 1$

8. $2y - x = 3$
$y - 6 = 2x$

9. $2y = x - 4$
$x + 4y = -2$

Questions to Think About

10. How is a system of equations different from an equation?

11. If a system of linear equations has no solution, what must be true about the graphs of the equations?

12. Maisha says that the solution to the system of equations $y = 3x + 4$ and $y = -2x + 9$ is $\{1, 7\}$. Explain the error in her solution.

7.6 Solving Systems of Linear Equations by Substitution

Some systems of equations, like the one below, can't be solved easily by graphing:

$$y = 21x$$
$$91x - y = 10$$

Even with the help of technology, the solution is not immediately clear (Figure 7.7):

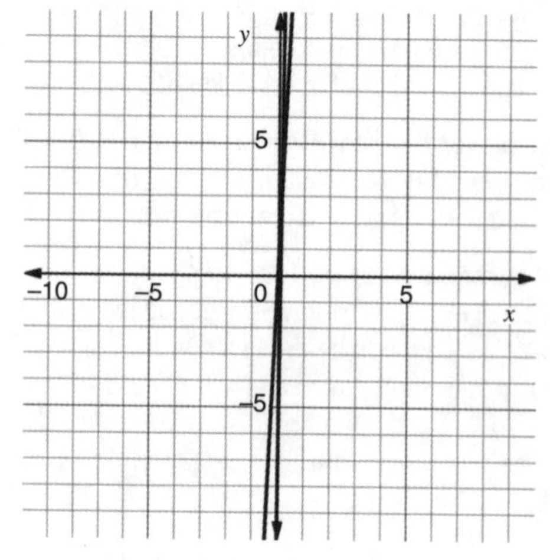

Figure 7.7 Some systems aren't solved easily by graphing.

Upon closer inspection, we can see that since one of the equations is already solved for a variable, we can substitute its equivalent into the second equation:

$91x - 21x = 10$	Substitute $21x$ for y in the second equation.
$70x = 10$	Combine like terms.
$\frac{70x}{70} = \frac{10}{70}$	Divide both sides by 70.
$x = \frac{10}{70} = \frac{1}{7}$	Solve for x.

We substitute this value into either equation to find the value of y. Since the first equation is already solved for y, substituting into this equation is easier: $y = \frac{1}{7}(21) = 3$, so the solution is $\left(\frac{1}{7}, 3\right)$.

This method of solving linear equations is called **substitution** since we are *substituting* an expression for one variable in another equation. Either equation can be solved for the variable, but it is easier when one or both equations is already solved for one of the variables.

Example 7.18 Solve the system of equations:

$$3m + 2n = 13$$

$$4n - 5 = m$$

Solution: To use substitution, one equation must be solved for one of the variables. Since the second equation is solved for m, we substitute $4n - 5$ for m in the first equation:

$3m + 2n = 13$	
$3(4n - 5) + 2n = 13$	Substitute $4n - 5$ for m.
$12n - 15 + 2n = 13$	Distributive property.
$14n - 15 = 13$	Combine like terms.
$\quad + 15 + 15$	Add 15 to both sides.
$14n = 28$	Combine like terms.
$\frac{14n}{14} = \frac{28}{14}$	Divide both sides by 14.
$n = 2$	Simplify.
$m = 4(2) - 5 = 3$	Substitute $n = 2$ into the second equation to find m.

The final answer is $m = 3$ and $n = 2$. In this example, we recommend *not* writing the solution as an ordered pair. Since the variables are not x and y, writing the solution as (3, 2) does not clearly indicate which variable has the value of 3 and which has the value of 2.

Exercises

Solve each system of equations by substitution.

1. $d = 4c$
$2c + d = 18$

2. $-q = 2p + 3$
$q = p$

3. $2x + 1 = y$
$4y = 8x - 8$

4. $2x - 5 = y$
$3x + 3y = 21$

5. $x = y + 10$
$2x = 3y + 14$

6. $x = 9 - 3y$
$4x + 12y = 36$

7. $r = 4t - 1$
$6t + r = 79$

8. $a = b + 2$
$3a + 4b = 20$

9. $q = 27 - 3p$
$2q = 3p$

7.7 Solving Systems of Linear Equations by Elimination

Some systems of equations can't be solved easily by substitution. Take, for example:

$$5x + 2y = 29$$

$$7x - 2y = 19$$

Solving the first equation for x gives us $x = -\frac{2}{5}y + \frac{29}{5}$. Substituting this messy equation into the second means having to simplify $7\left(-\frac{2}{5}y + \frac{29}{5}\right) - 2y = 19$! Fortunately, there's an easier way. Since the coefficients of $+2y$ and $-2y$ are opposites, we add the two equations to get an equivalent equation in one variable that we solve.

$$\begin{array}{r} 5x + 2y = 29 \\ + \, 7x - 2y = 19 \\ \hline \end{array}$$

$12x = 48$	Combine like terms.
$\frac{12x}{12} = \frac{48}{12}$	Divide both sides by 12.
$x = 4$	Solve for x.
$5(4) + 2y = 29$	Substitute $x = 4$ into the first equation.
$20 + 2y = 29$	Simplify.
$-20 \quad\quad - 20$	Subtract 20 from both sides.
$2y = 9$	Combine like terms.
$\frac{2y}{2} = \frac{9}{2}$	Divide both sides by 2.
$y = \frac{9}{2}$	Solve for y.

The solution is $\left(4, \frac{9}{2}\right)$, or $x = 4$ and $y = \frac{9}{2}$.

This method of solving systems is called **elimination** since one of the variables is *eliminated* when we add the equations.

In some systems, the coefficients of terms for one variable are not opposites.

Example 7.19 Solve the system of equations:

$$\mathbf{2x - y = 5}$$

$$\mathbf{3x + 3y = 21}$$

Solution: Adding the equations gives us $5x + 2y = 26$, which doesn't help us solve for x or y. If the coefficient of y in the first equation were -3, then we could add the equations to eliminate y. To do that, multiply *every term* in the first equation by 3:

$3(2x - y) = 3(5) \rightarrow$	$6x - 3y = 15$	Multiply both sides of the first equation by 3.
$3x + 3y = 21$	$\underline{+ \, 3x + 3y = 21}$	Leave the second equation unchanged.
	$9x = 36$	Add the two equations.
	$\frac{9x}{9} = \frac{36}{9}$	Divide both sides by 9.
	$x = 4$	Simplify.
	$3(4) + 3y = 21$	Substitute into an original equation to find y.

NOTE: You can also substitute into $2x - y = 5$ to find y.

$12 + 3y = 21$	Simplify.
$-12 \quad -12$	Subtract 12 from both sides.
$3y = 9$	Combine like terms.
$\frac{3y}{3} = \frac{9}{3}$	Divide both sides by 3.
$y = 3$	Solve for y.

The final answer is $x = 4$ and $y = 3$, or (4, 3).

In some equations, you'll have to multiply not one but both equations to eliminate a variable:

Example 7.20 Solve the system of equations:

$$\mathbf{4u + 2v = 14}$$

$$\mathbf{7u - 3v = -8}$$

Solution: We multiply each equation by an appropriate number so that the coefficients of one variable become opposites. Finding the least common multiple (using the same procedure described in Chapter 2 for finding the least common denominator or by using the lcm() function on your device) will help us use numbers that are smaller and easier to manage.

As with any system of equations, we can eliminate either variable (Table 7.3). Whether you eliminate u or v first, you should get the same answer by the end:

Table 7.3 We can eliminate *u* or *v*.

Eliminate *u*		Eliminate *v*	
$7(4u + 2v) = 7(14) \rightarrow$	$28u + 14v = 98$	$3(4u + 2v) = 3(14) \rightarrow$	$12u + 6v = 42$
$-4(7u - 3v) = -4(-8) \rightarrow$	$+ -28u + 12v = 32$	$2(7u - 3v) = 2(-8) \rightarrow$	$+ 14u - 6v = -16$
	$26v = 130$		$26u = 26$
	$\frac{26v}{26} = \frac{130}{26}$		$\frac{26u}{26} = \frac{26}{26}$
	$v = 5$		$u = 1$
$4u + 2(5) = 14$	Substitute $v = 5$ into an equation.	$4(1) + 2v = 14$	Substitute $u = 1$ into an equation.
$4u + 10 = 14$	Simplify.	$4 + 2v = 14$	Simplify.
$4u = 4$	Subtract 10 from both sides.	$2v = 10$	Subtract 4 from both sides.
$\frac{4u}{4} = \frac{4}{4}$	Divide both sides by 4.	$\frac{2v}{2} = \frac{10}{2}$	Divide both sides by 2.
$u = 1$	Final answer: $u = 1$ and $v = 5$	$v = 5$	Final answer: $u = 1$ and $v = 5$

Note that when eliminating u, we could have multiplied the first equation by -7 and the second equation by 4.

You can use any of the three methods we've discussed—graphing, substitution, or elimination—to solve any system of equations. However, each method works better for some systems. Table 7.4 summarizes these methods:

Table 7.4 Comparing solution methods for systems of equations.

Method	Ideal Situation	Example	Notes
Graphing	• Exact answer is not needed	$y = 2x + 6$ $y = 3x - 1$	• Can be done quickly with technology • Little algebraic manipulation required
Substitution	• One or both equations are solved for one of the variables.	$x = 9 - 3y$ $4x + 12y = 36$	• Yields an exact answer • Can get complicated if an equation isn't solved for one of the variables
Elimination	• Equations are not solved for one of the variables.	$4u + 2v = 14$ $7u - 3v = -8$	• Yields an exact answer • Often requires algebraic manipulation to get equations in the form $ax + by = c$

Exercises

Solve each system of equations by elimination.

1. $3x - y = 21$
   $2x + y = 4$
2. $3k + 5p = 9$
   $3k - p = -9$
3. $7a + t = 42$
   $3a - t = 8$
4. $2y - 7x = 2$
   $3x = 14 - y$
5. $3x - 4y = 8$
   $6x = 8y + 12$
6. $7r - 5s = 15$
   $2r - s = 9$
7. $3p + 6q = -12$
   $5p - 4q = -6$
8. $3k - 4d = 10$
   $4k - 5d = 14$
9. $6z + 9w = -9$
   $5z + 7w = -6$

Questions to Think About

Without solving for the variables, determine the method (graphing, substitution, or elimination) that is best used to answer each question. Explain your answer.

10. Solve for the variables:

$$2x - y = 5$$

$$5x + 2y = 17$$

11. How many solutions does the following system have?

$$y = 7x + 6$$
$$y = -3x + 4$$

12. Solve for the variables:

$$y = 2x - 7$$
$$4x + 5y = 17$$

7.8 Solving Systems of Linear Inequalities

In Chapter 5, we discussed how to solve linear inequalities in *one* variable (like $5x + 4 > 12$) and graph the solution on a number line. We can extend this concept into *two* variables on the coordinate plane.

Let's start by graphing an inequality like $x < 10$ on a number line. The numbers in the solution set—meaning the numbers that make this inequality true—are all less than 10. We represent the solution as in Figure 7.8:

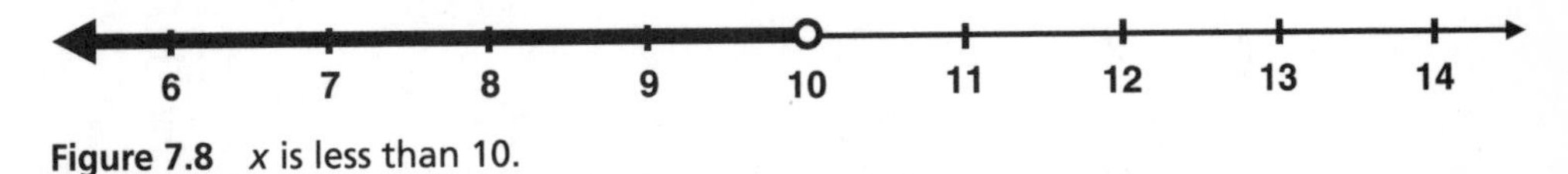

Figure 7.8 x is less than 10.

How can we represent $x < 10$ on a coordinate plane? The solution set is not a point on a line, but points like (9, 3) and (6, −2) that lie in a region on the coordinate plane. In this case, the solution set is all the points whose x-coordinates are less than 10. The boundary is the vertical line $x = 10$. We draw a dotted line since the inequality has a < symbol, so points on the boundary are not part of the solution. We shade to the left of the boundary since these points have x-coordinates that are less than 10. We mark the solution set by putting an S inside it, as shown in Figure 7.9:

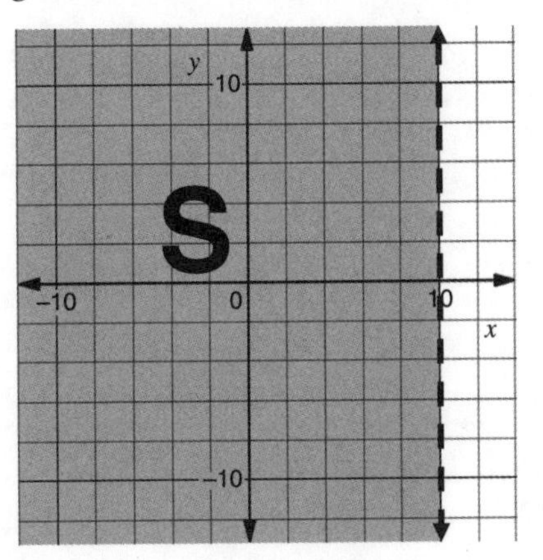

Figure 7.9 x is less than 10 on the coordinate plane.

We solve a system of inequalities by graphing each inequality on the same coordinate plane. The region where the solution sets overlap is the solution set to the system, as shown in Examples 7.21 and 7.22:

Example 7.21 Solve the following system of inequalities:

$$x > 3$$

$$y \geq -4$$

Solution:

1. For the first inequality, graph the boundary of the solution set: $x = 3$.
   - Draw a *dotted* line since the points on the line are not included in the solution set.
   - Since the inequality symbol is >, shade to the *right* of the boundary.

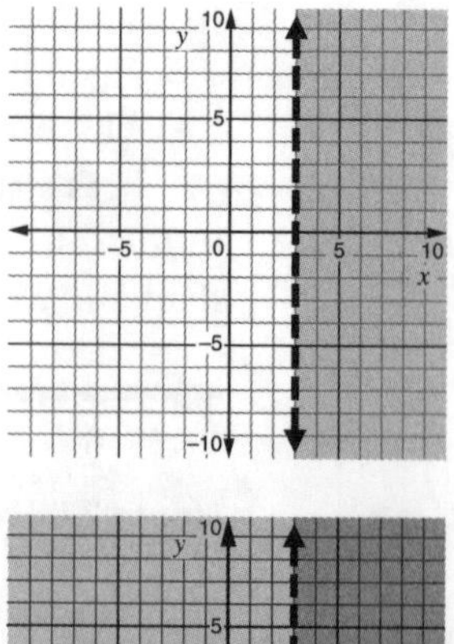

2. For the second inequality, graph the boundary of its solution set: $y = -4$.
   - Draw a *solid* line since the points on the line are included in the solution set.
   - Since the inequality symbol is ≥, shade *above* the boundary.

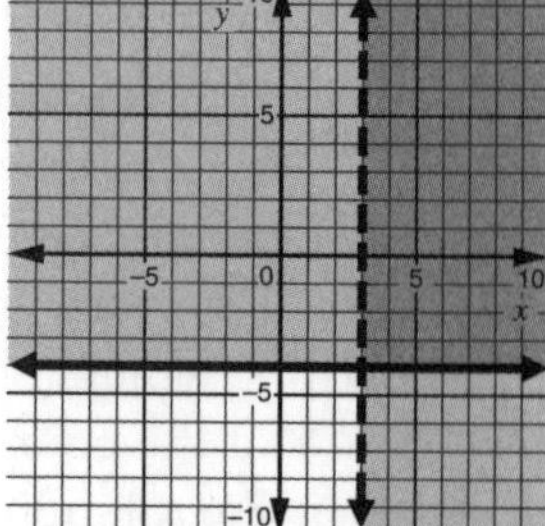

3. Label each graph with its inequality. Label the solution set with an S.

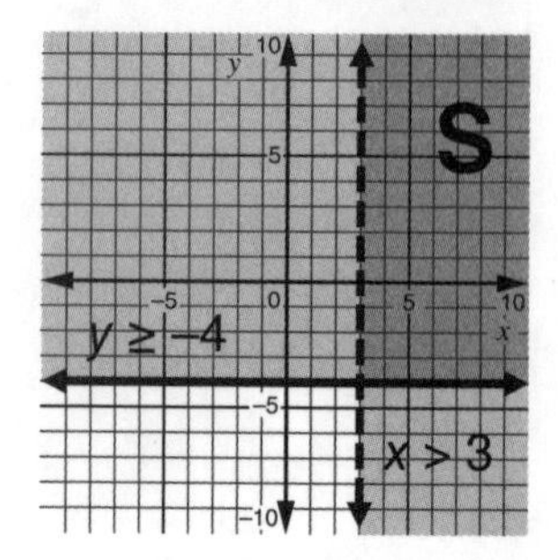

Example 7.22 Solve the following system of inequalities:

$$y \leq \frac{1}{2}x + 4$$

$$2x + y > 7$$

Solution:

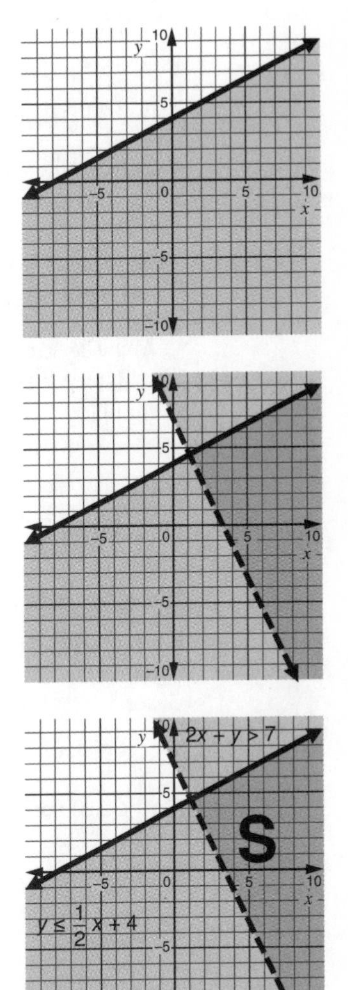

1. For the first inequality, graph the boundary of its solution set: $y = \frac{1}{2}x + 4$.
   - Draw a *solid* line since the points on the line are included in the solution set.
   - Since the inequality symbol is $\leq$, shade *below* the boundary.
2. For the second inequality, graph the boundary of its solution set: $2x + y = 7$.
   - Draw a *dotted* line since the points on the line are not included in the solution set.
   - Since the inequality symbol is $>$, shade *above* the boundary.
3. Label each graph with its inequality. Label the solution set with an S.

How to Solve a System of Inequalities

1. Graph the boundary for the first inequality. Draw a dotted line (if the inequality has a $>$ or $<$ symbol) or solid line (if the inequality has a $\geq$ or $\leq$ symbol).
2. Show the region for the first inequality. Shade above or to the right of the boundary (if the inequality has a $>$ or $\geq$ symbol) or below or to the left of the boundary (if the inequality has a $<$ or $\leq$ symbol).
3. Repeat steps 1 and 2 for each inequality.
4. Label the inequalities on the graph.
5. Label the solution set with an "S" inside the shaded region where the solutions for each inequality overlap.

Technology Tip

Many devices allow you to graph inequalities with the correct boundaries and shading simply by typing their expressions. Some devices require that you solve the inequality for *y* before graphing.

Table 7.5 summarizes the key differences between equations and inequalities in two variables:

Table 7.5 Comparing systems of equations and inequalities.

Characteristic	System of Linear Equations	System of Linear Inequalities
Graph	Two or more lines	Two or more regions bounded by a dotted or solid line
Method of solution	Graphically or algebraically (substitution or elimination)	Only graphically
Number of solutions	0 (when lines do not intersect), 1 (when lines intersect at a point), or infinite (when lines are the same)	0 (when regions do not intersect) or infinite (when regions intersect)
Expression of solution	Written as an ordered pair or set of equations	Shown as a region (typically marked with an S) on the coordinate plane

Exercises

Find the solution to each system of inequalities by graphing. Mark the solution set with an S.

1. $x > 5$
   $y > 3$
2. $x \leq -4$
   $y \leq -1$
3. $x > 5$
   $x > -2$
4. $x + y > 5$
   $y \leq 2x - 6$
5. $y > \frac{1}{2}x + 1$
   $y + 3x < -2$
6. $\frac{2}{3}x + y \leq 2$
   $y > x + 5$
7. $4x + y < -8$
   $-2x + 3y > 6$
8. $5x \geq -2y - 10$
   $-x + y \leq -2$
9. $x + 2y > 4$
   $3x - y \leq 6$

Questions to Think About

10. Can a system of inequalities have a solution even if their boundaries do not intersect? Explain your answer.
11. State the coordinates of a point that is in the solution set to the inequality $y > 3x + 2$. Explain your answer.
12. Is the point (2, −3) in the solution set to the inequality $y < -2x - 1$? Explain your answer.

7.9 Word Problems with Systems of Linear Equations and Inequalities

In this section, we discuss how we use systems of linear equations and inequalities to solve word problems. In these situations, we have two unknowns, so we need to write two equations or inequalities.

Some problems give us information that allow us to write equations representing the total number of unknown quantities ("the total number of dimes and nickels is 40") or the total value or cost of the quantities ("the total value of the coins is $3.50"). Here is an example:

Example 7.23 **On a recent school field trip, 130 students attended. The students traveled either by car or by van. A total of 24 vehicles transported students on the trip. Each car held 4 students, and each van held 6 students. How many cars and how many vans were used to transport students?**

Solution:

Step 1: *Identify what is given and what we need to find.*
130 students attended a trip.
24 cars and vans were used.
Each car held 4 students. Each van held 6 students.

Step 2: *Represent the unknown information.*
Let c = number of cars and v = number of vans. Then

$c + v = 24$ — 24 cars and vans were used.
$4c + 6v = 130$ — Each car held 4 students, so $4c$ students traveled in cars. Each van held 6 students, so $6v$ students traveled in vans. 130 students traveled in all (in a car or in a van).

Step 3: *Solve.*
Multiply the first equation by -4 to get c-terms to add up to zero:

$$-4(c + v) = -4(24) \quad \rightarrow -4c - 4v = -96$$
$$4c + 6v = 130 \quad \rightarrow \underline{+ 4c + 6v = 130}$$

$2v = 34$ — Add the equations.
$\frac{2v}{2} = \frac{34}{2}$ — Divide both sides by 2.
$v = 17$
$c + 17 = 24$ — Substitute $v = 17$ into an equation to find c.
$c = 7$ — Subtract 17 from both sides.

Seven cars and 17 vans were used.

Step 4: *Check.*

$$4(7) + 6(17) \stackrel{?}{=} 130$$

$$130 = 130$$

In another type of problem, we define two linear functions and use their graphs to find their intersection point. In context, this is the point where functions have equal value ("after how many minutes will the two rides have the same cost?"). Here are some tips for graphing real-world functions on the coordinate plane:

- If the variables make sense only if their values are non-negative, then put the origin in the lower left corner of the graph so that the axes extend up and to the right. This represents Quadrant I of the coordinate plane. We don't put arrows on the ends of functions if the values of the variable don't extend in those directions.
- When possible, use descriptive letters to represent variables, such as t for time or c for cost.

Example 7.24 **After expenses, Chris saves \$90 per month in his bank account. As of January, he had \$200 in savings. Since he delivers food and groceries by car to people who live in the city, where parking is scarce, he often must park illegally. As a result, he accumulates many parking tickets, which he pays from his savings. As of January, Chris already had \$400 in parking tickets from the city. Every month after January, he accumulates an additional \$70 in parking tickets and late fees.**

(a) Write a function A representing the amount, in dollars, in his bank account t months after January.

(b) Write a function P representing his total parking fines, in dollars, t months after January.

(c) Graph A and P on the coordinate plane.

(d) Use your graph to determine the number of months after January that Chris will have to wait before he has enough money to pay his parking fines.

Solution:

Step 1: *Identify what is given and what we need to find.*
Chris has \$200 in his account. He saves an additional \$90 per month.
Chris must pay \$400 in parking tickets. He accumulates an additional \$70 per month in fines after the first month.
How many months after January will it take Chris to pay his parking fines?

Step 2: *Represent the unknown information.*
Let t = number of months after January.

Step 3: *Solve.*

(a) Chris's savings are represented by $A(t) = 90t + 200$. (The starting value, at $t = 0$, is \$200. His savings increase by \$90 every month after January.)

(b) His parking fines are represented by $P(t) = 70t + 400$. (The starting value, at $t = 0$, is \$400. His fines increase by \$70 every month after January.)

(c) Since the parking fines can't be negative and we're only concerned with time after January, both variables have non-negative values. Our graph will focus on Quadrant I of the coordinate plane. A has a slope of 90 and passes through the point (0, 200). Our vertical scale is \$100 per box. We use technology to help us graph the functions, as shown below. We label the horizontal axis as time (months) and the vertical axis as amount ($).

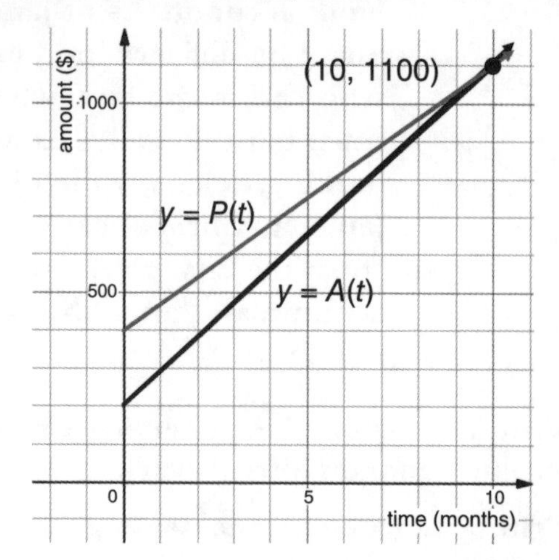

(d) From the graph (or from the function's table of values), the functions intersect at the point (10, 1100). This means that after 10 months, Chris will have just enough money to pay his parking fines. Unfortunately, by then, he will have to pay $1,100 in fines!

Step 4: *Check.*

Substitute (10, 1100) into both functions to see if they have equal outputs.

$$A(10) = 90(10) + 200 = 1{,}100.$$

$$P(10) = 70(10) + 400 = 1{,}100.$$

We use systems of inequalities to determine a range of values that are limited by certain constraints, as shown in Example 7.25.

Example 7.25 **An auditorium with a maximum capacity of 300 people is holding a concert. It plans to sell discounted tickets for $15 each and regular-price tickets for $40 each. The auditorium wants to sell at least $9,000 of tickets for each performance.**

(a) Write a system of inequalities that represents the given information.

(b) Graph the solution set to the system of inequalities.

(c) What is the largest number of discounted tickets that the auditorium can sell that will still enable it to meet its sales goal?

Solution:

Step 1: *Identify what is given and what we need to find.*
An auditorium holds no more than 300 people.
Discounted tickets sell for \$15 each. Regular-price tickets are \$40 each.
The auditorium wants to sell at least \$9,000 of tickets.

Step 2: *Represent the unknown information.*
Let d = number of discounted tickets sold and r = number of regular-price tickets sold.

Step 3: *Solve.*

(a) $d + r \leq 300$ (The number of discounted and regular-price tickets is at most 300.)

$15d + 40r \geq 9{,}000$ $15d$ is the sales, in dollars, from discounted tickets.
$40r$ is the sales, in dollars, from regular-price tickets.
The total ticket price sales $(15d + 40r)$ must be at least \$9,000.

$0 \leq r \leq 300$ The number of regular-price tickets sold must be between 0 and 300.

$0 \leq d \leq 300$ The number of discounted tickets sold must be between 0 and 300.

(b) We graph the first two inequalities on the coordinate plane. The last two inequalities provide constraints on r and d, so graph only within the rectangle bounded by the r-axis, d-axis, $r = 300$, and $d = 300$. Mark the solution set with the letter S.

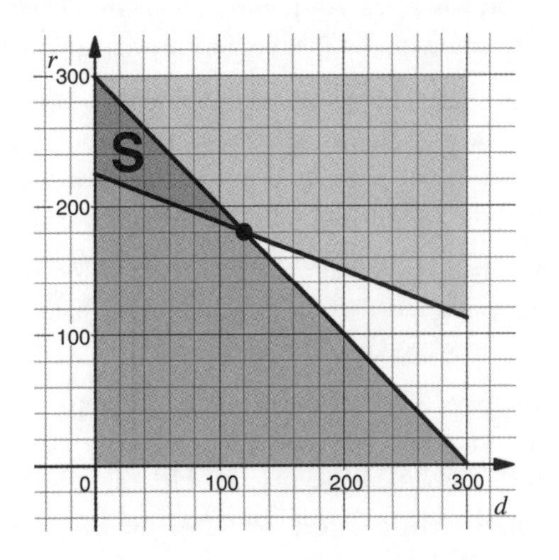

(c) Since d represents the number of discounted tickets, we need to find the maximum value of d that still lies within the region that defines the solution set. We see from the graph that the maximum value of d in the solution set is at the point (120, 180). The largest number of discounted tickets sold that will still enable the auditorium to meet its sales goal is 120.

Step 4: *Check.*

If 120 discounted tickets are sold, then $300 - 120 = 180$ regular-price tickets can be sold.

$$15(120) + 40(180) \stackrel{?}{=} 9{,}000$$

$$9{,}000 = 9{,}000$$

Watch Out!

In many real-world problems involving systems of inequalities, the solution set is only the set of points within the region labeled "S" *that have integer coordinates*, not the entire region labeled "S." (For example in Example 7.25, the auditorium isn't selling fractional tickets!) However, we can't draw them easily on the grid, so you'll have to keep this in mind as you interpret these graphs.

Exercises

1. The sum of two numbers is 28. Five more than three times the first number is equal to 47 less than the second. Find the two numbers.
2. Hyun has a bag containing $5.40 in nickels and quarters. If the bag contains 40 coins, how many nickels and how many quarters are in the bag?
3. A bus takes 10 hours to travel from a small town to Chicago and back again to the town along the same route. The bus's average speed going from Chicago was 72 miles per hour, and its average speed returning was 48 miles per hour. How long did the bus spend traveling in each direction?
4. At a bakery, Alicia paid $50.50 for 4 brownies and 5 cookies. Belinda bought 9 brownies and 8 cookies at the same bakery for $92.50. Determine the price of one brownie and one cookie.
5. Mia pays $25 to buy 4 pounds of apples and 4 pounds of pears from Yourway Supermarket. Nia pays $28.50 to buy 2 pounds of apples and 6 pounds of pears from the same supermarket. How much does one pound of apples cost at Yourway? How much does one pound of pears cost?

6. A hardware store sells both halogen bulbs and compact fluorescent light (CFL) bulbs. A shipment of 8 halogen bulbs and 12 CFL bulbs from the store costs \$74. A shipment of 5 halogen bulbs and 7 CFL bulbs from the same store costs \$45. What is the cost of each type of bulb?

7. In a large city, the DaisyTrip ride-sharing company charges \$0.60 per mile, with an initial fee of \$3.50 per ride. The GoRose ride-sharing company charges \$0.40 per mile, with an initial fee of \$5 per ride.
   (a) Write a function for D, which represents the cost, in dollars, of a DaisyTrip ride after x miles.
   (b) Write a function for G, which represents the cost, in dollars, of a GoRose ride after x miles.
   (c) Graph D and G on the coordinate plane.
   (d) Determine the length, in miles, of the trip that would cost the same using either GoRose or DaisyTrip.

8. To run his food truck, Amir spends \$450 in food and other fixed costs every day. He also spends \$20 per hour to pay his workers. He sells \$120 of food every hour.
   (a) Write a function E that models Amir's total daily expenses, in dollars, after t hours.
   (b) Write a function S that models Amir's total daily sales, in dollars, after t hours.
   (c) Graph E and S on the coordinate plane.
   (d) What is the minimum number of hours that the food truck needs to be open for Amir to cover his daily expenses?

9. Maria spends \$1,800 to edit and design a book with a self-publishing company. To print the book, the company charges \$1.75 per copy. The book will sell for \$8.
   (a) Write a function C that models her publishing costs, in dollars, for x books.
   (b) Write a function S that models her book sales, in dollars, for x books.
   (c) What is the minimum number of copies that Maria needs to sell to meet her publishing expenses?

10. The sum of two numbers x and y is more than 6. If three times x is subtracted from y, the difference is less than or equal to 4. Graphically determine the set of possible values for x and y.

11. A high school club plans to sell no more than 40 raffle tickets at the school carnival. Each raffle ticket comes with a slice of pizza or a burger. The club wants to make

at least \$100 in food sales. It will charge \$2 for a raffle ticket with a slice and \$5 for a ticket with a burger.
(a) Write a system of inequalities that can be used to represent this situation.

(b) Graph the system of inequalities that you wrote in part a on the coordinate plane.

(c) Determine if a combination of 35 tickets with pizza and 5 tickets with burgers will enable the club to meet its fundraising goal. Explain your answer.

12. Gina earns \$7 per hour working at a fast-food restaurant and \$14 per hour working at a clothing store. She can work no more than 50 hours per week. She needs to earn at least \$450 a week to meet her expenses.
(a) Write a system of inequalities that can be used to represent this situation.

(b) Graph the system of inequalities that you wrote in part a on the coordinate plane.

(c) Determine if Gina will meet her expenses if she works 20 hours at the fast-food restaurant and 25 hours at the clothing store. Explain your answer.

CHAPTER 7 TEST

1. What are the slope and y-intercept of the equation $y = -3x + 5$?

   (A) slope = −3, y-intercept = (0, 5)
   (B) slope = 3, y-intercept = (0, −5)
   (C) slope = 5, y-intercept = (0, −3)
   (D) slope = −5, y-intercept = (0, 3)

2. What is the slope of the line that passes through (8, 4) and (−6, 2)?

   (A) 7
   (B) 1
   (C) $\frac{1}{7}$
   (D) 6

3. What is an equation of the line that has a slope of 5 and passes through the point (7, −1)?

   (A) $y + 7 = 5(x - 1)$
   (B) $y - 7 = 5(x + 1)$
   (C) $y - 1 = 5(x + 7)$
   (D) $y + 1 = 5(x - 7)$

4. Which point is a solution to the equation $5 - y = 3x$?

   (A) (3, 4)
   (B) $\left(3, \frac{2}{3}\right)$
   (C) (3, −4)
   (D) $\left(3, \frac{3}{2}\right)$

5. What is the solution to the system of equations $y = 7x - 15$ and $y = 2x$?

   (A) (4, 8) (B) (8, 4) (C) (4, 13) (D) (3, 6)

6. Which graph represents the solution set to the system $x > 3$ and $y \leq 6$?

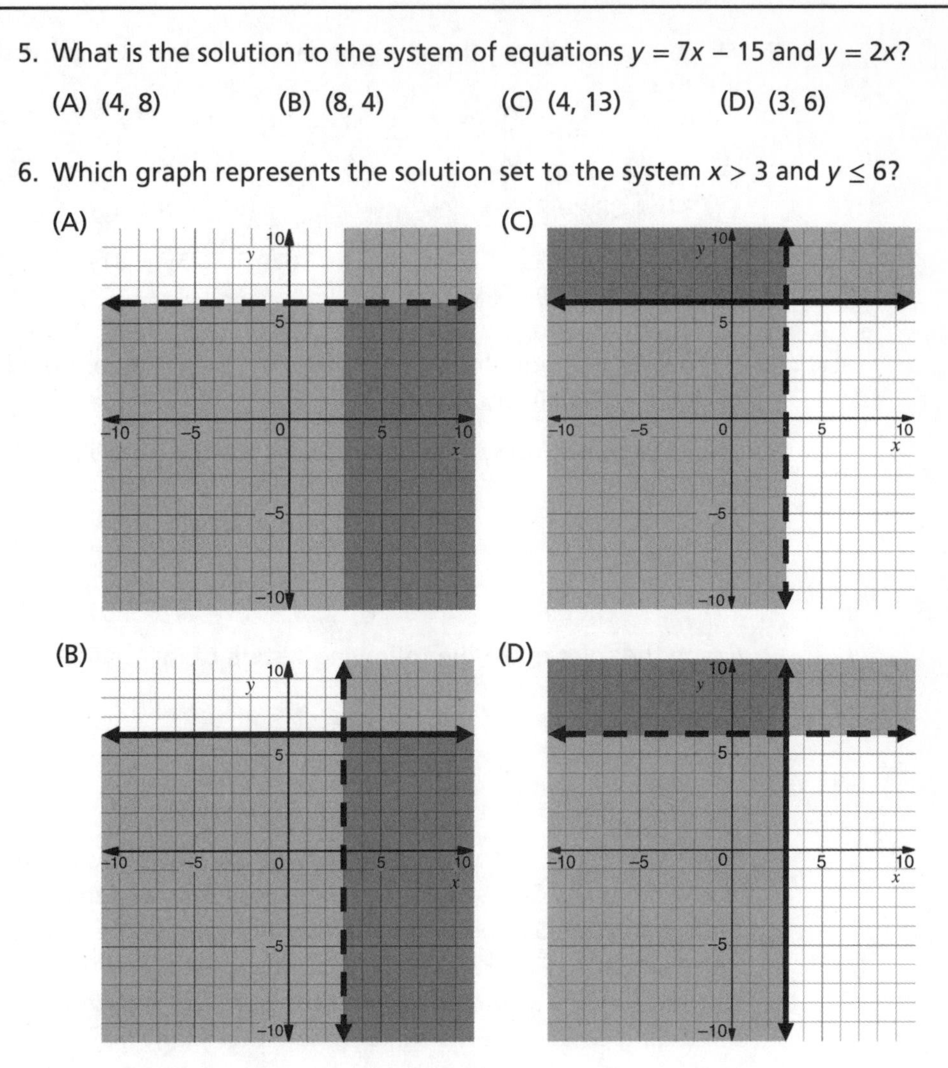

7. Which sentence best describes a line whose slope is 0?

   (A) As x increases, y increases, then decreases.

   (B) As x increases, y decreases, then increases.

   (C) As x increases, y is unchanged.

   (D) x remains unchanged.

8. Which line *cannot* be written in point-slope form?

   (A) $x = 4$ (B) $y = -4$ (C) $2x = y + 4$ (D) $y = 0$

9. Which system of equations has the same solution as the one shown below?

$$4x - 6y = 80$$
$$17x + 14y = 3$$

(A) $4x - 6y = 80$
$34x + 28y = 3$

(B) $2x - 3y = 40$
$51x + 42y = 9$

(C) $21x + 8y = 83$
$17x + 14y = 3$

(D) $5x - 5y = 81$
$7x + 4y = -7$

10. If the system of equations $y = ax + 8$ and $y = 3x - 4$ has no solution, find the value of a. Explain your work.

11. Solve the following system of equations for the variables:

$$x = 3y$$
$$2x - 7y = 4$$

12. Graph the solution to the following system of inequalities:

$$3x + 5y > 10$$
$$y \leq x - 2$$

13. Marcus needs a text messaging service for his music business. His mobile phone company offers two plans—the Basic Plan and the Essentials Plan. The Basic Plan costs \$20 per month and charges \$0.04 per message. The Essentials Plan costs \$30 per month and charges \$0.02 per message. How many text messages would Marcus need to send each month for his monthly text messaging costs to be the same using either plan?

14. Idi and Emma buy children's gifts from an online toy store. Idi bought 12 bells and 8 whistles for \$120. Emma bought 10 bells and 6 whistles for \$99. Determine the cost of one bell and the cost of one whistle.

15. An office supply factory wants to make small and large wooden file cabinets using 900 pounds of wood. The factory uses 65 pounds of wood to make each large file cabinet and 20 pounds of wood to make each small file cabinet. It has the capacity to make no more than 30 cabinets at a time. The factory wants to determine the number of cabinets of each size that it can make.

(a) Write a system of inequalities that models this situation.

(b) Graph the solution set to this system.

REVIEW TEST 1: CHAPTERS 1–7

1. Which number is equivalent to $\frac{3}{8}$?

   (A) $\frac{4}{9}$ (B) $\frac{2}{7}$ (C) $\frac{6}{16}$ (D) $-\frac{3}{8}$

2. 20 is 40% of what number?

   (A) 50 (B) 8 (C) 800 (D) 0.5

3. Over what interval is $f(x) = (x + 2)(x - 3)(x - 5)$ decreasing?

   (A) $\{x \mid -1 < x < 0\}$ (C) $\{x \mid -5 < x < -4\}$

   (B) $\{x \mid 4 < x < 5\}$ (D) $\{x \mid 1 < x < 2\}$

4. What is the solution set to the inequality $-2x + 5 > 11$?

   (A) $(-3, \infty)$ (B) $(-\infty, -3)$ (C) $(3, \infty)$ (D) $(-\infty, 3)$

5. Which point is a solution to the system of equations listed below?

$$y = 15x + 6$$

$$3x + y = 42$$

   (A) (1, 21) (B) (2, 35) (C) (2, 36) (D) (3, 33)

6. The function f is defined by the table below.

x	0	1	2	3	4
$f(x)$	−1	2	4	6	7

   What is $3f(2)$?

   (A) 3 (B) 6 (C) 12 (D) 34

7. Which equation is read as "2 added to one-third the quantity of x minus 3 equals 12"?

   (A) $2 + \frac{1}{3}x - 3 = 12$ (C) $\left(2 + \frac{1}{3}\right)(x - 3) = 12$

   (B) $2 + \frac{1}{3}(x - 3) = 12$ (D) $\left(2 + \frac{1}{3}x\right)(-3) = 12$

8. If $a = 2(hd + dw + hw)$, then what is h?

(A) $h = \dfrac{a}{2(d + w + dw)}$

(B) $h = \dfrac{a - 2dw}{2d + 2w}$

(C) $h = \dfrac{a}{2}(d + w + dw)$

(D) $h = \dfrac{a}{2}(dw) - d - h$

9. The foxglove tree can grow as much as 40 centimeters per month. What is this growth rate in inches per year? (2.54 centimeters = 1 inch)

(A) 8.47 (B) 15.75 (C) 101.6 (D) 188.98

10. Solve for the variable: $3(x + 1) = 4(6 - x)$.

11. Write the pronunciation of $\{x \mid x \in \mathbf{Z}\}$.

12. Is the relation {(1, 6), (2, 7), (3, –5), (4, –5)} a function? Explain.

13. The function $C(x) = 20 + 80x$ represents the total cost, in dollars, of attending a dance class after x months. State and interpret the slope and y-intercept of the function in context.

14. Solve the system of inequalities below. Mark the solution set with an S.

$$y > \frac{1}{2}x + 4$$

$$2x + y \leq 3$$

15. Michelle wants to buy a laptop that costs \$1,500. She has already saved \$200. She currently earns \$320 per week from her part-time job and plans to save 20% of her earnings to buy the laptop.

(a) How much money, in dollars, does she plan to save every week?

(b) Write an equation for E, the function that represents the total amount of money she saves x weeks after starting her job.

(c) After how many weeks will she have enough money to buy the laptop?

CHAPTER 7 SOLUTIONS

7.1. 1. $m = \frac{1}{4}$. As x increases by 4 units, y increases by 1 unit.

2. $m = -2$. As x increases by 2 units, y decreases by 1 unit.

3. $m = -\frac{1}{3}$. As x increases by 1 unit, y decreases by 3 units.

4. $m = 4$. As x increases by 1 unit, y increases by 4 units.

5. $m = 0$. As x increases by 1 unit, y does not change.

6. $m = 0$. As x increases by 1 unit, y does not change.

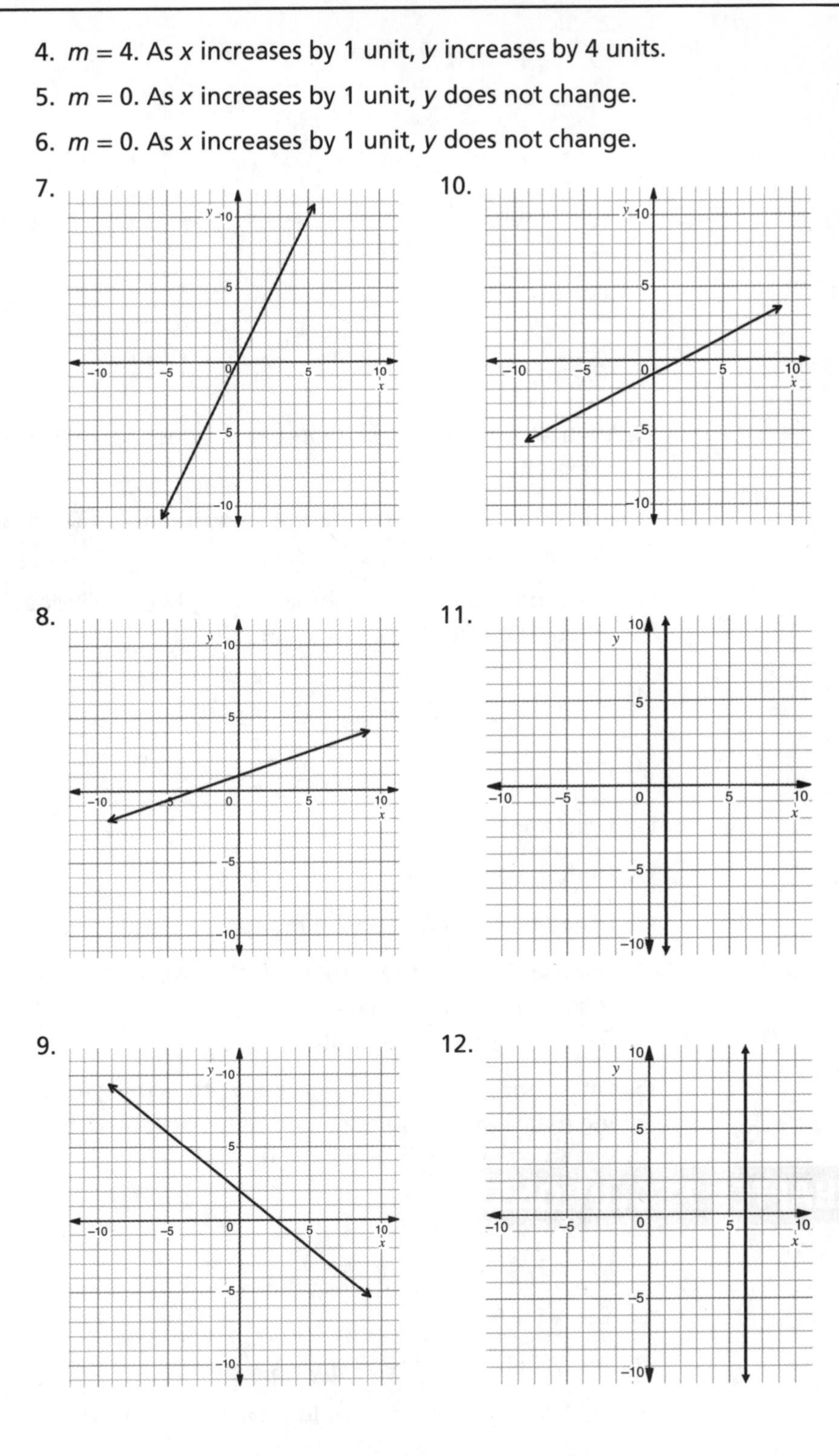

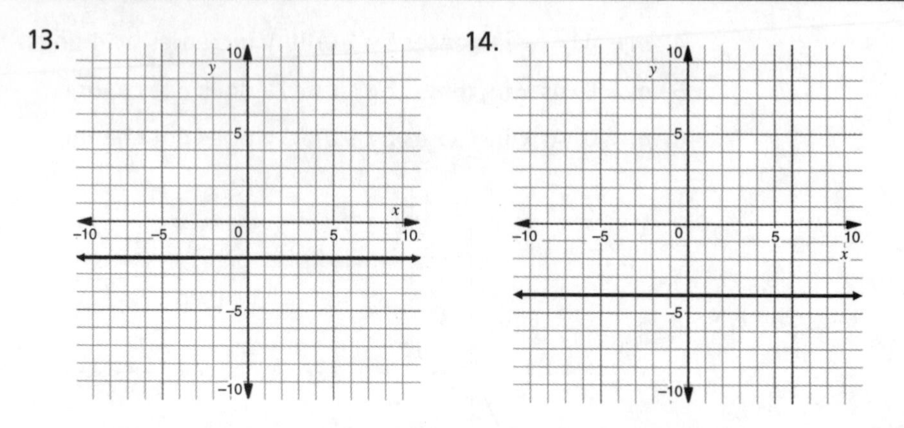

15. Vertical lines have an undefined slope, so they can't be written in slope-intercept form.

16. All points on the *x*-axis have a *y*-coordinate of 0. Thus, the equation of the *x*-axis is $y = 0$.

17. The term "no slope" can be confusing because it doesn't clearly indicate whether the slope is zero or undefined.

7.2. 1. $\frac{2}{3}$ 3. $\frac{4}{3}$ 5. $-\frac{5}{9}$ 7. -1 9. 0

2. $-\frac{1}{6}$ 4. $\frac{6}{5}$ 6. $-\frac{1}{2}$ 8. $-\frac{3}{5}$ 10. 0

11. Undefined

12. Undefined

13. "The quotient of *y*-sub-2 minus *y*-sub-1 and *x*-sub-2 minus *x*-sub-1."

14. Marissa flipped the order of the *x*-coordinates when subtracting. If $(x_1, y_1) = (3, 17)$ and $(x_2, y_2) = (-1, 4)$, then she should have subtracted $3 - (-1)$ in the denominator.

15. On a horizontal line, all points have the same *y*-coordinates, so subtracting them would result in a difference of 0. Thus, the calculation for slope would be $m = \frac{y_2 - y_1}{x_2 - x_1} = \frac{0}{x_2 - x_1} = 0$.

7.3. 1. Yes. $2(3) = 6$.

2. No. $7(2) + 3 = 17$, not 11.

3. Yes. $-1 + 5 = 4$.

4. Yes. When $x = 2$, the *x*-value is always 2.

5. No. When $y = 11$, the *y*-value is always 11, never 12.

6. Yes. $2(0) + 12 = 12$.
7. Yes. $15(-17) + 230 = -25$.
8. No. $3(40) - 4(20) = 40$, not 30.
9. No. $\frac{5}{2}(6) - 1 = 14$, not 15.
10. The graph represents all ordered pairs that are solutions to the equation.
11. All of the solutions to the equation are points on the graph.
12. Determine if the ordered pair is on the graph of the equation, or substitute the coordinates of the ordered pair into the equation to see if the resulting mathematical statement is true

7.4.

1. $y = -5x + 3$
2. $y = \frac{2}{7}x + 4$
3. $y = \frac{8}{3}x + \frac{1}{2}$
4. $y = 7$
5. $x = 2$
6. $x = -18$
7. $y = -2x + 7$ or $y - 11 = -2(x + 2)$
8. $y = \frac{4}{5}x - 2$ or $y - 6 = \frac{4}{5}(x - 10)$
9. $y = \frac{1}{8}x + 1$or $y - 2 = \frac{1}{8}(x - 8)$
10. $y = 5x - 16$ or $y + 1 = 5(x - 3)$
11. $y = \frac{3}{7}x + \frac{37}{7}$ or $y - 7 = \frac{3}{7}(x - 4)$
12. $y = 8$
13. $y = x + 6$, $y - 7 = x - 1$, or $y - 11 = x - 5$
14. $y = -x + 5$, $y - 7 = -(x + 2)$, or $y + 3 = -(x - 8)$
15. $y = \frac{1}{2}x + 9$, $y - 12 = \frac{1}{2}(x - 6)$, or $y - 8 = \frac{1}{2}(x + 2)$
16. $y = \frac{11}{10}x + 5$, $y - 5 = \frac{11}{10}x$, or $y - 16 = \frac{11}{10}(x - 10)$
17. $x = 18$
18. $y = -11$
19. $y = \frac{1}{5}x + 3$
20. $y = -3x + 2$
21. $y = -\frac{1}{2}x + 6$
22. $y = \frac{2}{5}x + 4$
23. $x = 7$
24. $y = -9$

25. No. Given only the slope, the line has an infinite number of possible y-intercepts.
26. We rearrange the second equation into slope-intercept form: $y + 13 = 3x + 18$, so $y = 3x + 5$. The equations represent the same line.
27. No. The equations of vertical lines do not have y-intercepts.

7.5. 1. 2.

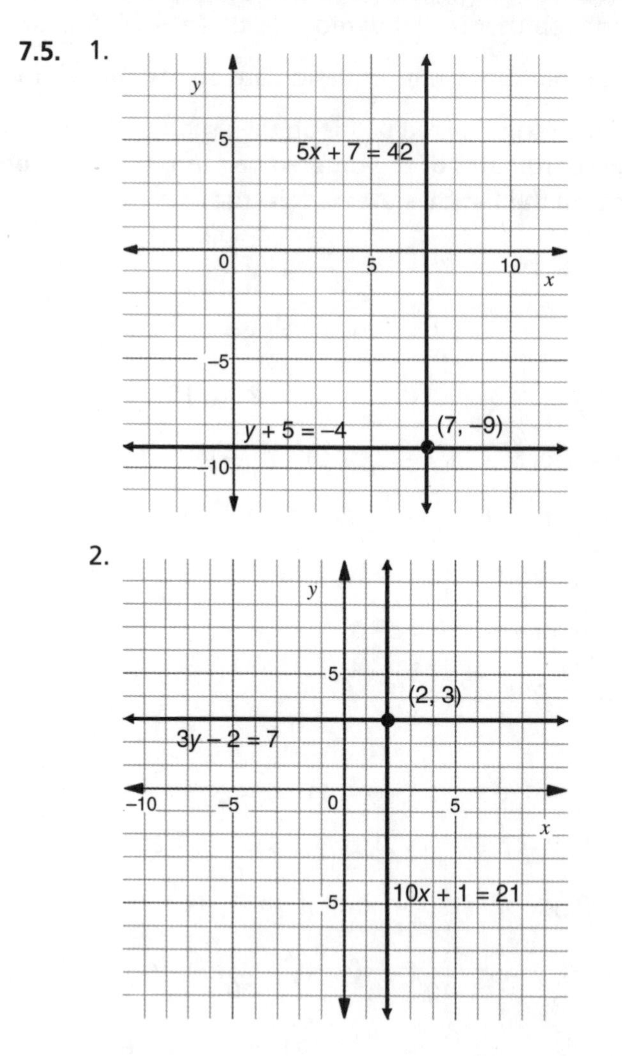

3.

4.

5.

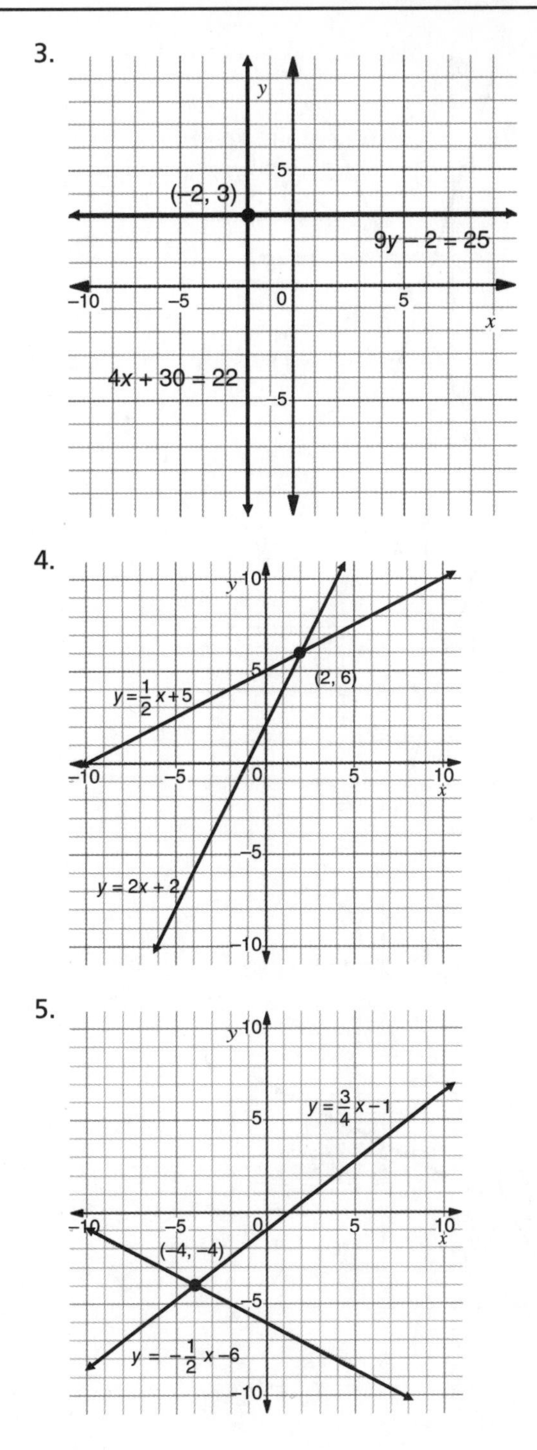
y
5
(−2, 3)
9y − 2 = 25
−10
−5
0
5
x
4x + 30 = 22
−5
y 10
y = 1/2 x + 5
(2, 6)
5
−10
−5
0
5
10
x
−5
y = 2x + 2
−10
y 10
y = 3/4 x − 1
5
−10
−5
0
5
10
x
(−4, −4)
−5
y = −1/2 x − 6
−10

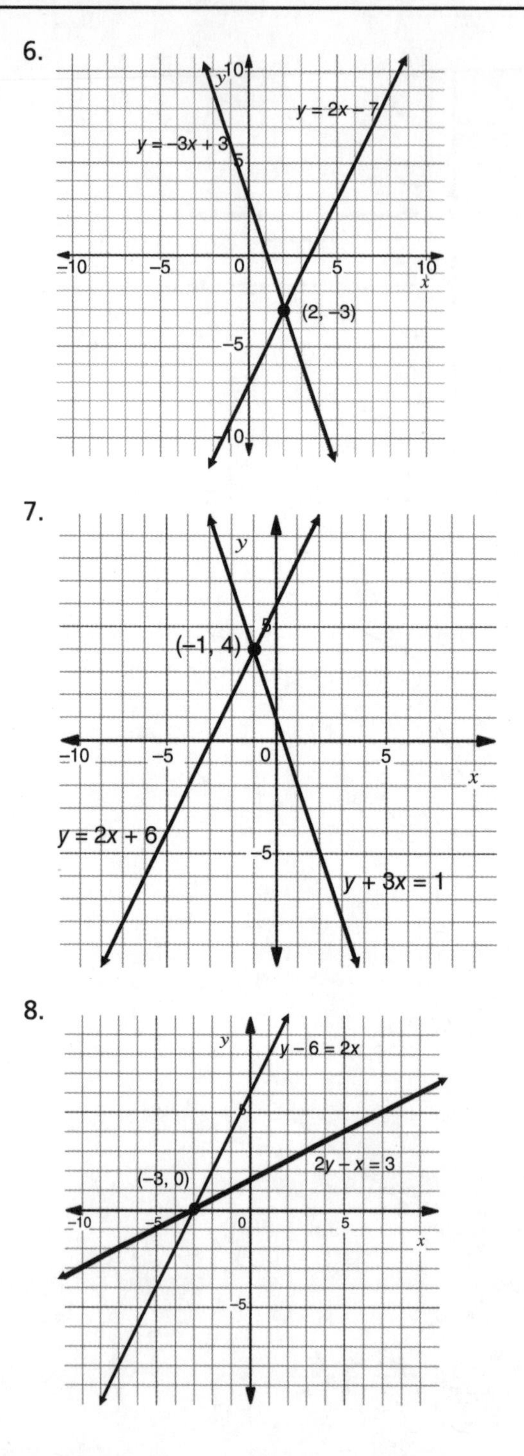

6.
y = −3x + 3
y = 2x − 7
(2, −3)
7.
(−1, 4)
y = 2x + 6
y + 3x = 1
8.
y − 6 = 2x
2y − x = 3
(−3, 0)

9.

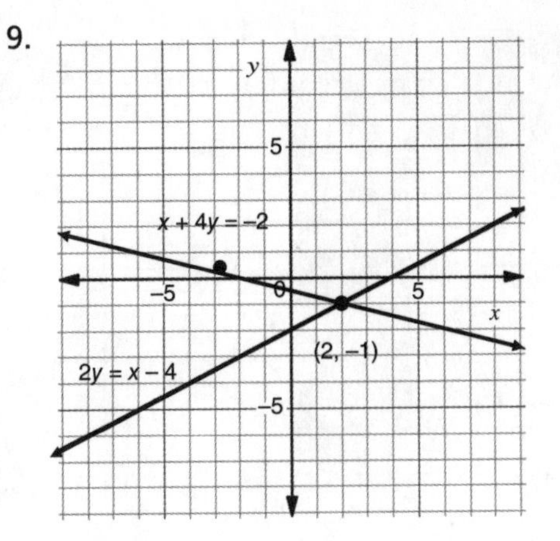

10. A system of equations is a collection of two or more equations that have the same variables.
11. The graphs do not intersect.
12. Her solution should be written as the ordered pair (1, 7). She should not have used curly brackets, which indicate an unordered set of numbers.

7.6.
1. $c = 3$ and $d = 12$
2. $p = -1$ and $q = -1$
3. No solution.
4. $x = 4$ and $y = 3$
5. $x = 16$ and $y = 6$
6. Infinitely many solutions
7. $t = 8$ and $r = 31$
8. $b = 2$ and $a = 4$
9. $p = 6$ and $q = 9$

7.7.
1. $x = 5$ and $y = -6$
2. $p = 3$ and $k = -2$
3. $a = 5$ and $t = 7$
4. $x = 2$ and $y = 8$
5. No solution
6. $r = 10$ and $s = 11$
7. $q = -1$ and $p = -2$
8. $d = 2$ and $k = 6$
9. $w = -3$ and $z = 3$
10. Elimination—the equations are not solved for a variable, and an exact answer is needed.
11. Graphing—we don't need the exact solution, only the number of solutions.
12. Substitution—one of the equations is solved for a variable.

7.8.

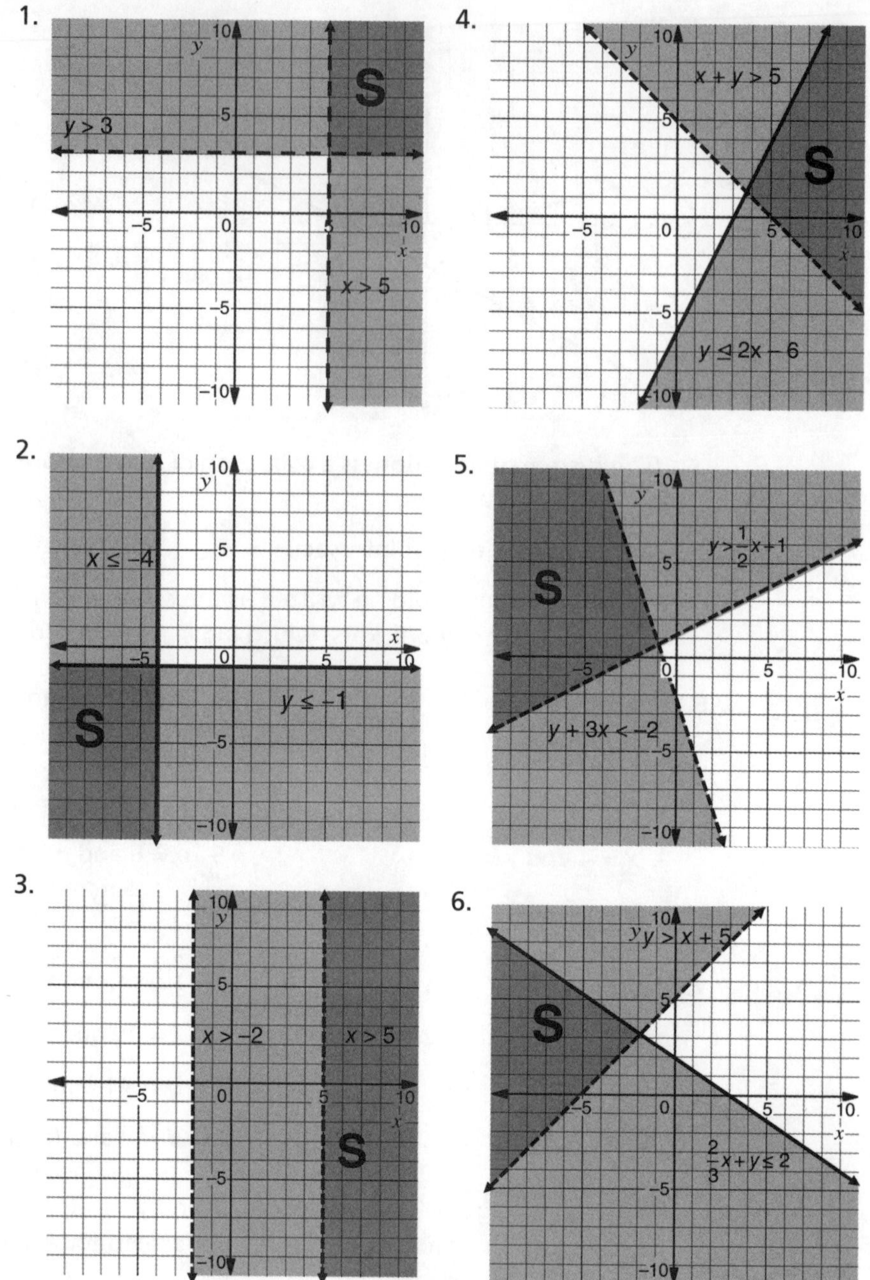

7. 8. 9.

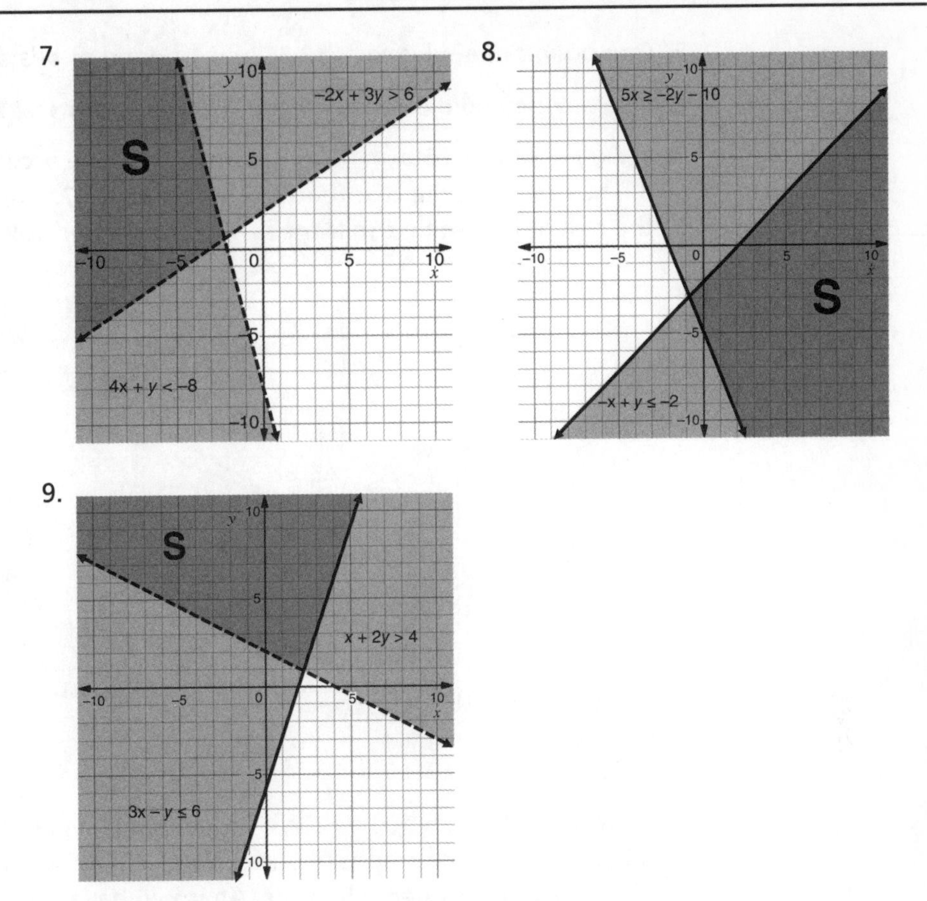

10. Yes, it can have a solution if the regions representing the solutions overlap somewhere on the coordinate plane. For example, the graphs of $x = 3$ and $x = 4$ are vertical lines that are parallel, so they do not overlap. However, the solution set to the system $x > 3$ and $x > 4$ is a region of points whose x-coordinates are greater than 4.

11. Answers may vary. For example, if $x = 0$, then $3(0) + 2 = 2$, so we need a point whose x-coordinate is 0 and y-coordinate is greater than 2. The point (0, 3) is in the solution set.

12. Substituting the coordinates $x = 2$ and $y = -3$ into the inequality gives us $-3 < -5$, which is not true. Therefore, the point is not in the solution set.

7.9.

1. −6 and 34
2. 23 nickels and 17 quarters.
3. The bus spent 4 hours driving from Chicago and 6 hours traveling to Chicago.
4. A brownie costs \$4.50 and a cookie costs \$6.50.

5. One pound of apples costs \$2.25 and 1 pound of pears costs \$4.
6. One halogen bulb costs \$5.50 and 1 CFL bulb costs \$2.50.
7. a. $D(x) = 3.50 + 0.60x$, b. $G(x) = 5 + 0.40x$, c. See accompanying graph, d. The graphs of the two functions intersect at (7.5, 8). This means that the cost of a 7.5-mile trip is the same using either GoRose or DaisyTrip.

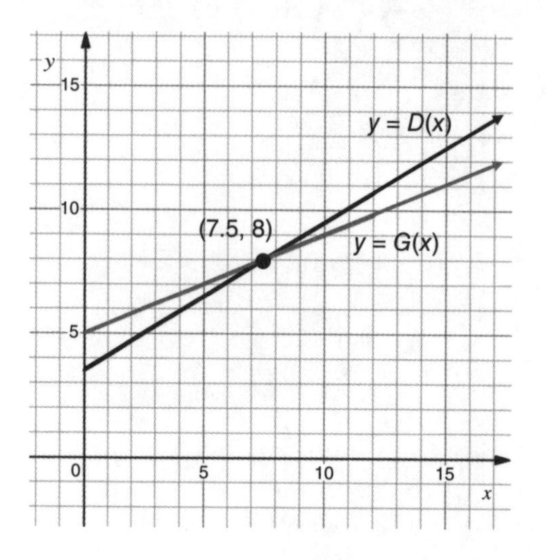

8. a. $E(t) = 450 + 20t$, b. $S(t) = 120t$, c. See accompanying graph, d. The intersection point of the graphs of $E(t)$ and $S(t)$ is (4.5, 540), which means that after the truck is open 4.5 hours, Amir will have made enough money to cover his daily expenses.

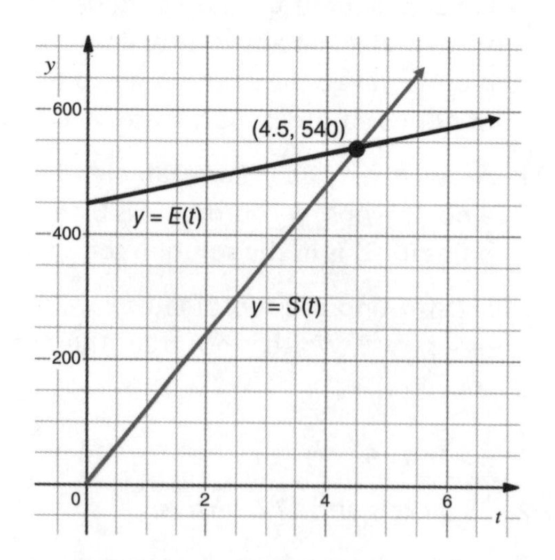

9. a. $C(x) = 1{,}800 + 1.75x$, b. $S(x) = 8x$, c. Her sales will equal her expenses when $C(x) = S(x)$, or when $1{,}800 + 1.75x = 8x$. Solving for x gives us $1{,}800 = 6.25x$, or when $x = 288$. She needs to sell a minimum of 288 copies.

10. 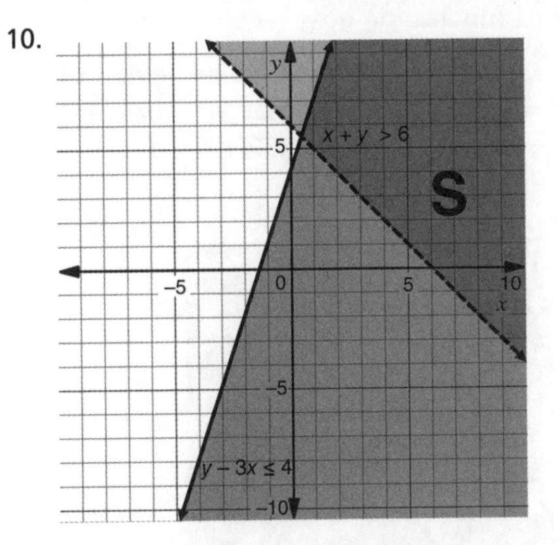

11. a. If p = number of tickets with pizza sold and b = number of tickets with burgers sold, then $p + b \leq 40$ and $2p + 5b \geq 100$. In addition, since only 40 raffle tickets are being sold, then $0 \leq p \leq 40$ and $0 \leq b \leq 40$, which means we are limited to the rectangle whose vertices are (0, 0), (0, 40), (40, 0), and (40, 40).

b. 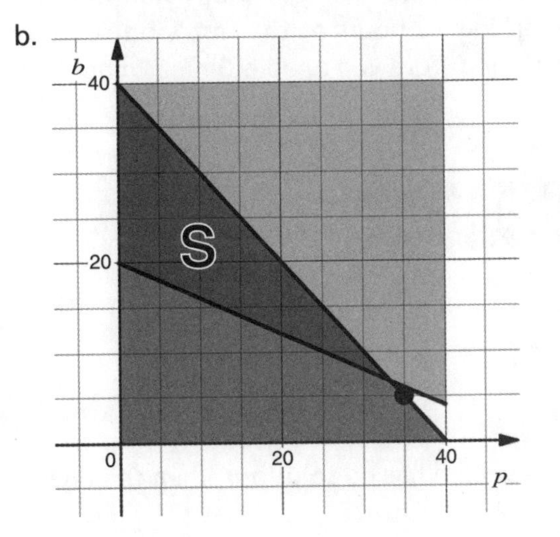

c. As shown in the graph from part *b*, the point (35, 5) is not in the region where the solution set is located, so a combination of 35 tickets with pizza and 5 tickets with burgers will not enable the club to meet its fundraising goal.

12. a. Let f = number of hours that Gina works at the fast-food restaurant and c = number of hours that she works at the clothing store. Then $f + c \leq 50$ and $7f + 14c \geq 450$. Also, $0 \leq f \leq 50$ and $0 \leq c \leq 50$.

b. 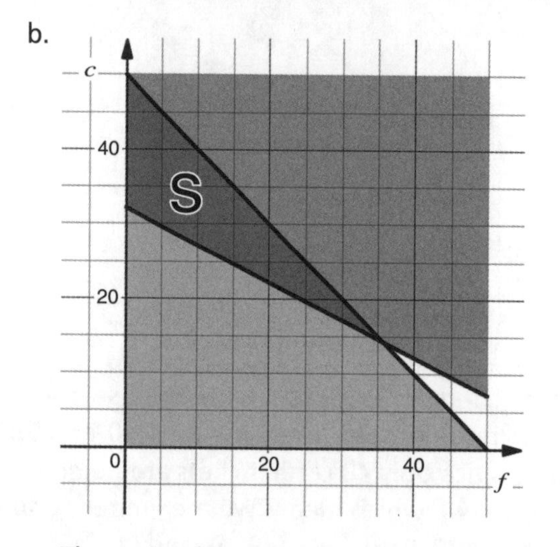

c. The point (20, 25) is in the region defined by the solution set, so working 20 hours at the fast-food restaurant and 25 hours in the clothing store will enable Gina to meet her expenses.

CHAPTER 7 TEST SOLUTIONS

1. (A)
2. (C)
3. (D)
4. (C)
5. (D)
6. (B)
7. (C)
8. (A)
9. (B)
10. $a = 3$. If the system has no solutions, then the lines have the same slope.
11. (–12, –4)

12.

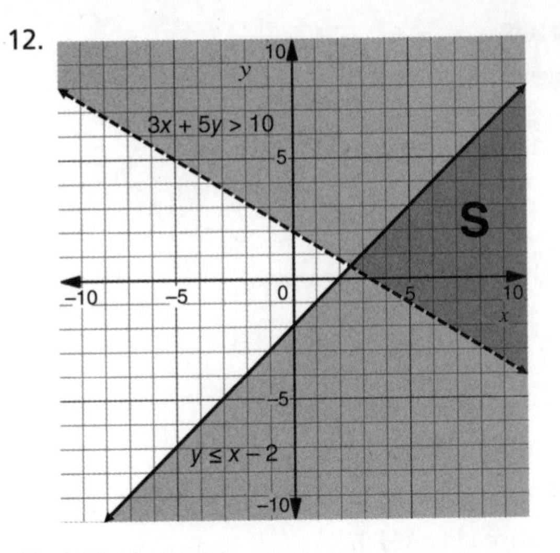

13. 500 messages.

14. One bell costs \$9 and one whistle costs \$1.50.

(a) Let x = number of large file cabinets that can be made and y = number of small file cabinets that can be made. Then $65x + 20y \leq 900$ and $x + y \leq 30$. Since $0 \leq x \leq 30$ and $0 \leq y \leq 30$, then we are limited to the rectangle whose corners are (0, 0), (0, 30), (30, 0) , and (30, 30).

(b)

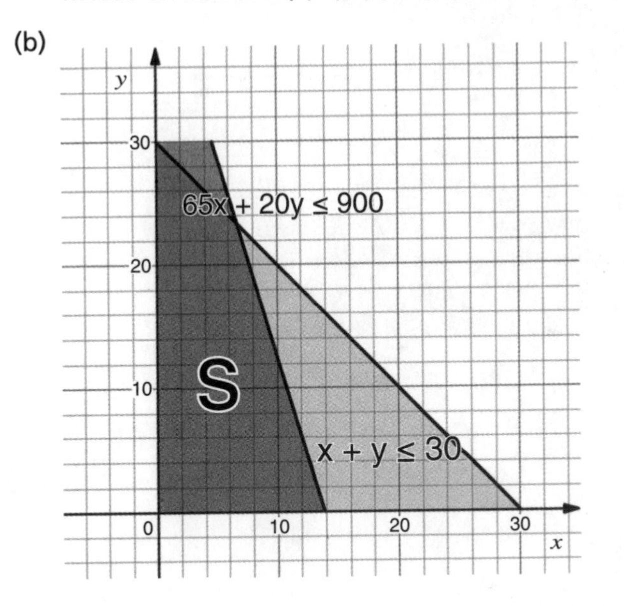

REVIEW TEST 1 SOLUTIONS

1. C
2. A
3. D
4. B
5. C
6. C
7. B
8. B
9. D
10. $x = 3$
11. "the set of all xs such that x is an integer"
12. Yes, since every input gets exactly one output.
13. slope: monthly cost of attending the dance class ($80), y-intercept: initial fee for attending the dance class ($20).
14. 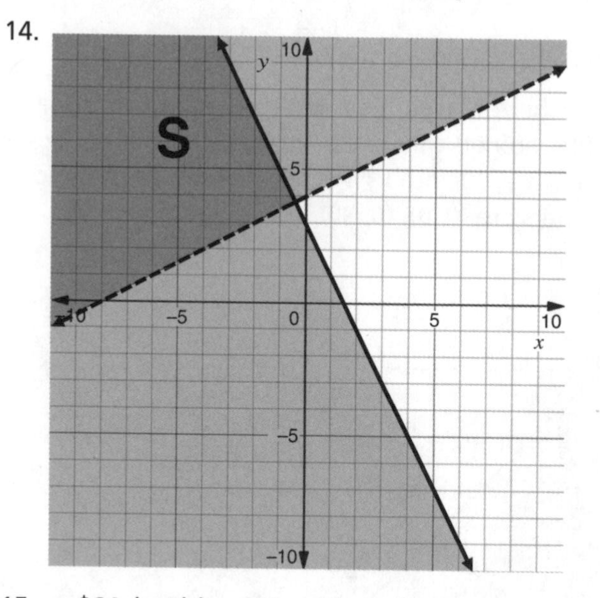

15. a. $64, b. $E(x) = 200 + 64x$, c. 21

8 OPERATIONS WITH POLYNOMIALS

So far, we've focused on linear expressions. In this chapter, we'll discuss operations (addition, subtraction, multiplication, and division) with non-linear expressions, whose variables have exponents that are greater than 1.

8.1 Adding and Subtracting Polynomials

Let's start by looking at an addition problem: 204 + 650. What's wrong with the following work?

$$\begin{array}{r} 24 \\ +65 \\ \hline 89 \end{array}$$

It ignores the importance of 0 as a placeholder for the tens in 204 and for the zeros in 650. If we rewrite 204 and 650 in expanded form (in other words, as powers of 10), we get the following:

$$\begin{array}{r} 2(10^2) + 0(10^1) + 4(10^0) \\ +6(10^2) + 5(10^1) + 0(10^0) \\ \hline 8(10^2) + 5(10^1) + 4(10^0) \end{array}$$

Writing the numbers in this way shows that we need to "line up" the digits with the same place value. Adding $4(10^0)$ and $5(10^1)$ to get $9(10^0)$ is incorrect. The number system that we use, which was developed in India and brought to Europe through the work of Middle Eastern mathematicians such as al-Khwārizmī and al-Kindi, is based on powers of 10. Since the bases in all of the powers here are 10, we call this a base-10 system. But it would apply just as well if we used a different base. (In fact, we use different bases all the time. Our division of hours into minutes and seconds is a product of the base-60 number system used by Mesopotamians thousands of years ago!)

If we replaced the numerical base with a variable like x, then we see that to add these expressions, we need to "line up" the like terms and add up their coefficients.

$$\begin{array}{r} 2x^2 + 0x + 4 \\ +6x^2 + 5x \\ \hline 8x^2 + 5x + 4 \end{array}$$

When talking about expressions with variables raised to a power greater than 1, we need some vocabulary:

- A **polynomial** is the sum or difference of one or more terms. Examples of polynomials include a **monomial** (which has *one* term), a **binomial** (which has *two* terms), and a **trinomial** (which has *three* terms).
- The **degree of a term** is the sum of the exponents of the variables in the term. The **degree of a polynomial** is the largest of all of these sums. Most of the time, you'll only encounter one variable, so in those cases you can think of the degree of a polynomial as the highest exponent of the variable. For example
  - $x^5 + 2x^2 + 4x$ has degree 5 (The polynomial has only one variable, whose highest exponent is 5.)
  - $x^2 + xyz + z^2$ has degree 3 (The terms have degrees 2, $1 + 1 + 1 = 3$, and 2, respectively. The highest of these degrees is 3.)
- We typically write polynomials in **standard form**, which means that terms decrease in degree from left to right and all like terms are combined.

Table 8.1 shows examples and non-examples of polynomials.

Table 8.1 Polynomials.

Example	Reason	Non-Example	Reason
$5a^2 - 3a - 7$	This polynomial is a trinomial—the sum of three terms.	$\frac{q-5}{3q}$	This expression cannot be rewritten as a product of numbers and variables. There is a variable in the denominator.
6	A polynomial may consist of only one term—in this case, the number 6.	$n^2 + 7n + 3 + \frac{1}{n+6}$	There is a variable in the denominator of $\frac{1}{n+6}$.
$\frac{x}{7}$	This can be rewritten as $\frac{1}{7}x$, which is a monomial—the product of the number $\frac{1}{7}$ and the variable x.	$\frac{7}{x}$	There is a variable in the denominator.

Watch Out!

As said in Chapter 1, we read x^2 as "x-squared," "x to the second power," or "x to the second," not "x-two." Similarly, we read x^3 as "x-cubed," "x to the third power," or "x to the third," not "x-three."

Example 8.1 Express $1+9x^3-x+34x^2-7x^3+5x+3x^3+1+52x-4$ in standard form, and state the degree of the polynomial.

Solution: First, we group the like terms and arrange them so that the exponents of their variables are in descending order. Then we add the like terms to write the polynomial in standard form:

$$\begin{array}{llll} +9x^3 & +34x^2 & -x & +1 \\ -7x^3 & & +5x & -4 \\ \underline{+3x^3} & & \underline{+52x} & \underline{+1} \\ 5x^3 & +34x^2 & +56x & -2 \end{array}$$

In standard form, the polynomial is $5x^3 + 34x^2 + 56x - 2$. Since the highest power of the exponent is 3, this is a third-degree polynomial.

Example 8.2 Express $5xy + 3x^2y - 2 - 6x^2y - 7xy^2 + 8 + 10xy + 4xy^2$ in standard form and state its degree.

Solution: Arrange the terms in order of descending powers of one of the variables (in this case, y). Then add the like terms.

$$\begin{array}{l} -7xy^2 - 6x^2y + 10xy - 2 \\ \underline{+4xy^2 + 3x^2y + 5xy \ \ +8} \\ -3xy^2 - 3x^2y + 15xy + 6 \end{array}$$

The degree of the polynomial is the largest sum of all exponents of x and y. In the term $-3xy^2$, the exponent of x is 1 and the exponent of y^2 is 2, so the degree is $1 + 2 = 3$. Note that we could have written this in standard form by descending powers of x: $-3x^2y - 3xy^2 + 15xy + 6$.

Example 8.3 Express $(2x^3 + 4x^2 - 6) - (3x^3 + 2x - 2)$ in standard form.

Solution: Arrange the like terms vertically and subtract the coefficients. We add $+0x$ and $+0x^2$ as placeholders. Recall that subtracting two numbers is equivalent to adding the opposite of the second number.

$$\begin{array}{lcl} 2x^3 + 4x^2 - 6 & \rightarrow & 2x^3 + 4x^2 + 0x - 6 \\ \underline{-\ (3x^3 + 2x \ \ - 2)} & \rightarrow & \underline{+\ -3x^3 + 0x^2 - 2x + 2} \\ & & -x^3 + 4x^2 - 2x - 4 \end{array}$$

Our answer is $-x^3 + 4x^2 - 2x - 4$.

Exercises

Express each polynomial in standard form and state its degree.

1. $1 + 9x + 3x^2 - 4x^3$
2. $7 + 2y^4 - 8y + 3y^3 - 3y^2$
3. $9 + 2m^3 - 1 + 5m + 6m^2 - 7m^3$
4. $10b^3 - b^2 + 9b^5 + b^3 + 8b + b + 2$
5. $-3 + 4x^3 - 2xy + 5x^2$
6. $4mn^3 + m^4 + 6m^2n^2 + 4m^3n + n^4$

Express each sum or difference in standard form.

7. $(4x^2 - 3x + 5) + (2x^2 + 4x - 7)$
8. $(8y^2 + y - 6) + (9y^2 - 3y - 2)$
9. $(2a^2 - 10a - 3) + (-4a^2 + 5a + 11)$
10. $(3y^2 + 2y - 1) - (6y^2 - 5y + 4)$
11. $(5b^2 - 5b - 1) - (2b^2 + 9b - 2)$
12. $(-3z^2 + 2z - 4) - (-5z^2 - 6z + 7)$
13. $(4x^2 - 3 + 2x) + (5x - 2x^2)$
14. $(k^3 + 8 + 2k) + (3k - 1 - 4k^3)$
15. $5r^2 + 8r^3 + 7 + (7r^2 + 1 + 3r - 17r^3)$
16. $(9z + 6z^2 - 1) - (3 - z^2 - 4z)$
17. $(6n^4 - 7n^3 + 2n^2 - 9) - (n^3 - 4n^2)$
18. $(7x^3 - 3x + 1) - (x^3 + 4x^2 - 2)$

Questions to Think About

19. Explain how the method for adding and subtracting polynomials is similar to the method for adding and subtracting integers.
20. Is $\frac{14}{7}x - 12$ a polynomial? Explain.
21. Is the polynomial $14x^3 - 6x^3 + 2x^2 - 1$ in standard form? Explain.

8.2 Multiplying Monomials: Rules of Exponents

So far, we've talked about how we add and subtract polynomials. To multiply polynomials, we must first discuss how we multiply monomials.

As we said in Chapter 1, raising a number to an exponent means multiplying that number by itself several times. For example, $2^3 = 2(2)(2) = 8$. Let's see what happens when we multiply a number like 2^3 (which we call a power) by another number with the same base:

- $2^3 2^2 = (2)(2)(2) \cdot (2)(2) = (2)(2)(2)(2)(2) = 2^5$ (or $2^3 2^2 = 8 \cdot 4 = 32$)
- $5^1 5^3 = (5) \cdot (5)(5)(5) = (5)(5)(5)(5) = 5^4$ (or $5^1 5^3 = 5 \cdot 125 = 625$)
- $3^2 3^4 = (3)(3) \cdot (3)(3)(3)(3) = (3)(3)(3)(3)(3)(3) = 3^6$ (or $3^2 3^4 = 9 \cdot 81 = 729$)

These examples illustrate the following rule: **to multiply powers with the same base, keep the base the same and add the exponents.** We express this using variables as follows: $\boldsymbol{x^a(x^b) = x^{a+b}}$.

This rule only works if the bases are the same. For example, $2^3(5^1)$, which is $(2)(2)(2)(5) = 40$, is not equal to $(2{\cdot}5)^{3+1} = 10^4 = 10{,}000$ or $(2 + 5)^{3+1} = 7^4 = 2{,}401$.

Example 8.4 Express $(-4a^3)(5a^5)$ in standard form.

Solution: Expressing this product in standard form means that all like terms are combined.

$(-4a^3)(5a^5) = (-4 \cdot 5)(a^3 \cdot a^5)$	Commutative and associative properties of multiplication.
$= -20(a^3 \cdot a^5)$	Multiply the coefficients of the monomials.
$= -20a^8$	$x^a(x^b) = x^{a+b}$.

Watch Out!

When we multiply numbers and positive variables without a multiplication symbol, we write them with no spaces in between (such as $5x$, pronounced "five x"). If the variable is negative, we put parentheses around the variable (such as $5(-x)$, pronounced "five times negative x"). We never write $5{+}x$ or $5{-}x$ to mean multiplication since this looks like addition or subtraction.

We can also raise a power to a power. Recall from Section 1.1 that we use the word power to refer to both an exponent (such as the 4 in 2^4) or a number raised to a number (such as the entire expression 2^4). Here are some examples:

- $(2^3)^4 = (2^3)(2^3)(2^3)(2^3) = (2)(2)(2) \cdot (2)(2)(2) \cdot (2)(2)(2) \cdot (2)(2)(2) = 2^{12}$
- $(3^5)^2 = (3^5)(3^5) = (3)(3)(3)(3)(3) \cdot (3)(3)(3)(3)(3) = 3^{10}$
- $((-5)^2)^3 = (-5)^2(-5)^2(-5)^2 = (-5)(-5) \cdot (-5)(-5) \cdot (-5)(-5) = (-5)^6$

These examples illustrate the following rule: **to raise a power to a power, keep the base the same and multiply the exponents.** We express this using variables as follows: $(x^a)^b = x^{ab}$.

To raise a product to a power, we raise each factor to that power: $(x^a y^b)^c = x^{ac} y^{bc}$.

Example 8.5 Express $(-6a^2)^3$ in standard form.

Solution:

$(-6a^2)^3 = (-6a^2)(-6a^2)(-6a^2)$	Definition of exponents.
$= (-6)(-6)(-6) \cdot (a^2)(a^2)(a^2)$	Commutative and associative properties of multiplication.
$= (-6)^3 \cdot a^6$	Definition of exponents.
$= -216a^6$	Simplify.

Exercises

Write each expression in simplest form.

1. $(2m^4)(-6m^2)$
2. $(x^2)(x^3)(4x^5)$
3. $(8p^2)(10p^3)$
4. $(-2a)(7ab)(a^4b^2)$
5. $(3a^2b^3)(4a^3b^4)$
6. $(-5x^2y^3)(-2xy^2)$
7. $(d^2)^6$
8. $(3k^3)^4$
9. $(-4q^5)^3$
10. $(-2m^2)^3(3n^3)^4$
11. $(x^3)^4(x^2)$
12. $(3a^2b^3)^4$

8.3 Dividing Monomials: Rules of Exponents

When we divide powers with the same base, we see another pattern:

- $\frac{2^5}{2^3} = \frac{(2)(2)(2)(2)(2)}{(2)(2)(2)} = (2)(2) = 2^2 \left(\text{or } \frac{2^5}{2^3} = \frac{32}{8} = 4\right)$
- $\frac{3^4}{3^1} = \frac{(3)(3)(3)(3)}{(3)} = (3)(3)(3) = 3^3 \left(\text{or } \frac{3^4}{3^1} = \frac{81}{3} = 27\right)$
- $\frac{5^6}{5^2} = \frac{(5)(5)(5)(5)(5)(5)}{(5)(5)} = (5)(5)(5)(5) = 5^4 \left(\text{or } \frac{5^6}{5^2} = \frac{15{,}625}{25} = 625\right)$

These examples illustrate the following rule: **to divide powers with the same nonzero base, keep the base the same and subtract the exponents.** (The base cannot be 0 since we would then be dividing by 0, which is not allowed.) We express this using variables as follows: $\frac{x^a}{x^b} = x^{a-b}$, **where** $x \neq 0$.

Example 8.6 Express $\frac{36y^6}{-15y^2}$ in standard form.

Solution:

$\frac{36y^6}{-15y^2} = \frac{36}{-15} \cdot \frac{y^6}{y^2}$	$\left(\frac{a}{b}\right)\left(\frac{c}{d}\right) = \frac{ac}{bd}$, where $b \neq 0$ and $d \neq 0$.
$= -\frac{12}{5} \cdot \frac{y^6}{y^2}$	Simplify $\frac{36}{45}$.
$= -\frac{12}{5}y^4$	$\frac{x^a}{x^b} = x^{a-b}$, where $x \neq 0$.

If the base is a fraction, we see the following pattern, as shown by the following examples:

- $\left(\frac{1}{4}\right)^3 = \left(\frac{1}{4}\right)\left(\frac{1}{4}\right)\left(\frac{1}{4}\right) = \frac{1^3}{4^3}$
- $\left(\frac{2}{3}\right)^4 = \left(\frac{2}{3}\right)\left(\frac{2}{3}\right)\left(\frac{2}{3}\right)\left(\frac{2}{3}\right) = \frac{2^4}{3^4}$
- $\left(-\frac{1}{5}\right)^3 = \left(-\frac{1}{5}\right)\left(-\frac{1}{5}\right)\left(-\frac{1}{5}\right) = \frac{(-1)^3}{(5)^3}$

These examples illustrate the following rule: **to raise a fraction with a nonzero denominator to an exponent, raise the numerator and the denominator to that exponent.** In symbols, we write the following: $\left(\frac{x}{y}\right)^a = \frac{x^a}{y^a}$ **, where $y \neq 0$.**

If we arrange powers with the same base in descending order by their exponents, an interesting pattern emerges. For example, let's look at powers of 3:

$3^5 = 243$ $\quad 3^4 = 81$ $\quad 3^3 = 27$ $\quad 3^2 = 9$ $\quad 3^1 = 3$

We see that as the exponents decrease by 1, the result is divided by 3: $\frac{243}{3} = 81$, $\frac{81}{3} = 27$, $\frac{27}{3} = 9$, and $\frac{9}{3} = 3$. What should 3^0 be? Following this pattern, the next number should be $\frac{3}{3} = 1$. This illustrates the following rule: **any nonzero number raised to the 0 power equals 1.** (The base cannot be 0 since 0 raised to any positive exponent is 0, so the powers of 0 don't fit this pattern.) In symbols, we write: $\boldsymbol{x^0 = 1}$**, where $x \neq 0$**. There is nothing "natural" about this rule. Mathematicians have defined x^0 to be 1 (for $x \neq 0$) so that our other rules will still work.

What happens for negative exponents? If we divide 3^0 and subsequent powers by 3, we get the following:

$3^{-1} = 1 \div 3 = \frac{1}{3}$ $\quad 3^{-2} = \frac{1}{3} \div 3 = \frac{1}{3} \cdot \frac{1}{3} = \frac{1}{3^2}$ $\quad 3^{-3} = \frac{1}{9} \div 3 = \frac{1}{3^2} \cdot \frac{1}{3} = \frac{1}{3^3}$

These examples illustrate the following rule: **a number raised to a negative exponent equals the reciprocal of the number raised to the exponent's opposite.** In symbols, we write: $\boldsymbol{x^{-a} = \frac{1}{x^a}}$**, where $x \neq 0$.**

When working with quotients, we may get expressions with negative exponents. We typically don't include negative or zero exponents in our final answer. Instead, we prefer positive exponents, using fractions if necessary. For example, we express x^2y^{-3} as $\frac{x^2}{y^3}$.

When multiplication, division, and powers of monomials occur together, we follow the order of operations (explained in Section 1.2): we evaluate powers first and then perform multiplication and division.

Example 8.7 Express $\left(\frac{30m^3}{15m^{18}}\right)(5m^2)^4$ in standard form or without negative exponents.

Solution:

$$\left(\frac{30m^3}{15m^{18}}\right)(5m^2)^4$$

$= \left(\frac{30m^3}{15m^{18}}\right)(5^4)(m^2)^4$	$(x^a y^b)^c = x^{ac}y^{bc}$.
$= \left(\frac{30m^3}{15m^{18}}\right)(5^4)(m^8)$	$(x^a)^b = x^{ab}$.
$= (2m^{-15})(5^4)(m^8)$	$\frac{x^a}{x^b} = x^{a-b}$, where $x \neq 0$.
$= (2)(5^4)(m^{-15})(m^8)$	Commutative and associative properties of multiplication.
$= 1{,}250m^{-7}$	$x^a(x^b) = x^{a+b}$.
$= \frac{1{,}250}{m^7}$	$x^{-a} = \frac{1}{x^a}$, where $x \neq 0$.

Table 8.2 summarizes the rules of exponents:

Exercises

Write each expression in standard form or without negative exponents.

1. $\frac{x^8}{x^6}$

2. $\frac{c^{10}}{c^7}$

3. $\frac{m^{12}}{m^4}$

4. $\frac{-48y^4}{8y^2}$

5. $\frac{24a^6}{-3a^2}$

6. $\frac{-52b^5}{-4b^7}$

7. $\left(\frac{a}{b}\right)^4$

8. $\left(\frac{3x}{4y}\right)^2$

Table 8.2 Rules of exponents.

Rule in Words	Rule in Symbols	Example
To multiply powers with the same base, keep the base the same and add the exponents.	$x^a(x^b) = x^{a+b}$	$5^1 5^3 = (5)\cdot(5)(5)(5) = 5^4$
To raise a power to a power, keep the base the same and multiply the exponents.	$(x^a)^b = x^{ab}$ $(x^a y^b)^c = x^{ac} y^{bc}$	$(3^3)^2 = (3)(3)(3)\cdot(3)(3)(3) = 3^6$ $(ax^2)^3 = (ax^2)(ax^2)(ax^2) = a^3x^6$
To divide powers with the same nonzero base, keep the base the same and subtract the exponents.	$\frac{x^a}{x^b} = x^{a-b},\ x \neq 0$	$\frac{2^5}{2^3} = \frac{2(2)(2)(2)(2)}{(2)(2)(2)} = 2^2$
To raise a fraction with a nonzero denominator to an exponent, raise the numerator and the denominator to that exponent.	$\left(\frac{x}{y}\right)^a = \frac{x^a}{y^a},\ y \neq 0$	$\left(\frac{2}{3}\right)^4 = \left(\frac{2}{3}\right)\left(\frac{2}{3}\right)\left(\frac{2}{3}\right)\left(\frac{2}{3}\right)$ $= \frac{2^4}{3^4}$
Any nonzero number raised to the 0 power equals 1.	$x^0 = 1, x \neq 0$	$3^0 = 1$
A number raised to a negative exponent equals the reciprocal of the number raised to the exponent's opposite.	$x^{-a} = \frac{1}{x^a},\ x \neq 0$	$3^{-2} = \frac{1}{3^2}$

9. $\left(\frac{-4m}{5p}\right)^3$
10. $(5t^4)(7t^{-5})$
11. $(9u^6)(-9u^{-6})$
12. $(3d^4v^2)(-5d^0v^{-3})$
13. $\frac{17a^2y^3}{17a^2y^2}$
14. $\frac{28m^3n}{-7mn^3}$
15. $\frac{-22j^7k^4}{11j^3k^2}$
16. $\frac{16q^6}{2q^3}\cdot 3q^4$
17. $\left(\frac{2r^0}{4r^2}\right)^3\cdot 2r^3$
18. $\frac{10n^5}{10n^5}\cdot(2n^2)^3$

8.4 Multiplying Polynomials

In Section 8.1, we saw that the method for adding and subtracting polynomials is similar to the method for adding and subtracting multi-digit numbers. In the same way, we start our discussion of multiplying polynomials by examining how we multiply whole numbers.

In Section 1.1, we discussed how we can use rectangles to represent multiplication. For example, we can represent 3(34) as a rectangle whose length is 34 squares and width is 3 squares (Figure 8.1):

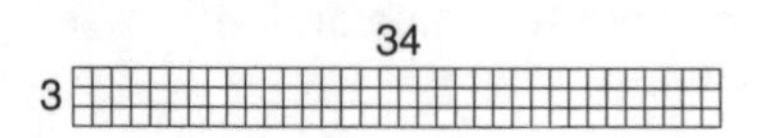

Figure 8.1 Multiplication as a rectangle.

Counting all the squares in the rectangle takes too much work! Instead, we can group the squares into smaller rectangles whose areas are easier to calculate. Since 34 = 30 + 4, we can divide this rectangle into two smaller rectangles (Figure 8.2) whose areas are 30(3) = 90 and 4(3) = 12. Adding them gives the total area of 90 + 12 = 102.

30 4
3 90 12

Figure 8.2 Multiplication as addition of smaller rectangles.

We can simplify this process even more by eliminating the squares entirely and representing the multiplication as a rectangle with one row and two columns (Figure 8.3):

	30	4
3	90	12

Figure 8.3 Multiplication with the area model.

We write our work symbolically as follows:

3(34)	
= 3(30 + 4)	Write 34 as 30 + 4.
= 3(30) + 3(4)	Divide the large rectangle into 2 smaller rectangles.
= 90 + 12	Find the area of each smaller rectangle.
= 102	Add the areas of the smaller rectangles.

Does this look familiar? It's the distributive property, which we discussed in Section 3.7: $a(b + c) = ab + ac$.

Because this method of multiplication uses boxes to represent the area of rectangles, it's often called the area model or the "box" method. The length and width of the rectangle are factors of the product since they multiply together to form the area. We use a rectangle whose dimensions are determined by the number of terms in the rows and columns.

Example 8.8 Express $ab(2a - 5b)$ in standard form.

Solution: Here, we multiply a binomial by a monomial, so we use a rectangle with 2 rows and 1 column.

The product is the sum of the terms expressed in order of descending powers of a: $2a^2b - 5ab^2$.

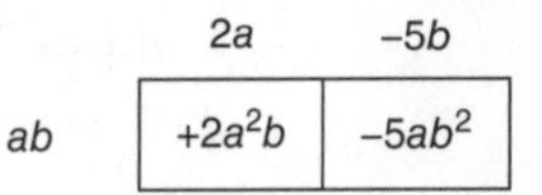

To multiply a binomial by a binomial, we add another row to our table:

Example 8.9 Express $(3x + 2)(2x - 5)$ in standard form.

Solution: We draw a table with 2 rows and 2 columns and multiply the row and column header for each cell:

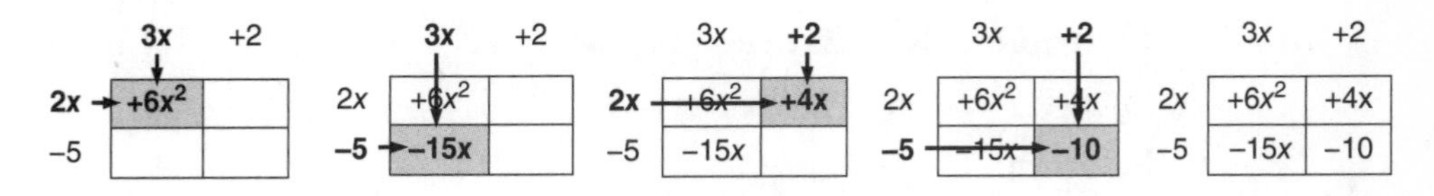

The product is the sum of the terms inside the table: $6x^2 + 4x - 15x - 10 = 6x^2 - 11x - 10$.

Example 8.10 Express $(p + 7)^2$ in standard form.

Solution: $(p + 7)^2 = (p + 7)(p + 7)$

The product is $p^2 + 7p + 7p + 49 = p^2 + 14p + 49$.

	p	$+7$
p	$+p^2$	$+7p$
$+7$	$+7p$	$+49$

Watch Out!

Many students make the mistake of thinking that $(p + 7)^2 = p^2 + 49$. However, Example 8.10 shows that this is not correct. In general,

$(x + y)^2 = x^2 + 2xy + y^2$, **not** $x^2 + y^2$.

$(x - y)^2 = x^2 - 2xy + y^2$, **not** $x^2 - y^2$.

These polynomials are examples of **perfect square trinomials**—they represent the square of a binomial and have three terms.

Example 8.11 Express $(3z - 2)(3z^2 + 4z - 3)$ in standard form.

Solution: We use a table with two rows (for $3z - 2$) and three columns (for $3z^2 + 4z - 3$).

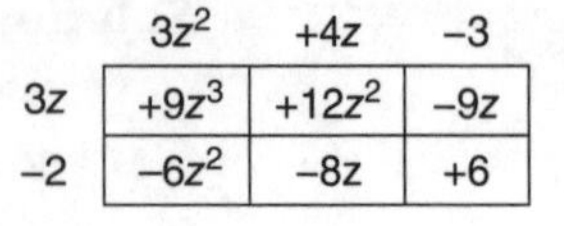

	$3z^2$	$+4z$	-3
$3z$	$+9z^3$	$+12z^2$	$-9z$
-2	$-6z^2$	$-8z$	$+6$

The product is $9z^3 + (-6z^2 + 12z^2) + (-8z - 9z) + 6 = 9z^3 + 6z^2 - 17z + 6$.

The like terms are located in the four diagonals in the rectangle. The first diagonal represents the z^3 terms, the second diagonal represents the z^2 terms, the third represents the z terms, and the fourth represents the z^0 terms. We see from this example that the number of diagonals equals the number of terms in the final product.

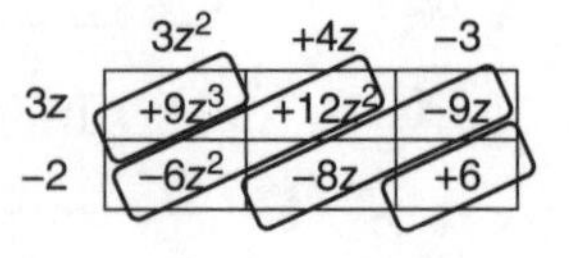

	$3z^2$	$+4z$	-3
$3z$	$+9z^3$	$+12z^2$	$-9z$
-2	$-6z^2$	$-8z$	$+6$

Example 8.12 Express $3u(2u^2 + 6u - 5) + 2(u - 4)^2$ in standard form.

Solution:

$3u(2u^2 + 6u - 5) + 2(u - 4)^2$

$= 3u(2u^2) + 3u(6u) + 3u(-5) + 2(u - 4)^2$ Distributive property.

$= 6u^3 + 18u^2 - 15u + 2(u - 4)^2$ Multiply monomials.

$= 6u^3 + 18u^2 - 15u + 2 \cdot$

	u	-4
u	$+u^2$	$-4u$
-4	$-4u$	$+16$

Order of operations: evaluate exponents before multiplication.

$= 6u^3 + 18u^2 - 15u + 2(u^2 - 8u + 16)$ Combine like terms.

$= 6u^3 + 18u^2 - 15u + 2u^2 - 16u + 32$ Distributive property.

$= 6u^3 + 20u^2 - 31u + 32$ Combine like terms.

Exercises

Write each expression in standard form.

1. $(k + 3)(k - 7)$
2. $(x - 5)(x + 4)$
3. $(5 + b)(2 - b)$
4. $(2a - 3)(a + 2)$
5. $(3c + 2)(4c - 5)$
6. $(x^2 - 2)(2x^2 + 1)$
7. $(a + 6)^2$
8. $(b - 3)^2$
9. $(2x + 1)^2$
10. $(3k + 4)^2$
11. $(3 - 2p)^2$
12. $(p + k)^2$

13. $(r^2 - 3r + 5)(5r - 2)$
14. $(k^2 + 2k - 3)(k - 4)$
15. $(2a + 1)(2a^2 + 3a - 4)$
16. $2(y + 1)^2 - 2(y + 1) + 5$
17. $3(2q - 7)^2 + 4(q - 2)$
18. $(x^2 + 1)(x + 3) + 3(2x - 1)$

Questions to Think About

19. Explain how the area model can be used to illustrate the commutative property of multiplication.
20. The work below shows two ways to multiply 32(21) – the vertical method (on the left) and the area model (on the right). Explain how the two methods are similar.

Vertical Method	Area Model
$\begin{array}{r} 31 \\ \times 23 \\ \hline 93 \\ +620 \\ \hline 713 \end{array}$	$\begin{array}{c|c|c} & \textbf{30} & \textbf{1} \\ \hline \textbf{20} & 600 & 20 \\ \hline \textbf{3} & 90 & 3 \end{array}$ $= (30 + 1)(20 + 3) = 600 + 90 + 20 + 3 = 713$

21. Jenna tries to multiply $3x^2 + 5$ and $2x^2 + 3x$ using the area model, but notices that the like terms do not line up in her diagonals. How can she rearrange her work so that the like terms appear in the diagonals?

8.5 Dividing Polynomials

Since division is the inverse of multiplication, we would expect that we divide polynomials by doing the opposite of what we do when multiplying. Fortunately, the area model discussed in the previous section allows us to do exactly that! Let's start by dividing a polynomial by a monomial:

Example 8.13 Calculate $(12x^5 - 27x^4 + 18x^3 - 3x^2 + 6x) \div 3x$, $x \neq 0$.

Solution: Draw a rectangle with 1 row and 5 columns, one for each term in the dividend. Then divide each term of the polynomial by the monomial.

	$+4x^4$	$-9x^3$	$+6x^2$	$-x$	$+2$
$3x$	$+12x^5$	$-27x^4$	$+18x^3$	$-3x^2$	$+6x$

We can also write this without a rectangle as follows: $\frac{12x^5-27x^4+18x^3-3x^2+6x}{3x} = 4x^4 - 9x^3 + 6x^2 - x + 2$.

To divide a polynomial by a binomial, we reverse the process of multiplication. We know the total area represented by the rectangle and one of its dimensions, so we need to find the other dimension.

Example 8.14 Calculate $(6x^2 + 7x - 20) \div (3x - 4)$.

Solution:

1. Create a rectangle with 2 rows and 2 columns (a 2 x 2 box has 3 diagonals, one for each term in the dividend). Write the dividend below the rectangle and the divisor on the left.

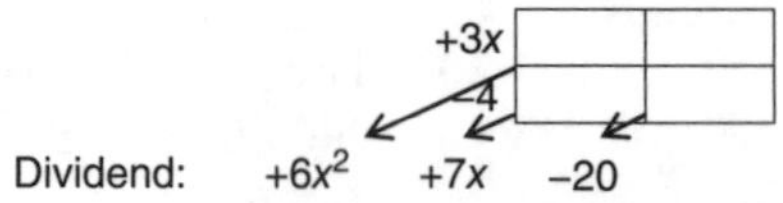

$+3x$		
-4		

Dividend: $+6x^2$ $+7x$ -20

2. Write the first term of the dividend in the top left box of the rectangle.

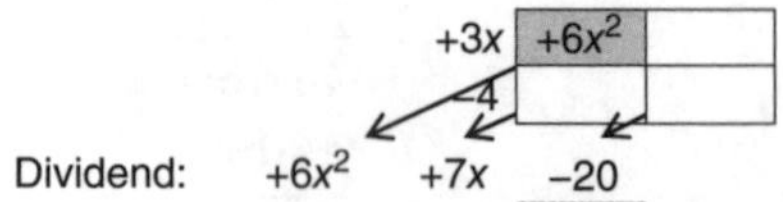

$+3x$	$+6x^2$	
-4		

Dividend: $+6x^2$ $+7x$ -20

3. DIVIDE: What number multiplied by $+3x$ equals $+6x^2$? Since $+6x^2 \div +3x = +2x$, write $+2x$ above the $+6x^2$.

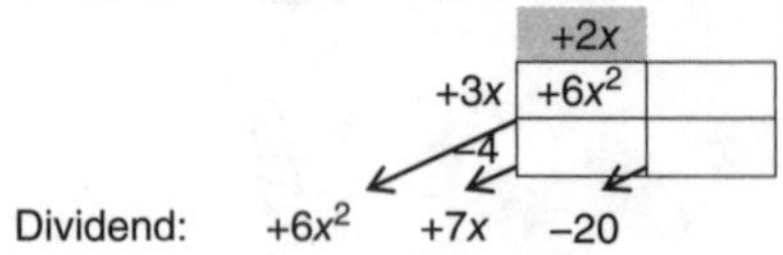

	$+2x$	
$+3x$	$+6x^2$	
-4		

Dividend: $+6x^2$ $+7x$ -20

4. MULTIPLY: Multiply $2x(-4) = -8x$, which we write in the lower left box.

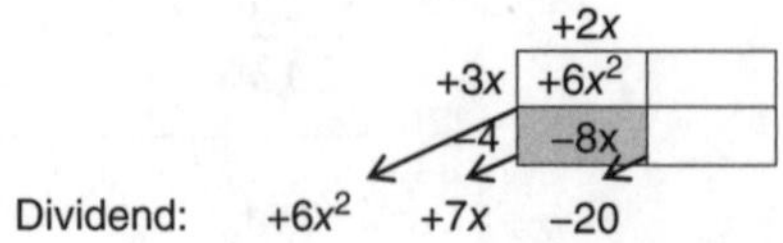

	$+2x$	
$+3x$	$+6x^2$	
-4	$-8x$	

Dividend: $+6x^2$ $+7x$ -20

5. SUBTRACT: What number added to $-8x$ equals $+7x$? Since $+7x - (-8x) = +15x$, write $+15x$ in the top right box.

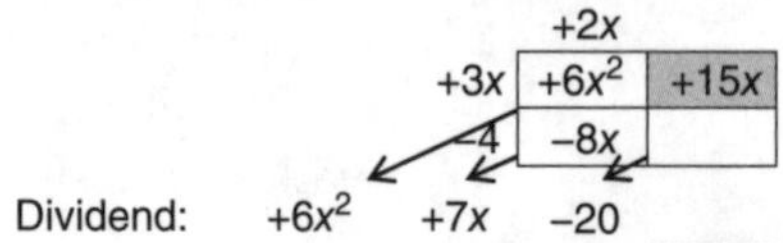

	$+2x$	
$+3x$	$+6x^2$	$+15x$
-4	$-8x$	

Dividend: $+6x^2$ $+7x$ -20

6. DIVIDE: What number multiplied by $+3x$ equals $+15x$? Since $+15x \div +3x = +5$, write $+5$ above the $+15x$.

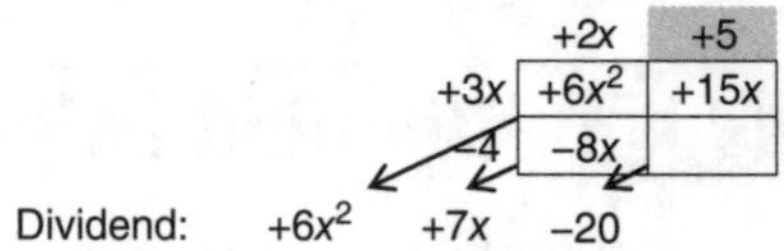

	$+2x$	$+5$
$+3x$	$+6x^2$	$+15x$
-4	$-8x$	

Dividend: $+6x^2$ $+7x$ -20

7. MULTIPLY: Multiply $+5(-4) = -20$, which we write in the lower right box.

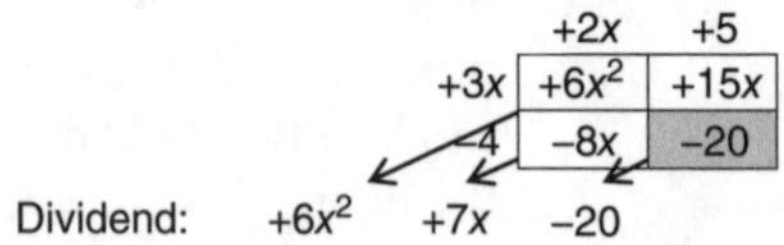

	$+2x$	$+5$
$+3x$	$+6x^2$	$+15x$
-4	$-8x$	-20

Dividend: $+6x^2$ $+7x$ -20

FINAL ANSWER: $(+6x^2 + 7x - 20) \div (3x - 4) = 2x + 5$.

The quotient is $2x + 5$.

We add more rows or columns to our table as necessary:

Example 8.15 Calculate $(9x^3 - 6x^2 - 30x - 8) \div (3x + 4)$.

Solution: We need a rectangle with 4 diagonals (one for each term in the dividend), so we need 3 columns and 2 rows, one for each term in the divisor. The first term, $9x^3$, goes in the top left of the rectangle.

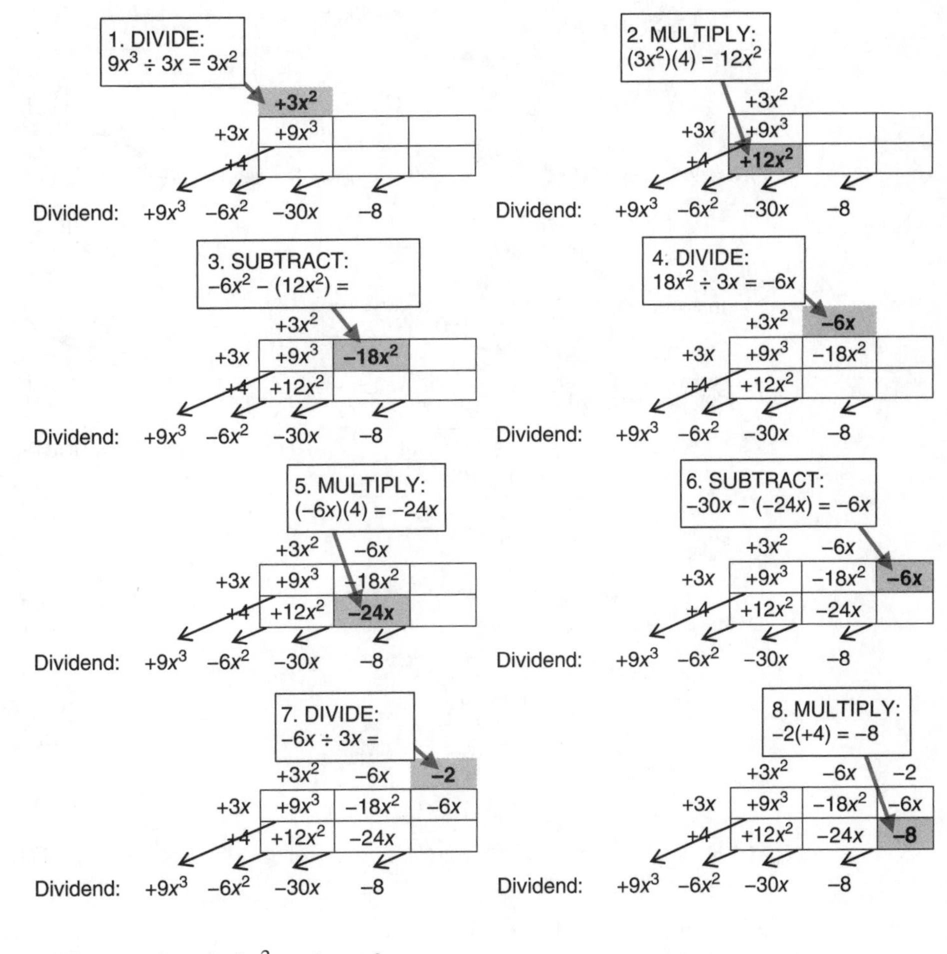

The quotient is $3x^2 - 6x - 2$.

If the dividend or divisor has a term with a 0 coefficient, include a placeholder for it so that the like terms line up in the diagonals:

Example 8.16 Calculate $(9x^3 - 49x + 40) \div (3x + 4)$.

Solution: Rewrite the dividend as $9x^3 + 0x^2 - 49x + 40$ before dividing.

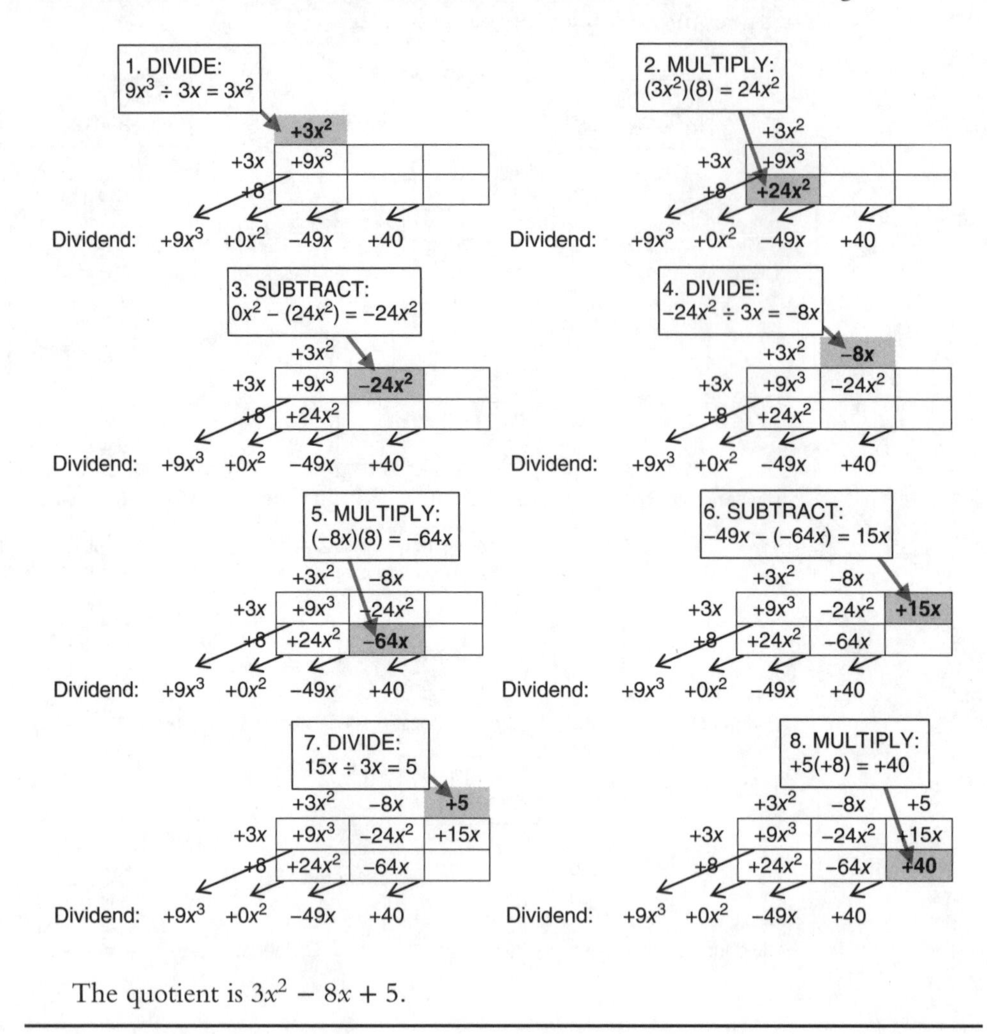

The quotient is $3x^2 - 8x + 5$.

Sometimes a divisor will not go evenly into a dividend, as shown in the following example:

Example 8.17 Calculate $(8x^3 - 2x^2 - 7x + 5) \div (2x - 1)$.

Solution:

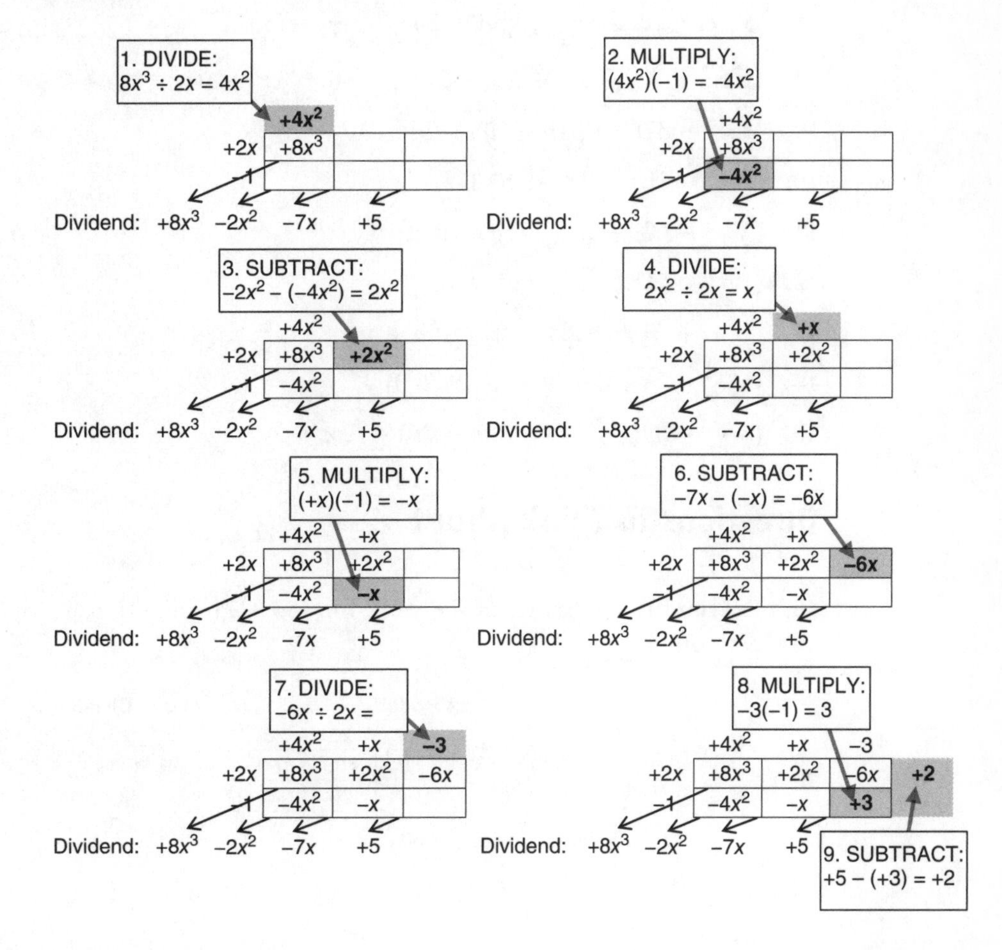

We express the remainder as the numerator of a fraction whose denominator is the divisor, so our final answer is $4x^2 + x - 3 + \frac{2}{2x-1}$. We do something similar when we get a remainder when dividing whole numbers. For example, $\frac{13}{4} = 3 + \frac{1}{4}$.

Exercises

Calculate each quotient.

1. $(36x^4 - 60x^3 + 72x^2 - 18x) \div (6x)$
2. $(16y^4 + 80y^3 - 24y^2 + 64y) \div (-8y)$
3. $(4q^6 + 40q^5 - 52q^4 - 28q^3) \div (2q^2)$
4. $(10x^2 - 31x - 14) \div (5x + 2)$
5. $(21x^2 - 32x + 12) \div (7x - 6)$
6. $(34u^2 + u - 8) \div (2u + 1)$
7. $(12d^3 - 13d^2 - 31d + 5) \div (4d + 5)$
8. $(3a^3 + 16a^2 + 9a + 20) \div (a + 5)$
9. $(45b^3 - 37b^2 - 24b + 18) \div (5b - 3)$
10. $(35c^3 + 71c^2 - 18) \div (5c + 3)$
11. $(27p^3 - 63p - 20) \div (3p + 4)$
12. $(18t^3 - 137t - 24) \div (3t + 8)$
13. $(10n^3 + 37n^2 + 22n - 1) \div (5n + 1)$
14. $(18k^4 - k^3 + 65k + 5) \div (2k + 3)$
15. $(6m^4 - 9m^3 - 7m^2 + 16m - 26) \div (2m + 3)$

Questions to Think About

16. Explain the similarities in the work for $483 \div 21$ using the area model and long division.

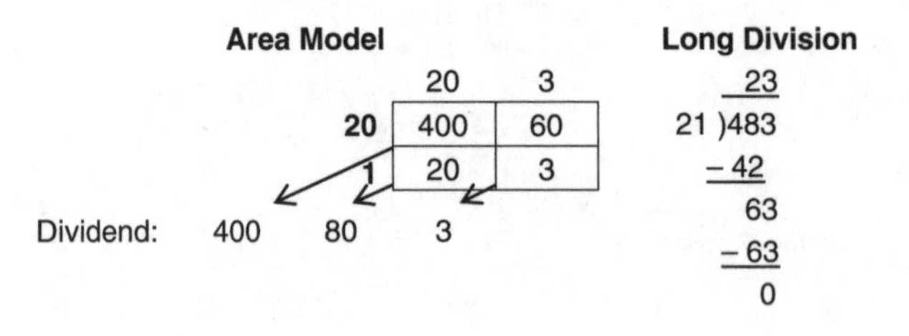

17. If a polynomial in standard form with nonzero coefficients and degree n is divided by a first-degree binomial, how many rows and columns should be in the rectangle used for division? Explain.

18. When Elise divides $8x^3 + 58x^2 + 15x + 9$ by $x + 7$, she gets $8x^2 + 2x + 1 + \frac{2}{x+7}$. How can she check her answer?

CHAPTER 8 TEST

1. How many terms are in the polynomial $4m^4 + 2m^2 + 5$?

(A) 2 (B) 3 (C) 5 (D) 6

2. What is the degree of $-2x^3 + 7x^5 + 6x$?

(A) 3 (B) 5 (C) 9 (D) 11

3. Find the sum of $x^2 - 9$ and $3x^2 + x + 8$.

(A) $4x^2 - 8x - 1$ (B) $4x^2 + x - 1$ (C) $4x^2 + x^- + 1$ (D) $4x^2 + x - 17$

4. Subtract $-a^2 + 4a - 6$ from $-3a^2 - 5a + 3$.

(A) $-2a^2 - 9a - 3$ (B) $-2a^2 - a - 3$ (C) $-2a^2 + 9a + 9$ (D) $-2a^2 - 9a + 9$

5. Which expression is equivalent to $3r^5(-8r^2)$?

(A) $-5r^7$ (B) $-24r^{10}$ (C) $-24r^7$ (D) $-5r^3$

6. Which expression is equivalent to $(6y^3)^2$?

(A) $36y^6$ (B) $12y^6$ (C) $12y^9$ (D) $36y^9$

7. Which expression is equivalent to $(xy^0)(x^{-2}y^4)$?

(A) $\frac{1}{x}$ (B) $\frac{1}{x^2}$ (C) $\frac{y^4}{x^2}$ (D) $\frac{y^4}{x}$

8. If $x^4x^2 = x^mx^{16}$, then what is the value of m?

(A) $-\frac{3}{8}$ (B) 2 (C) 10 (D) -10

9. If $y^5 = m$ and $y^2 = 6x$, then which expression is equivalent to x?

(A) $\frac{m}{6y^3}$ (B) $\frac{1}{6y^3}$ (C) $\frac{m}{y^3}$ (D) $\frac{6y^3}{m}$

10. Calculate $(11m^3k^2 + 33mk^3 - 22mk) \div 11mk$.

11. Is $5x^2 + 3x^2 + \frac{1}{x}$ a polynomial? Explain.

12. Express $5b^3 + 1 + 8b^2 - 8b - 2(7 - 3b^2 - 3b^3)$ in standard form.

13. Express $4(2z - 1)^2$ in standard form.

14. Express $(3k^3 + 2k^2 - 5)(6k - 7) + 4$ in standard form.
15. Calculate $(20j^4 - 38j^3 + 22j^2 - 9j + 7) \div (5j - 2)$.

CHAPTER 8 SOLUTIONS

8.1.
1. $-4x^3 + 3x^2 + 9x + 1$, degree = 3
2. $2y^4 + 3y^3 - 3y^2 - 8y + 7$, degree = 4
3. $-5m^3 + 6m^2 + 5m + 8$, degree = 3
4. $9b^5 + 11b^3 - b^2 + 9b + 2$, degree = 5
5. $4x^3 + 5x^2 - 2xy - 3$
6. $m^4 + 4m^3n + 6m^2n^2 + 4mn^3 + n^4$
7. $6x^2 + x - 2$
8. $17y^2 - 2y - 8$
9. $-2a^2 - 5a + 8$
10. $-3y^2 + 7y - 5$
11. $3b^2 - 14b + 1$
12. $2z^2 + 8z - 11$
13. $2x^2 + 7x - 3$
14. $-3k^3 + 5k + 7$
15. $-9r^3 + 12r^2 + 3r + 8$
16. $7z^2 + 13z - 4$
17. $6n^4 - 8n^3 + 6n^2 - 9$
18. $6x^3 - 4x^2 + 3x + 3$
19. Just as quantities with the same place value are grouped together and added or subtracted, like terms are grouped together and their coefficients are added or subtracted.
20. Yes, since it is the difference of two terms.
21. No. Although the terms are in decreasing order from left to right, $14x^3$ and $-6x^3$ are like terms that are not combined. In standard form, this polynomial would be $8x^3 + 2x^2 - 1$.

8.2.
1. $-12m^6$
2. $4x^{10}$
3. $80p^5$
4. $-14a^6b^3$
5. $12a^5b^7$
6. $10x^3y^5$
7. d^{12}
8. $81k^{12}$
9. $-64q^{15}$
10. $-648m^6n^{12}$
11. x^{14}
12. $81a^8b^{12}$

8.3.
1. x^2
2. c^3
3. m^8
4. $-6y^2$
5. $-8a^4$
6. $\frac{13}{b^2}$
7. $\frac{a^4}{b^4}$
8. $\frac{9x^2}{16y^2}$
9. $-\frac{64m^3}{125p^3}$
10. $\frac{35}{t}$
11. -81
12. $\frac{-15d^4}{v}$
13. y
14. $-\frac{4m^2}{n^2}$
15. $-2j^4k^2$
16. $24q^7$
17. $\frac{1}{4r^3}$
18. $8n^6$

8.4.
1. $k^2 - 4k - 21$
2. $x^2 - x - 20$
3. $-b^2 - 3b + 10$
4. $2a^2 + a - 6$
5. $12c^2 - 7c - 10$
6. $2x^4 - 3x^2 - 2$
7. $a^2 + 12a + 36$
8. $b^2 - 6b + 9$
9. $4x^2 + 4x + 1$
10. $9k^2 + 24k + 16$
11. $4p^2 - 12p + 9$
12. $p^2 + 2pk + k^2$
13. $5r^3 - 17r^2 + 31r - 10$
14. $k^3 - 2k^2 - 11k + 12$
15. $4a^3 + 8a^2 - 5a - 4$
16. $2y^2 + 2y + 5$
17. $12q^2 - 80q + 139$
18. $x^3 + 3x^2 + 7x$
19. If you rotate the rectangle so that the length and width are reversed, the area of the rectangle remains unchanged, so $ab = ba$.
20. Both methods generate the same partial products, which are added together to get the final product.
21. Some of the terms in the expressions that she is multiplying have coefficients of 0. To get the like terms to line up in her diagonals, she should rewrite the expressions as $3x^2 + 0x + 5$ and $2x^2 + 3x + 0$.

8.5.
1. $6x^3 - 10x^2 + 12x - 3$
2. $-2y^3 - 10y^2 + 3y - 8$
3. $2q^4 + 20q^3 - 26q^2 - 14q$
4. $2x - 7$
5. $3x - 2$
6. $17u - 8$
7. $3d^2 - 7d + 1$
8. $3a^2 + a + 4$
9. $9b^2 - 2b - 6$
10. $7c^2 + 10c - 6$

11. $9p^2 - 12p - 5$

12. $6t^2 - 16t - 3$

13. $2n^2 + 7n + 3 - \dfrac{4}{5n+1}$

14. $9k^3 - 14k^2 + 21k + 1 + \dfrac{2}{2k+3}$

15. $3m^3 - 9m^2 + 10m - 7 - \dfrac{5}{2m+3}$

16. In both methods, digits from the dividend (483) are systematically divided by the divisor (21). Note that the calculations within each method are the same, but they are written differently.

17. A polynomial with degree n has $n + 1$ terms (one with the variable raised to the 0th power, 1st power, 2nd power, ..., nth power). Each diagonal is used for terms with the same degree. To have $n + 1$ diagonals, the rectangle needs 2 rows and n columns.

18. If she multiplies $8x^2 + 2x + 1$ by $x + 7$ and adds 2 to the result, she should get the original quotient of $8x^3 + 58x^2 + 15x + 9$.

CHAPTER 8 TEST SOLUTIONS

1. (B)
2. (B)
3. (B)
4. (D)
5. (C)
6. (A)
7. (D)
8. (D)
9. (A)
10. $m^2k + 3k^2 - 2$.
11. No, since $\frac{1}{x}$ cannot be expressed as the product of a number and a variable, so it is not a polynomial.
12. $11b^3 + 14b^2 - 8b - 13$
13. $16z^2 - 16z + 4$
14. $18k^4 - 9k^3 - 14k^2 - 30k + 39$
15. $4j^3 - 6j^2 + 2j - 1 + \dfrac{5}{5j-2}$.

9 QUADRATIC FUNCTIONS

In Chapter 8, we started our discussion of polynomials whose degree was greater than 1. We used rectangles to represent multiplication and division since multiplying the length and width to get the area of a rectangle is similar to multiplying factors to get a product. The area model allows us to link algebra and geometry—a connection that mathematicians around the world recognized thousands of years ago! In this chapter, we focus on **quadratic** polynomials, whose degree is 2. (The word "quadratic" comes from the Latin word *quadratum*, or "square." It relates the geometric shape with four congruent sides to the algebraic concept of a second-degree polynomial.)

9.1 Factoring a Monomial from a Polynomial

In Chapter 8, we encountered quadratic polynomials in two situations (Figure 9.1):

- MULTIPLICATION: Given two first-degree factors, find the product.
- DIVISION: Given a second-degree product and a first-degree factor, find the other factor.

Figure 9.1 Find the missing factor or product.

What if we're only given the product and we need to find the missing factors?

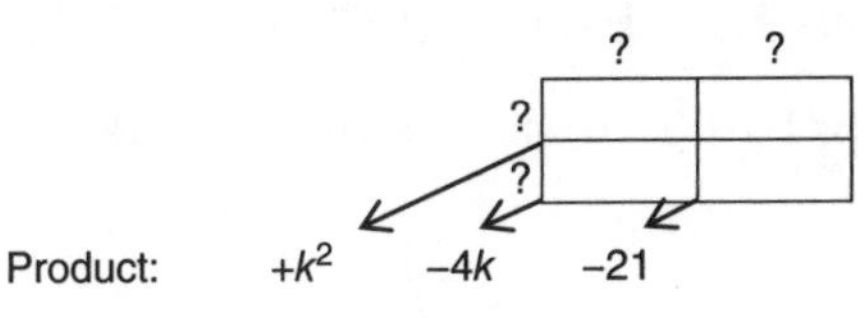

Figure 9.2 Find the missing factors.

Recall from Section 2.2 that factoring is expressing a quantity as a product of factors. Since the process of factoring is the reverse of multiplication, then multiplication, division, and factoring are related. Table 9.1 summarizes this relationship:

Table 9.1 Vocabulary for multiplication, division, and factoring.

Process	Vocabulary	Example				
Multiplication	factor • factor = product	Given two factors,		find		the product.
		3(7)		=		21
		$a(a + 3)$		=		$a^2 + 3a$
Division	product ÷ factor = factor (dividend ÷ divisor = quotient)	Given the product	and	a factor,	find	the other factor.
		21	÷	3	=	7
		$(a^2 + 3a)$	÷	a	=	$a + 3$
Factoring	product = factor • factor	Given a product,		find		the factors.
		21		=		3(7)
		$a^2 + 3a$		=		$a(a + 3)$

We are generally interested in factoring into primes. Just as a prime number has no other factors except 1 and itself, a **prime polynomial** cannot be factored any further—the common factor of all terms is 1. A monomial cannot be factored further since it is already considered prime. Factoring a polynomial into primes is called **factoring completely**, although in practice, we usually refer to it as "factoring."

We start by identifying common monomial factors—any numbers and variables that are common to all terms in the polynomial. As with factoring numbers, when factoring polynomials, we want to find the greatest common factor (GCF).

Example 9.1 Factor $40k^5 - 4k^4 + 12k^3$.

Solution: First, write each term in factored form and identify factors common to each term:

$$\begin{aligned} &40k^5 - 4k^4 + 12k^3 \\ &= 2\cdot2\cdot2\cdot5\cdot k\cdot k\cdot k\cdot k\cdot k - 2\cdot2\cdot k\cdot k\cdot k\cdot k + 2\cdot2\cdot3\cdot k\cdot k\cdot k \end{aligned}$$

The GCF is the product of all factors that are common to each term: $2 \cdot 2 \cdot k \cdot k \cdot k = 4k^3$. We then divide each term by the GCF: $\frac{40k^5}{4k^3} - \frac{4k^4}{4k^3} + \frac{12k^3}{4k^3} = 10k^2 - k + 3$.

Since factoring an expression means writing an expression as a product of prime factors, our final answer is the GCF multiplied by all remaining factors: $4k^3(10k^2 - k + 3)$. We can also represent this with the area model:

	$4k^3$
$+10k^2$	$40k^5$
$-k$	$-4k^4$
$+3$	$+12k^3$

Example 9.2 Factor $9x^3 + 3x^2y - 6xy^2$.

Solution: First, write each term in factored form and identify factors common to each term.

$$\begin{array}{ccccc} 9x^3 & + & 3x^2y & - & 6xy^2 \\ = 3 \cdot 3 \cdot x \cdot x \cdot x & + & 3 \cdot x \cdot x \cdot y & - & 2 \cdot 3 \cdot x \cdot y \cdot y \end{array}$$

The GCF is $3x$. (Even though y is a common factor of $3x^2y$ and $-6xy^2$, it is not a factor of $9x^3$. Thus, y is not part of the GCF.) Dividing our original expression by the GCF gives us $\frac{9x^3}{3x} + \frac{3x^2y}{3x} - \frac{6xy^2}{3x} = 3x^2 + xy - 2y^2$. Our final answer is $3x(3x^2 + xy - 2y^2)$. We can also represent this with the area model:

	$3x$
$+3x^2$	$9x^3$
$+xy$	$+3x^2y$
$-2y^2$	$-6xy^2$

Examples 9.1 and 9.2 illustrate the following shortcut: **to factor a monomial from a polynomial, find the greatest common numerical factor and the smallest exponent of each variable that appear in each term of the polynomial.** In Example 9.1, the GCF of 40, −4, and 12 is 4, and the GCF of k^5, k^4, and k^3 is k^3 (the smallest exponent of k), so the GCF of the original expression is $4k^3$.

Technology Tip

Use the gcf() function on your device to find the greatest common factor of the coefficients of terms. Some devices can even find the GCF of expressions with variables—check your device's manual for details.

Exercises

Factor each polynomial, if possible.

1. $3a + 3b$
2. $2x^2 - 6xy$
3. $x^2 - 3x^3$
4. $r + ry - r^2$
5. $5k^3 - 15k^2 + 35k$
6. $y^3 - 2y + 5$
7. $10m^3 - 4m^2 + 2m$
8. $5k^2 - 2k + 1$
9. $32a^3 + 8a^2 + 20a$
10. $p^4q + p^3q + p^2q$
11. $10x^3y^2 - 5x^2y^3 + 15x^2y^2$
12. $a^2\,bc - ab^2\,c + abc^2$

Questions to Think About

13. What is a prime polynomial?
14. What is meant by factoring completely?
15. Heeyun claims that $9x^2 + 4y^2$ is not factored completely since 9 and 4 are not prime. Explain the error in her thinking.

9.2 Factoring by Grouping

Some polynomials, such as $4m^3 + 8m^2 + 3m + 6$, don't have a common monomial factor. However, we can group this polynomial into two binomials (Figure 9.3), each of which has a different monomial GCF:

	$4m^2$
m	$+4m^3$
$+2$	$+8m^2$

	$+3$
m	$+3m$
$+2$	$+6$

Figure 9.3 Factoring binomials with a common binomial factor.

Since these binomials have the same width $(m + 2)$, we can combine these two rectangles into a larger rectangle (Figure 9.4):

	$6m^2$	$+3$
m	$+6m^3$	$+3m$
$+2$	$-12m^2$	$+6$

Figure 9.4 Factor by grouping.

Algebraically, we write this as follows:

$4m^3 + 8m^2 + 3m + 6$	
$= (4m^3 + 8m^2) + (3m + 6)$	Group the polynomial into two binomials.
$= 4m^2(m + 2) + 3(m + 2)$	Factor the GCF from each binomial—the remaining factor should be the same.
$= (4m^2 + 3)(m + 2)$	Distributive property.

If we had grouped the first and third terms together, we would have found the same common factor:

$$(4m^3 + 3m) + (8m^2 + 6)$$
$$= m(4m^2 + 3) + 2(4m^2 + 3)$$
$$= (m + 2)(4m^2 + 3)$$

When factoring terms with negative coefficients, we need to be careful with our signs, as shown in Example 9.3:

Example 9.3 Factor $3m^3 - 15m^2 - m + 5$.

Solution: We can represent the factoring with a rectangle, grouping the first two terms and the last two terms:

$+3m^3$	$-m$
$-15m^2$	$+5$

The GCF of the first two terms is $3m^2$, which makes the remaining factor $m - 5$:

	$3m^2$	
m	$+3m^3$	$-m$
-5	$-15m^2$	$+5$

To find the missing term, we divide $\frac{-m}{m}$ and $\frac{5}{-5}$ and get -1:

	$3m^2$	-1
m	$+3m^3$	$-m$
-5	$-15m^2$	$+5$

We write this algebraically as follows:

$3m^3 - 15m^2 - m + 5$	
$= (3m^3 - 15m^2) + (-m + 5)$	Group the polynomial into binomials.
$= 3m^2(m - 5) + (-1)(m - 5)$	Factor the GCF from each binomial—the remaining factor should be the same.
$= (3m^2 - 1)(m - 5)$	Distributive property.

Exercises

Factor each polynomial.

1. $12m^3 + 18m^2 + 10m + 15$
2. $15p^3 + 21p^2 + 10p + 14$
3. $d^3 + 5d^2 + 2d + 10$
4. $121y^3 - 11y^2 - 11y + 1$
5. $6u^3 - 15u^2 - 8u + 20$
6. $21q^3 - 28q^2 - 6q + 8$
7. $6a^4 - 27a^3 - 8a + 36$
8. $4z^3 - 6z^2d - 2zd + 3d^2$
9. $2a^2b^2 + 10ab^2c - abc - 5bc^2$

Questions to Think About

10. Explain why the polynomial $6n^3 + 4n^2 + 3n + 3$ cannot be factored by grouping.
11. Group $7x^3 + 2x^2 + 7x + 2$ in two different ways to factor.

9.3 Factoring Trinomials

As we saw in the previous section, factoring by grouping allows us to factor many polynomials with four terms. However, we often encounter polynomials that have three terms, such as $x^2 + 7x + 12$. We can factor many of these trinomials by grouping, but to do so we need to split one of the terms into two so that we can factor each binomial.

To see how we factor trinomials, we start by looking at several examples of multiplying binomials using the area model (Figure 9.5):

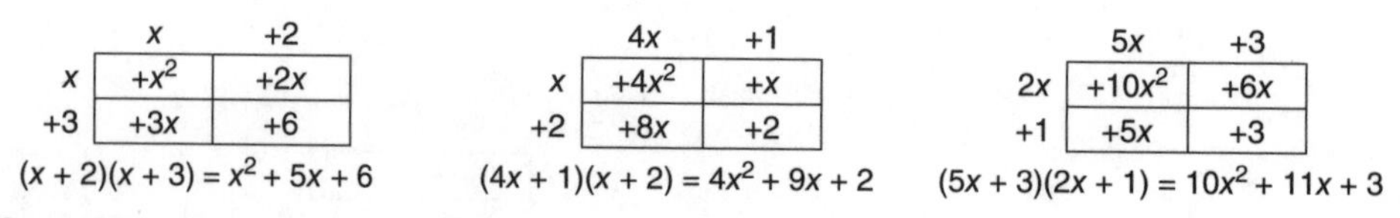

Figure 9.5 Examples of multiplying binomials.

If we look at the coefficients of each trinomial and the monomials in each rectangle, we see several patterns (Figure 9.6):

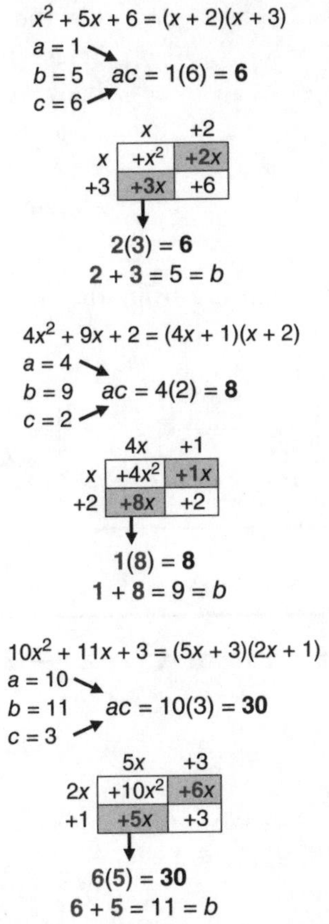

$x^2 + 5x + 6 = (x + 2)(x + 3)$

$a = 1$
$b = 5$ $ac = 1(6) = \mathbf{6}$
$c = 6$

	x	$+2$
x	$+x^2$	$+2x$
$+3$	$+3x$	$+6$

2(**3**) = **6**
2 + **3** = 5 = b

$4x^2 + 9x + 2 = (4x + 1)(x + 2)$

$a = 4$
$b = 9$ $ac = 4(2) = \mathbf{8}$
$c = 2$

	$4x$	$+1$
x	$+4x^2$	$+1x$
$+2$	$+8x$	$+2$

1(**8**) = **8**
1 + **8** = 9 = b

$10x^2 + 11x + 3 = (5x + 3)(2x + 1)$

$a = 10$
$b = 11$ $ac = 10(3) = \mathbf{30}$
$c = 3$

	$5x$	$+3$
$2x$	$+10x^2$	$+6x$
$+1$	$+5x$	$+3$

6(**5**) = **30**
6 + **5** = 11 = b

Trinomial ($ax^2 + bx + c$)	Factored Form	a	b	c	ac
$x^2 + 5x + 6$	$(x + 2)(x + 3)$	1	5	6	1(6) = 6
$4x^2 + 9x + 2$	$(4x + 1)(x + 2)$	4	9	2	4(2) = 8
$10x^2 + 11x + 3$	$(5x + 3)(2x + 1)$	10	11	3	10(3) = 30

Figure 9.6 Factoring with the *ac* method.

These examples illustrate that for quadratic trinomials in standard form ($ax^2 + bx + c$), the coefficients of the terms in the diagonal are the factors of ac that add up to b. (You'll prove this in one of the exercises in this section.) We use this fact to factor quadratic trinomials, as shown in Examples 9.4, 9.5, and 9.6:

Example 9.4 Factor $x^2 + 8x + 12$.

Solution:

$a = 1$, $b = 8$, $c = 12$	Identify values of a, b, and c.
$ac = 1(12) = 12$	Calculate ac.

1 and 12: $1 + 12 \neq 8$	List the number pairs that are factors of $ac = 12$.
2 and 6: $2 + 6 = 8$	Select the pair that adds up to $b = 8$.
$x^2 + \mathbf{2x} + \mathbf{6x} + 12$	Rewrite the original trinomial using the selected pair of *ac*'s factors.
$= x(x + 2) + 6(x + 2)$	Factor by grouping.
$= (x + 2)(x + 6)$	Distributive property.

We can represent this using the area model as follows:

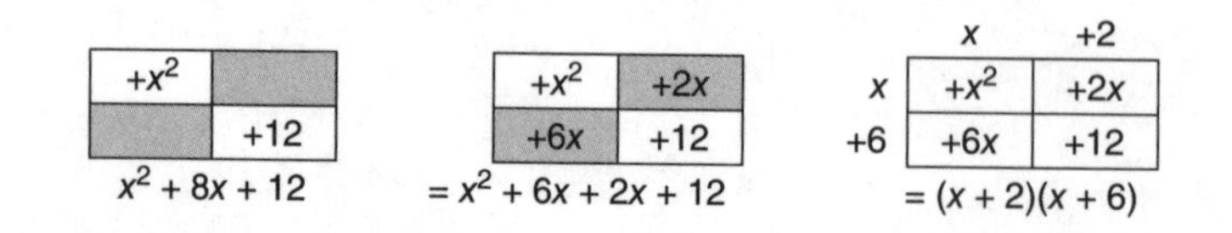

Example 9.5 Factor $x^2 - 2x - 24$.

Solution:

$a = 1, b = -2, c = -24$		Identify values of *a*, *b*, and *c*.
$ac = 1(-24) = -24$		Calculate *ac*. (Since *ac* is negative, one factor is positive and the other is negative.)
1 and −24:	$1 + (-24) \neq -2$	List the number pairs that are factors of $ac = -24$.
2 and −12:	$2 + (-12) \neq -2$	
3 and −8:	$3 + (-8) \neq -2$	
4 and −6:	$4 + (-6) = -2$	Select the pair that adds up to $b = -2$.
$x^2 + \mathbf{4x} - \mathbf{6x} - 24$		Rewrite the original trinomial using the selected pair of *ac*'s factors.
$= x(x + 4) - 6(x + 4)$		Factor by grouping.
$= (x - 6)(x + 4)$		Distributive property.

We can represent this using the area model as follows:

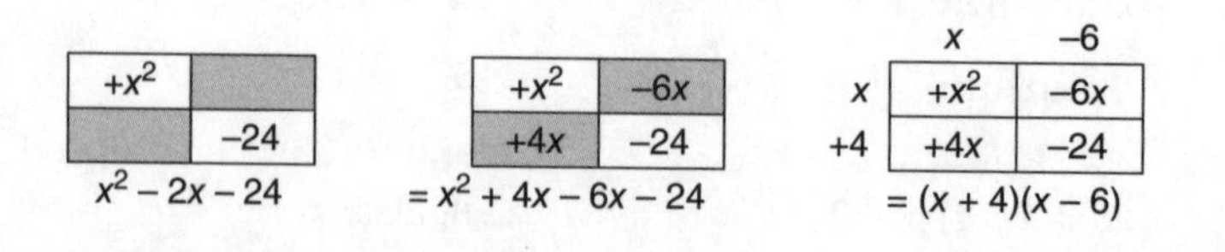

Example 9.6 Factor $4x^2 - 8x - 21$.

Solution:

$a = 4$, $b = -8$, $c = -21$		Identify values of a, b, and c.
$ac = 4(-21) = -84$		Calculate ac. (Since ac is negative, one factor is positive and the other is negative.)
1 and -84:	$1 + (-84) \neq -8$	List the number pairs that are factors of $ac = -84$.
2 and -42:	$2 + (-42) \neq -8$	
3 and -28:	$3 + (-28) \neq -8$	
4 and -21:	$4 + (-21) \neq -8$	
6 and -14:	$6 + (-14) = -8$	Select the pair that adds up to $b = -8$.
$4x^2 + \mathbf{6x - 14x} - 21$		Rewrite the original trinomial using the selected pair of ac's factors.
$= 2x(2x + 3) - 7(2x + 3)$		Factor by grouping.
$= (2x - 7)(2x + 3)$		Distributive property.

We can represent this using the area model as follows:

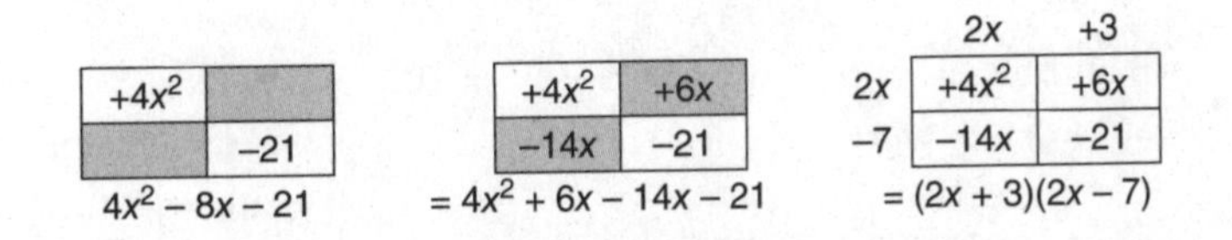

Because this method of factoring trinomials relies on the product ac, it is often called the ***ac*** **method of factoring** (pronounced "*a-c*," not "ack").

Exercises

Factor each polynomial.

1. $m^2 + 13m + 22$
2. $r^2 + 10r + 24$
3. $y^2 + 17y + 72$
4. $q^2 - 5q + 6$
5. $x^2 - 12x + 32$
6. $z^2 - 14z + 48$
7. $a^2 + a - 12$
8. $x^2 - 2x - 8$
9. $k^2 - k - 20$
10. $b^2 + 5b - 14$
11. $x^2 - 5x - 6$
12. $p^4 - p^2 - 30$
13. $2x^2 + x - 15$
14. $3a^2 - 2a - 5$
15. $2a^2 - 9a + 4$
16. $5c^2 + 17c + 14$
17. $6a^2 + a - 5$
18. $9y^2 + 3y - 2$

Questions to Think About

19. If $ax^2 + bx + c = (mx + n)(px + q)$, show that the product $(mx + n)(px + q)$ can be written as a four-term polynomial whose two middle terms' coefficients have a product of ac and a sum of b.

20. What are some of the challenges of using the ac method?

21. Use the ac method to verify that $x^2 + x + 4$ is a prime polynomial.

9.4 Special Cases of Factoring

In this section, we discuss several cases of factoring that occur frequently enough to deserve special mention. These problems can be solved using one or more of the methods described earlier.

Example 9.7 Factor $x^2 + 10x + 25$.

Solution:

$a = 1,\ b = 10,\ c = 25$		Identify values of a, b, and c.
$ac = 1(25) = 25$		Calculate ac.
1 and 25:	$1 + 25 \neq 10$	List the number pairs that are factors of $ac = 25$.
5 and 5:	$5 + 5 = 10$	Select the pair that adds up to $b = 10$.
$x^2 + \mathbf{5x} + \mathbf{5x} + 25$		Rewrite the original trinomial using the selected pair of ac's factors.
$= x(x + 5) + 5(x + 5)$		Factor by grouping.
$= (x + 5)(x + 5)$		Distributive property.
$= (x + 5)^2$		Express as a perfect square.

We can represent this with the area model as follows:

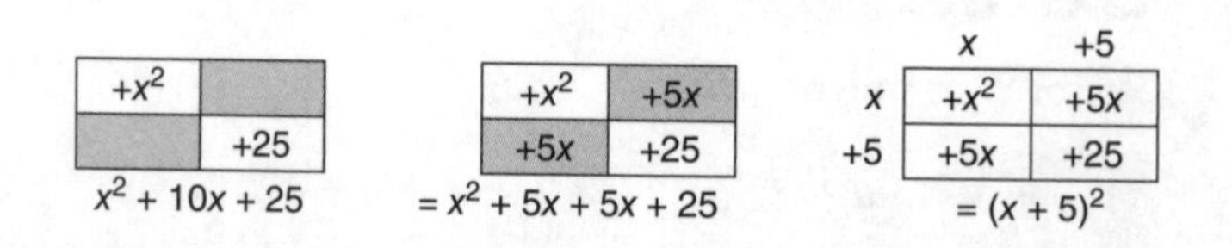

As we said in Section 8.4, expressions like the one in Example 9.7 are perfect square trinomials, which have the form $(x + y)^2 = x^2 + 2xy + y^2$. Here are other examples of perfect square trinomials in factored form:

1. $y^2 + 2y + 1 = (y + 1)^2$
2. $a^2 - 6a + 9 = (a - 3)^2$
3. $k^2 + 14km + 49m^2 = (k + 7m)^2$

Dividing the middle coefficient b by 2 gives us the last term in the binomial that is squared. In symbols, $x^2 + bx + c = \left(x + \frac{b}{2}\right)^2$.

Example 9.8 Factor $4x^2 - 25$.

Solution: Rewrite this as a trinomial in which $b = 0$: $4x^2 + 0x - 25$.

$a = 4,\ b = 0,\ c = -25$	Identify values of a, b, and c.
$ac = 4(-25) = -100$	Calculate ac.
10 and −10	List the number pairs that are factors of $ac = -100$.
$10 + -10 = 0$	Select the pair that adds up to $b = 0$.
$4x^2 + \mathbf{10x} - \mathbf{10x} - 25$	Rewrite the original trinomial using the selected pair of ac's factors.
$= 2x(2x + 5) - 5(2x + 5)$	Factor by grouping.
$= (2x - 5)(2x + 5)$	Distributive property: $ab + ac = a(b + c)$.

We can represent this with the area model as follows:

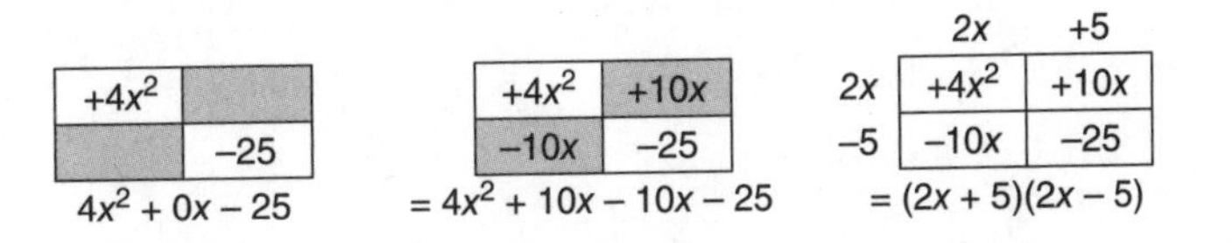

A binomial like the one in Example 9.8 is a **difference of two squares:** $\boldsymbol{x^2 - y^2 = (x + y)(x - y)}$.

The polynomials we've worked with so far in this chapter have required us to factor only once. However, some polynomials must be factored more than once to get prime polynomials. To emphasize this idea, we will now not just *factor* but *factor completely*, as in Example 9.9:

Example 9.9 Factor $3y^4 + 9y^3 - 12y^2 - 36y$ completely.

Solution:

$3y^4 + 9y^3 - 12y^2 - 36y$	
$= 3y(y^3 + 3y^2 - 4y - 12)$	Factor the GCF from each term.
$= 3y([y^3 + 3y^2] + [-4y - 12])$	Group the polynomial into binomials.
$= 3y(y^2[y + 3] + [-4][y + 3])$	Factor the GCF from each binomial—the remaining factor is the same.
$= 3y(y^2 - 4)(y + 3)$	Distributive property.
$= 3y(y + 2)(y - 2)(y + 3)$	Factor the difference of two squares: $x^2 - y^2 = (x + y)(x - y)$.

How to Factor Polynomials

1. Factor the monomial GCF.
2. Factor the remaining polynomial:
   - If it has 2 terms, try factoring using the *ac* method or the difference of two squares.
   - If it has 3 terms, try factoring using the *ac* method.
   - If it has 4 terms, try factoring by grouping.
3. Repeat step 2 until all factors are prime.

Exercises

Factor each polynomial completely.

1. $k^2 + 12k + 36$
2. $a^2 - 16a + 64$
3. $x^2 + 20x + 100$
4. $9z^2 + 6z + 1$
5. $x^4 - 4x^2 + 4$
6. $25x^2 + 20x + 4$
7. $16 - y^2$
8. $25x^2 - 64$
9. $36 - a^2$
10. $a^2 - b^4$
11. $1 - 49r^4$
12. $p - 25$
13. $c^2 - 9y^2$
14. $1 - a^2b^2$
15. $4a^2x^2 - 121z^2$
16. $10x^2 + 12x + 2$
17. $14ah^2 + 20ah + 6a$
18. $15x^2 - 12x - 3$
19. $6a^3 + 8a^2 + 2a$
20. $5x^2y^2 + 10xy^2 + 5y^2$
21. $3ax^2 + 3ax - 18a$
22. $y^4 - 2y^3 - 9y^2 + 18y$
23. $6m^4 + 3m^3 - 150m^2 - 75m$
24. $9u^4 + 45u^3 - 4u^2 - 20u$

Questions to Think About

25. Is $x^2 + 2x + 1$ a perfect square trinomial? Explain.
26. A difference of two perfect squares $x^2 - y^2$ can be factored as $(x + y)(x - y)$. Does a similar formula for the sum of two perfect squares exist? In other words, does $x^2 + y^2 = (x + y)(x + y)$? Explain.
27. Julian factored $x^3 + 7x^2 + 12x$ as $x(x^2 + 7x + 12)$. Explain why he did not factor completely.

9.5 Solving Quadratic Equations by Factoring

We now have all the tools needed to solve quadratic equations by factoring.

How to Solve Quadratic Equations by Factoring

1. Write the quadratic equation in standard form: $ax^2 + bx + c = 0$, where $a \neq 0$.
2. Factor the quadratic expression into linear factors.
3. Set each factor containing the variable equal to 0. (Recall from Section 1.1 that any number multiplied by 0 becomes 0.) This leads to an important property called the **zero-product property: if a product equals 0, then at least one of its factors must equal 0.** We write it in symbols as follows: if $ab = 0$, then $a = 0$, $b = 0$, or $a = 0$ and $b = 0$.)
4. Solve the resulting linear equations.

Example 9.10 Solve for the variable: $m^2 - 18m + 81 = 0$.

Solution:

$m^2 - 18m + 81 = 0$	The quadratic equation is already in standard form.
$a = 1, b = -18, c = 81$	Identify values of a, b, and c.
$ac = 1(81) = 81$	Calculate ac.
-9 and -9:	List the number pairs that are factors of $ac = 81$.
$-9 + (-9) = -18$	Select the pair that adds up to $b = -18$.
$m^2 - 9m - 9m + 81 = 0$	Rewrite the original trinomial using the selected pair of ac's factors.
$m(m - 9) - 9(m - 9) = 0$	Factor by grouping.
$(m - 9)(m - 9) = 0$	Distributive property.
$m - 9 = 0$ or $m - 9 = 0$	Zero-product property.
$m = 9$	Solve each linear equation for the variable.

In Example 9.10, $m = 9$ is an example of a **double root**—a root that appears twice in the solution of an equation. When writing the solution, we don't list it twice. We can simply say $m = 9$. If we use curly brackets, we say that the solution set is $\{9\}$.

Watch Out

A quadratic equation must be in standard form so that we can factor the polynomial and use the zero-product property. Sometimes, you'll need to rearrange terms, as shown in Example 9.11.

Example 9.11 Solve for the variable: $3x^2 + 13x = 10$.

Solution:

$3x^2 + 13x \quad = 10$	
$\quad - 10 - 10$	Subtract 10 from both sides.
$3x^2 + 13x - 10 = 0$	Write the quadratic equation in standard form.
$a = 3, b = 13, c = -10$	Identify values of a, b, and c.
$ac = 3(-10) = -30$	Calculate ac.
15 and -2:	List the number pairs that are factors of $ac = -30$.
$15 + (-2) = 13$	Select the pair that adds up to $b = 13$.
$3x^2 + 15x - 2x - 10 = 0$	Rewrite the original trinomial using the selected pair of ac's factors.
$3x(x + 5) - 2(x + 5) = 0$	Factor by grouping.
$(3x - 2)(x + 5) = 0$	Distributive property.
$3x - 2 = 0$ or $x + 5 = 0$	Zero-product property.
$x = \frac{2}{3}$ or $x = -5$	Solve each linear equation for the variable.

Here are some important notes about solving quadratic equations:

- Unlike linear equations (which have only one solution), quadratic equations have zero, one, or two solutions—one for each linear factor. Verify each solution by substituting it into the original equation.
- Write the solution to a quadratic equation in any of the following ways:
  - Listing each solution in its own equation, linked by "or": $x = \frac{2}{3}$ or $x = -5$
  - Listing each solution in its own equation, linked by commas: $x = \frac{2}{3}, x = -5$
  - Listing the solutions using curly brackets: $\left\{\frac{2}{3}, -5\right\}$. When we use set notation, we don't state the variable. We don't say $x = \left\{\frac{2}{3}, -5\right\}$. Instead, we say the solution set is $\left\{\frac{2}{3}, -5\right\}$.
- Since the order in which we list the solutions of a quadratic equation doesn't matter, we don't use parentheses, which indicate an *ordered* pair of *two* variables. Note the difference: $\{3, 7\}$ means $x = 3$ *or* $x = 7$, but $(3, 7)$ means $x = 3$ *and* $y = 7$.

Example 9.12 Solve for the variable: $7k^2 = 35k$.

Solution:

$7k^2 = 35k$	
$-35k \quad -35k$	Subtract $35k$ from both sides.
$7k^2 - 35k = 0$	Write the quadratic equation in standard form.
$7k(k - 5) = 0$	Factor the GCF from each term on the left side.
$7k = 0$ or $k - 5 = 0$	Zero-product property.
$k = 0$ or $k = 5$	Solve each linear equation for the variable.

Note that if we divide both sides of the equation in Example 9.12 by $7k$, we would get $\frac{7k^2}{7k} = \frac{35k}{7k}$, or $k = 5$. We would "lose" the root $k = 0$! Since $k = 0$, dividing by $7k$ is essentially dividing by 0, which as we said in Section 1.1 is meaningless.

Exercises

Solve each equation for the variable.

1. $x^2 + 4x - 21 = 0$
2. $x^2 + 6x + 8 = 0$
3. $4x^2 - 7x - 2 = 0$
4. $a^2 - 16a + 64 = 0$
5. $q^2 - 12q + 36 = 0$
6. $z^2 + 4z + 4 = 0$
7. $x^2 - 9 = 0$
8. $w^2 - 100 = 0$
9. $4x^2 - 25 = 0$
10. $9x^2 - 6x = -1$
11. $3x^2 = x$
12. $2y^2 - 5y = 25$
13. $k^2 - 4k = 0$
14. $y^2 - 2y = 0$
15. $4x^2 = 28x$
16. $8k^2 - 64k = 0$
17. $3a^2 - 18a = 0$
18. $2x^2 + 3x = 0$

Questions to Think About

19. Keiko solved the equation $x^2 - 10x = 24$ using the following work. Explain the error in her reasoning.

$$x(x - 10) = 24$$

$$x = 24 \text{ or } x - 10 = 24$$

$$x = 24 \text{ or } x = 34$$

20. Explain the difference between the solution set $\{2, 4\}$ and the coordinate $(2, 4)$.

21. Miles wrote a "proof" that $1 = 2$. Explain the conceptual error in his work.
Let $a = b$.

1. $a^2 = ab$	Multiply both sides by a.
2. $a^2 - b^2 = ab - b^2$	Subtract b^2 from both sides.
3. $(a + b)(a - b) = b(a - b)$	Factor both sides.
4. $a + b = b$	Divide by $a - b$ on both sides.
5. $a + a = a$	Substitute a for b in the equation.
6. $2a = a$	Combine like terms.
7. $2 = 1$	Divide both sides by a.

9.6 Radical Expressions

So far, we've solved quadratic equations by factoring. What if the trinomial can't be factored? For example, let's look at the equation $x^2 = 2$. When we write it in standard form ($x^2 - 2 = 0$), we see that no two factors of -2 add up to $b = 0$. However, in the same way we solve $2x = 6$ by dividing or we solve $x + 2 = 6$ by subtracting, we can "undo" the exponent in the equation by performing the inverse of squaring. To do this, we need to define a new operation called the **square root** of a number a, which is the number r such that $r^2 = a$.

This definition is somewhat ambiguous since the square root of a number could be positive or negative. For example, the square root of 9 could be $+3$ (since $3^2 = 9$) or -3 (since $(-3)^2 = (-3)(-3) = 9$). To avoid confusion, we define the **principal square root** of a number n as the *positive* number r such that $r^2 = a$.

We write the principal square root of a as $\sqrt[2]{a}$ and pronounce it as "the square root of a" or "radical a." This expression has several parts (Figure 9.7):

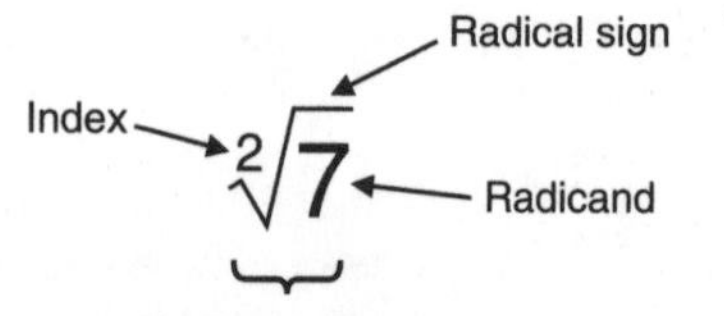

Figure 9.7 Vocabulary for radicals.

- The $\sqrt{\ }$ is called a **radical sign**.
- The number 2 is called the **index**. (Square roots have an index of 2, so the index 2 in $\sqrt[2]{a}$ represents the exponent in the equation $r^2 = a$. Since square roots occur so often, we usually write them as $\sqrt{a}$, without the index.)
- The expression underneath the radical symbol is called the **radicand**.
- The entire expression is called a **radical expression**.

We can generalize the concept for powers with exponents greater than 2 with the following definition: the ***n*th root** of a number a is the number r such that $r^n = a$.

Here are some examples:

- $\sqrt{9} = 3$ because $3^2 = 9$. Note that $-\sqrt{9}$ (read as "negative square root of 9") equals -3 since it is the opposite of $\sqrt{9}$.
- $\sqrt[3]{8} = 2$ because $2^3 = 8$. Radical expressions with an index of 3 are called **cube roots**, so $\sqrt[3]{8}$ is read "the cube root of 8."
- $\sqrt[3]{-8} = -2$ because $(-2)^3 = -8$.
- $\sqrt[4]{16} = 2$ because $2^4 = 16$.

Here are some important notes about radical expressions:

- Do not confuse a radicand's index with its coefficient. For example, $\sqrt[3]{64} = 4$, but $3\sqrt{64} = 3(8) = 24$.
- When we add, subtract, or multiply a radical by a number, we always put the radical last. This avoids confusion about what numbers are in the radicand. For example, the product of $\sqrt{5}$ and 2 is $2\sqrt{5}$ (never $\sqrt{5}2$, which could be mistaken for $\sqrt{52}$), and the sum of 2 and $\sqrt{5}$ is $2 + \sqrt{5}$ (not $\sqrt{5} + 2$, which could be mistaken for $\sqrt{5+2}$).

Since we are focusing on second-degree equations right now, we will focus on square roots in this chapter.

Just as we often simplify fractions, we can also simplify radicals. A radical expression is in **simplest form** if all of the following are true:

- The radicand has no perfect square factors (except 1).
- The radicand is not a fraction.
- The denominator contains no radicals.

A **perfect square** is a number that can be expressed as the product of two equal numbers. Table 9.2 shows some common perfect squares.

Table 9.3 shows examples of radicals that are and are not in simplest form.

In order to simplify radicals, we also need to know how to add, subtract, multiply, and divide radicals. Table 9.4 summarizes the rules for operations with radicals. All radicands are nonnegative numbers.

Table 9.2 Perfect squares.

Number (x)	Perfect Square (x^2)
2	4
3	9
4	16
5	25
6	36
7	49
8	64
9	81
10	100
11	121
12	144
13	169
14	196
15	225

Table 9.3 Examples and non-examples of radicals in simplest form.

Example	Reason	Non-Example	Reason
$\sqrt{17}$	Radicand has no perfect square factors except 1.	$\sqrt{40}$	Radicand has a perfect square factor since $40 = 4(10)$ and 4 is a perfect square.
$\sqrt{38}$	Radicand has no perfect square factors except 1.	$\sqrt{\frac{21}{5}}$	Radicand is a fraction.
$\frac{\sqrt{21}}{5}$	Equivalent to $\frac{1}{5}\sqrt{21}$ - radicand is not a perfect square and is not a fraction	$\frac{1}{\sqrt{5}}$	Denominator contains a radical.

Table 9.4 Operations with radicals.

Operation	In Words	In Symbols	Example
Addition	To add radicals with the same radicand, add their coefficients and leave the radicands unchanged.	$a\sqrt{x} + b\sqrt{x} = (a+b)\sqrt{x}$	$7\sqrt{5} + 4\sqrt{5} = 11\sqrt{5}$

Table 9.4 (*Continued*)

Operation	In Words	In Symbols	Example
Subtraction	To subtract radicals with the same radicand, subtract their coefficients and leave the radicands unchanged.	$a\sqrt{x} - b\sqrt{x} = (a - b)\sqrt{x}$	$7\sqrt{5} - 4\sqrt{5} = 3\sqrt{5}$
Multiplication	To multiply radicals, multiply their coefficients by the square root of the product of their radicands.	$a\sqrt{x}(b\sqrt{y}) = ab\sqrt{xy}$	$12\sqrt{3}(2\sqrt{5}) = 24\sqrt{15}$
Division	To divide radicals, multiply the quotient of their coefficients by the square root of the quotient of their radicands.	$\frac{a\sqrt{x}}{b\sqrt{y}} = \frac{a}{b}\sqrt{\frac{x}{y}}$ $(b \neq 0, y \neq 0)$	$\frac{12\sqrt{24}}{2\sqrt{8}} = 6\sqrt{3}$

Example 9.13 Express $5\sqrt{48}$ in simplest form.

Solution:

$5\sqrt{48}$

$= 5\sqrt{16 \cdot 3}$ — Find the largest perfect square factor in the radicand. List all factor pairs of the radicand: $48 = 1 \cdot 48 = 2 \cdot 24 = \underline{3 \cdot \mathbf{16}} = 4 \cdot 12 = 6 \cdot 8$

($3 \cdot 16$ has the largest perfect square factor.)

$= 5\sqrt{16}\sqrt{3}$ — $a\sqrt{x}(b\sqrt{y}) = ab\sqrt{xy}$.

$= 5 \cdot 4\sqrt{3}$ — Simplify the perfect square radicand.

$= 20\sqrt{3}$ — Multiply coefficients.

Finding the largest perfect square factor saves steps, as shown here:

$5\sqrt{48}$	$5\sqrt{48}$
$= 5\sqrt{16 \cdot 3}$	$= 5\sqrt{4 \cdot 12}$
$= 5\sqrt{16}\sqrt{3}$	$= 5\sqrt{4}\sqrt{12}$
$= 5 \cdot 4\sqrt{3}$	$= 5 \cdot 2\sqrt{12}$
$= 20\sqrt{3}$	$= 5 \cdot 2\sqrt{4}\sqrt{3}$
	$= 5 \cdot 2 \cdot 2\sqrt{3}$
	$= 20\sqrt{3}$

Example 9.14 Express $3\sqrt{8} - \sqrt{50} + 6\sqrt{32}$ in simplest form.

Solution:

$3\sqrt{8} - \sqrt{50} + 6\sqrt{32}$

$= 3\sqrt{4 \cdot 2} - \sqrt{25 \cdot 2} + 6\sqrt{16 \cdot 2}$ Find the largest perfect square factor in each radicand.
List all factor pairs of each radicand:
$8 = 1 \cdot 8 = \underline{2 \cdot 4}$
($2 \cdot 4$ has the largest perfect square factor.)
$50 = 1 \cdot 50 = \underline{2 \cdot 25} = 5 \cdot 10$
($2 \cdot 25$ has the largest perfect square factor.)
$32 = 1 \cdot 32 = \underline{2 \cdot 16} = 4 \cdot 8$
($2 \cdot 16$ has the largest perfect square factor.)

$= 3\sqrt{4}\sqrt{2} - \sqrt{25}\sqrt{2} + 6\sqrt{16}\sqrt{2}$ $a\sqrt{x}(b\sqrt{y}) = ab\sqrt{xy}$.

$= 3 \cdot 2\sqrt{2} - 5\sqrt{2} + 6 \cdot 4\sqrt{2}$ Simplify the perfect square radicands.

$= 6\sqrt{2} - 5\sqrt{2} + 24\sqrt{2}$ Multiply coefficients.

$= 25\sqrt{2}$ $a\sqrt{x} + b\sqrt{x} = (a+b)\sqrt{x}$.

Example 9.15 Express $\sqrt{6} \cdot \sqrt{15}$ in simplest form.

Solution:

$\sqrt{6} \cdot \sqrt{15}$

$= \sqrt{90}$ $a\sqrt{x}(b\sqrt{y}) = ab\sqrt{xy}$.

$= \sqrt{9 \cdot 10}$ Find the largest perfect square factor in the radicand.

$= \sqrt{9}\sqrt{10}$ $a\sqrt{x}(b\sqrt{y}) = ab\sqrt{xy}$.

$= 3\sqrt{10}$ Simplify the perfect square radicand.

If the denominator of a fraction has a radical expression that can't be simplified, then we often **rationalize** the denominator, meaning that we multiply the numerator and denominator by the same quantity to make the denominator rational.

Example 9.16 Express $\sqrt{\frac{20}{147}}$ in simplest form.

Solution:

$$\sqrt{\frac{20}{147}}$$

$= \dfrac{\sqrt{20}}{\sqrt{147}}$ — $\dfrac{a\sqrt{x}}{b\sqrt{y}} = \dfrac{a}{b}\sqrt{\dfrac{x}{y}}$, $b \neq 0$, $y \neq 0$.

$= \dfrac{2\sqrt{5}}{7\sqrt{3}}$ — Simplify numerator ($\sqrt{20} = \sqrt{4}\sqrt{5} = 2\sqrt{5}$) and denominator ($\sqrt{147} = \sqrt{49}\sqrt{3} = 7\sqrt{3}$).

$= \dfrac{2\sqrt{5}}{7\sqrt{3}} \cdot \dfrac{\sqrt{3}}{\sqrt{3}}$ — Multiply numerator and denominator by $\sqrt{3}$ to rationalize the denominator.

(We multiply by $\sqrt{3}$ since $\sqrt{3}\sqrt{3} = 3$, which makes the denominator rational.)

$= \dfrac{2\sqrt{5}\sqrt{3}}{7\sqrt{3}\sqrt{3}}$ — $\left(\frac{a}{b}\right)\left(\frac{c}{d}\right) = \frac{ac}{bd}$, where $b \neq 0$ and $d \neq 0$.

$= \dfrac{2\sqrt{15}}{21}$ — $a\sqrt{x}(b\sqrt{y}) = ab\sqrt{xy}$.

Finally, we consider another property of square roots. First, recall from Section 1.3 that a rational number can be expressed as the quotient of an integer and a nonzero integer. We define an **irrational number** as one that *cannot* be expressed in this way. Together, the rational and irrational numbers form what we call the **real numbers**, which we denote as **R** or $\mathbb{R}$. Figure 9.8 summarizes the relationship between the different types of numbers we discuss in this book:

Did You Know?

Irrational numbers have been studied for thousands of years. Mesopotamian mathematicians estimated square roots as early as 2000–1500 BCE. By around 750 BCE, Indian mathematical texts included problems involving square roots. Many historians attribute the first formal proof of the existence of irrational numbers to followers of the Greek philosopher Pythagoras, who lived around 500 BCE. The Egyptian Abū Kāmil Shujāʿ ibn Aslam (9th and 10th centuries) was the first mathematician to accept irrational numbers as solutions to quadratic equations.

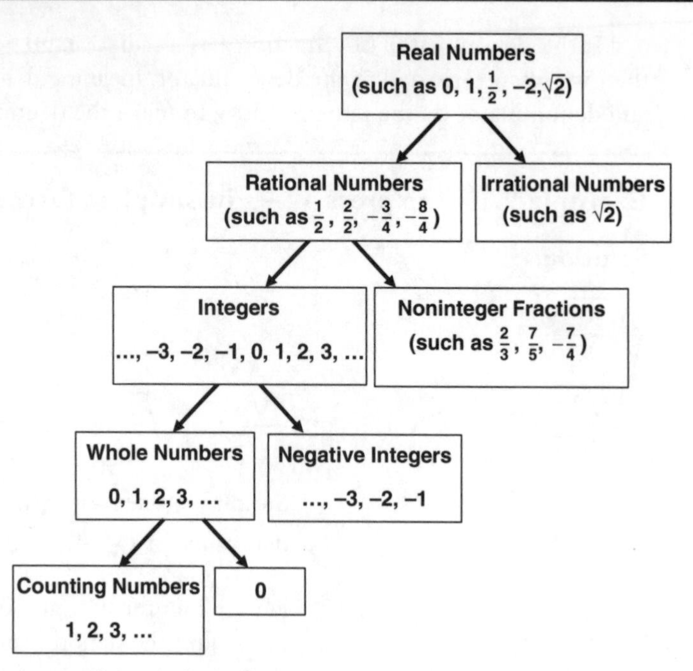

Figure 9.8 Subsets of the real numbers.

Our work so far in this section illustrates the following property:

- **If the radicand of a square root is a perfect square, then the square root is rational.**
- **If the radicand of a square root is *not* a perfect square, then the square root is irrational.**

Here are some examples:

- $\sqrt{36}$ is rational since 36 is a perfect square ($6^2 = 36$).
- $\sqrt{\frac{16}{25}}$ is rational since $\frac{16}{25}$ is a perfect square $\left(\left(\frac{4}{5}\right)^2 = \frac{16}{25}\right)$.
- $\sqrt{2}$ is irrational since 2 is not a perfect square ($\sqrt{2}$ is between 1 and 2 since $1^2 = 1$ and $2^2 = 4$. In fact, $\sqrt{2}$ is approximately 1.41421.)

When we perform operations with rational and irrational numbers, the result can be rational or irrational, as summarized here:

- **The sum, difference, or product of two rational numbers is rational.**
- **The sum, difference, or product of a rational and irrational number is irrational.**

- **The sum, difference, or product of two irrational numbers can be rational or irrational.** In these situations, simplify the expression and examine the result. For example:
  - $\sqrt{5} + \sqrt{6}$ is irrational but $\sqrt{5} - \sqrt{5}$ is rational (it equals 0).
  - $\sqrt{5}\sqrt{7}$ is irrational but $\sqrt{12}\sqrt{3}$ is rational (it equals $\sqrt{36} = 6$).

Exercises

Write the pronunciation of each expression.

1. $3\sqrt{30}$
2. $7\sqrt{7}$
3. $\sqrt[3]{16}$
4. $\sqrt[4]{8}$
5. $2\sqrt[3]{5}$
6. $3\sqrt[5]{81}$

Simplify each expression.

7. $\sqrt{49}$
8. $\sqrt{64}$
9. $2\sqrt{100}$
10. $5\sqrt{144}$
11. $\sqrt[3]{27}$
12. $\sqrt[4]{16}$
13. $3\sqrt[3]{64}$
14. $5\sqrt[4]{81}$
15. $\sqrt{\frac{1}{4}}$
16. $\sqrt{\frac{9}{16}}$
17. $6\sqrt{\frac{169}{25}}$
18. $\frac{5}{3}\sqrt{\frac{49}{64}}$
19. $\sqrt{52}$
20. $\sqrt{125}$
21. $4\sqrt{27}$
22. $3\sqrt{128}$
23. $\sqrt{\frac{2}{3}}$
24. $\sqrt{\frac{5}{7}}$
25. $\sqrt{\frac{11}{18}}$
26. $\sqrt{\frac{32}{45}}$
27. $7\sqrt{7} + 4\sqrt{7} - 3\sqrt{7}$
28. $\sqrt{8} - 2\sqrt{8} + 7\sqrt{8}$
29. $4\sqrt{7} - \sqrt{28} - \sqrt{63}$
30. $\sqrt{75} + \sqrt{48} + \sqrt{27}$
31. $3\sqrt{6} \cdot 5\sqrt{3}$
32. $\sqrt{2}\sqrt{5}\sqrt{6}$
33. $2\sqrt{6} \cdot 3\sqrt{8}$
34. $\sqrt{7}\sqrt{28}$
35. $\dfrac{12\sqrt{8}}{6\sqrt{2}}$
36. $\dfrac{15\sqrt{48}}{5\sqrt{6}}$
37. $\dfrac{2\sqrt{3}}{14\sqrt{27}}$
38. $\dfrac{2\sqrt{4}}{3\sqrt{20}}$

Determine if each expression is rational or irrational.

39. $1 + \sqrt{2}$

40. $4\sqrt{5} + 2\sqrt{5}$

41. $\sqrt{121} - 6$

42. $4(3\sqrt{7} - 2\sqrt{7})$

43. $\sqrt{3}(\sqrt{12} + \sqrt{27})$

44. $\sqrt{2}(\sqrt{8} + \sqrt{64})$

Questions to Think About

45. Arvin simplifies $\sqrt{80}$ as follows: $\sqrt{80} = \sqrt{4}\sqrt{20} = 2\sqrt{20}$. Explain the error in his reasoning.

46. In general, does $\sqrt{a} + \sqrt{b} = \sqrt{a+b}$? Explain.

47. Is $\sqrt{\frac{1}{2}}$ in simplest form? Explain.

48. Is $\frac{7}{\sqrt{3}}$ in simplest form? Explain.

49. Let $\frac{a}{b}$ and $\frac{c}{d}$ be two rational numbers, where a, b, c, and d are integers, $b \neq 0$, and $d \neq 0$. Show that their product $\frac{a}{b} \cdot \frac{c}{d}$ is a rational number. (HINT: Use the fact that the sum, difference, or product of integers is an integer.)

50. Let $\frac{a}{b}$ and $\frac{c}{d}$ be two rational numbers, where a, b, c, and d are integers, $b \neq 0$, and $d \neq 0$. Show that their sum $\frac{a}{b} + \frac{c}{d}$ is a rational number. (HINT: Use the fact that the sum, difference, or product of integers is an integer.)

9.7 Solving Quadratic Equations by Completing the Square

In Section 9.6, we discussed how the square root "undoes" squaring. We use this knowledge to solve quadratic equations without factoring, as shown here:

$x^2 = 4$

$\sqrt{x^2} = \sqrt{4}$ Take the square root of both sides.

$|x| = \sqrt{4}$ In this equation, x has two possible values. If $x \geq 0$, then taking the square root of its square gives us the nonnegative number we started with. (If $x = 2$, then $\sqrt{2^2} = 2$.) However, if $x < 0$, then taking the square root of its square gives us the *opposite* of what we started with—a positive number. (If $x = -2$, then $\sqrt{(-2)^2} = +2$, not -2.) We express this more succinctly by saying that $\sqrt{x^2} = |x|$ (recall from Section 1.1 that the absolute value of a number is never negative).

$x = \sqrt{4}$ or $x = -\sqrt{4}$	Since $\lvert +\sqrt{4}\rvert$ and $\lvert -\sqrt{4}\rvert = \sqrt{4}$, we say that if $\lvert x\rvert = a$, then $x = \pm a$.
$x = 2$ or $x = -2$	Final answer, which we can also write as ± 2 (read as "positive or negative 2," not "plus or minus 2").

Watch Out!

Make sure that you write the signs of radical expressions outside the radical. In other words, write $\pm\sqrt{4}$, not $\sqrt{\pm 4}$.

The work above shows that the solution to $x^2 = p$ is $x = \pm\sqrt{p}$. We extend it to solve any quadratic equation.

Example 9.17 Solve for the variable: $x^2 + 6x + 4 = 0$.

Solution:

$x^2 + 6x + 4 = 0$	
$\quad -4 \; -4$	Subtract the constant (4) from both sides.
$x^2 + 6x = -4$	Combine like terms.

	x	
x	$+x^2$	$+3x$
	$+3x$	

$= -4$

Express the left side ($x^2 + 6x$) as part of a perfect square. Divide $+6x$ into two equal parts ($+3x$ and $+3x$).

	x	$+3$
x	$+x^2$	$+3x$
$+3$	$+3x$	

$= -4$

Divide to find the missing term in the dimensions of the square: $\frac{3x}{x} = 3$.

	x	$+3$
x	$+x^2$	$+3x$
$+3$	$+3x$	**+9**

$= -4$ **+ 9**

Add $3^2 = 9$ to both sides to complete the square.

$(x + 3)^2 = 5$	Express the left side as a perfect square.
$\sqrt{(x+3)^2} = \sqrt{5}$	Take the square root of both sides.
$\lvert x + 3\rvert = \sqrt{5}$	$\sqrt{x^2} = \lvert x\rvert$.
$x + 3 = \pm\sqrt{5}$	If $\lvert x\rvert = a$, then $x = \pm a$.
$\quad -3 \quad -3$	Subtract 3 from both sides.
$x = -3 \pm \sqrt{5}$	Final answer, read as "negative 3 plus or minus the square root of 5."

Watch Out!

The meaning and pronunciation of the $\pm$ symbol depends on whether a quantity appears before it:

- 8 ± 3 ("8 plus or minus 3") means $8 + 3$ or $8 - 3$, meaning 11 or 5.
- ± 3 ("positive or negative 3") means $+3$ or -3.

See Section 1.1 for a discussion of the difference between operations (like addition and subtraction) and signs (positive or negative).

Example 9.18 Solve for the variable: $3x^2 - 4x - 1 = 0$.

Solution: We divide the equation by the leading coefficient of 3 in order to make the equation similar to the one in Example 9.17.

$$3x^2 - 4x - 1 = 0$$

$$\frac{3x^2}{3} - \frac{4x}{3} - \frac{1}{3} = \frac{0}{3}$$ Divide both sides by the leading coefficient.

$$x^2 - \frac{4}{3}x - \frac{1}{3} = 0$$ Simplify.

$$+\frac{1}{3} \qquad +\frac{1}{3}$$ Add the constant $\left(\frac{1}{3}\right)$ to both sides.

$$x^2 - \frac{4}{3}x = \frac{1}{3}$$ Combine like terms.

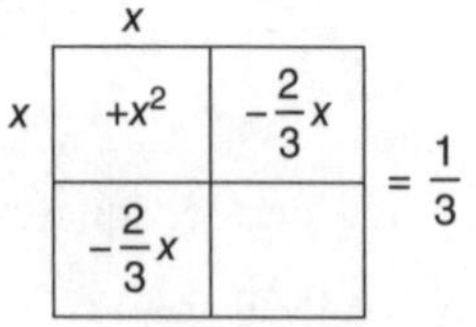

	x	
x	$+x^2$	$-\frac{2}{3}x$
	$-\frac{2}{3}x$	

$= \frac{1}{3}$

Express the left side $\left(x^2 - \frac{4}{3}x\right)$ as part of a perfect square. Divide $-\frac{4}{3}x$ into two equal parts ($-\frac{2}{3}x$ and $-\frac{2}{3}x$).

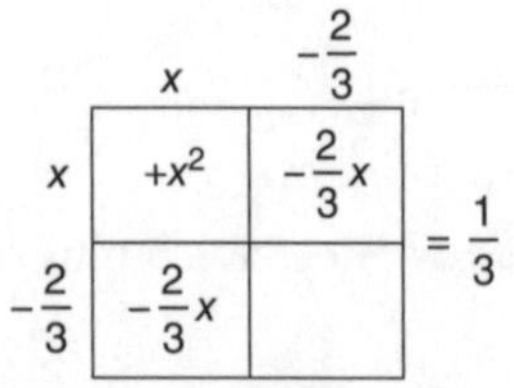

	x	$-\frac{2}{3}$
x	$+x^2$	$-\frac{2}{3}x$
$-\frac{2}{3}$	$-\frac{2}{3}x$	

$= \frac{1}{3}$

Divide to find the missing term in the dimensions of the square: $\frac{-\frac{2}{3}x}{x} = -\frac{2}{3}$.

	x	$-\frac{2}{3}$
x	$+x^2$	$-\frac{2}{3}x$
$-\frac{2}{3}$	$-\frac{2}{3}x$	$+\frac{4}{9}$

$= \frac{1}{3} + \frac{4}{9}$

Add $\left(-\frac{2}{3}\right)^2 = \frac{4}{9}$ to both sides to complete the square.

$$\left(x - \frac{2}{3}\right)^2 = \frac{1}{3} + \frac{4}{9}$$

Express the left side as a perfect square.

$$\left(x - \frac{2}{3}\right)^2 = \frac{7}{9}$$

Simplify the right side: $\frac{1}{3} + \frac{4}{9} = \frac{3}{9} + \frac{4}{9} = \frac{7}{9}$.

$$\sqrt{\left(x - \frac{2}{3}\right)^2} = \sqrt{\frac{7}{9}}$$

Take the square root of both sides.

$$\left|x - \frac{2}{3}\right| = \sqrt{\frac{7}{9}}$$

$\sqrt{x^2} = |x|$.

$$x - \frac{2}{3} = \pm\sqrt{\frac{7}{9}}$$

If $|x| = a$, then $x = \pm a$.

$$x - \frac{2}{3} = \pm\frac{\sqrt{7}}{3}$$

Simplify the right side: $\sqrt{\frac{7}{9}} = \frac{\sqrt{7}}{\sqrt{9}} = \frac{\sqrt{7}}{3}$.

$$+\frac{2}{3} \quad +\frac{2}{3}$$

Add $\frac{2}{3}$ to both sides.

$$x = \frac{2}{3} \pm \frac{\sqrt{7}}{3}$$

Final answer, which can also be written as $x = \frac{2 \pm \sqrt{7}}{3}$ or $\left\{\frac{2-\sqrt{7}}{3}, \frac{2+\sqrt{7}}{3}\right\}$.

In these examples, we changed the left side from the binomial $x^2 + bx$ to the perfect square $\left(x + \frac{b}{2}\right)^2$.

Because this method involves rewriting a quadratic equation to form a perfect square, we call it **completing the square**, which was popularized by al-Khwārizmī.

How to Solve Quadratic Equations by Completing the Square

1. Rearrange the equation into the form $ax^2 + bx = -c$.
2. If $a \neq 1$, divide both sides of the equation by a.

3. Use the area model to complete the square (or add $\left(\frac{b}{2}\right)^2$, the square of half the coefficient of x, to both sides of the equation).
4. Write the trinomial as the square of a binomial.
5. Take the square root of both sides.
6. Solve the resulting two equations for the variable.

While only some quadratic equations can be solved by factoring, any quadratic equation can be solved by completing the square. However, some equations are solved more easily using one method (you'll examine this in the exercises).

Exercises

Solve each equation for the variable by completing the square.

1. $w^2 - 20w + 96 = 0$
2. $q^2 - 18q - 144 = 0$
3. $d^2 + 14d - 72 = 0$
4. $m^2 - 4m + 1 = 0$
5. $y^2 - 8y + 5 = 0$
6. $c^2 + 10c + 23 = 0$
7. $x^2 + 2x - 5 = 0$
8. $p^2 - 6p + 2 = 0$
9. $u^2 + 12u + 7 = 0$
10. $2r^2 + 4r - 8 = 0$
11. $2a^2 - 3a - 1 = 0$
12. $5x^2 + 8x + 2 = 0$

Questions to Think About

13. Solve the equation $x^2 + 6x + 8 = 0$ by factoring and by completing the square. Which method is easier for you? Why?
14. Solve the equation $x^2 + 10x - 96 = 0$ by factoring and by completing the square. Which method is easier for you? Why?
15. Tamika believes that since the solutions to $x^2 = 9$ are $+3$ and -3, then $\sqrt{9} = \pm 3$. Explain the error in her reasoning.

9.8 Solving Quadratic Equations Using the Quadratic Formula

As said in the previous section, we can complete the square to solve any quadratic equation. However, the steps can get tricky and tedious, as shown in Table 9.5:

Table 9.5 Examples of completing the square.

$2x^2 + 11x + 3 = 0$	$3x^2 + 8x + 2 = 0$
$\frac{2x^2}{2} + \frac{11x}{2} + \frac{3}{2} = \frac{0}{2}$	$\frac{3x^2}{3} + \frac{8x}{3} + \frac{2}{3} = \frac{0}{3}$
$x^2 + \frac{11x}{2} = -\frac{3}{2}$	$x^2 + \frac{8x}{3} = -\frac{2}{3}$
$x^2 + \frac{11x}{2} + \frac{121}{16} = \frac{121}{16} - \frac{3}{2}$	$x^2 + \frac{8x}{3} + \frac{64}{36} = \frac{64}{36} - \frac{2}{3}$
$\left(x + \frac{11}{4}\right)^2 = \frac{121}{16} - \frac{3}{2}$	$\left(x + \frac{8}{6}\right)^2 = \frac{64}{36} - \frac{2}{3}$
$\left\lvert x + \frac{11}{4} \right\rvert = \frac{121}{16} - \frac{3}{2}$	$\left\lvert x + \frac{8}{6} \right\rvert = \frac{64}{36} - \frac{2}{3}$
$\sqrt{\left(x + \frac{11}{4}\right)^2} = \sqrt{\frac{121}{16} - \frac{3}{2}}$	$\sqrt{\left(x + \frac{8}{6}\right)^2} = \sqrt{\frac{64}{36} - \frac{2}{3}}$
$x + \frac{11}{4} = \pm\sqrt{\frac{121}{16} - \frac{3}{2}}$	$x + \frac{8}{6} = \pm\sqrt{\frac{64}{36} - \frac{2}{3}}$
$x = -\frac{11}{4} \pm \sqrt{\frac{121}{16} - \frac{24}{16}}$	$x = -\frac{8}{6} \pm \sqrt{\frac{64}{36} - \frac{24}{36}}$
$x = -\frac{11}{4} \pm \frac{\sqrt{121 - 24}}{\sqrt{16}}$	$x = -\frac{8}{6} \pm \frac{\sqrt{64 - 24}}{\sqrt{36}}$
$x = -\frac{11}{4} \pm \frac{\sqrt{121 - 24}}{4}$	$x = -\frac{8}{6} \pm \frac{\sqrt{64 - 24}}{6}$

Fortunately, though, we see several patterns here. The denominator is twice the leading coefficient, and the square of the coefficient of the middle term appears in the radicand. If we replace the coefficients with variables, we derive what we call the **quadratic formula: if $ax^2 + bx + c = 0$ and $a \neq 0$, then $x = \frac{-b \pm \sqrt{b^2 - 4ac}}{2a}$.**

Compared to completing the square, using the quadratic formula reduces the steps involved in solving equations.

Example 9.19 Solve for the variable: $3j^2 - 9j + 5 = 0$.

Solution:

$a = 3$, $b = -9$, $c = 5$ — Determine the values of a, b, and c.

$j = \frac{-(-9) \pm \sqrt{(-9)^2 - 4(3)(5)}}{2(3)}$ — Substitute into the quadratic formula.

$= \dfrac{9 \pm \sqrt{21}}{6}$ Final answer. The answer can also be written as:

- $j = \frac{3}{2} + \frac{\sqrt{21}}{6}$ or $j = \frac{3}{2} - \frac{\sqrt{21}}{6}$
- $\left\{ \frac{3}{2} - \frac{\sqrt{21}}{6}, \frac{3}{2} + \frac{\sqrt{21}}{6} \right\}$

Technology Tip

When substituting into the quadratic formula, use technology to calculate the radicand. Type parentheses carefully when substituting negative numbers. For Example 9.19, type $(-9)^2 - 4(3)(5)$ (note the parentheses around -9), not $-9^2 - 4(3)(5)$.

Like completing the square, the quadratic formula can be used to solve any quadratic equation. However, the equation must be in standard form.

Example 9.20 Solve for the variable: $2x^2 + 4x = 7$.

Solution:

$2x^2 + 4x = 7$

$\quad\quad -7 \quad -7$ Subtract 7 from both sides.

$2x^2 + 4x - 7 = 0$ Write the equation in standard form.

$a = 2,\ b = 4,\ c = -7$ Determine the values of a, b, and c.

$x = \dfrac{-4 \pm \sqrt{4^2 - 4(2)(-7)}}{2(2)}$ Substitute into the quadratic formula.

$= \dfrac{-4 \pm \sqrt{72}}{4} = \dfrac{-4 \pm 6\sqrt{2}}{4}$ Simplify $\left(\sqrt{72} = \sqrt{36}\sqrt{2} = 6\sqrt{2}\right)$.

$= \dfrac{2(-2 \pm 3\sqrt{2})}{2(2)}$ Factor the GCF from the numerator and the denominator.

$= \dfrac{-2 \pm 3\sqrt{2}}{2}$ Simplify. The answer can also be written as $-1 \pm \frac{3\sqrt{2}}{2}$ or $\left\{-1 - \frac{3\sqrt{2}}{2}, -1 + \frac{3\sqrt{2}}{2}\right\}$.

The quadratic formula enables us to prove that **if $ax^2 + bx + c = 0$ and $a \neq 0$, then the roots have a sum of $-\frac{b}{a}$ and a product of $\frac{c}{a}$.**

Example 9.21 Without solving for the variable, find the sum and product of the roots of the equation $3x^2 - 7x + 4 = 0$.

Solution: From the equation, $a = 3$, $b = -7$, and $c = 4$. The roots have a sum of $-\frac{b}{a} = -\frac{-7}{3} = \frac{7}{3}$ and a product of $\frac{4}{3}$.

We can check our answer by solving the equation for x.

$3x^2 - 7x + 4 = 0$	
$(3x - 4)(x - 1) = 0$	Factor the left side.
$3x - 4 = 0$ or $x - 1 = 0$	Zero-product property.
$x = \frac{4}{3}$ or $x = 1$	Solve each linear equation for the variable.

The roots have a sum of $\frac{4}{3} + 1 = \frac{4}{3} + \frac{3}{3} = \frac{7}{3}$ and a product of $\frac{4}{3}(1) = \frac{4}{3}$.

Exercises

Solve each equation for the variable using the quadratic formula.

1. $x^2 - 5x + 6 = 0$
2. $a^2 - 25 = 0$
3. $r^2 + 10r + 21 = 0$
4. $y^2 - 5y - 14 = 0$
5. $2x^2 + 11x + 6 = 0$
6. $m^2 - 7m + 8 = 0$
7. $x^2 - 9x + 6 = 0$
8. $p^2 - 5p + 1 = 0$
9. $k^2 - 4k - 8 = 0$
10. $2j^2 - 8j + 1 = 0$
11. $3h^2 - 6h - 5 = 0$
12. $4w^2 + 2w - 3 = 0$

Without solving the equation, determine the sum and product of the roots.

13. $2x^2 - 5x - 6 = 0$
14. $4x^2 + 9x + 1 = 0$
15. $7k^2 + k - 8 = 0$
16. $3m^2 - 5m + 2 = 0$

Questions to Think About

17. Let $ax^2 + bx + c = 0$ (where $a \neq 0$). Use the method of completing the square to show that $x = \frac{-b \pm \sqrt{b^2 - 4ac}}{2a}$.
18. Use the quadratic formula to show that the roots of $ax^2 + bx + c = 0$ have a sum of $-\frac{b}{a}$ and a product of $\frac{c}{a}$.
19. To solve the equation $2x^2 - 7x + 1 = 0$, Noah substituted into the quadratic formula as follows $x = \frac{-7 \pm \sqrt{-7^2 - 4(2)(1)}}{2(2)} = \frac{-7 \pm \sqrt{-49 - 8}}{4} = \frac{-7 \pm \sqrt{-57}}{4}$. Explain the errors in his work.

9.9 Graphing Quadratic Functions

Let's look at equations in the form $x^2 = m$, where m is a number. We see the following:

- If $m > 0$, $x^2 = m$ has 2 distinct real solutions: $\pm\sqrt{m}$. For example, the solutions to $x^2 = 4$ are $x = \sqrt{4} = 2$ and $x = -\sqrt{4} = -2$.
- If $m = 0$, $x^2 = m$ has 1 distinct real solution: the double root $x = 0$.
- If $m < 0$, $x^2 = m$ has no real solutions. For example, no real number is a solution to the equation $x^2 = -1$ since no real number multiplied by itself equals -1. (In an advanced algebra course, you'll learn how to work with such equations, but that's beyond the scope of this book.)

These observations suggest that the graph of the equation $y = x^2$ is symmetrical. We verify this by creating a table of values and a graph (Figure 9.9):

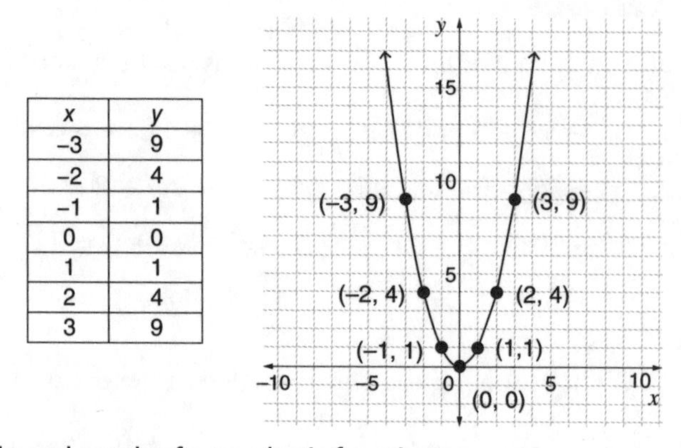

x	y
–3	9
–2	4
–1	1
0	0
1	1
2	4
3	9

Figure 9.9 Table and graph of a quadratic function.

The graph of a quadratic function is not linear. It is a curve called a **parabola** (pronounced puh-RA-buh-luh). Here are some important characteristics of parabolas:

- A parabola has an **axis of symmetry**, which is a line that divides it into two identical parts. The graph is decreasing on one side of the parabola and increasing on the other.
- The axis of symmetry passes through the **vertex**, the point where the graph changes from decreasing to increasing or increasing to decreasing (also called a **turning point**).
- Although a parabola is often described as U-shaped, its sides are not strictly vertical like a U but flare outward slightly like a V.
- We draw arrows on the ends of the parabola to show that it extends infinitely in those directions.

We write the equation of a parabola in three different ways:

- Standard form: $y = ax^2 + bx + c$
- Factored form: $y = a(x - x_1)(x - x_2)$, where x_1 and x_2 are the roots of $ax^2 + bx + c = 0$
- Vertex form: $y = a(x - h)^2 + k$, where (h, k) is the vertex

We use these forms to determine the characteristics of the parabola's graph. Figure 9.10 shows the graph of the equation $y = x^2 - 4x - 12$:

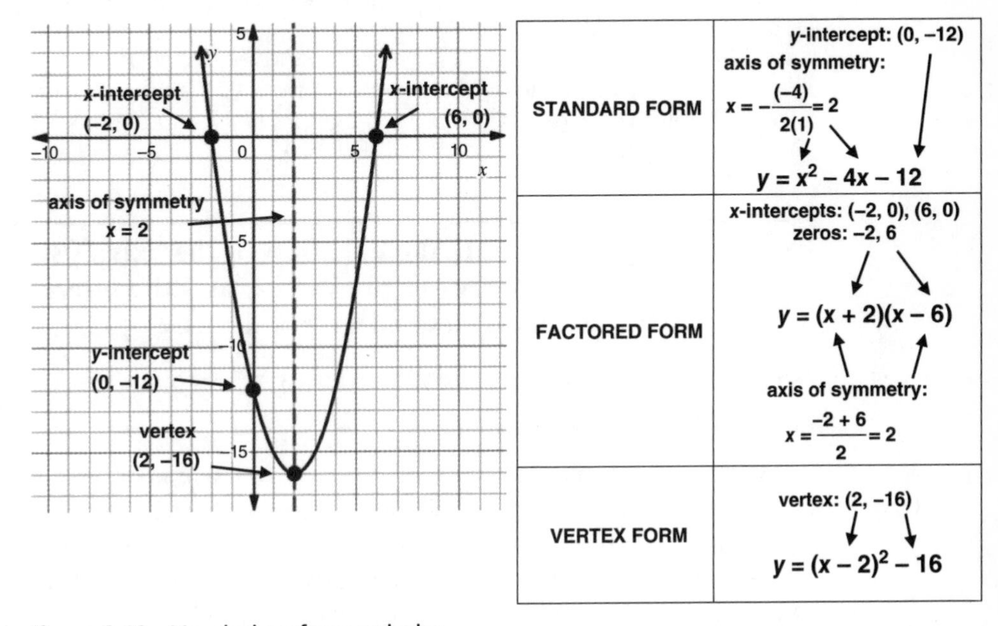

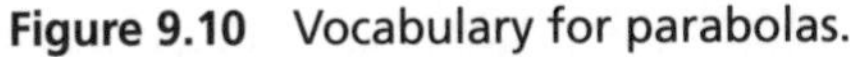

Figure 9.10 Vocabulary for parabolas.

Here are some things to keep in mind about the graphs of quadratic equations:

- The axis of symmetry is midway between the x-intercepts.
- We find the equation of the axis of symmetry from the standard form by using the formula $x = -\frac{b}{2a}$ or from the factored form by using the formula $x = \frac{x_1+x_2}{2}$, which is the mean, or average, of the zeros. (We talk about the mean more in Section 13.3.)
- The coordinates of the vertex can be found in two ways:
  - From the standard form of the equation, determine the equation of the axis of symmetry. This is the x-coordinate of the vertex. Then substitute the x-value into the equation to find the y-coordinate of the vertex.
  - Rewrite the equation in vertex form by completing the square.

Watch Out!

Make sure you know the difference between x-intercepts, zeros, and roots:

- The x-intercepts of the graph of f are the points where the graph intersects the x-axis—in other words, the *ordered pairs* where $f(x) = 0$.
- The zeros of f are the x-values of the x-intercepts—in other words, the *x-values* where $f(x) = 0$.
- The roots of an equation are the *values* that make the equation true.

Graphs have x-intercepts, *functions* have zeros, and *equations* have roots. Thus, the x-coordinates of the x-intercepts of the graph of $y = (x - x_1)(x - x_2)$ are also the zeros of the function $f(x) = (x - x_1)(x - x_2)$ and the roots of the equation $(x - x_1)(x - x_2) = 0$.

Example 9.22 Determine the equation of the axis of symmetry, the coordinates of the vertex, and the intercepts of the parabola defined by $y = x^2 - 8x - 33$.

Solution: Axis of symmetry: Use the formula $x = -\frac{b}{2a}$. Since $a = 1$ and $b = -8$, then the equation of the axis of symmetry is $x = -\frac{(-8)}{2(1)} = \frac{8}{2}$, or $x = 4$.

Vertex: From the equation of the axis of symmetry, we know that the x-coordinate of the vertex is 4. To find the y-coordinate, substitute $x = 4$ into the equation: $4^2 - 8(4) - 33 = -49$. The coordinates of the vertex are $(4, -49)$.

x-intercepts: To find the x-intercepts, solve the equation $y = 0$: $x^2 - 8x - 33 = 0$, so $(x + 3)(x - 11)$. The x-intercepts are $(-3, 0)$ and $(11, 0)$.

y-intercept: To find the y-intercept, find the value of c when the equation is written in standard form. Since $c = -33$, then the y-intercept is $(0, -33)$.

Example 9.23 Determine the equation of the axis of symmetry, the coordinates of the vertex, the intercepts, and the zeros of the parabola defined by $f(x) = (x - 5)(x + 21)$.

Solution: Axis of symmetry: we could write the equation in standard form by multiplying the binomials and then using the equation $x = -\frac{b}{2a}$. However, this requires extra work. We can use the factored form given here. Since the axis of symmetry is midway between the x-intercepts, we find its location by taking the mean of the zeros.

$(x-5)(x+21)=0$ — Set $f(x)=0$.

$x-5=0$ or $x+21=0$ — Zero-product property.

$x=5$ or $x=-21$ — Solve each linear equation for the variable.

$$\frac{5+(-21)}{2}=-\frac{16}{2}=-8$$

Find the mean of the roots. The equation of the axis of symmetry is $x=-8$.

Vertex: The vertex has an x-coordinate of -8. To find its y-coordinate, substitute -8 into the equation: $(-8-5)(-8+21)=(-13)(13)=-169$. The vertex has coordinates $(-8,-169)$.

x-intercepts: Find the roots of the equation $y=0$ that we found earlier. The x-intercepts are $(5, 0)$ and $(-21, 0)$.

Zeros: The zeros are 5 and -21.

y-intercept: Find the point where $x=0$ by substituting $x=0$ into the equation: $(0-5)(0+21)=(-5)(21)=-105$. The y-intercept is $(0, -105)$.

Example 9.24 Write the equation $y=x^2-2x-8$ in vertex form and determine the coordinates of the parabola's vertex.

Solution: We write the equation in vertex form by completing the square.

$$y=x^2-2x-8$$
$$\underline{+8\qquad\qquad +8}$$
$$y+8=x^2-2x$$

Add the constant (8) to both sides.
Combine like terms.

$y+8=$

	x	
x	$+x^2$	$-x$
	$-x$	

Express the right side (x^2-2x) as a perfect square. Divide $-2x$ into two equal parts ($-x$ and $-x$).

$y+8=$

	x	-1
x	$+x^2$	$-x$
-1	$-x$	

Divide to find the missing term in the dimensions of the square: $\frac{-x}{x}=-1$.

$y+8+1=$

	x	-1
x	$+x^2$	$-x$
-1	$-x$	$+1$

Add $(-1)^2=1$ to both sides to complete the square.

$y+9=(x-1)^2$ — Express the right side as a perfect square.

$y=(x-1)^2-9$ — Subtract 9 from both sides to write the equation in vertex form.

Writing the equation of a parabola in vertex form allows us to see that the vertex is $(1, -9)$.

How to Graph Quadratic Equations on Paper

1. Create a table of values with at least five points:
   - the vertex
   - the x- and y-intercepts
   - at least one additional point on each side of the axis of symmetry
2. Plot the points on the coordinate plane. If necessary, adjust the scale of your axes so that the parabola is roughly centered on the plane. It should be symmetrical and continuous (meaning that it should not have any "gaps").
3. Connect the points smoothly. Make sure that the curve that you draw does not have vertical sides like a U or come to a sharp point like a V.
4. Draw arrows at the ends of the parabola to indicate that it continues forever in those directions.

Avoid these common mistakes when graphing parabolas (Figure 9.11):

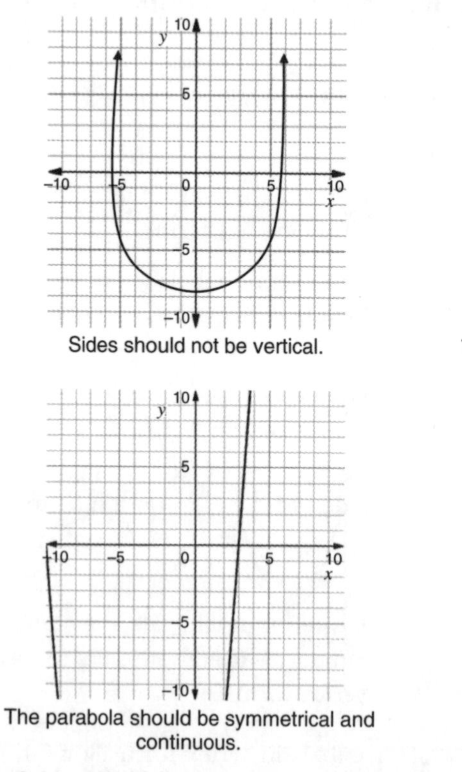

Sides should not be vertical.

The parabola should be symmetrical and continuous.

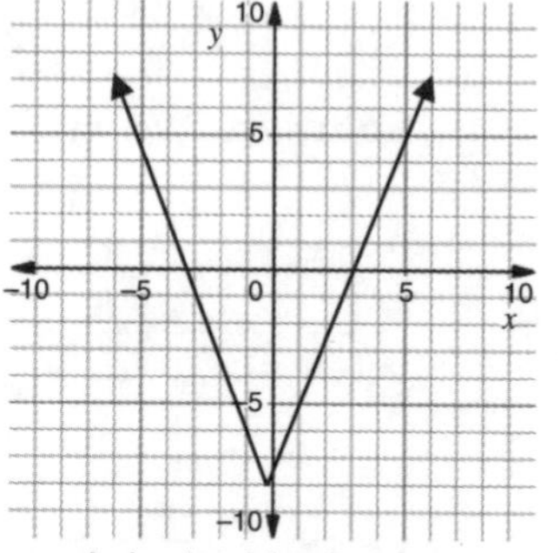

The parabola should not come to a point.

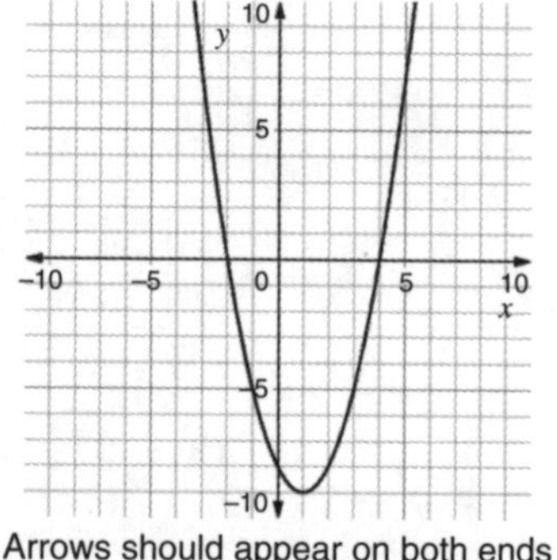

Arrows should appear on both ends (unless the domain is restricted).

Figure 9.11 Mistakes to avoid when graphing parabolas.

Example 9.25 Graph the equation $y = x^2 - 4x - 5$ on the coordinate plane.

Solution: Determine the location of the axis of symmetry: $x = -\frac{b}{2a} = -\frac{(-4)}{2(1)} = -(-2) = 2$. Since $x^2 - 4x - 5 = (x - 5)(x + 1)$, then the x-intercepts are (5, 0) and (−1, 0). To determine the y-intercept, find the value of c when the equation is written in standard form. Since $c = -5$, the y-intercept is (0, −5). Then create a table of values that is centered around $x = 2$ and contains the x- and y-intercepts. Note the symmetry in the table around $x = 2$:

x	y
5	0
4	$4^2 - 4(4) - 5 = -5$
3	$3^2 - 4(3) - 5 = -8$
2	$2^2 - 4(3) - 5 = -9$
1	$1^2 - 4(1) - 5 = -8$
0	$0^2 - 4(0) - 5 = -5$
−1	0

Use the ordered pairs from the table to graph the parabola.

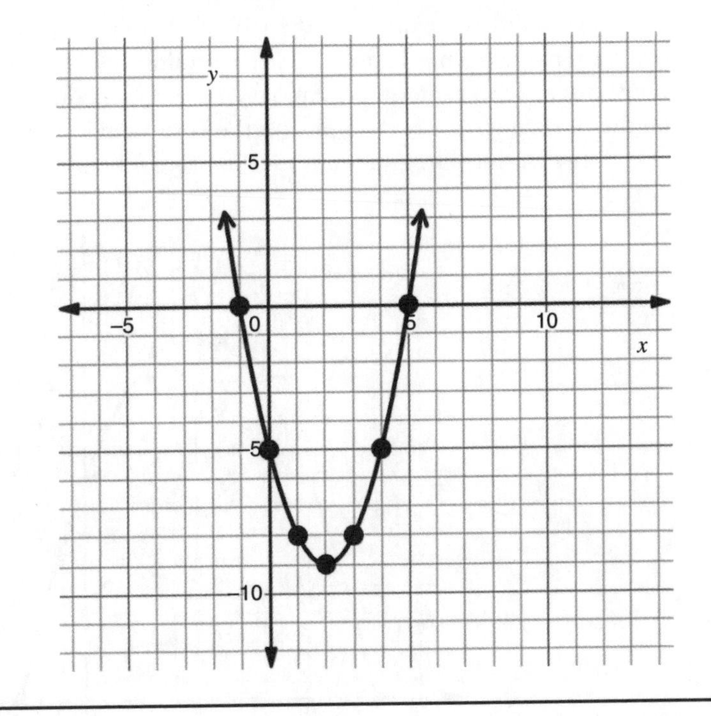

Example 9.26 Graph the function $f(x) = -x^2 - 10x + 39$ on the coordinate plane.

Solution: Find the axis of symmetry: $x = -\frac{b}{2a} = -\frac{(-10)}{2(-1)} - \frac{(-10)}{(-2)} = -5$. Since $-x^2 - 10x + 39 = -1(x^2 + 10x - 39) = -1(x + 13)(x - 3)$, then the x-intercepts are $(-13, 0)$ and $(3, 0)$. Since $c = 39$, then the y-intercept is $(0, 39)$. Then create a table of values that is centered around $x = -5$ and contains the x- and y-intercepts. Since the parabola is symmetrical and we know that $(0, 39)$ (which is 5 units to the right of the axis of symmetry) is on the graph, then we also know that the point $(-10, 39)$ (which is 5 units to the left of the axis of symmetry) is also on the graph.

x	$f(x)$
-13	0
-10	39
-6	$-(-6)^2 - 10(-6) + 39 = 63$
-5	$-(-5)^2 - 10(-5) + 39 = 64$
-4	$-(-4)^2 - 10(-4) + 39 = 63$
0	39
3	0

Use the ordered pairs from the table to graph the parabola. Since the x-values in our table range from -13 to 0 and the y-values in our table range from 0 to 64, we adjust our scale to center the parabola on the coordinate plane. Each box equals 1 unit on the horizontal axis and 5 units on the vertical axis.

Note that since the equation is written as a function, we label the vertical axis $f(x)$, not y.

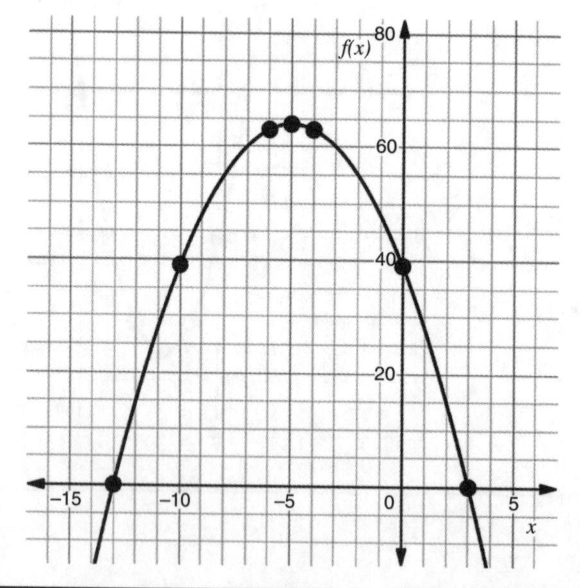

These examples show that for a parabola whose equation is $y = ax^2 + bx + c$:

- If $a > 0$, the parabola **opens upward** (it extends infinitely in the positive-y direction, like the parabola in Example 9.25) and the vertex is a minimum.
- If $a < 0$, the parabola **opens downward** (it extends infinitely in the negative-y direction, like the parabola in Example 9.26) and the vertex is a maximum.

Table 9.6 summarizes the properties of parabolas that we can determine from their equations:

Table 9.6 Properties determined from equations of parabolas.

Form	Equation	Characteristic
Standard Form	$y = ax^2 + bx + c$	axis of symmetry: $x = -\frac{b}{2a}$
		vertex: $\left(-\frac{b}{2a},\ a\left(-\frac{b}{2a}\right)^2 + b\left(-\frac{b}{2a}\right) + c\right)$
		y-intercept: $(0, c)$
Factored Form	$y = a(x - x_1)(x - x_2)$	x-intercepts: $(x_1, 0)$, $(x_2, 0)$
		zeros: x_1, x_2
		axis of symmetry: $x = \frac{x_1 + x_2}{2}$
Vertex Form	$y = a(x - h)^2 + k$	vertex: (h, k)

Exercises

For the parabola defined by each equation, determine the equation of the axis of symmetry, the coordinates of the vertex, and the intercepts of the parabola.

1. $y = x^2 + 8x + 12$
2. $y = x^2 - 6x + 8$
3. $y = (x - 10)(x + 20)$
4. $y = (2x + 1)(x - 4)$

Write each equation in vertex form and determine the coordinates of the parabola's vertex.

5. $y = x^2 + 8x + 6$
6. $y = x^2 - 2x - 6$
7. $y = x^2 + 6x + 4$
8. $y = x^2 - 10x + 11$
9. $y = x^2 - 16x + 75$
10. $y = x^2 + 14x + 51$

Graph each equation on the coordinate plane.

11. $y = x^2 + 4x - 5$
12. $y = x^2 + 2x - 8$
13. $y = x^2 - 4x + 3$
14. $y = 4x^2 + 7x - 2$
15. $y = -x^2 + 2x + 15$
16. $y = -2x^2 + 9x - 4$

Questions to Think About

17. If the x-intercepts of a quadratic equation are $(p, 0)$ and $(q, 0)$, how do we determine the equation of the axis of symmetry?

18. The graph of a quadratic function has a vertex at (3, 7). The function increases over the interval $2 < x < 3$ and decreases over the interval $3 < x < 4$. Is the vertex a relative maximum or a relative minimum? Explain.

19. Ilham graphed the equation $y = x^2 - 9x - 70$ as follows. Explain the errors in her graph.

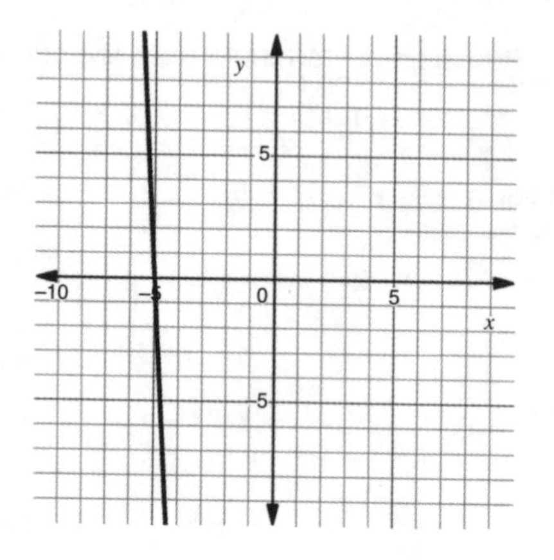

9.10 Solving Quadratic Equations by Graphing

We use the techniques that we discussed in Section 9.9 to solve quadratic equations by graphing. Graphing can be helpful when we only need an approximate answer, when we want to avoid algebraic work, or when we want to check an answer obtained through another method.

To solve quadratic equations in the form $ax^2 + bx + c = 0$ graphically, we do the following:

1. Graph the equation $y = ax^2 + bx + c$.
2. Label the x-intercepts of the graph. The x-coordinates of the x-intercepts are the roots of the equation $ax^2 + bx + c = 0$.

> **Technology Tip**
>
> Use technology to find the intercepts for the graphs of quadratic equations. Once you find the intercepts, you can factor the expression or solve the equation, depending on the problem.

Example 9.27 Solve the equation $x^2 - 2x - 3 = 0$ by graphing. Use the graph to factor $x^2 - 2x - 3$ if possible.

Solution: Graph the equation $y = x^2 - 2x - 3$ on the coordinate plane:

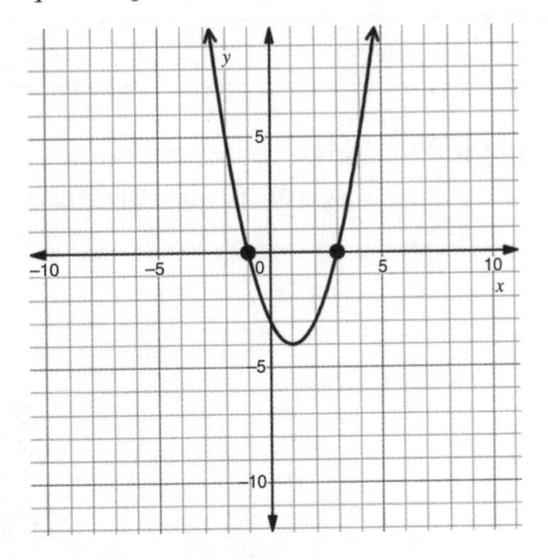

Since the graph has x-intercepts $(-1, 0)$ and $(3, 0)$, then the equation $x^2 - 2x - 3 = 0$ has two roots: $x = -1$ and $x = 3$.

If $x = -1$ is a root, then $x + 1$ is a factor. If $x = 3$ is a root, then $x - 3$ is a factor. The equation in factored form is $(x + 1)(x - 3) = 0$, so the factored form of $x^2 - 2x - 3$ is $(x + 1)(x - 3)$.

Example 9.28 Solve the equation $x^2 - 8x + 16 = 0$ by graphing. Use the graph to factor $x^2 - 8x + 16$ if possible.

Solution: Graph the equation $y = x^2 - 8x + 16$ on the coordinate plane. Here, we adjusted the scale of our graphs to fit the intercepts and show the parabola's symmetry:

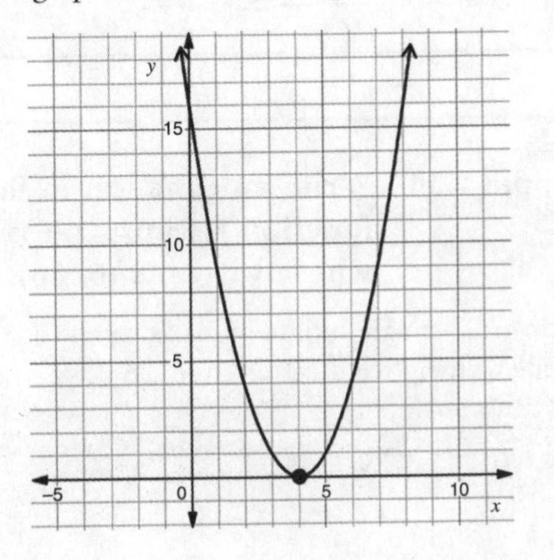

Since the graph has an x-intercept at (4, 0), then the equation $x^2 - 8x + 16 = 0$ has one double root: $x = 4$.

If $x = 4$ is a double root, then $x - 4$ is a perfect square factor. The equation in factored form is $(x - 4)^2 = 0$, so the factored form of $x^2 - 8x + 16$ is $(x - 4)^2$.

Example 9.29 Solve the equation $x^2 + 5x + 7 = 0$ by graphing. Use the graph to factor $x^2 + 5x + 7$ if possible.

Solution: Graph the equation $y = x^2 + 5x + 7$ on the coordinate plane:

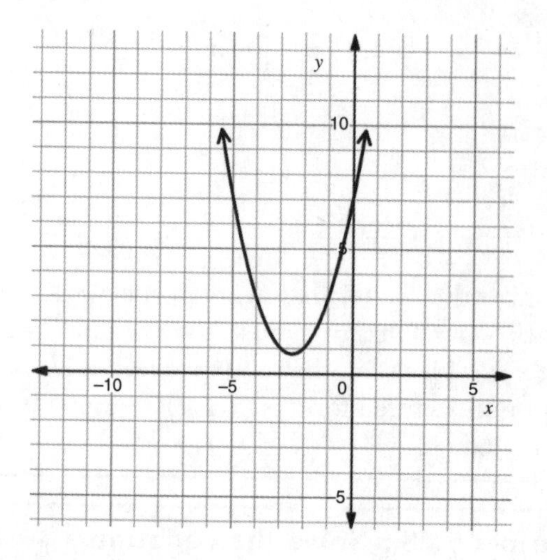

Since this graph has no x-intercepts, then the equation $x^2 + 5x + 7 = 0$ has no real roots. We could confirm this by solving the equation algebraically using the quadratic formula: $x = \frac{-5 \pm \sqrt{5^2 - 4(1)(7)}}{2(1)} = \frac{-5 \pm \sqrt{-3}}{2}$.

Example 9.30 Write an equation in factored form of the quadratic function f whose zeros are $x = 3$ and $x = 9$ and whose vertex is (6, 36).

Solution: There are infinitely many parabolas whose zeros are $x = 3$ and $x = 9$, but only one of them has a vertex of (6, 36):

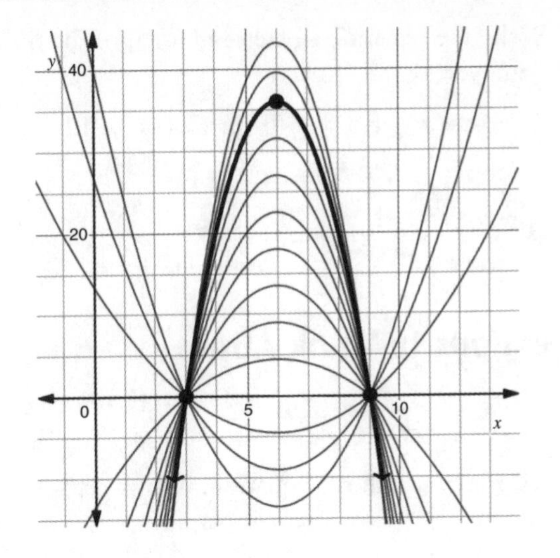

All of these equations have the form $f(x) = a(x - 3)(x - 9)$. We need to find the value of a.

$f(x) = a(x - 3)(x - 9)$	
$36 = a(6 - 3)(6 - 9)$	Substitute $x = 6$ and $f(6) = 36$ (the vertex coordinates) into the equation.
$36 = -9a$	Simplify.
$\frac{36}{-9} = \frac{-9a}{-9}$	Divide both sides by -9.
$-4 = a$	The equation is $f(x) = -4(x - 3)(x - 9)$.

Examples 9.27 through 9.30 show that for a quadratic function $y = f(x)$:

- If its graph crosses the x-axis at two points, then the equation $f(x) = 0$ has *two real and unequal roots* at x_1 and x_2, and $f(x) = a(x - x_1)(x - x_2)$.
- If its graph touches the x-axis at one point but doesn't cross the axis, then the equation $f(x) = 0$ has *one double root* at x_1, and $f(x) = a(x - x_1)^2$.
- If its graph doesn't intersect the x-axis, then the equation $f(x) = 0$ has *no real roots*.

Exercises

Solve each equation by graphing and write the equation in factored form if possible.

1. $x^2 + 5x + 4 = 0$

2. $x^2 + 2x - 8 = 0$

3. $x^2 - 3x = 0$

4. $x^2 - 6x + 9 = 0$

5. $x^2 - 4x + 5 = 0$

6. $x^2 + 10x + 25 = 0$

7. $-x^2 - 5x - 6 = 0$

8. $-x^2 - 22x - 105 = 0$

9. $-x^2 + 2x + 80 = 0$

Write the equation in factored form of the quadratic function $f(x)$ with the given zeros and vertex.

10. Zeros: $x = -4$, $x = -8$; vertex $(-6, -12)$

11. Zeros: $x = 2$, $x = 10$; vertex $(6, -32)$

12. Zeros: $x = -1$, $x = -5$; vertex $(-3, 2)$

13. Zeros: $x = -2$, $x = 4$; vertex $(1, 3)$

Questions to Think About

14. The vertex of the graph of $y = -x^2 + bx + c$ is $(3, 2)$. How many real roots does the equation $-x^2 + bx + c = 0$ have? Explain.

15. A parabola whose equation is $y = ax^2 + bx + c$ has a vertex located at its x-intercept. How many roots does the equation $ax^2 + bx + c = 0$ have? Explain.

16. A parabola whose equation is $y = ax^2 + bx + c$ has a vertex of $(1, 4)$ and a y-intercept of $(0, 5)$. How many real roots does the equation $ax^2 + bx + c = 0$ have? Explain.

9.11 Solving Quadratic Equations by the Mean-Product Method

In this section, we discuss another method for solving quadratic equations, which combines completing the square with graphing quadratic equations. It incorporates techniques used by the ancient Babylonians and Greeks as well as strategies described by François Viète (17th century), math teacher John Savage (1989), and mathematician Po-Shen Loh (2019).

Like completing the square, this method, which we call the **mean-product method** for solving quadratic equations, can be used to solve *any* quadratic equation. It avoids the tricky substitutions that can arise when using the quadratic formula and the frustrating trial-and-error that can occur when factoring. This method also allows you to draw a picture to help you see the steps.

The mean-product method is based on the idea that the axis of symmetry is midway between the x-intercepts of the parabola (Figure 9.12). For an equation in the form $x^2 + bx + c = 0$ whose roots are x_1 and x_2:

- If the equation of the axis of symmetry is $x = m$, then m is the mean of the roots and $m = -\frac{b}{2a} = -\frac{b}{2(1)} = -\frac{b}{2}$.
- If the product of the roots is $\frac{c}{a} = \frac{c}{1} = c$, then the roots are $x = m - u$ and $x = m + u$, where u represents the distance from the axis of symmetry to one of the roots.

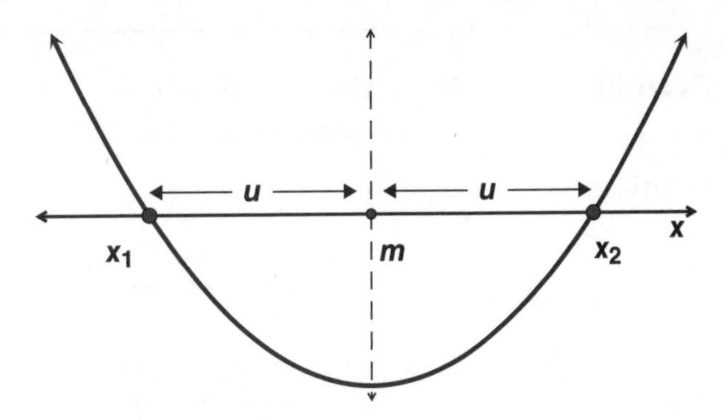

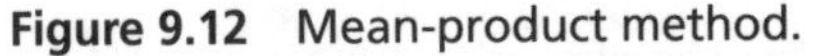

Figure 9.12 Mean-product method.

How to Solve Quadratic Equations Using the Mean-Product Method

1. Write the quadratic equation in the form $x^2 + bx + c = 0$. (If necessary, divide the equation by the leading coefficient.)
2. Find the mean of the roots, m, using the equation of the axis of symmetry: $m = -\frac{b}{2}$.
3. Write an equation representing the product of the roots: $(m + u)(m - u) = c$. Solve for u.
4. The roots of the equation are $x = m + u$ and $x = m - u$.

Example 9.31 Solve the equation $x^2 - 18x + 32 = 0$ using the mean-product method.

Solution:

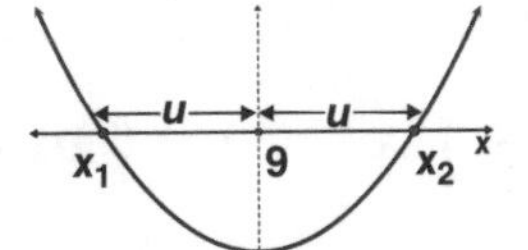

$b = -18$, so the equation of the axis of symmetry is $x = -\frac{b}{2} = -\frac{(-18)}{2} = 9$.

$x_1 x_2 = c = 32$	Product of roots.
$x_1 = 9 - u,\ x_2 = 9 + u$	The axis of symmetry is midway between the roots.
$(9 - u)(9 + u) = 32$	Substitute into the equation $x_1x_2 = 32$.
$81 - u^2 = 32$	Multiply the left side.
$-32 + u^2 \quad -32 + u^2$	Subtract 32 from and add u^2 to both sides.
$49 = u^2$	Simplify.
$\sqrt{49} = 7 = u$	Take the square root of both sides. (Since the parabola is symmetrical, we only need the principal square root.)

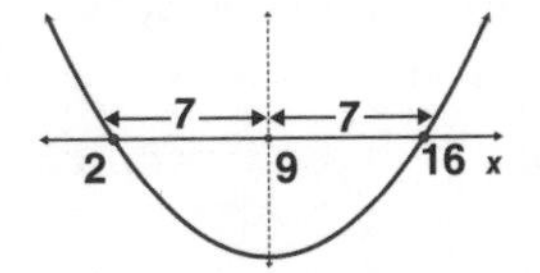

Determine the roots: $x_1 = 9 - 7 = 2$, $x_2 = 9 + 7 = 16$.

Example 9.32 Solve the equation $x^2 + 21x + 80 = 0$ using the mean-product method.

Solution:

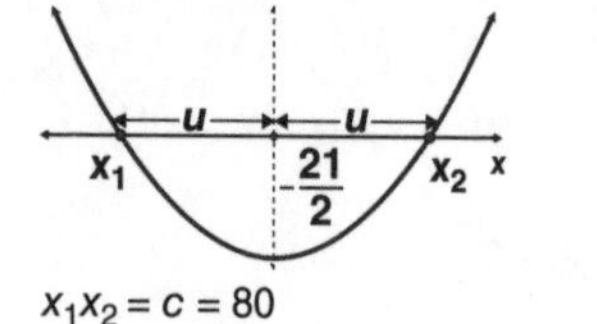

$b = 21$, so the equation of the axis of symmetry is $x = -\frac{b}{2} = -\frac{21}{2}$.

$x_1 x_2 = c = 80$ — Product of roots.

$x_1 = -\frac{21}{2} - u,\ x_2 = -\frac{21}{2} + u$ — The axis of symmetry is midway between the roots.

$\left(-\frac{21}{2} + u\right)\left(-\frac{21}{2} - u\right) = 80$ — Substitute into the equation $x_1 x_2 = 80$.

$\frac{441}{4} - u^2 = 80$ — Multiply the left side.

$-80 + u^2 \quad -80 + u^2$ — Subtract 80 from and add u^2 to both sides.

$\frac{121}{4} = u^2$ — Simplify.

$\sqrt{\frac{121}{4}} = \frac{\sqrt{121}}{\sqrt{4}} = \frac{11}{2} = u$ — Take the square root of both sides. (Since the parabola is symmetrical, we only need the principal square root.)

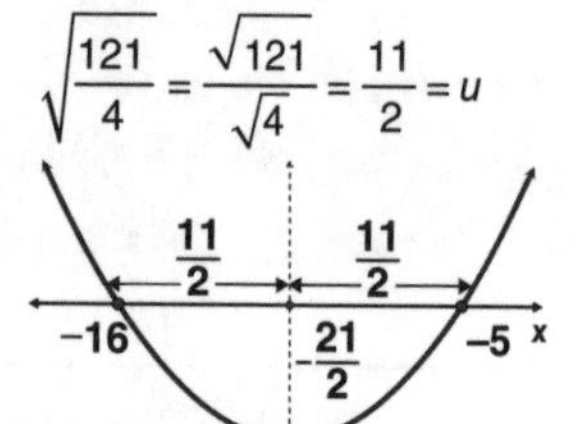

Determine the roots: $x = -\frac{21}{2} - \frac{11}{2} = -\frac{32}{2} = -16,$

$x_2 = -\frac{21}{2} + \frac{11}{2} = -\frac{10}{2} = -5.$

Example 9.33 Solve the equation $3x^2 - 12x - 20 = 0$ using the mean-product method.

Solution:

$\frac{3x^2}{3} - \frac{12x}{3} - \frac{20}{3} = 0$ — Divide the equation by the leading coefficient.

$x^2 - 4x - \frac{20}{3} = 0$ — Simplify.

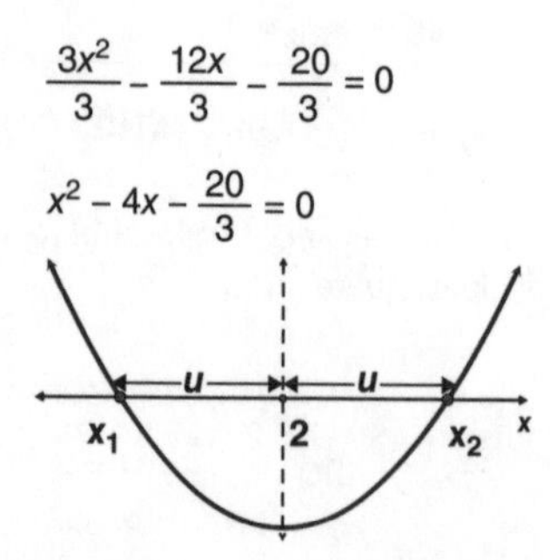

$b = -4$, so the equation of the axis of symmetry is

$x = -\frac{b}{2a} = -\frac{(-4)}{2} = 2.$

$x_1 x_2 = c = -\frac{20}{3}$	Product of roots.
$x_1 = 2 - u,\ x_2 = 2 + u$	The axis of symmetry is midway between the roots.
$(2-u)(2+u) = -\frac{20}{3}$	Substitute into the equation $x_1x_2 = 80$.
$4 - u^2 = -\frac{20}{3}$	Multiply the left side.
$+\frac{20}{3} + u^2 + \frac{20}{3} + u^2$	Add $\frac{20}{3}$ and u^2 to both sides.
$\frac{32}{3} = u^2$	Simplify.
$\sqrt{\frac{32}{3}} = \frac{\sqrt{32}}{\sqrt{3}} = \frac{4\sqrt{2}}{\sqrt{3}} = u$	Take the square root of both sides. (Since the parabola is symmetrical, we only need the principal square root.)
$\frac{4\sqrt{2}\sqrt{3}}{\sqrt{3}\sqrt{3}} = \frac{4\sqrt{6}}{3} = u$	Rationalize the denominator.

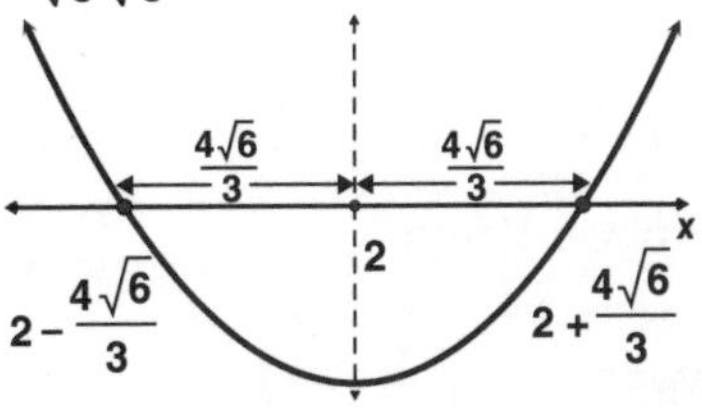

Determine the roots: $x_1 = 2 - \frac{4\sqrt{6}}{3}$, $x_2 = 2 + \frac{4\sqrt{6}}{3}$.

Technology Tip

Use technology to add and subtract complicated expressions with fractions, such as $80 - \frac{441}{4}$.

In this book, we've discussed five methods for solving quadratic equations: factoring, completing the square, the quadratic formula, graphing, and the mean-product method. Each method has its advantages and disadvantages, which we summarize in Table 9.7:

Table 9.7 Methods for Solving Quadratic Equations.

COMPARING METHODS FOR SOLVING QUADRATIC EQUATIONS

Method	Advantages	Disadvantages
Factoring	• Requires few steps	• "Guess-and-check" can be frustrating • Difficult if numbers are large • Can't be used for some quadratic equations

(Continued)

Table 9.7 (*Continued*)

COMPARING METHODS FOR SOLVING QUADRATIC EQUATIONS		
Method	**Advantages**	**Disadvantages**
Completing the Square	• Can be used to solve any quadratic equation • Best if $a = 1$ and b is even	• Difficult if b is odd or $a \neq 1$ • Requires several algebraic steps • Remembering steps can be difficult
Quadratic Formula	• Can be used to solve any quadratic equation • Requires few steps	• Substituting into formula can be difficult, especially if values are large or negative
Graphing	• Can be used to solve any quadratic equation with real roots • Can be done easily with technology • Requires no algebra • Can be used to verify work from another method • Best if roots are integers	• Can't be used to find irrational roots exactly
Mean-Product Method	• Can be used to solve any quadratic equation • Sketch makes remembering steps easier	• Requires several algebraic steps • Requires knowledge of formulas for the sum and product of the roots

As you can see, no one technique is best for every equation. We suggest the following:

- Learn how to solve by graphing. You can do it quickly with technology and verify a solution found by another method.
- Learn one or more of the techniques that can be used to solve any quadratic equation: completing the square, the quadratic formula, or the mean-product method.
- As you become more familiar with solving quadratic equations, learn to recognize quadratic expressions that can be factored so you can solve by factoring when possible.

Exercises

For each equation, use the mean-product method to find the mean of the roots (m), the distance from the axis of symmetry to a root (u), and the roots.

1. $x^2 - 20x + 96 = 0$
2. $x^2 + 22x + 120 = 0$
3. $x^2 - 6x - 135 = 0$
4. $x^2 - 4x - 32 = 0$
5. $x^2 - 19x + 60 = 0$
6. $x^2 + 27x + 92 = 0$
7. $x^2 + 9x - 90 = 0$
8. $x^2 - 3x - 18 = 0$
9. $x^2 - 10x + 18 = 0$
10. $x^2 - 12x - 20 = 0$
11. $x^2 + 7x + 11 = 0$
12. $x^2 - 5x + 3 = 0$
13. $4x^2 + 16x + 15 = 0$
14. $4x^2 - 16x + 13 = 0$
15. $5x^2 + 8x - 2 = 0$
16. $3x^2 - 6x - 5 = 0$

Questions to Think About

17. Explain how the mean and product of the roots are used in the mean-product method.
18. If the mean of the roots is 5 and the product of the roots is 6, what are the roots? Explain.
19. Use the mean-product method to show that the roots of the equation $x^2 + bx + c = 0$ are $x = -\frac{b}{2} \pm \frac{\sqrt{b^2-4c}}{2}$. (This shows that the mean-product method is equivalent to completing the square and the quadratic formula.)

9.12 Solving Quadratic-Linear Systems

Just as we solved systems of linear equations, we can also solve systems with quadratic equations. In this section, we solve **quadratic-linear systems**, which have one quadratic and one linear equation, both algebraically and graphically. Here are some examples:

Example 9.34 Solve the system of equations

$$\mathbf{x^2 - 6x + 8 = y}$$
$$\mathbf{y = 3x - 6}$$

Solution:

METHOD 1: Solve algebraically by substitution.

$x^2 - 6x + 8 = 3x - 6$	Substitute $x - 1$ for y in the first equation.
$\quad - 3x + 6 - 3x + 6$	Subtract $3x$ and add 6 to both sides.
$x^2 - 9x + 14 = 0$	Write the equation in standard form.
$(x - 7)(x - 2) = 0$	Factor the left side.

$x - 7 = 0$ or $x - 2 = 0$	Zero-product property.
$x = 7$ or $x = 2$	Solve each linear equation for the variable.
When $x = 7$, $y = 3(7) - 6 = 15$.	Substitute each x-value into an equation to find its corresponding y-value.
When $x = 2$, $y = 3(2) - 6 = 0$.	

METHOD 2: Solve graphically.

Graph both equations on the same coordinate plane. The points where the graphs intersect are the solution set to the system.

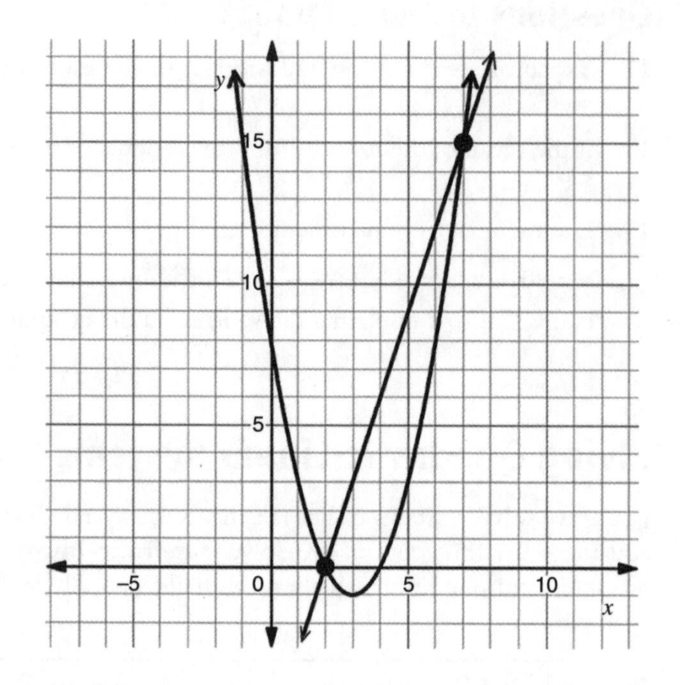

The solution set can be written in one of the following ways:

- (2, 0) and (7, 15)
- {(2, 0), (7, 15)}
- $x = 2$ and $y = 0$, $x = 7$ and $y = 15$

Example 9.35 Solve the system of equations

$$-2x^2 + 5x + 3 = y$$

$$11 = 3x + y$$

Solution:

METHOD 1: Solve algebraically by substitution.

$11 = 3x + y$	Solve the second equation for y.
$\underline{-3x \quad -3x}$	Subtract $3x$ from both sides.
$-3x + 11 = y$	
$-2x^2 + 5x + 3 = -3x + 11$	Substitute $-3x + 11$ for y in the first equation.
$\underline{\quad +3x - 11 \quad +3x - 11}$	Add $3x$ to and subtract 11 from both sides.
$-2x^2 + 8x - 8 = 0$	Write the equation in standard form.
$x^2 - 4x + 4 = 0$	Divide by the GCF (-2).
$(x - 2)(x - 2) = 0$	Factor the left side.
$x - 2 = 0$ or $x - 2 = 0$	Zero-product property.
$x = 2$ or $x = 2$	Solve each linear equation for the variable: $x = 2$ is a double root.
When $x = 2$, $y = -3(2) + 11 = 5$.	Substitute $x = 2$ into an equation to find y.

METHOD 2: Solve graphically.

Graph both equations on the same coordinate plane.

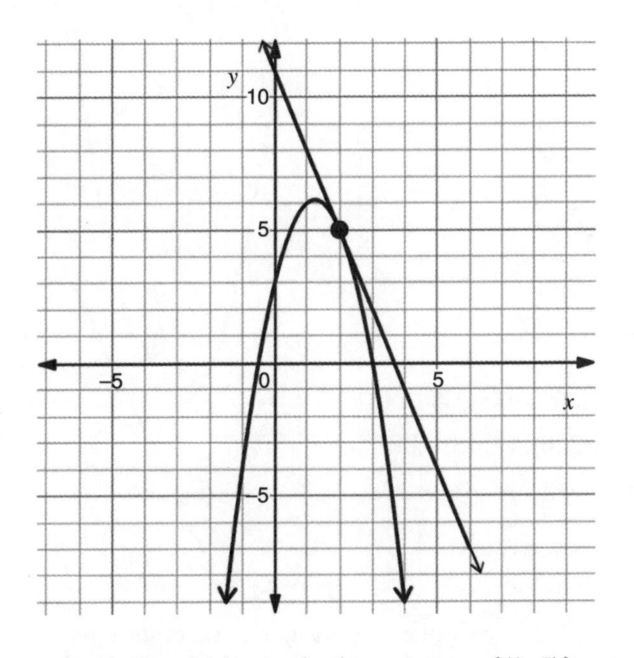

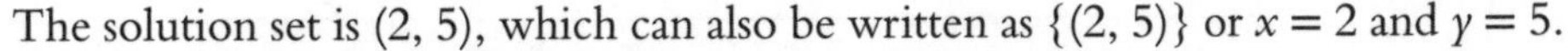
The solution set is $(2, 5)$, which can also be written as $\{(2, 5)\}$ or $x = 2$ and $y = 5$.

Example 9.36 Solve the system of equations

$$y = (x + 1)^2 - 4$$

$$2x - 5 = y$$

Solution:

METHOD 1: Solve algebraically by substitution.

$y = x^2 + 2x - 3$	Write the first equation in standard form: $(x + 1)^2 - 4 = (x^2 + 2x + 1) - 4 = x^2 + 2x - 3$
$2x - 5 = x^2 + 2x - 3$	Substitute $2x - 5$ for y in the first equation.
$-2x + 5 \qquad -2x + 5$	Subtract $2x$ from and add 5 to both sides.
$0 = x^2 + 2$	Write the equation in standard form.
$-2 \qquad -2$	Subtract 2 from both sides.
$-2 = x^2$	We see that since x^2 is negative, then the equation has no real solutions.

METHOD 2: Solve graphically.

Graph both equations on the same coordinate plane. The graphs do not intersect.

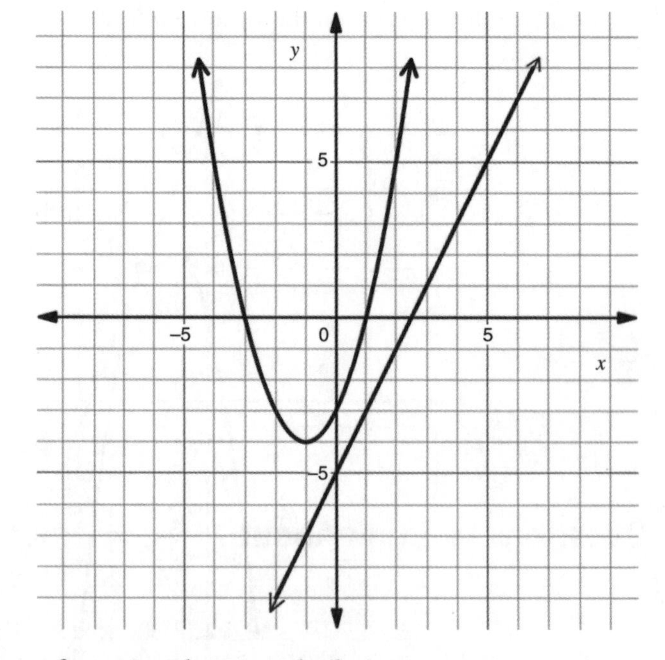

The system of equations has no real solutions.

As we can see from Examples 9.34, 9.35, and 9.36, a quadratic-linear system can have 2, 1, or 0 distinct solutions:

- If the system has two distinct solutions, then the parabola and line will intersect at two distinct points.
- If the system has one solution, then the parabola and line will intersect at only one point.
- If the system has no solutions, then the parabola and line will not intersect.

Technology Tip

If you're using technology to solve a quadratic-linear system, remember that the graphs may have two intersection points. Adjust your window if necessary to find them.

Exercises

Solve each system of quadratic-linear equations.

1. $y = x^2 - 6x + 10$
$y = -x + 4$

2. $y = x^2 + 6x + 5$
$y = 2x + 5$

3. $y = x^2 - 4x + 8$
$y = -2x + 7$

4. $y = x^2 + 5x + 6$
$x + y = 1$

5. $y = x^2 - 7x + 6$
$3x + y = 1$

6. $y = (x - 5)^2 - 8$
$y + 2x = 10$

7. $y = x^2 + \frac{1}{2}x - 6$
$y + 2 = \frac{1}{2}x$

8. $y = 2x^2 - x - 6$
$24 + y = 11x$

9. $y = (x - 4)^2 - 9$
$5x + y = 1$

Questions to Think About

10. To solve a quadratic-linear system. Aileen uses a calculator to graph both equations, resulting in the accompanying graph. She concludes that the system has one solution. In fact, the system has two distinct solutions. Explain the error in her reasoning.

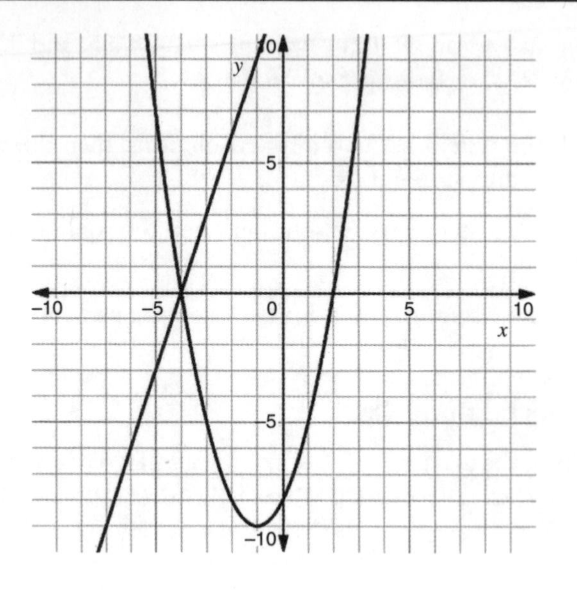

11. To solve the equation $ax^2 + bx + c = 0$, Selina graphs the parabola $y = ax^2 + bx + c$. She claims that the x-coordinates of the x-intercepts represent the solutions to the equation. Use your knowledge of solving quadratic-linear systems graphically to explain why her method works. (HINT: Recall that the equation of the x-axis is $y = 0$.)

12. If a, b, c, m, and n are real numbers, how many solutions can the quadratic-linear system $y = ax^2 + bx + c$ and $x = m$ have? Explain. (HINT: Note that the domain of the quadratic function $f(x) = ax^2 + bx + c$ is the set of all real numbers.)

9.13 Using Quadratic Equations to Solve Word Problems

Quadratic functions can be used to solve word problems. Example 9.37 is based on a problem that appeared in a Babylonian text from 2000–1500 BC.

Example 9.37 The length of a rectangle exceeds its width by 22. If the area of the rectangle is 240, determine its length and width.

Solution:

Step 1: Identify what is given and what we need to find.

The length of a rectangle exceeds its width by 22.

The area is 240.

Find the length and width.

Step 2: *Represent the unknown information.*
Let w = width of rectangle.
Then $w + 22$ = length.

The area of the rectangle is 240.	$w(w + 22) = 240$

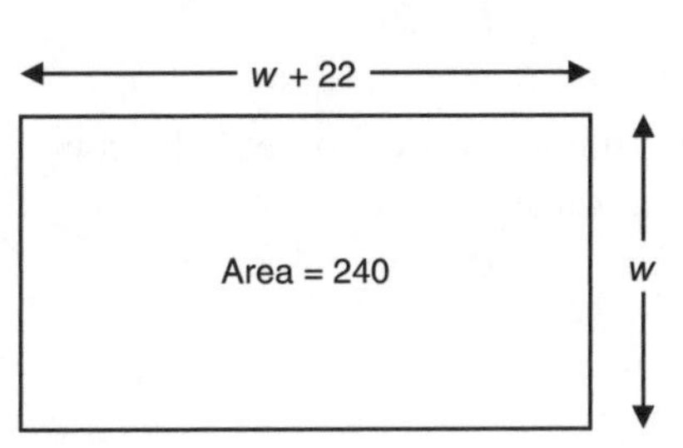

Step 3: *Solve.*

$$w(w + 22) = 240$$

$w^2 + 22w = 240$	
$\underline{\quad -240 \quad -240}$	Subtract 240 from both sides.
$w^2 + 22w - 240 = 0$	Write the quadratic equation in standard form.
$(w + 30)(w - 8) = 0$	Factor the left side.
$w + 30 = 0$ or $w - 8 = 0$	Zero-product property.
$w = -30$ or $w = 8$	Solve each linear equation for the variable. Since the width cannot be negative, we reject $w = -30$.

Step 4: *Check.*

$8(8 + 22) \stackrel{?}{=} 240$	Substitute $w = 8$ into the original equation. $240 = 240$ The width is 8 and the length is $22 + 8 = 30$.

Example 9.38 is based on a problem from al-Khwārizmī's *Al-Jabr*.

Example 9.38 Sama thinks of two numbers that add up to 10. She squares the two numbers and adds the squares. To that sum, she adds the difference of the two numbers. Her result is 54. What numbers could Sama be thinking of?

Solution:

Step 1: *Identify what is given and what we need to find.*
Two numbers add up to 10.
The squares of the numbers are added to make a sum.

The numbers are subtracted to make a difference.
The sum and difference are added to get 54.
Find the two numbers.

Step 2: Represent the unknown information.

Let x = the first number.
Then $10 - x$ = the second number since $x + (10 - x) = 10$.

She squares the two numbers and adds the squares.	$x^2 + (10 - x)^2$
To that sum, she adds	$+$
the difference of the two numbers,	$x - (10 - x)$
The result is 54.	$= 54$

Step 3: Solve.

$$x^2 + (10 - x)^2 + x - (10 - x) = 54$$

$x^2 + 100 - 20x + x^2 + x - 10 + x = 54$	Multiply and distribute.
$2x^2 - 18x + 90 = 54$	Combine like terms.
$\quad - 54 - 54$	Subtract 54 from both sides.
$2x^2 - 18x + 36 = 0$	Write the quadratic equation in standard form.
$x^2 - 9x + 18 = 0$	Divide the equation by the GCF (2).
$(x - 6)(x - 3) = 0$	Factor the left side.
$x - 6 = 0$ or $x - 3 = 0$	Zero-product property.
$x = 6$ or $x = 3$	Solve each linear equation for the variable.
	If $x = 6$, then the other number is $10 - 6 = 4$.
	If $x = 3$, then the other number is $10 - 3 = 7$.

Step 4: Check.

$6^2 + 4^2 + 6 - (10 - 6) \stackrel{?}{=} 54$ Substitute $x = 6$ into the original equation.

$$54 = 54$$

$3^2 + 7^2 + 3 - (10 - 3) \stackrel{?}{=} 54$ Substitute $x = 3$ into the original equation.

$$54 = 54$$

Sama could be thinking of 6 and 4 or 3 and 7.

We use parabolas to model the height of objects thrown in the air. Since objects rise and fall after being thrown, their paths are modeled by parabolas that open downward, and so they have negative leading coefficients.

Example 9.39 Juanita throws a ball into the air. The ball's height, in feet, t seconds after being thrown can be modeled using the function $h(t) = -16t^2 + 32t + 5$.

(a) Explain what the point (0, 5) means in context.

(b) When does the ball reach its maximum height?

(c) What is the maximum height that it reaches?

(d) To the nearest tenth of a second, determine when the ball hits the ground.

Solution: We graph the function to help us visualize the problem. The ball's height can't be negative (that would mean that it goes underground, which makes no sense here), and time can't be negative, so we limit the graph to Quadrant I. Since the graph doesn't extend infinitely in both directions, we don't put arrows on either end of the parabola. The graph is asymmetrical.

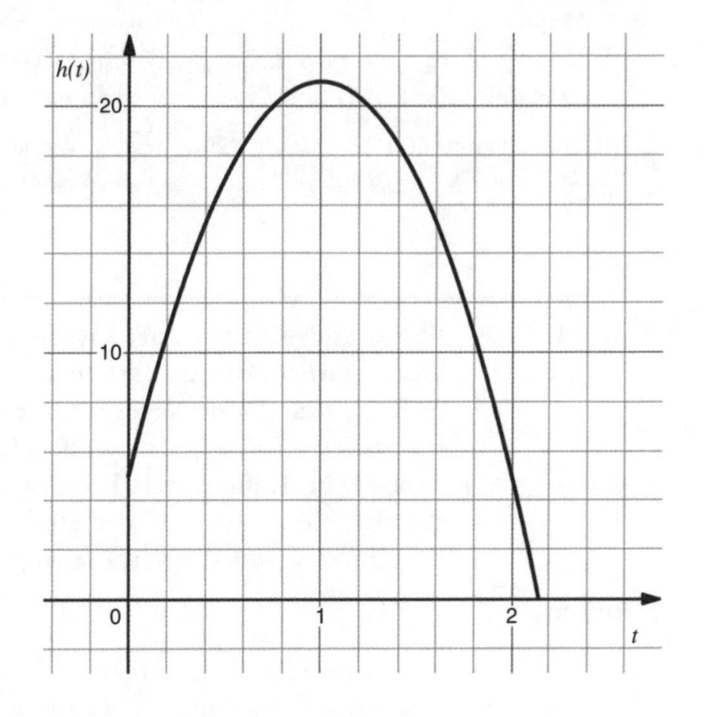

(a) At (0, 5), $t = 0$ (0 seconds after being thrown) and $h(t) = 5$ (the height of the ball is 5 feet). In context, this means that Juanita throws the ball from a height of 5 feet.

(b) The ball reaches its maximum height at the vertex. Since we need the time when the ball reaches its maximum, we need the t-value of the vertex. Using the formula for the axis of symmetry, $t = -\frac{b}{2a} = -\frac{32}{2(-16)} = 1$. The ball reaches its maximum height 1 second after being thrown.

(c) The maximum height is determined from the vertex of the parabola. Since $h(1) = 21$, the ball reaches a maximum height of 21 feet. We can answer parts b and c by graphing the function and interpreting the coordinates of the vertex (1, 21).

(d) The ball hits the ground when $h(t) = 0$, at the t-intercept. Using the trace or intercept function in technology, we see that the t-intercept is approximately (2.146, 0). The ball hits the ground in approximately 2.1 seconds.

Technology Tip

When using technology, do any of the following to make the graph of a quadratic function fit into your window:

- Zoom out until we can see the general shape of the parabola, including the vertex and intercepts.
- If the function is limited to Quadrant I, adjust the window settings so that the vertical and horizontal axes start at 0 (typically, this means setting Xmin = 0 and Ymin = 0).
- Adjust the maximum vertical value, maximum horizontal value, or both, increasing each by 50 or 100 at a time (typically, this means changing the Xmax and Ymax values).

Example 9.40 **Jiwoo and Kendra start saving money to buy a new video game system. The total amount of money, in dollars, that Jiwoo saves after x weeks is modeled by the function $j(x) = 2x^2 + 20$. The total amount of money, in dollars, that Kendra saves is modeled by the function $k(x) = 10x + 60$. During what week will they have saved the same amount of money?**

Solution:

Step 1: *Identify what is given and what we need to find.*
Jiwoo's savings after x weeks is modeled by $j(x) = 2x^2 + 20$.
Kendra's savings is modeled by $k(x) = 10x + 60$.
When will their savings be equal?

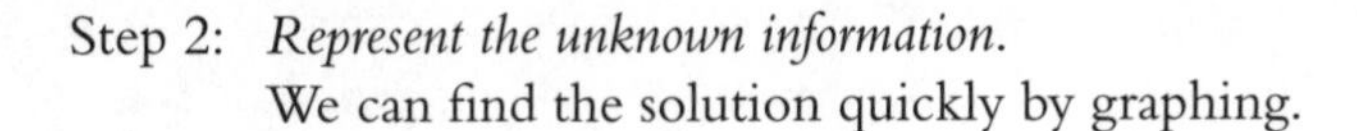

Step 2: *Represent the unknown information.*
We can find the solution quickly by graphing.

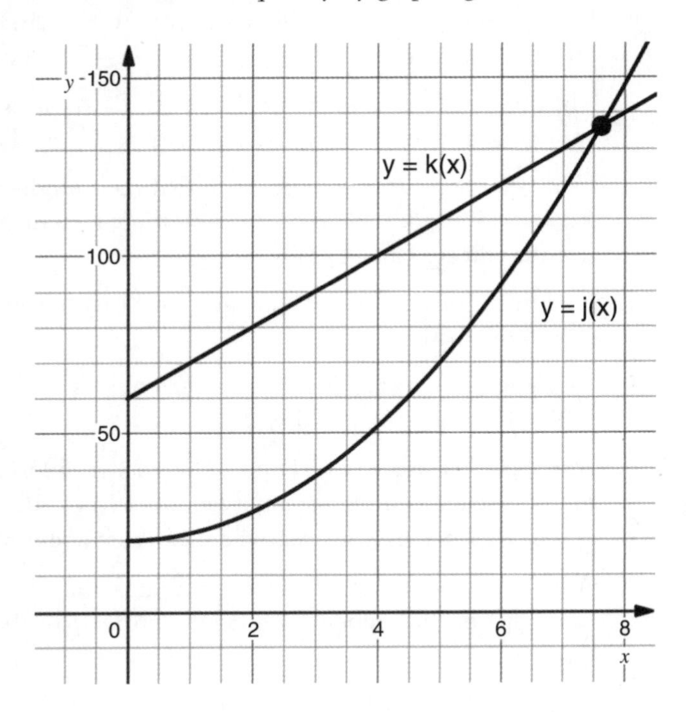

The functions representing Jiwoo's and Kendra's savings will be equal at the point where their graphs intersect.

Step 3: *Solve.*

Using the trace or intercept functions in technology, we find the intersection point is approximately (7.623, 136.235), or when $x \approx 7.623$. In context, this means that Jiwoo's and Kendra's savings will be equal during the seventh week. Note that rounding 7.623 to 8 makes no sense in this problem since that implies that their savings are equal during the eighth week.

Step 4: *Check.*

Substitute $x = 7.623$ into each function to see if the amounts are approximately equal.

$$j(7.623) = 2(7.623)^2 + 20 \approx \$136.22$$

$$k(7.623) = 10(7.623) + 60 \approx \$136.23$$

$$\$136.22 \approx \$136.23$$

Exercises

1. If 16 times a negative number is added to the square of the number, the result is 36. Find the number.
2. The number 10 is divided into two parts. Each part is then squared. If the squares are subtracted from each other, the result is 40. Find the two parts.
3. Find two consecutive integers such that their product is 130 more than 20 times the second. (HINT: Review Section 3.5, which discusses how to represent consecutive integers algebraically.)
4. Find three consecutive even integers such that the product of the first and second is 52 more than the third.
5. The length of a rectangle is 3 more than twice its width. The area of the rectangle is 275. Find the length and width.
6. A triangle has an area of 54 and a base that is 15 less than three times the height. Find the lengths of the base and height. (HINT: Recall that the formula for the area of a triangle is $A = \frac{1}{2}bh$, where b = base and h = height.)
7. A ball is thrown from a platform above ground. The accompanying graph shows h, a function that models the height (in feet) of the ball t seconds after being thrown.

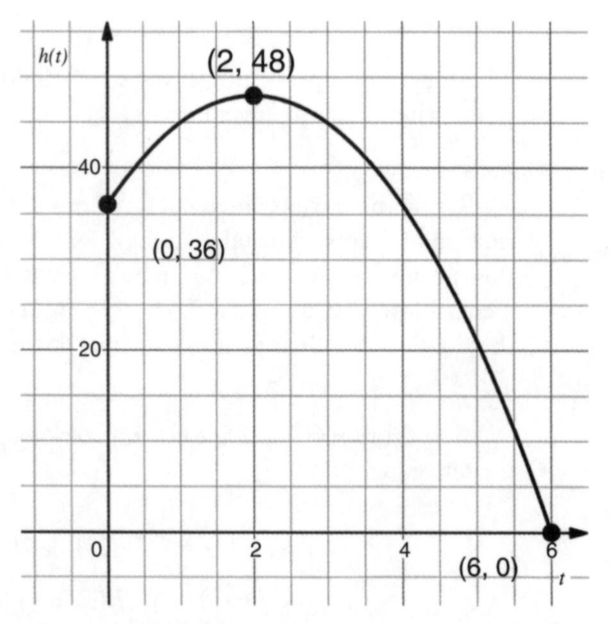

(a) How many feet above the ground is the platform?

(b) After how many seconds does the ball reach its maximum height?

(c) After how many seconds does the ball fall to the ground?

(d) Write an equation for h.

8. The height (in feet) of a diver jumping off a springboard can be modeled by the function $h(x) = -x^2 + 6x + 16$, where x is the horizontal distance from the springboard.
(a) How high, in feet, is the springboard?

(b) What is the maximum height that the diver reaches?

(c) What is an appropriate domain for this function?

9. The height of an object thrown from a building after t seconds can be represented by the function $f(t) = -16t^2 + 48t + 64$.
(a) After how many seconds does the object reach its maximum height?

(b) What is the maximum height, in feet, that the object achieves?

(c) Over what interval, in seconds, is the object's height decreasing?

10. The predicted price (in dollars) of the Equis Corporation's stock can be modeled by the function $E(x) = \frac{1}{2}x^2 - 6x + 25$, where x is the number of days after December 31. (For example, $x = 2$ represents January 2.) The predicted price (in dollars) of the Zeta Corporation's stock can be modeled by the function $Z(x) = \frac{3}{2}x + 33$.
(a) According to the model, on what date will the prices of the two stocks be equal?

(b) What will be the predicted price of the stocks on that date?

CHAPTER 9 TEST

1. What is the correct factorization of $6x^3 + 4x^2 - 15x - 10$?

(A) $(2x^2 + 5)(-3x - 2)$
(B) $(2x^2 + 5) - (3x + 2)$
(C) $(2x^2 + 5)(3x + 2)$
(D) $(2x^2 - 5)(3x + 2)$

2. What are the sum and product of the roots of the equation $5x^2 - 7x - 2 = 0$?

(A) sum $= -\frac{7}{5}$, product $= -\frac{2}{5}$
(B) sum $= \frac{7}{5}$, product $= -\frac{2}{5}$
(C) sum $= -\frac{2}{5}$, product $= \frac{7}{5}$
(D) sum $= \frac{2}{5}$, product $= \frac{7}{5}$

3. Which equation represents the correct factorization of a difference of two perfect squares?

   (A) $9x^2 - 49 = (3x + 7)(3x - 7)$
   (B) $36x^2 - 229x + 25 = (4x - 25)(9x - 1)$
   (C) $9x^2 + 49 = (3x + 7)(3x + 7)$
   (D) $9x^2 - 49 = (3x - 7)(3x - 7)$

4. Which expressions are rational?

   (A) $\sqrt{50} + \sqrt{50}$ and $\sqrt{50} - \sqrt{50}$
   (B) $\sqrt{50} \cdot \sqrt{2}$ and $\dfrac{\sqrt{50}}{\sqrt{2}}$
   (C) $\sqrt{50} + \sqrt{2}$ and $\sqrt{50} - \sqrt{2}$
   (D) $50\sqrt{50}$ and $-50\sqrt{50}$

5. A parabola has an equation of $y = (x + 3)(x - 8)$. What is the equation of its axis of symmetry?

   (A) $x = -\dfrac{11}{2}$ (B) $x = -\dfrac{11}{2}$ (C) $x = -\dfrac{5}{2}$ (D) $x = \dfrac{5}{2}$

6. Which equation has no real roots?

   (A) $2x^2 + 8x + 12 = 0$
   (B) $2x^2 + 8x + 8 = 0$
   (C) $2x^2 + 8x = 0$
   (D) $2x^2 + 8x - 8 = 0$

7. The vertex of a parabola is (4, 6). One of the parabola's x-intercepts is (10, 0). What is the other x-intercept?

   (A) (−1, 0) (B) (14, 0) (C) (−2, 0) (D) $\left(0, \dfrac{10}{3}\right)$

8. A quadratic function $y = f(x)$ is shown in the accompanying graph. Which points can be used to determine the zeros of the function?

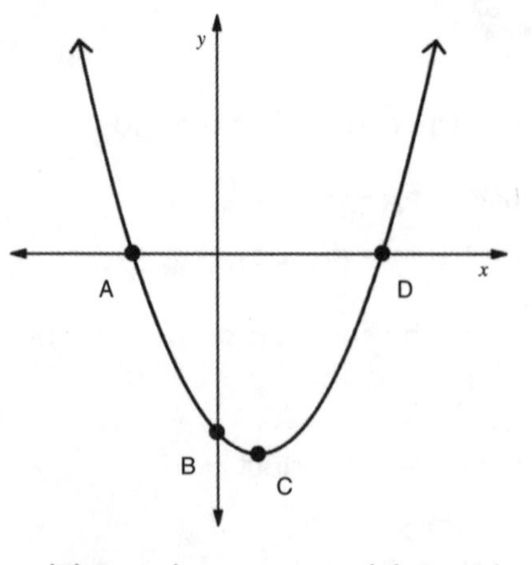

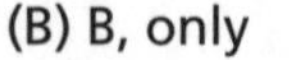

   (A) A, only (B) B, only (C) C, only (D) A and D

9. Which statement about the function $f(x) = -3(x + a)^2 + b$, where $a > 0$ and $b > 0$, *must* be true?

   (A) The function must have exactly 2 real zeros.

   (B) The function must have 1 or 2 real zeros.

   (C) The function has no real zeros.

   (D) The function can have 0, 1, or 2 real zeros.

10. Factor $16r^2q^3 - 8r^2q^2 - 15r^2q$ completely.

11. Solve $x^2 + 12x - 189 = 0$ for x.

12. Solve $2x^2 - 8x - 7 = 0$ for x.

13. Solve for x and y:

$$x^2 - 12x + 20 = y$$

$$x + y = 2$$

14. The number 25 is divided into two positive parts such that the square of one part is equal to 20 more than 16 times the other part. Find the two parts.

15. The height of a ball t seconds after being thrown in the air can be modeled with the equation $h(t) = -16t^2 + 64t + 3$.

   (a) Determine the height, in feet, at which the ball was thrown.

   (b) How many seconds does the ball spend traveling upwards?

   (c) What is the maximum height, in feet, reached by the ball?

   (d) How many seconds after being thrown does the ball take to return to its starting height? Round your answer to the nearest whole number.

CHAPTER 9 SOLUTIONS

9.1.

1. $3(a + b)$
2. $2x(x - 3y)$
3. $x^2(1 - 3x)$
4. $r(1 + y - r)$
5. $5k(k^2 - 3k + 7)$
6. Cannot be factored (prime)
7. $2m(5m^2 - 2m + 1)$
8. Cannot be factored (prime)
9. $4a(8a^2 + 2a + 5)$
10. $p^2q(p^2 + p + 1)$

11. $5x^2 y^2(2x - y + 3)$ 12. $abc(a - b + c)$

13. A prime polynomial cannot be factored any further—it has no other factors except 1 and itself. A monomial cannot be factored further since it is already considered prime.

14. Factoring completely means factoring a polynomial into primes.

15. Although the coefficients 9 and 4 are not prime, $9x^2$ and $4y^2$ have no factors in common, so the polynomial is prime.

9.2. 1. $(6m^2 + 5)(2m + 3)$ 4. $(11y - 1)(11y^2 - 1)$ 7. $(3a^3 - 4)(2a - 9)$

2. $(3p^2 + 2)(5p + 7)$ 5. $(3u^2 - 4)(2u - 5)$ 8. $(2z^2 - d)(2z - 3d)$

3. $(d^2 + 2)(d + 5)$ 6. $(7q^2 - 2)(3q - 4)$ 9. $(ab + 5bc)(2ab - c)$

10. There is no common factor among the pairs of binomials: $6n^3 + 4n^2 + 3n + 3 = 2n^2(3n + 2) + 3(n + 1)$.

11. $(7x^3 + 2x^2) + (7x + 2) = x^2(7x + 2) + 1(7x + 2) = (x^2 + 1)(7x + 2)$ or $(7x^3 + 7x) + (2x^2 + 2) = 7x(x^2 + 1) + 2(x^2 + 1) = (7x + 2)(x^2 + 1)$.

9.3. 1. $(m + 2)(m + 11)$ 7. $(a + 4)(a - 3)$ 13. $(2x - 5)(x + 3)$

2. $(r + 4)(r + 6)$ 8. $(x - 4)(x + 2)$ 14. $(3a - 5)(a + 1)$

3. $(y + 8)(y + 9)$ 9. $(k - 5)(k + 4)$ 15. $(2a - 1)(a - 4)$

4. $(q - 3)(q - 2)$ 10. $(b + 7)(b - 2)$ 16. $(5c + 7)(c + 2)$

5. $(x - 8)(x - 4)$ 11. $(x - 6)(x + 1)$ 17. $(6a - 5)(a + 1)$

6. $(z - 6)(z - 8)$ 12. $(p^2 - 6)(p^2 + 5)$ 18. $(3y + 2)(3y - 1)$

19. Multiplying $(mx + n)(px + q)$ gives us $mpx^2 + npx + mqx + nq = mpx^2 + (np + mq)x + nq$. Since this equals $ax^2 + bx + c$, then $a = mp$, $b = np + mq$, and $c = nq$. The two coefficients np and mq have a product of $(np)(mq) = npmq = mpnq = ac$ and a sum of $np + mq = b$.

20. Answers may vary. For example, if the product ac has many factors or if ac is negative, finding the right factor pair can be tedious.

21. $a = 1$ and $c = 4$, so $ac = 1(4) = 4$. $b = 1$, but $1(4) \neq 1$ and $2(2) \neq 1$, so we cannot split the middle term into two terms for factoring by grouping.

9.4. 1. $(k + 6)^2$ 4. $(3z + 1)^2$

2. $(a - 8)^2$ 5. $(x^2 - 2)^2$

3. $(x + 10)^2$ 6. $(5x + 2)^2$

7. $(4 + y)(4 - y)$
8. $(5x + 8)(5x - 8)$
9. $(6 + a)(6 - a)$
10. $(a + b^2)(a - b^2)$
11. $(1 + 7r^2)(1 - 7r^2)$
12. $(p^2 + 5)(p^2 - 5)$
13. $(c + 3y)(c - 3y)$
14. $(1 + ab)(1 - ab)$
15. $(2ax + 11z)(2ax - 11z)$
16. $2(5x + 1)(x + 1)$
17. $2a(7h + 3)(h + 1)$
18. $3(5x + 1)(x - 1)$
19. $2a(3a + 1)(a + 1)$
20. $5y^2(x + 1)^2$
21. $3a(x + 3)(x - 2)$
22. $y(y - 2)(y + 3)(y - 3)$
23. $3m(2m + 1)(m + 5)(m - 5)$
24. $u(u + 5)(3u + 2)(3u - 2)$
25. Yes, since $x^2 + 2x + 1 = (x + 1)(x + 1) = (x + 1)^2$.
26. In general, $x^2 + y^2 \neq (x + y)(x + y)$ since $(x + y)(x + y) = x^2 + 2xy + y^2$.
27. Julian didn't factor completely since $x^2 + 7x + 12$ is not prime. In fact, $x^3 + 7x^2 + 12x = x(x^2 + 7x + 12) = x(x + 4)(x + 3)$.

9.5.

1. $x = -7$ or $x = 3$
2. $x = -2$ or $x = -4$
3. $x = 2$ or $x = -\frac{1}{4}$
4. $a = 8$
5. $q = 6$
6. $z = -2$
7. $x = 3$ or $x = -3$
8. $w = 10$ or $w = -10$
9. $x = \frac{5}{2}$ or $x = -\frac{5}{2}$
10. $x = \frac{1}{3}$
11. $x = \frac{1}{3}$ or $x = 0$
12. $y = -\frac{5}{2}$ or $y = 5$
13. $k = 0$ or $k = 4$
14. $y = 0$ or $y = 2$
15. $x = 0$ or $x = 7$
16. $k = 0$ or $k = 8$
17. $a = 0$ or $a = 6$
18. $x = 0$ or $x = -\frac{3}{2}$
19. She mistakenly thought that if a product equals 24, then at least one of the factors must be 24. This is incorrect. For example, $6(4) = 24$, but $6 \neq 24$ and $4 \neq 24$.
20. {2, 4} is a set that represents all possible values of one variable, such as $x = 2$ or $x = 4$. The coordinate (2, 4) represents an ordered pair representing two variables, such as $x = 2$ and $y = 4$.
21. The conceptual error occurs in step 4. Since $a = b$, then dividing by $a - b$ is equivalent to dividing by $a - a$ or 0, which is meaningless.

9.6.

1. "3 times the square root of 30" or "3 times radical 30"
2. "7 times the square root of 7" or "7 times radical 7"

3. "the cube root of 16"
4. "the fourth root of 8"
5. "2 times the cube root of 5"
6. "3 times the fifth root of 81"
7. 7
8. 8
9. $2(10) = 20$
10. $5(12) = 60$
11. 3
12. 2
13. $3(4) = 12$
14. $5(3) = 15$
15. $\frac{1}{2}$
16. $\frac{3}{4}$
17. $6\left(\frac{13}{5}\right) = \frac{78}{5}$
18. $\frac{5}{3}\left(\frac{7}{8}\right) = \frac{35}{24}$
19. $2\sqrt{13}$
20. $5\sqrt{5}$
21. $12\sqrt{3}$
22. $24\sqrt{2}$
23. $\frac{\sqrt{6}}{3}$
24. $\frac{\sqrt{35}}{7}$
25. $\frac{\sqrt{22}}{6}$
26. $\frac{4\sqrt{10}}{15}$
27. $8\sqrt{7}$
28. $12\sqrt{2}$
29. $-\sqrt{7}$
30. $12\sqrt{3}$
31. $45\sqrt{2}$
32. $2\sqrt{15}$
33. $24\sqrt{3}$
34. 14
35. 4
36. $6\sqrt{2}$
37. $\frac{1}{21}$
38. $\frac{2\sqrt{5}}{15}$
39. Irrational $(1 + \sqrt{2})$
40. Irrational $(6\sqrt{5})$
41. Rational (5)
42. Irrational $(4\sqrt{7})$
43. Rational (15)
44. Irrational $(4 + 8\sqrt{2})$
45. Arvin did not factor the largest perfect square (16) from the radicand. He should have factored as follows: $\sqrt{80} = \sqrt{16}\sqrt{5} = 4\sqrt{5}$.
46. No. For example, $\sqrt{4} + \sqrt{9} = 2 + 3 = 5$, but $\sqrt{4+9} = \sqrt{13} \neq 5$.
47. No because the radicand $\left(\frac{1}{2}\right)$ is a fraction.

48. No because the denominator has a radical expression.

49. First, we find the product of two rational numbers: $\frac{a}{b} \cdot \frac{c}{d} = \frac{ac}{bd}$. We know that ac and bd are integers since the product of two integers is an integer. Since $b \neq 0$ and $d \neq 0$, then $bd \neq 0$. Thus, we know that $\frac{ac}{bd}$ is the quotient of two integers with a nonzero denominator, so it is rational.

50. First, we find the sum of the two rational numbers: $\frac{a}{b} + \frac{c}{d} = \frac{ad}{bd} + \frac{bc}{bd} = \frac{ad+bc}{bd}$. We know that ad and bc are integers since the product of two integers is an integer. We know that $ad + bc$ is an integer since the sum of two integers is an integer. Since b and d are integers, then their product bd is an integer. Since $b \neq 0$ and $d \neq 0$, then $bd \neq 0$. Thus, we know that $\frac{ad+bc}{bd}$ is the quotient of two integers with a nonzero denominator, so it is rational.

9.7.

1. $\{8, 12\}$
2. $\{24, -6\}$
3. $\{4, -18\}$
4. $2 \pm \sqrt{3}$
5. $4 \pm \sqrt{11}$
6. $-5 \pm \sqrt{2}$
7. $-1 \pm \sqrt{6}$
8. $3 \pm \sqrt{7}$
9. $-6 \pm \sqrt{29}$
10. $-1 \pm \sqrt{5}$
11. $\frac{3 \pm \sqrt{17}}{4}$
12. $\frac{-4 \pm \sqrt{6}}{5}$
13. The solution set is $\{-4, -2\}$. Reasons may vary. Factoring may be quicker, and completing the square may be tedious.
14. The solution set is $\{-16, 6\}$. Reasons may vary. Factoring may be difficult since -96 has many possible factor pairs.
15. Tamika is correct in thinking that the solutions to $x^2 = 9$ are $+3$ and -3. However, $\sqrt{9}$ is the principal square root, which is $+3$ by definition. Thus, $x = \pm 3$, but $\sqrt{9} = +3$.

9.8.

1. $\{2, 3\}$
2. $\{-5, 5\}$
3. $\{-3, -7\}$
4. $\{7, -2\}$
5. $\frac{-11 \pm \sqrt{73}}{4}$
6. $\frac{7 \pm \sqrt{17}}{2}$
7. $\frac{9 \pm \sqrt{57}}{2}$
8. $\frac{5 \pm \sqrt{21}}{2}$
9. $2 \pm 2\sqrt{3}$
10. $\frac{4 \pm \sqrt{14}}{2}$
11. $\frac{3 \pm 2\sqrt{6}}{3}$

12. $\frac{-1 \pm \sqrt{13}}{4}$

13. sum $= \frac{5}{2}$, product $= -3$

14. sum $= -\frac{9}{4}$, product $= \frac{1}{4}$

15. sum $= -\frac{1}{7}$, product $= -\frac{8}{7}$

16. sum $= \frac{5}{3}$, product $= \frac{2}{3}$

17. Subtract c on both sides: $ax^2 + bx = -c$. Divide both sides by a: $x^2 + \frac{b}{a} = -\frac{c}{a}$. Add $\left(\frac{b}{2a}\right)^2$ to each side to complete the square: $x^2 + \frac{b}{a} + \left(\frac{b}{2a}\right)^2 = \left(\frac{b}{2a}\right)^2 - \frac{c}{a}$. Rewrite the left side as a perfect square: $\left(x + \frac{b}{2a}\right)^2 = \left(\frac{b}{2a}\right)^2 - \frac{c}{a}$. Take the square root of both sides: $x + \frac{b}{2a} = \pm\sqrt{\left(\frac{b}{2a}\right)^2 - \frac{c}{a}}$. Subtract $\frac{b}{2a}$ from both sides: $x = -\frac{b}{2a} \pm \sqrt{\left(\frac{b}{2a}\right)^2 - \frac{c}{a}}$. Simplify the right side: $x = -\frac{b}{2a} \pm \sqrt{\frac{b^2}{4a^2} - \frac{4ac}{4a^2}} = -\frac{b}{2a} \pm \sqrt{\frac{b^2-4ac}{4a^2}} = -\frac{b}{2a} \pm \frac{\sqrt{b^2-4ac}}{\sqrt{4a^2}} = -\frac{b}{2a} \pm \frac{\sqrt{b^2-4ac}}{2a} = \frac{-b\pm\sqrt{b^2-4ac}}{2a}$.

18. From the quadratic formula, the sum of the roots is $\frac{-b+\sqrt{b^2-4ac}}{2a} + \frac{-b-\sqrt{b^2-4ac}}{2a} = \frac{-b+\sqrt{b^2-4ac}+(-b)-\sqrt{b^2-4ac}}{2a} = \frac{-2b}{2a} = -\frac{b}{a}$. The product is $\left(\frac{-b+\sqrt{b^2-4ac}}{2a}\right)\left(\frac{-b-\sqrt{b^2-4ac}}{2a}\right) = \frac{(-b)(-b)-(b^2-4ac)}{(2a)(2a)} = \frac{b^2-(b^2-4ac)}{4a^2} = \frac{b^2-b^2+4ac}{4a^2} = \frac{4ac}{4a^2} = \frac{c}{a}$.

19. Noah substituted $b = -7$ into the quadratic formula incorrectly. Since b is negative, then $x = \frac{-(-7)\pm\sqrt{(-7)^2-4(2)(1)}}{2(2)} = \frac{7\pm\sqrt{49-8}}{4} = \frac{7\pm\sqrt{41}}{4}$ (note the parentheses around -7).

9.9. 1. axis of symmetry: $x = -4$; vertex: $(-4, -4)$; x-intercepts: $(-6, 0)$, $(-2, 0)$; y-intercept: $(0, 12)$

2. axis of symmetry: $x = 3$; vertex: $(3, -1)$; x-intercepts: $(2, 0)$, $(4, 0)$; y-intercept: $(0, 8)$

3. axis of symmetry: $x = -5$; vertex: $(-5, -225)$; x-intercepts: $(10, 0)$, $(-20, 0)$; y-intercept: $(0, -200)$

4. axis of symmetry: $x = \frac{7}{4}$; vertex: $\left(\frac{7}{4}, -\frac{81}{8}\right)$; x-intercepts: $\left(-\frac{1}{2}, 0\right)$, $(4, 0)$; y-intercept: $(0, -4)$

5. $y = (x + 4)^2 - 10$, $(-4, -10)$

6. $y = (x - 1)^2 - 7$, $(1, -7)$

7. $y = (x + 3)^2 - 5$, $(-3, -5)$

8. $y = (x - 5)^2 - 14$, $(5, -14)$

9. $y = (x - 8)^2 + 11$, $(8, 11)$

10. $y = (x + 7)^2 + 2$, $(-7, 2)$

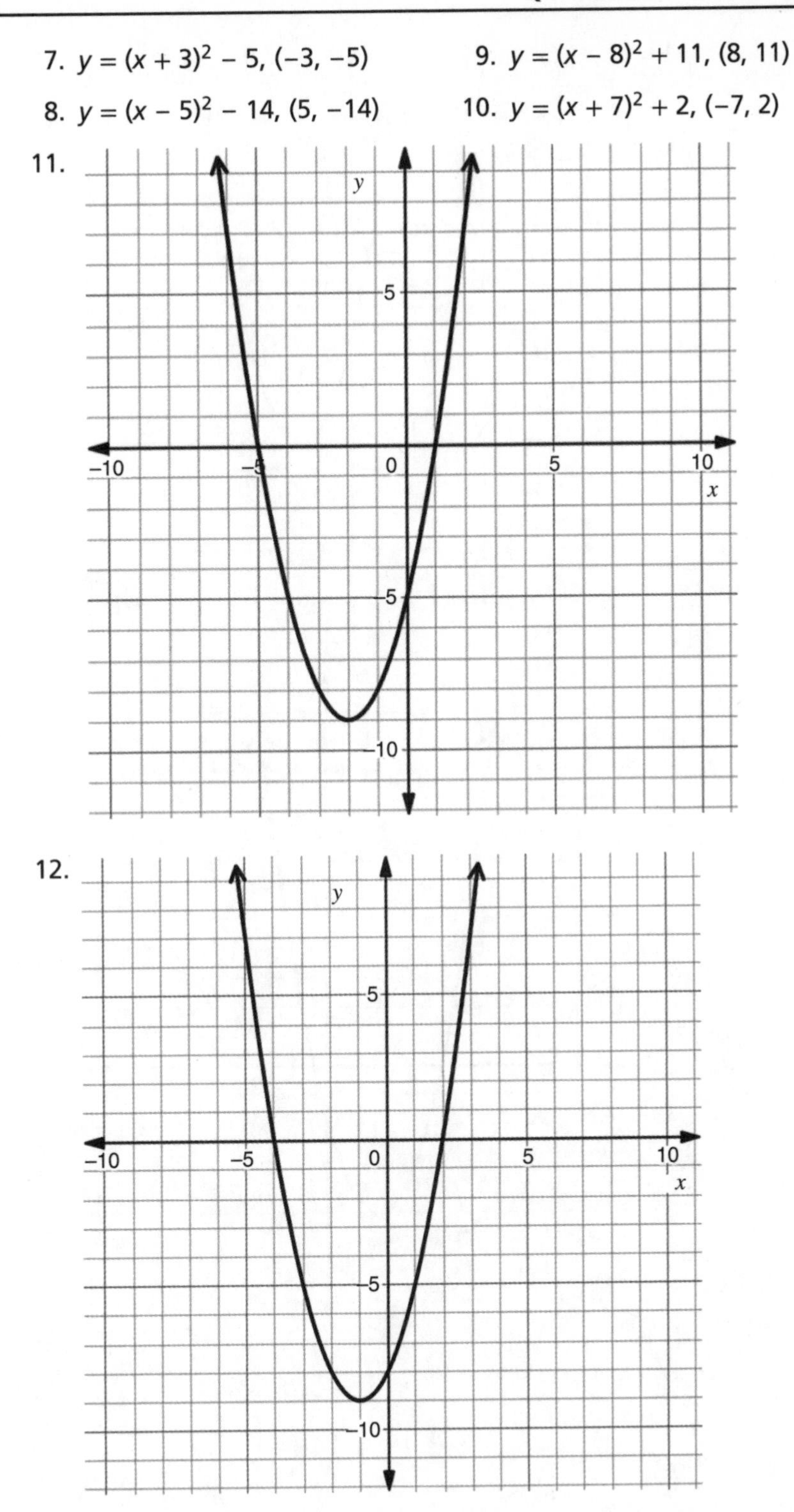

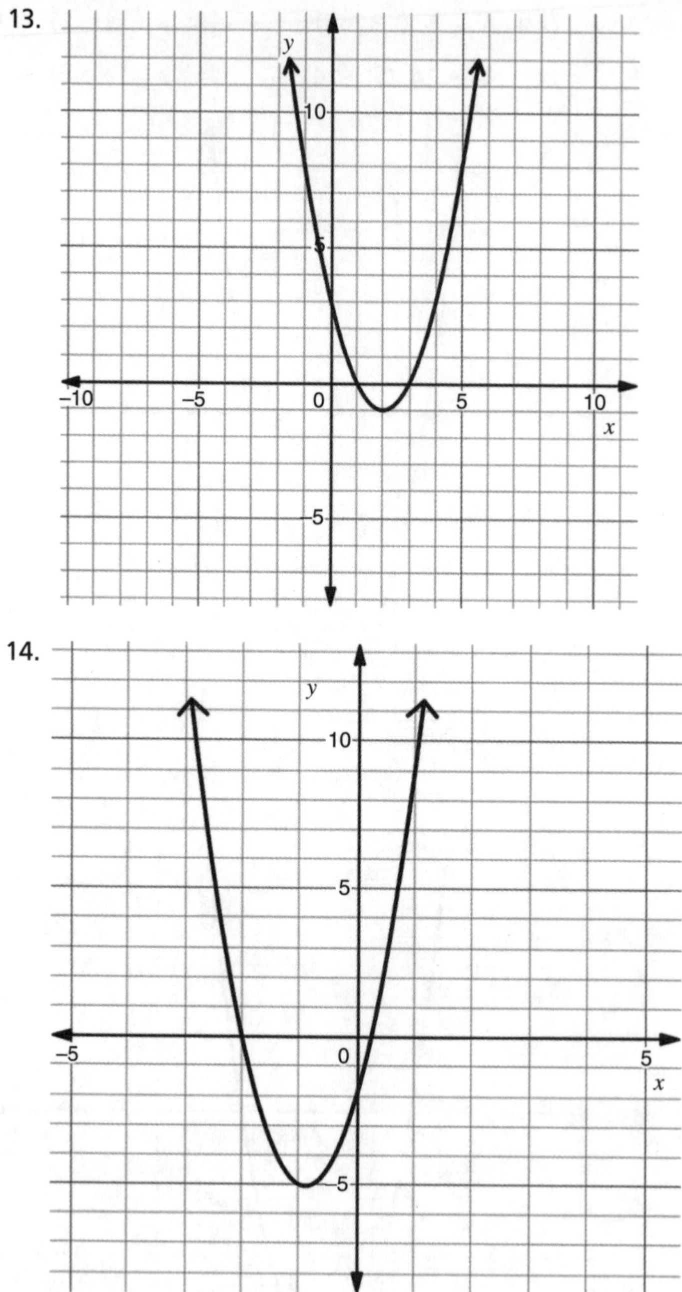
13.
y
10
5
−10
−5
0
5
10
x
−5
14.
y
10
5
−5
0
5
x
−5

15\.

16\.

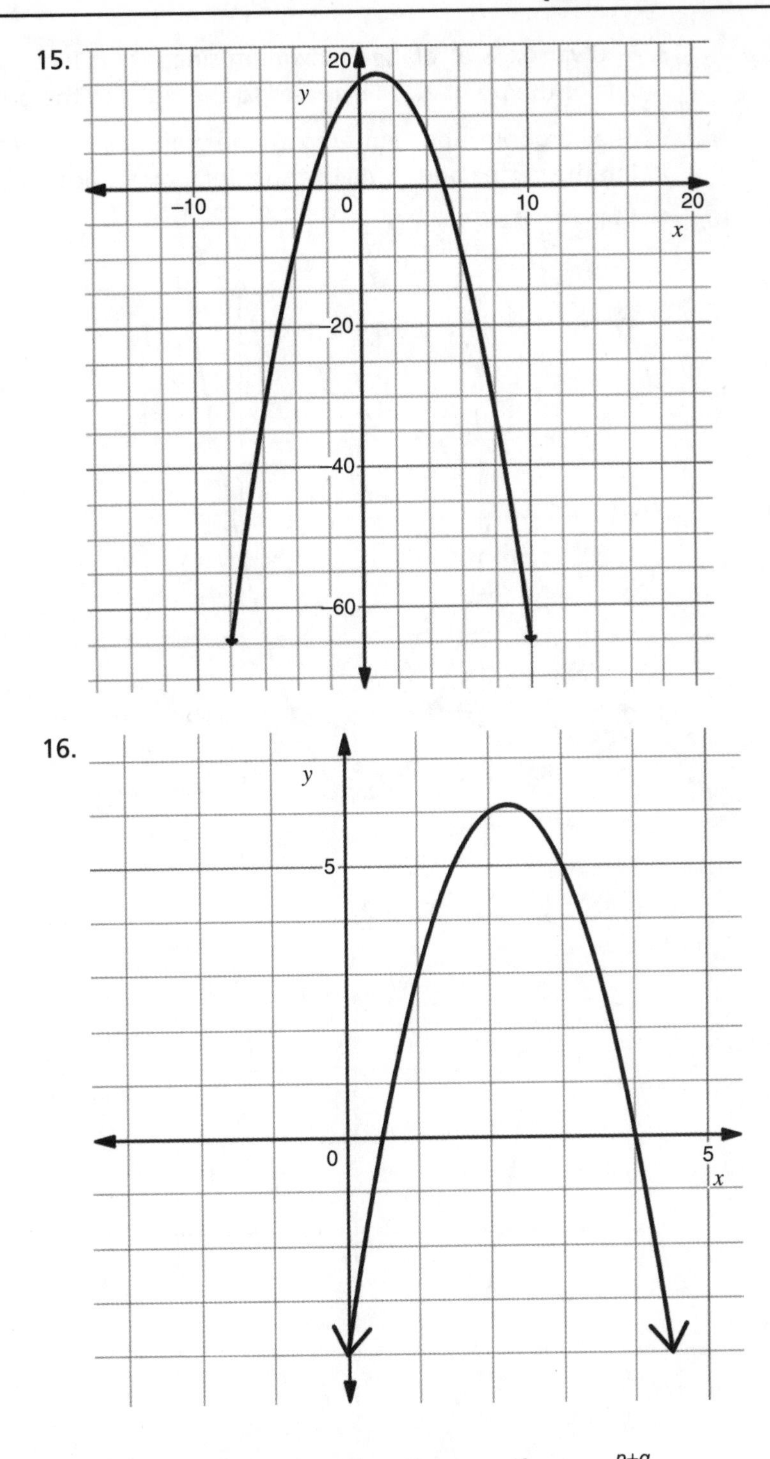

17\. The axis of symmetry has the equation $x = \frac{p+q}{2}$.

18. The vertex is a relative maximum since the function increases to the left of the point and decreases to the right of the point.
19. The parabola is not centered on the graph, so the overall shape (including the vertex and some of the intercepts) is not visible.

9.10. 1. $\{-4, -1\}$, $y = (x + 4)(x + 1)$

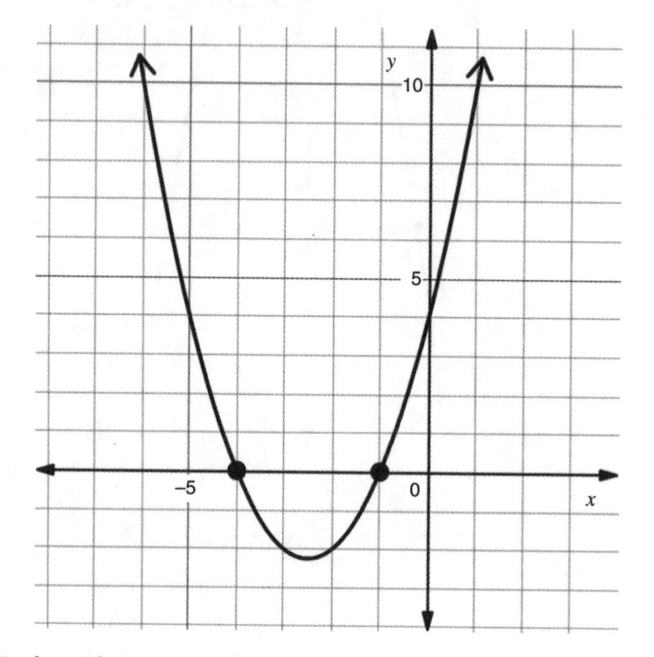

2. $\{-4, 2\}$, $y = (x + 4)(x - 2)$

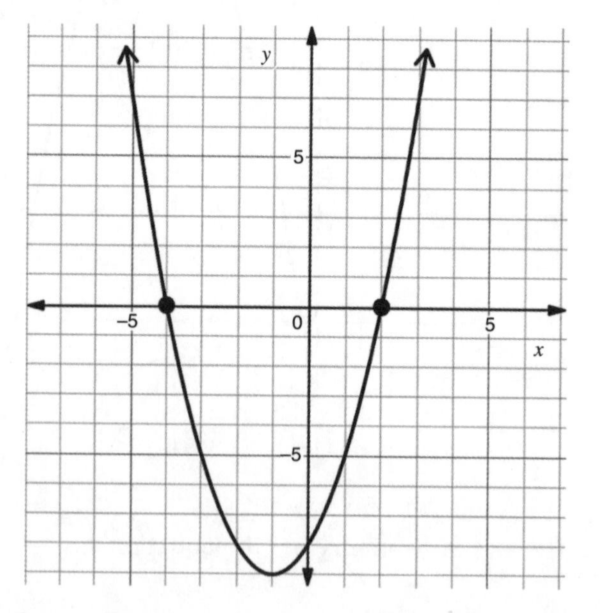

3. $\{0, 3\}$, $y = x(x - 3)$

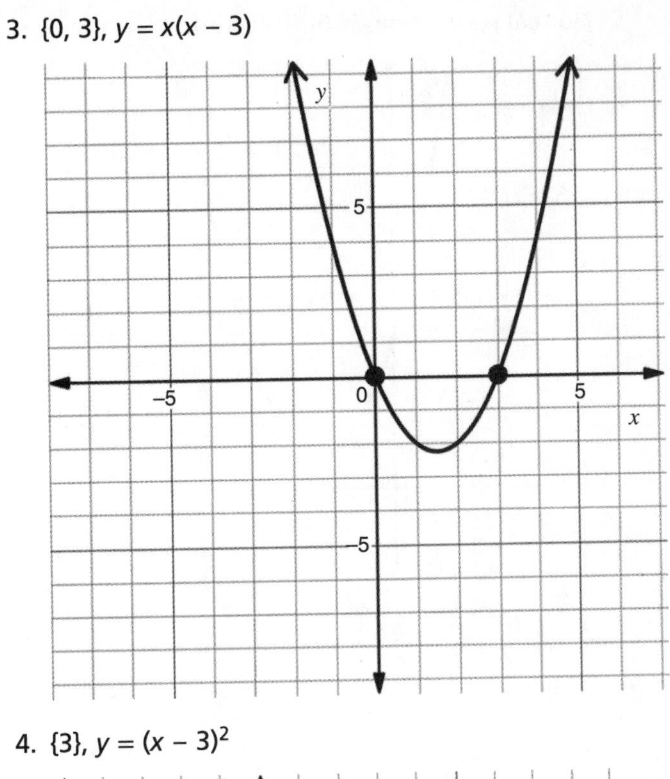

4. $\{3\}$, $y = (x - 3)^2$

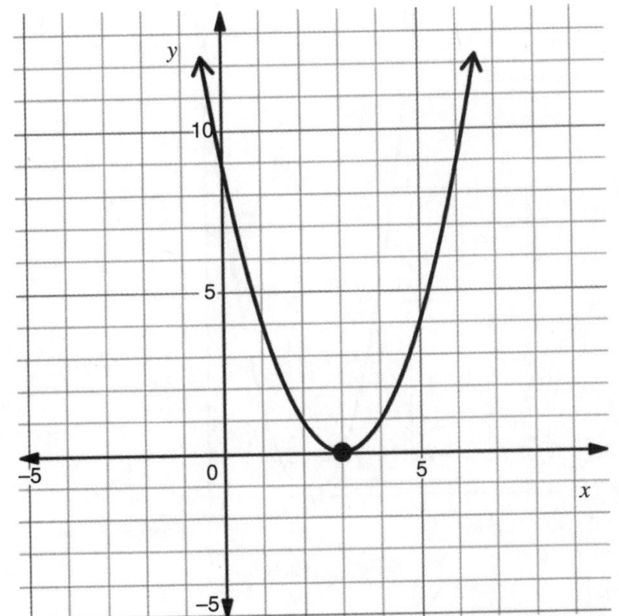

5. No real roots, cannot be factored

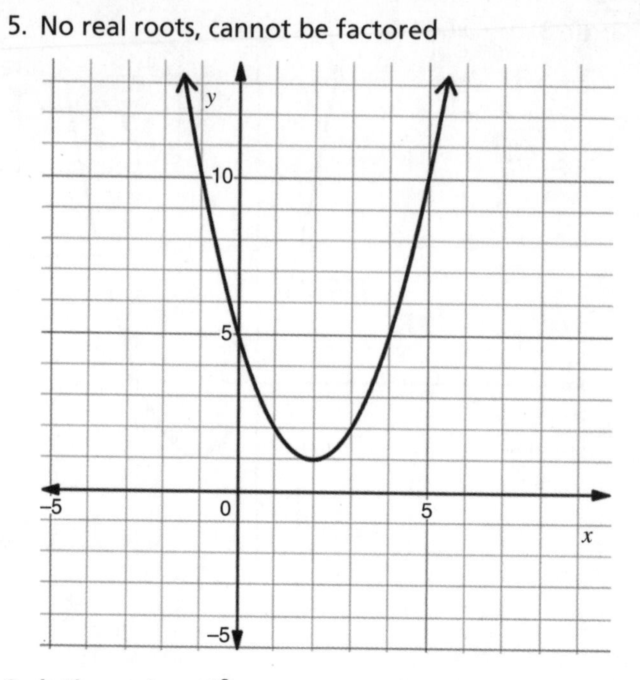

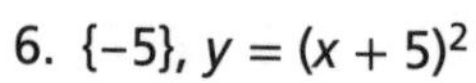
6. $\{-5\}$, $y = (x + 5)^2$

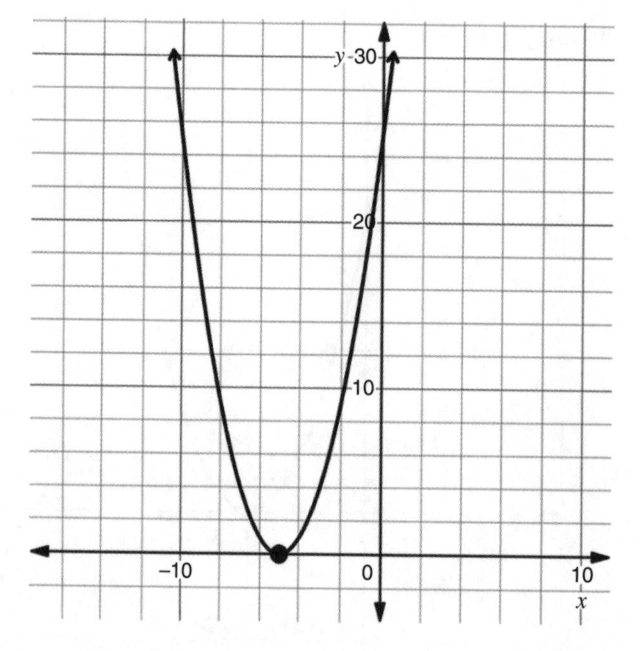

7. $\{-3, -2\}$, $y = -(x + 3)(x + 2)$

8. $\{-15, -7\}$, $y = -(x + 15)(x + 7)$

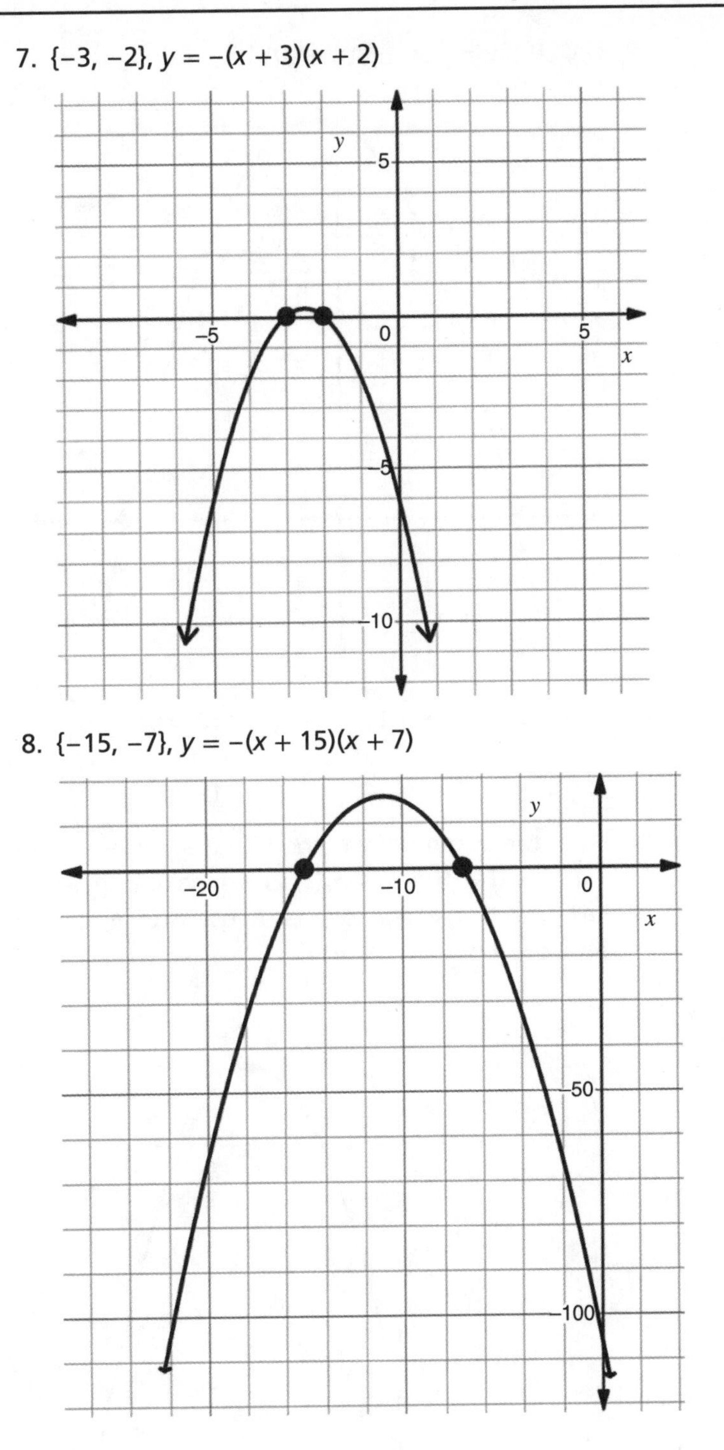

9. $\{-8, 10\}$, $y = -(x + 8)(x - 10)$

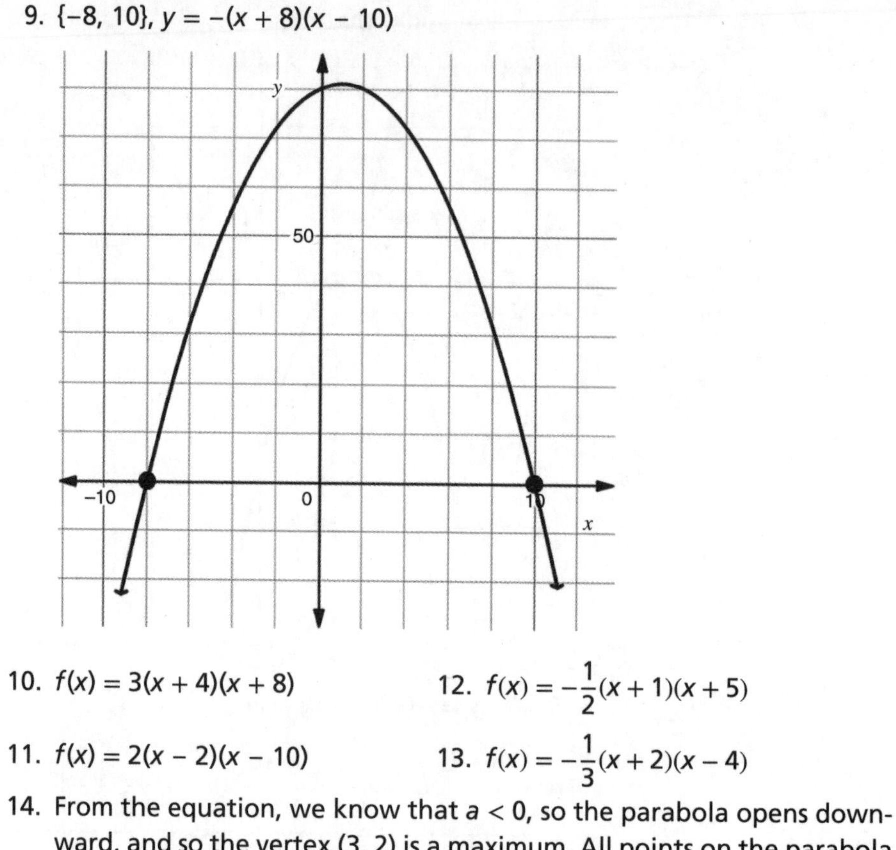

10. $f(x) = 3(x + 4)(x + 8)$

11. $f(x) = 2(x - 2)(x - 10)$

12. $f(x) = -\frac{1}{2}(x + 1)(x + 5)$

13. $f(x) = -\frac{1}{3}(x + 2)(x - 4)$

14. From the equation, we know that $a < 0$, so the parabola opens downward, and so the vertex (3, 2) is a maximum. All points on the parabola lie below the vertex, so it has two x-intercepts. Thus, the equation $-x^2 + bx + c = 0$ has two roots.

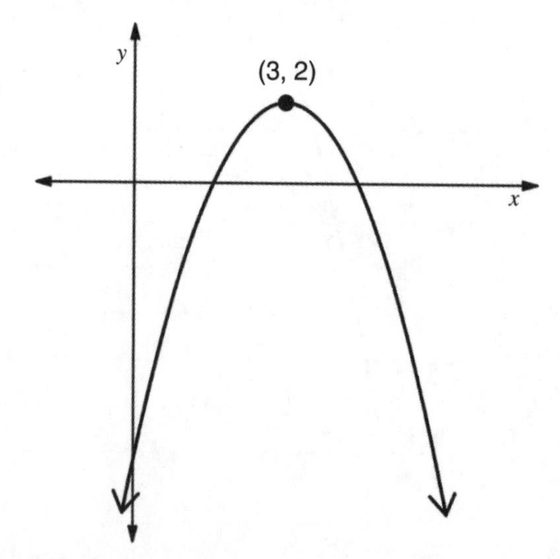

15. If the vertex and x-intercept are located at the same point, then the equation $ax^2 + bx + c = 0$ has one double root at the x-intercept.

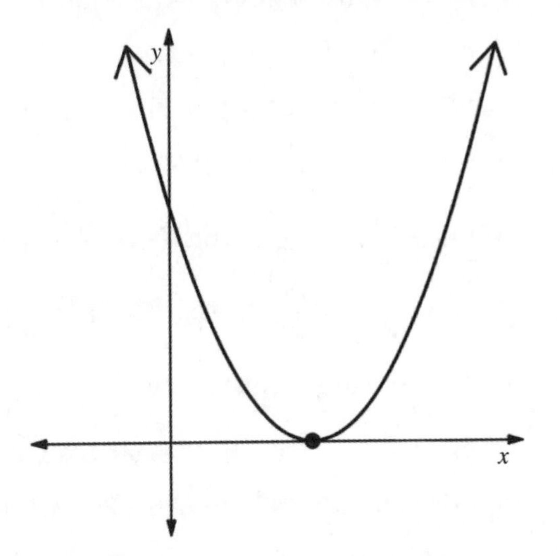

16. The vertex (1, 4) is either a maximum or a minimum of the parabola. Since (0, 5), whose y-value is greater than the vertex's y-value, is also a point on the parabola, then (1, 4) must be a minimum. Thus, the parabola does not intercept the x-axis, so $ax^2 + bx + c = 0$ has no real roots.

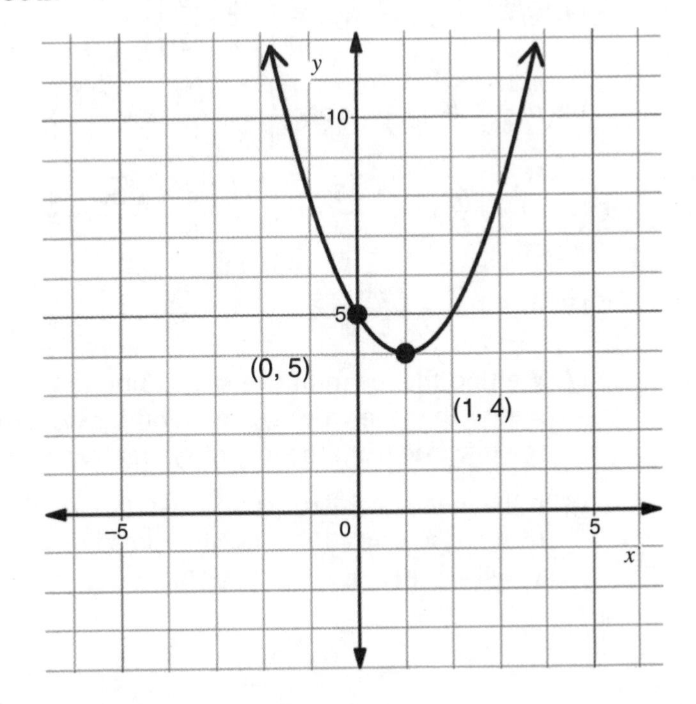

9.11. 1. $m = 10$, $u = 2$, roots: {8, 12}

2. $m = -11$, $u = 1$, roots: {−12, −10}

3. $m = 3$, $u = 12$, roots: {−9, 15}

4. $m = 2$, $u = 6$, roots: {−4, 8}

5. $m = \frac{19}{2}$, $u = \frac{11}{2}$, roots: {4, 15}

6. $m = -\frac{27}{2}$, $u = \frac{19}{2}$, roots: {−23, −4}

7. $m = -\frac{9}{2}$, $u = \frac{21}{2}$, roots: {−15, 6}

8. $m = \frac{3}{2}$, $u = \frac{9}{2}$, roots: {−3, 6}

9. $m = 5$, $u = \sqrt{7}$, roots: $\{5 - \sqrt{7}, 5 + \sqrt{7}\}$

10. $m = 6$, $u = 2\sqrt{14}$, roots: $\{6 - 2\sqrt{14}, 6 + 2\sqrt{14}\}$

11. $m = -\frac{7}{2}$, $u = \frac{\sqrt{5}}{2}$, roots: $\left\{-\frac{7}{2} - \frac{\sqrt{5}}{2}, -\frac{7}{2} + \frac{\sqrt{5}}{2}\right\}$

12. $m = \frac{5}{2}$, $u = \frac{\sqrt{13}}{2}$, roots: $\left\{\frac{5}{2} - \frac{\sqrt{13}}{2}, \frac{5}{2} + \frac{\sqrt{13}}{2}\right\}$

13. $m = -2$, $u = \frac{1}{2}$, roots: $\left\{-\frac{5}{2}, -\frac{3}{2}\right\}$

14. $m = 2$, $u = \frac{\sqrt{3}}{2}$, roots: $\left\{2 - \frac{\sqrt{3}}{2}, 2 + \frac{\sqrt{3}}{2}\right\}$

15. $m = -\frac{4}{5}$, $u = \frac{\sqrt{26}}{5}$, roots: $\left\{-\frac{4}{5} - \frac{\sqrt{26}}{5}, -\frac{4}{5} + \frac{\sqrt{26}}{5}\right\}$

16. $m = 1$, $u = \frac{2\sqrt{6}}{3}$, roots: $\left\{1 - \frac{2\sqrt{6}}{3}, 1 + \frac{2\sqrt{6}}{3}\right\}$

17. We find the mean of the roots using the formula for the axis of symmetry. Then we write an equation for the product of the roots to find the distance from the axis of symmetry to a root.

18. If the mean of the roots is 5, then $(5 - u)(5 + u) = 6$, so $25 - u^2 = 6$, so $u^2 = 19$, so $u = \sqrt{19}$. Thus, the roots are $x_1 = 5 - u$ and $x_2 = 5 + u$, or $x_1 = 5 - \sqrt{19}$ and $x_2 = 5 + \sqrt{19}$.

19. If $x^2 + bx + c = 0$, then the roots have a mean of $x = -\frac{b}{2}$ and a product of $\left(-\frac{b}{2} - u\right)\left(-\frac{b}{2} + u\right) = c$. Then $\frac{b^2}{4} - u^2 = c$, so $u^2 = \frac{b^2}{4} - c$. Taking the square root of both sides gives us $u = \sqrt{\frac{b^2}{4} - c} = \sqrt{\frac{b^2}{4} - \frac{4c}{4}} = \sqrt{\frac{b^2-4c}{4}} = \frac{\sqrt{b^2-4c}}{\sqrt{4}} = \frac{\sqrt{b^2-4c}}{2}$. The roots are $x = -\frac{b}{2} \pm u = -\frac{b}{2} \pm \frac{\sqrt{b^2-4c}}{2}$.

9.12.

1. {(2, 2), (3, 1)}
2. {(−4, −3), (0, 5)}
3. {(1, 5)}
4. {(−5, 6), (−1, 2)}
5. No real solutions
6. {(1, 8), (7, −4)}
7. {(2, −1), (−2, −3)}
8. {(3, 9)}
9. No real solutions
10. If Aileen had zoomed out, she would have seen that the line eventually intersects with the parabola at a second point in Quadrant I.
11. The system of equations $ax^2 + bx + c = 0$ and $y = 0$ has solutions whose y-coordinates are 0. These points are the x-intercepts of the graph of $y = ax^2 + bx + c$.
12. Since the domain of the quadratic function is all real numbers, then every x-value has exactly 1 y-value. Since $x = m$ is a vertical line, then the parabola will intersect the line at exactly 1 point, which has coordinates $(m, f(m))$. Thus, the system has 1 solution.

9.13.

1. −18
2. 3 and 7
3. 25 and 26
4. −8, −6, −4
5. width = 11, length = 25
6. height = 9, base = 12
7. a. 36 (height at $x = 0$), b. 2, c. 6, d. $h(t) = -3(t - 6)(t + 2) = -3t^2 + 12t + 36$
8. a. 16 feet (since $h(0) = 16$), b. 25 feet (since the vertex is (3, 25)), c. $0 \leq x \leq 8$
9. a. 1.5 seconds, b. 100 feet, c. 1.5 to 4 seconds
10. a. January 16, b. $57

CHAPTER 9 TEST SOLUTIONS

1. (D)
2. (B)
3. (A)
4. (B)
5. (D)
6. (A)
7. (C)
8. (D)
9. (A)
10. $r^2q(4q - 5)(4q + 3)$
11. $\{-21, 9\}$
12. $\left\{2 - \frac{\sqrt{30}}{2}, 2 + \frac{\sqrt{30}}{2}\right\}$ or $\frac{4 \pm \sqrt{30}}{2}$
13. $\{(2, 0), (9, -7)\}$
14. 11 and 14
15. a. 3, b. 2, c. 67, d. 4

10 EXPONENTIAL FUNCTIONS

So far, we have worked with polynomial functions. In this chapter, we briefly introduce another important type of function called an **exponential function**, which has its variable in the exponent. These functions are incredibly useful because they can model a great deal of real-world behavior, from the population of bacteria to the amount of money in your bank account.

10.1 Graphing Exponential Functions

A famous problem, first known to have been recorded by the 13th-century Muslim scholar Ibn Khallikān, has been used to illustrate exponential growth for hundreds of years: if an 8x8 chessboard has grains of wheat placed on each of its 64 squares such that 1 grain is placed on the first square, 2 on the second, 4 on the third, and so on, doubling the number of grains on each subsequent square, how many grains end up on the chessboard?

If we make a table (Table 10.1) that relates each square (x) to the number of grains of wheat on it (y), we see a pattern:

Table 10.1 Number of grains of wheat per square.

Square (x)	Calculation	Number of grains (y)
1	$= 1$	1
2	$= 2 = 2^1$	2
3	$= 2 \bullet 2 = 2^2$	4
4	$= 2 \bullet 2 \bullet 2 = 2^3$	8
5	$= 2 \bullet 2 \bullet 2 \bullet 2 = 2^4$	16
6	$= 2 \bullet 2 \bullet 2 \bullet 2 \bullet 2 = 2^5$	32
7	$= 2 \bullet 2 \bullet 2 \bullet 2 \bullet 2 \bullet 2 = 2^6$	64
...	...	...
x	$= 2 \bullet 2 \bullet \ldots \bullet 2 = 2^{x-1}$	2^{x-1}

We can summarize this pattern with the equation $y = 2^{x-1}$, meaning that the xth square has 2^{x-1} grains of wheat on it. The total number of grains of wheat on the chessboard is $1 + 2 + 4 + 8 + \ldots + 2^{63}$, or $2^{64} - 1 = 18{,}446{,}744{,}073{,}709{,}551{,}615$.

Reading and Writing Tip

Since reading ordinal words with variables (such as *x*th or *t*th) is difficult to say, we often pronounce them without the -th suffix, so we pronounce 2^x as "2 to the *x* power" and 2^t as "2 to the *t* power."

To get a better understanding of an exponential function, we can make a graph using a table of values.

Example 10.1 Graph $f(x) = 2^x$ using a table of values from $-3 \leq x \leq 3$.

Solution: To make a table of values, recall from Section 8.3 that $x^{-a} = \frac{1}{x^a}$, so $2^{-3} = \frac{1}{2^3} = \frac{1}{8}$, $2^{-2} = \frac{1}{2^2} = \frac{1}{4}$, and $2^{-1} = \frac{1}{2^1} = \frac{1}{2}$. From the table, we see that 2 acts as a "multiplier" between output values in the table. We then plot the points from the table on the coordinate plane and connect them to get a smooth curve. We put arrows at each end to indicate that the graph extends infinitely in those directions.

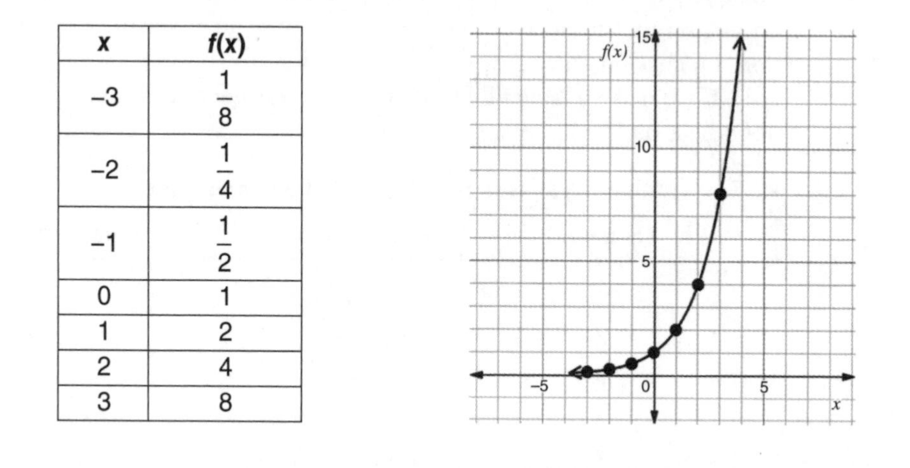

x	f(x)
−3	$\frac{1}{8}$
−2	$\frac{1}{4}$
−1	$\frac{1}{2}$
0	1
1	2
2	4
3	8

Example 10.2 Graph $f(x) = 3(2)^x$ using a table of values from $-3 \leq x \leq 3$.

Solution: Recall that $3(2)^x \neq 6^x$ since 3 is not being raised to a power, so $3(2)^{-3} = 3\left(\frac{1}{2^3}\right) = \frac{3}{8}$. From the table, we see that our "multiplier" is 2.

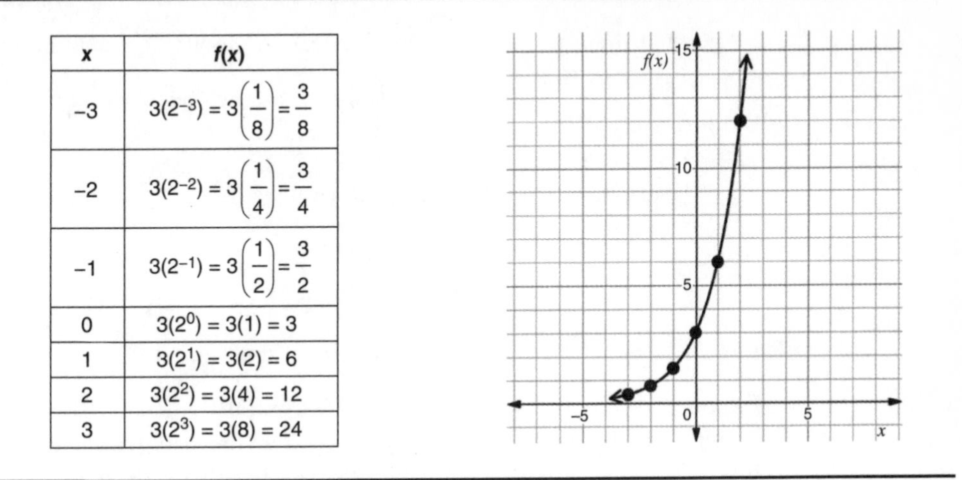

x	f(x)
−3	$3(2^{-3}) = 3\left(\frac{1}{8}\right) = \frac{3}{8}$
−2	$3(2^{-2}) = 3\left(\frac{1}{4}\right) = \frac{3}{4}$
−1	$3(2^{-1}) = 3\left(\frac{1}{2}\right) = \frac{3}{2}$
0	$3(2^{0}) = 3(1) = 3$
1	$3(2^{1}) = 3(2) = 6$
2	$3(2^{2}) = 3(4) = 12$
3	$3(2^{3}) = 3(8) = 24$

In Examples 10.1 and 10.2, the functions increase as x increases (if necessary, review the definitions of increasing functions in Section 6.3). Other exponential functions decrease as x increases, as shown here.

Example 10.3 Graph $f(x) = \left(\frac{1}{2}\right)^x$ using a table of values from $-3 \le x \le 3$.

Solution: We use the rules for multiplying powers and raising a power to a power (see Sections 8.2 and 8.3 if you need to refresh your memory). From the table, we see that our "multiplier" is $\frac{1}{2}$.

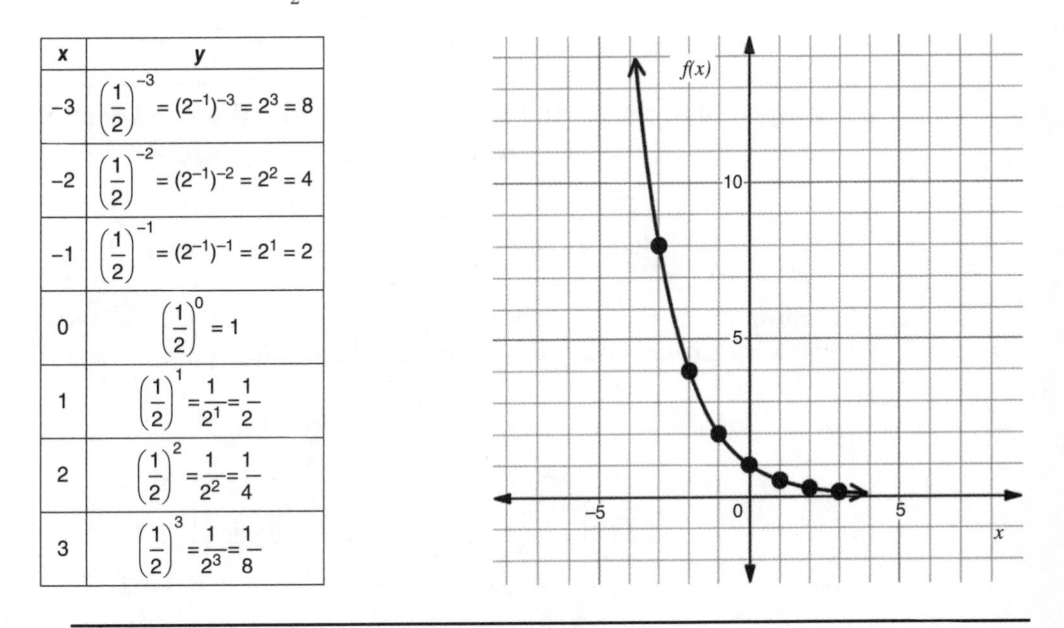

x	y
−3	$\left(\frac{1}{2}\right)^{-3} = (2^{-1})^{-3} = 2^3 = 8$
−2	$\left(\frac{1}{2}\right)^{-2} = (2^{-1})^{-2} = 2^2 = 4$
−1	$\left(\frac{1}{2}\right)^{-1} = (2^{-1})^{-1} = 2^1 = 2$
0	$\left(\frac{1}{2}\right)^{0} = 1$
1	$\left(\frac{1}{2}\right)^{1} = \frac{1}{2^1} = \frac{1}{2}$
2	$\left(\frac{1}{2}\right)^{2} = \frac{1}{2^2} = \frac{1}{4}$
3	$\left(\frac{1}{2}\right)^{3} = \frac{1}{2^3} = \frac{1}{8}$

Figure 10.1 summarizes what we've learned about exponential functions so far:

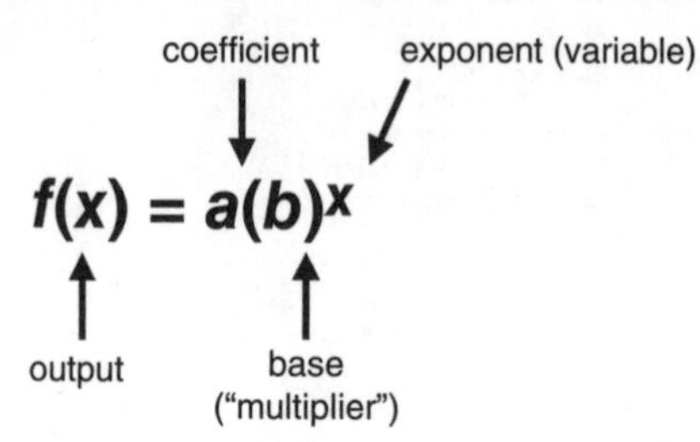

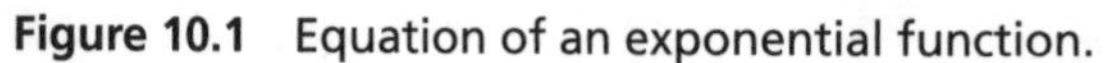
Figure 10.1 Equation of an exponential function.

- Exponential functions have the form $f(x) = a(b)^x$, where a is a coefficient and b (our "multiplier" between outputs whose x-values are consecutive integers) is called the **base**. (We use the word "base" because this number is similar to the base of a power like 2^4.)
- The y-intercept of f is $(0, a)$.
- If $a > 0$ and $b > 1$, then the function represents **exponential growth**—it is always increasing. If $a > 0$ and $0 < b < 1$, then the function represents **exponential decay**—it is always decreasing. The base cannot be a negative number or 0.
- Unlike polynomial graphs, exponential graphs approach the x-axis. We call the x-axis an **asymptote**, meaning that the graph of the exponential function gets closer and closer to the x-axis. The graph of $f(x) = a(b)^x$ has no x-intercepts and never goes below the x-axis.
- The domain of f is $(-\infty, \infty)$. The range of f is $(0, \infty)$.
- Unlike quadratic graphs, exponential graphs are not symmetrical.

Example 10.4 A partial table of values for the exponential function f is shown here. Write an equation for f.

x	−3	−2	−1	0	1	2
$f(x)$	32	16	8	4	2	1

Solution: Exponential functions have the form $f(x) = a(b)^x$, so we need to find a and b. From the table of values, we can identify the y–intercept: $x = 0$ and $f(0) = 4$.

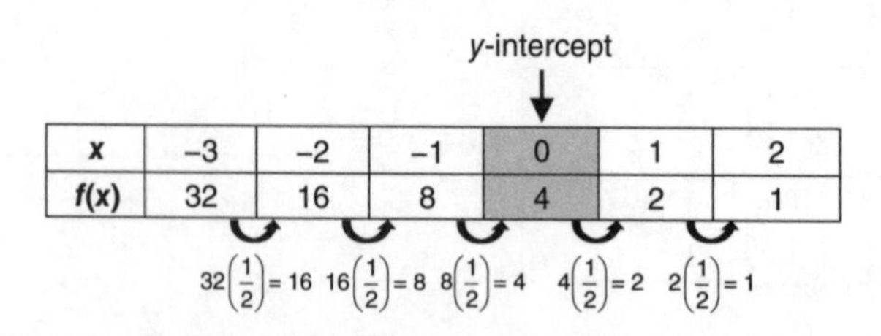

x	−3	−2	−1	0	1	2
f(x)	32	16	8	4	2	1

Substituting these values into the equation gives us $4 = a(b)^0$. Since any nonzero number raised to the 0 power equals 1, then $4 = a(1) = a$, so $a = 4$. To find the base b, look for a pattern among consecutive outputs. (In the next chapter, we'll discuss a more systematic way to find b.) Since $16\left(\frac{1}{2}\right) = 8$, $8\left(\frac{1}{2}\right) = 4$, and so on, then $b = \frac{1}{2}$. Thus, $a = 4$ and $b = \frac{1}{2}$, so the equation is $f(x) = 4\left(\frac{1}{2}\right)^x$.

Exercises

Graph each function.

1. $f(x) = 3^x$
2. $f(x) = 5^x$
3. $f(x) = 2(3)^x$
4. $f(x) = 3(4)^x$
5. $f(x) = \left(\frac{1}{3}\right)^x$
6. $f(x) = \left(\frac{1}{4}\right)^x$

A partial table of values for the exponential function f is shown. Write an equation for f.

7.

x	−3	−2	−1	0	1	2	3
f(x)	$\frac{1}{216}$	$\frac{1}{36}$	$\frac{1}{6}$	1	6	36	216

8.

x	−3	−2	−1	0	1	2	3
f(x)	$\frac{1}{125}$	$\frac{1}{25}$	$\frac{1}{5}$	1	5	25	125

9.

x	−3	−2	−1	0	1	2	3
f(x)	$\frac{5}{8}$	$\frac{5}{4}$	$\frac{5}{2}$	5	10	20	40

10.

x	−3	−2	−1	0	1	2	3
f(x)	$\frac{1}{32}$	$\frac{1}{8}$	$\frac{1}{2}$	2	8	32	128

11.

x	−3	−2	−1	0	1	2	3
$f(x)$	343	49	7	1	$\frac{1}{7}$	$\frac{1}{49}$	$\frac{1}{343}$

12.

x	−3	−2	−1	0	1	2	3
$f(x)$	729	81	9	1	$\frac{1}{9}$	$\frac{1}{81}$	$\frac{1}{729}$

Questions to Think About

13. Explain why a function $f(x) = a(b)^x$, where $b = 1$, is not exponential.

14. Explain why a function $f(x) = a(b)^x$, where $b < 0$, does not represent exponential growth or decay. (HINT: Let b equal a negative number. Substitute different values for x, such as 1, 2, 3, and 4. What do you notice?)

15. Solve for x in the equation $2(3)^x = 3(2)^x$.

10.2 Using Exponential Functions to Solve Word Problems

As we said at the beginning of this chapter, exponential functions model real-world behavior. When writing these functions, we use an important shortcut that is illustrated in Example 10.5:

Example 10.5 **Ben goes out to a restaurant to eat. His restaurant bill, including taxes, is \$80. He wants to add a percentage of his bill for tips. Calculate his total bill if this percentage is:**

(a) 15%

(b) 20%

(c) 25%

Solution: Convert each percentage to a decimal, multiply the bill by the decimal, and add the result to the initial amount.

(a) $\$80 + \$80(0.15)$

$= \$80(1 + 0.15)$	Distributive Property.
$= \$80(1.15)$	Simplify.
$= \$92$	Multiply.

(b) $\$80 + \$80(0.20) = \$80(1 + 0.20) = \$80(1.20) = \$96$

(c) $\$80 + \$80(0.25) = \$80(1 + 0.25) = \$80(1.25) = \$100$

Example 10.5 shows the following: if A_0 (pronounced "A-sub-zero") is the initial amount and r is the percentage (expressed as a decimal) being added to the initial amount per time period, then the final amount at the end of the time period is $A = A_0(1 + r)$.

Example 10.6 Desiree puts \$600 in a bank account that earns 1% interest per year. Assuming she makes no deposits or withdrawals to the account, determine *to the nearest cent* the account balance after 20 years.

Solution: We use the formula $A = A_0(1 + r)$ repeatedly to calculate the account balance at the end of each year. This enables us to find a pattern, as shown in the following table:

Year	Beginning Balance	Ending Balance
1	$\$600$	$\$600(1.01) = \$600(1.01)^1$
2	$\$600(1.01)^1$	$\$600(1.01)^1 \cdot (1.01) = \$600(1.01)^2$
3	$\$600(1.01)^2$	$\$600(1.01)^2 \cdot (1.01) = \$600(1.01)^3$
4	$\$600(1.01)^3$	$\$600(1.01)^3 \cdot (1.01) = \$600(1.01)^4$
...	...	...
n	$\$600(1.01)^{n-1}$	$\$600(1.01)^{n-1} \cdot (1.01) = \$600(1.01)^n$

Repeating this calculation 20 times would be tedious! We see from the table that the balance after 20 years is $\$600(1.01)^{20} \approx \$732.11$.

Example 10.7 Lin buys a vintage action figure for \$50. The action figure increases in value by 2.3% each year.

(a) Write a function *V* that models the value, in dollars, of the action figure *t* years after purchase.

(b) Use the equation for *V* to determine the value, to the nearest dollar, of the action figure 10 years after purchase.

(c) To the nearest tenth, determine the number of years that Lin would have to wait after buying the action figure before it is worth \$70.

Solution:

(a) The initial value is \$50, and the rate of change is $2.3\% = 0.023$, so $V(t) = 50(1 + 0.023)^t$. Simplifying the expression in parentheses gives us $V(t) = 50(1.023)^t$.

(b) To the nearest dollar, the value of the action figure 10 years after Lin buys it is $V(10) = 50(1.023)^{10} = \$62.766273 \approx \$63$.

(c) We use technology to graph $y = V(t)$ and $V(t) = 70$ on the same coordinate plane. Use the trace function or click to find the point where the two graphs intersect:

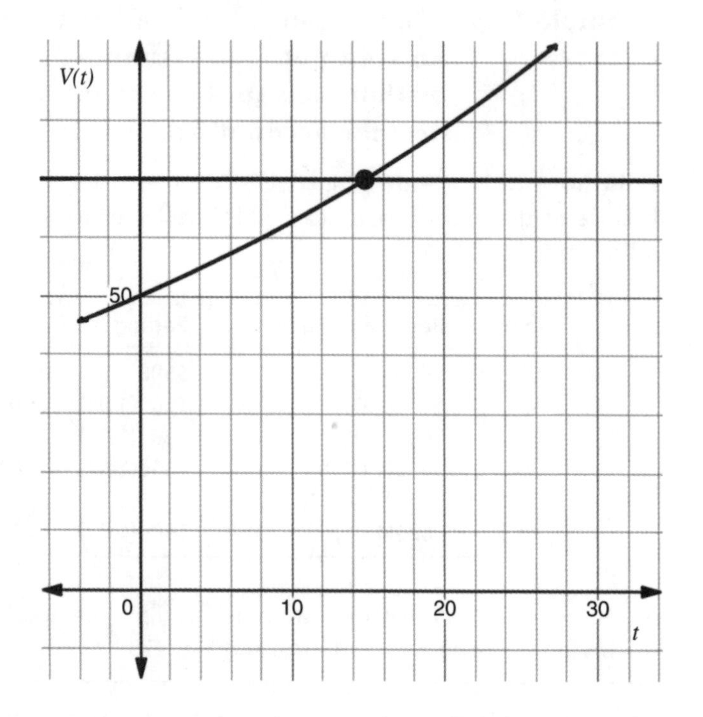

Since the two graphs intersect at approximately (14.797, 70), then the action figure is worth \$70 after approximately 14.8 years.

Example 10.8 The function $M(x) = 14(0.937)^x$ models the population, in thousands, of Megaville x years after 2000.

(a) Determine Megaville's population in 2000.

(b) Has the population of Megaville been increasing or decreasing since 2000?

(c) Explain what 0.937 represents in context.

Solution:

(a) Since x represents the population x years after 2000, then $M(0)$ represents the population in that year (in other words, 0 years after 2000). $M(0) = 14(0.937)^0 = 14(1) = 14$. Since M is expressed in thousands, then the population is 14,000.

(b) The base of M is between 0 and 1, so the function is decreasing. We confirm this by using technology to graph M. The graph shows that the population is decreasing:

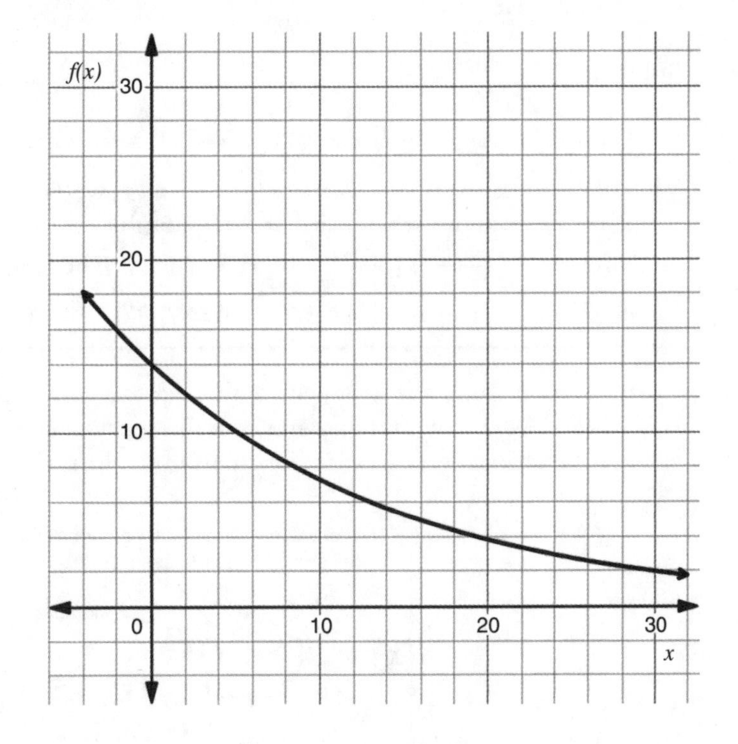

(c) If r is the rate of change in population (expressed as a decimal), then $1 + r = 0.937$. Solving for r gives us $r = 0.937 - 1 = -0.063$, or -6.3%. Since 2000, Megaville's population has decreased at a rate of 6.3% per year.

A common type of exponential decay problem deals with radioactive substances. The **half-life** of a radioactive substance is the time required for one-half of it to decay. For example, if a 100-gram radioactive sample has a half-life of 1 hour, then it would have $100\left(\frac{1}{2}\right) = 50$ grams of radioactive material after 1 hour, $100\left(\frac{1}{2}\right)^2 = 25$ grams after 2 hours, $100\left(\frac{1}{2}\right)^3 = 12.5$ grams after 3 hours, and so on.

Example 10.9 **The function $A(h) = 700(0.5)^h$ models the number of grams left in a sample of a radioactive element after h half-lives. The half-life of the element is 20 seconds.**

(a) How many grams were in the original sample?

(b) To the nearest tenth of a gram, how much of the element remains after 2 minutes?

Solution:

(a) The sample originally had $A(0) = 700(0.5)^0 = 700$ grams.

(b) Since $A(h)$ is expressed in terms of half-lives, we need to convert 2 minutes to half-lives to know what value of h to substitute into the function. 2 minutes = 2 minutes $\left(\frac{1 \textit{ half-life}}{20 \textit{ seconds}}\right)\left(\frac{60 \textit{ seconds}}{1 \textit{ minute}}\right) = 6$ half-lives. The number of grams remaining after 2 minutes, or 6 half-lives, is $A(6) = 700(0.5)^6 = 10.9375 \approx 10.9$ grams.

In Section 10.1, we represented the equation of an exponential function as $f(x) = a(b)^x$. As we saw in the examples in this section, when we deal with real-world exponential change, we can write functions in the format shown in Figure 10.2:

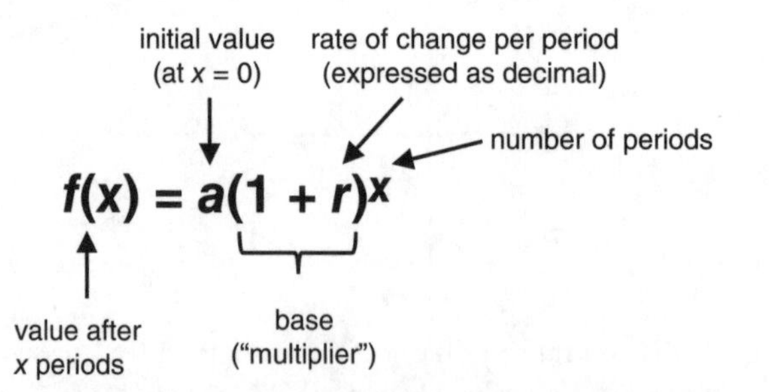

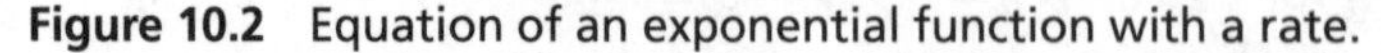

Figure 10.2 Equation of an exponential function with a rate.

- The expression $1 + r$ is equivalent to the base b in the equation $f(x) = a(b)^x$ that we discussed in Section 10.1.
- If $a > 0$ and $r > 0$, then $f(x)$ represents **exponential growth**. If $a > 0$ and $r < 0$, then $f(x)$ represents **exponential decay**.
- If the rate of change is expressed as a percentage, then we substitute its decimal form for r into the equation.
- The quantity a is the initial, or starting, value. The value of $f(x)$ is the value after x events.

If an exponential function is not in the form $f(x) = a(b)^x$, we use the rules of exponents (which we discussed in Sections 8.2 and 8.3) to rewrite it as an equivalent function. Two functions are **equivalent** if their expressions are equivalent. For example, $f(x) = 0.5^x$ and $f(x) = \left(\frac{1}{2}\right)^x$ are equivalent because $0.5 = \frac{1}{2}$.

Example 10.10 Rewrite each function as an equivalent function in the form $f(x) = a(b)^x$. Round all coefficients to the nearest thousandth.

(a) $f(x) = (0.501)^{x+3}$

(b) $f(x) = 95(1.242)^{x-4}$

(c) $f(x) = 20(1.005)^{52x}$

(d) $f(x) = 6(0.975)^{\frac{x}{12}}$

Solution:

(a) $f(x) = (0.501)^{x+3}$

$f(x) = (0.501)^x(0.501)^3$ $\quad x^a(x^b) = x^{a+b}$.

$f(x) = 0.126(0.501)^x$ $\quad$ Simplify.

(b) $f(x) = 95(1.242)^{x-4}$

$f(x) = 95\frac{(1.242)^x}{(1.242)^4}$ $\quad \frac{x^a}{x^b} = x^{a-b}, x \neq 0$.

$f(x) = 39.924(1.242)^x$ $\quad$ Simplify.

(c) $f(x) = 20(1.005)^{52x}$

$f(x) = 20(1.005^{52})^x$ $\quad (x^a)^b = x^{ab}$.

$f(x) = 20(1.296)^x$ $\quad$ Simplify.

(d) $f(x) = 6(0.975)^{\frac{x}{12}}$

$f(x) = 6\left(0.975^{\frac{1}{12}}\right)^x$ $\quad (x^a)^b = x^{ab}$.

$f(x) = 6(0.998)^x$ $\quad$ Simplify.

Technology Tip

Remember that you can check if two functions are equivalent by graphing them using technology. If the graphs overlap completely, then they are equivalent.

Exercises

Determine whether each function represents exponential growth or decay. State the rate of change as a percentage.

1. $f(x) = 198(1.05)^x$
2. $P(x) = 6(1.12)^x$
3. $A(x) = 42(0.87)^x$
4. $D(t) = 13(1.132)^t$
5. $g(x) = 5.1(1.098)^x$
6. $W(t) = 907(0.096)^t$
7. The balance, in dollars, in an investment account x years after the account was opened is modeled by the function $f(x) = 1{,}500(1.0092)^x$.

   (a) How much was originally invested in the account?

   (b) What is the account's yearly interest rate?

   (c) To the nearest dollar, determine how much money will be in the account after 12 years.

8. The number of milligrams of a drug dose in the body after t minutes can be modeled by the function $A(t) = 500(0.7)^t$.

   (a) Is the amount of drug in the body increasing or decreasing over time? Explain.

   (b) To the nearest tenth, determine the number of milligrams of the drug dose left in the body after 6 minutes.

9. A student loan balance, in thousands in dollars, is modeled by the function $B(x) = 85(1.045)^x$, where x represents the number of years after the loan was first taken out.

   (a) What was the original amount of the loan, in dollars?

   (b) Explain in context what 0.045 represents.

   (c) To the nearest thousand dollars, what will the loan balance be after 10 years? (Assume that payments to the loan are deferred, which means that no payments are made and there is no penalty for doing so.)

10. A checking account with an initial deposit of $900 charges a 2.5% monthly fee, which is taken from the account.

    (a) Write a function f that represents the account balance x months after the initial deposit.

    (b) Assuming no other withdrawals or deposits are made, determine the amount, to the nearest cent, that is left in the account after 2 years.

11. The population of a large city, which was 2,800,000 in 2000, has increased 1.9% per year since then.

(a) Write a function C that models the city's population, in millions, x years after 2000.

(b) Use the model to determine the city's population, to the nearest ten thousand, in 2010.

(c) According to the model, in what year the city's population reached above 3,800,000?

12. A 50-gram sample of iodine-124 has a half-life of 4 days.

(a) Write a function I that models the number of grams of iodine-124 remaining in the sample after x half-lives.

(b) Determine, to the nearest hundredth, the number of grams of iodine-124 in the sample that remain after 30 days.

(c) Determine, to the nearest whole number, the number of days required for the sample to contain 3 grams of iodine-124.

Rewrite each function as an equivalent function in the form $f(x) = a(b)^x$. Round all coefficients to the nearest thousandth.

13. $f(x) = 5.1(1.195)^{x+2}$

14. $f(x) = 62(0.571)^{x-5}$

15. $f(x) = 7(1.571)^{7x}$

16. $f(x) = 20(0.9)^{3x}$

17. $f(x) = 6(1.8)^{\frac{x}{365}}$

18. $f(x) = 15(1.25)^{\frac{x}{12}}$

CHAPTER 10 TEST

1. The function $P(t) = 839(1.026)^t$ models the population of Newtown t years after 2000. What was the population of Newtown in 2000?

(A) 818 (B) 839 (C) 860 (D) 861

2. Which function is equivalent to $f(x) = 4(0.7)^{2x}$?

(A) $f(x) = (1.96)^x$

(B) $f(x) = 4(0.14)^x$

(C) $f(x) = 4(0.49)^x$

(D) $f(x) = (7.84)^x$

3. What type of change is represented by the function $Q(t) = 265(1.024)^t$?

(A) 1.024% growth
(B) 2.4% growth
(C) 102.4% growth
(D) 271.36% growth

4. What type of change is represented by the function $P(x) = (0.825)^x$?

(A) 17.5% decay
(B) 0.825% decay
(C) 82.5% growth
(D) 0.825% growth

5. A partial table of values for f is shown here. What is the equation of f?

x	−3	−2	−1	0	1	2	3
$f(x)$	$\frac{3}{32}$	$\frac{3}{8}$	$\frac{3}{2}$	6	24	96	384

(A) $f(x) = 6^x$
(B) $f(x) = 4(6)^x$
(C) $f(x) = 6(4)^x$
(D) $f(x) = 24^x$

6. What is the x-intercept of the function $f(x) = 2(3)^x$?

(A) (1, 6)
(B) (6, 0)
(C) (0, 2)
(D) There is no x-intercept.

7. Over what interval is the function $f(x) = 0.5(2)^x$ increasing?

(A) $(-\infty, \infty)$
(B) $(0, \infty)$
(C) $(1, \infty)$
(D) The function is never increasing.

8. If $P(x) = b^x$ represents exponential decay, which statement must be true?

(A) $b < 0$
(B) $-1 < b < 0$
(C) $0 < b < 1$
(D) $0 < b \leq 1$

9. If $P(x) = b^x$ is an exponential function and $b \neq 1$, which statement must be true?

(A) $b > 0$
(B) $b > 1$
(C) $0 < b < 1$
(D) $0 \leq b < 1$

10. The value, in dollars, of an investment account x years after 2015 is modeled by the function $V(x) = 825(1.014)^x$. Determine the y-intercept of V and explain in context what it represents.

11. The ordered pair (2, 16) is a solution to the equation $y = b^x$. What is the value of b? Explain.

12. The number of followers, in thousands, that a celebrity has on a popular social media site m months after creating an account is modeled by the function $f(m) = 4(1.4)^m$. To the nearest thousand, use the model to determine the celebrity's number of followers 3 years after creating an account.

13. Graph $f(x) = 2(0.5)^x$ on the coordinate plane.

14. The function $C(x) = 45.5(0.94)^x$ represents the value of a new car, in thousands of dollars, x years after purchase.

    (a) What was the value of the car, in dollars, when it was purchased?

    (b) Describe in context what 0.94 represents.

15. The population of a large town, which was 26,000 in 2000, has increased at a rate of 6% per year since then.

    (a) Write a function P that represents the town's population, in thousands, x years since 2000.

    (b) Graph P over the interval $0 \le x \le 20$.

    (c) In what year the town's population began to exceed 40,000? Explain.

CHAPTER 10 SOLUTIONS

10.1. 1. 2.

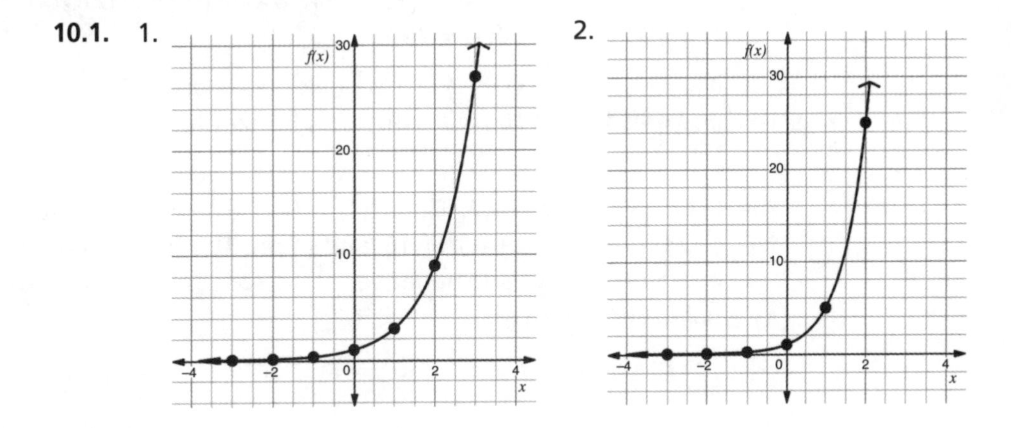

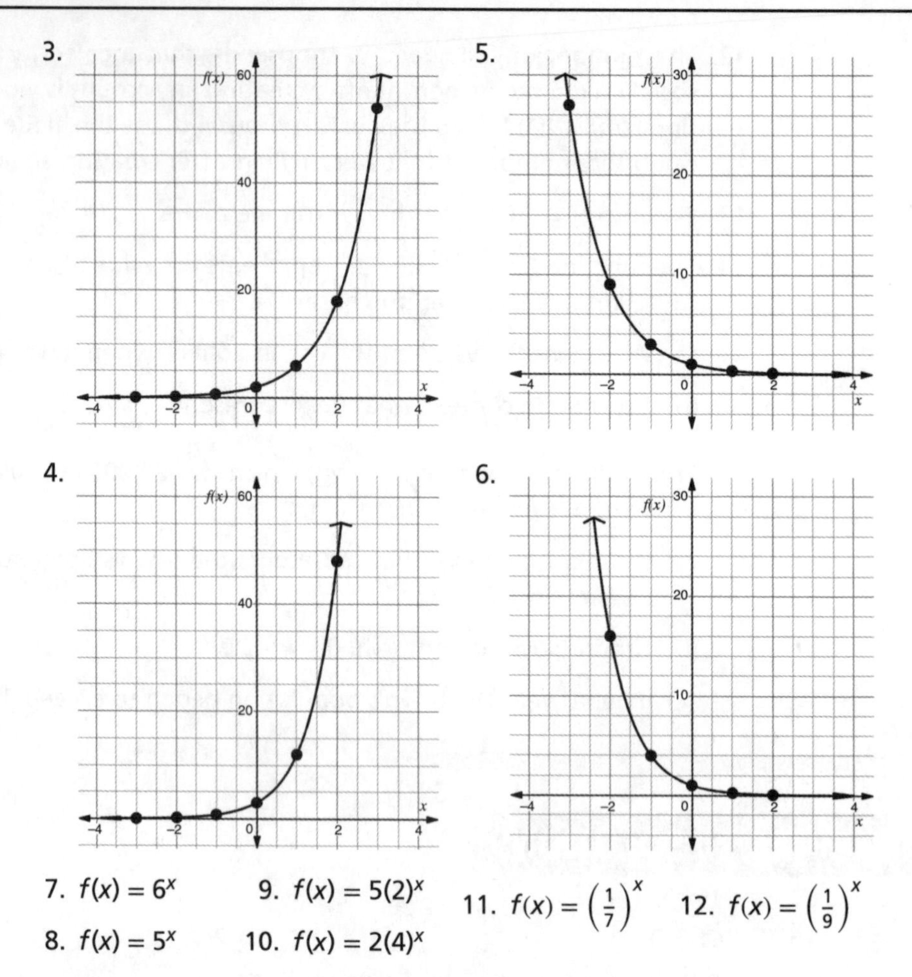

7. $f(x) = 6^x$

8. $f(x) = 5^x$

9. $f(x) = 5(2)^x$

10. $f(x) = 2(4)^x$

11. $f(x) = \left(\frac{1}{7}\right)^x$

12. $f(x) = \left(\frac{1}{9}\right)^x$

13. If $b = 1$, then $f(x) = a(b)^x$ becomes $f(x) = a(1)^x$. Since 1 raised to any power is 1, then $f(x) = a$, which is a horizontal line, not an exponential function.

14. Let $a = 1$ and $b = -2$. Then $f(1) = (-2)^1 = -2$, $f(2) = (-2)^2 = 4$, $f(3) = (-2)^3 = -8$, $f(4) = (-2)^4 = 16$, and so on. The function would alternate between positive and negative values, so it doesn't represent exponential growth or decay.

15. The graphs of $y = 2(3)^x$ and $y = 3(2)^x$ intersect at (1, 6), where $x = 1$.

10.2. 1. 5% growth

2. 12% growth

3. 13% decay

4. 13.2% growth

5. 9.8% growth

6. 90.4% decay

7. a. \$1,500, b. 0.92%, c. $x = 12$, so $1,500(1.0092)^{12} \approx \$1,674$.
8. a. decreasing since the base is less than 1 (or the graph shows decay), b. $500(0.7)^6 \approx 58.8$.
9. a. \$85,000; b. Student loan balance increases by 4.5% per year; c. $85(1.045)^{10} \approx \$132,000$.
10. a. $f(x) = 900(0.975)^x$, b. 2 years = 24 months, so $900(0.975)^{24} \approx \$490.18$.
11. a. $C(x) = 2.8(1.019)^x$, b. $2.8(1.019)^{10} \approx 3.3799$ or 3,380,000, c. the graphs of $y = C(x)$ and $y = 3.8$ intersect at $x \approx 16.225$, which represents the year 2016.
12. a. $I(x) = 50(0.5)^x$, b. 30 days = 7.5 half-lives so $I(7.5) = 50(0.5)^{7.5} \approx 0.28$, c. the graphs of $y = I(x)$ and $y = 3$ intersect at $x = 4.059$ half-lives, or $4.059(4) \approx 16$ days.
13. $f(x) = 7.283(1.195)^x$
14. $f(x) = 1,021.436(0.571)^x$
15. $f(x) = 7(23.617)^x$
16. $f(x) = 20(0.729)^x$
17. $f(x) = 6(1.002)^x$
18. $f(x) = 15(1.019)^x$

CHAPTER 10 TEST SOLUTIONS

1. (B)
2. (C)
3. (B)
4. (A)
5. (C)
6. (D)
7. (A)
8. (C)
9. (A)
10. (0, 825). The value of the account in 2015 was \$825.
11. Since $b^2 = 16$, then $b = 4$ or $b = -4$. Since the base of a power can't be negative, then $b = 4$.
12. 3 years = 36 months, so $f(36) = 4(1.4)^{36} \approx 729,000$.
13.

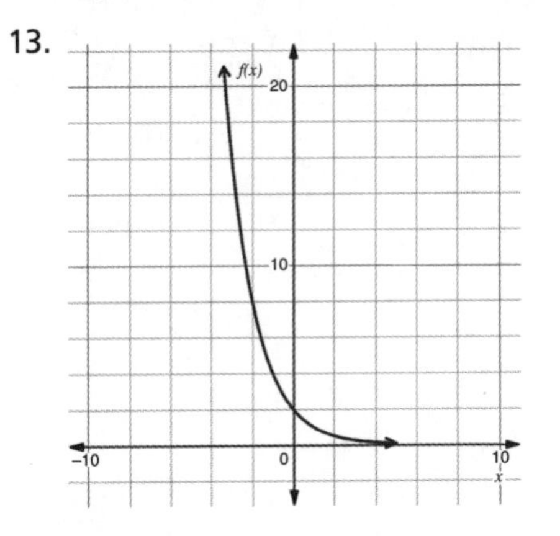

14. a. $45,500, b. The car loses 6% of its value each year.

15. a. $P(x) = 26(1.06)^x$

b.

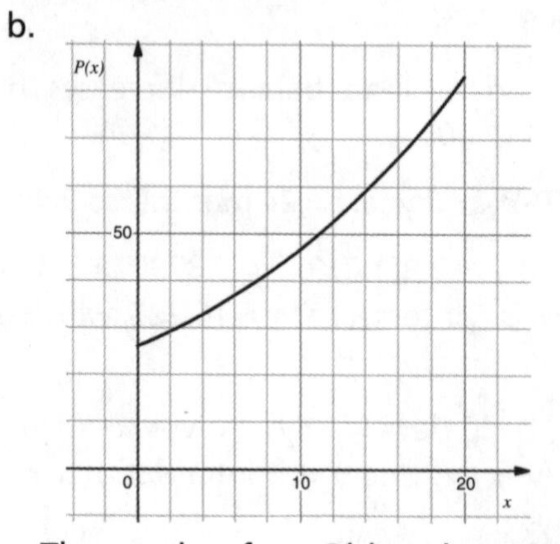

c. The graphs of $y = P(x)$ and $y = 40$ intersect at the point (7.393, 40), which means the population exceeds 40,000 during the 7th year after 2000, or 2007.

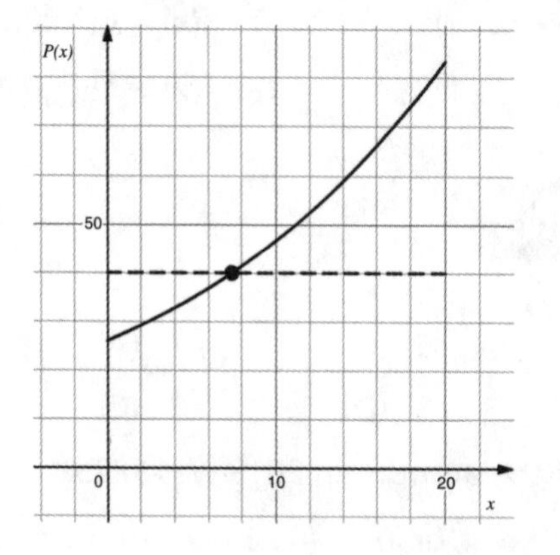

11 SEQUENCES

In math, we often work with patterns. In this chapter, we discuss a special type of numerical pattern called a **sequence**, which is a list of ordered numbers called **terms**. We also define a sequence as a function whose domain is the set of whole numbers or counting numbers.

11.1 Writing Recursive Formulas for Sequences

Let's start by looking at the following sequence: 3, 7, 11, 15, 19, 23, … What is the next number? How can we describe this sequence in words? We see that the numbers in the sequence follow a pattern. In this case, we add 4 to each term to get the next one: $3 + 4 = 7$, $7 + 4 = 11$, $11 + 4 = 15$, and so on. However, describing this sequence as "add 4 to a term to get the next one" won't uniquely identify it. This defines an infinite number of sequences, such as:

- 0, 4, 8, 12, 16, 20, …
- 1, 5, 9, 13, 17, 21, …
- 2, 6, 10, 14, 18, 22, …

To distinguish between these sequences, we also have to state the first term (our starting number). Thus, we say, "Start at 3 and then add 4 to get the next number."

In math, we use symbols to represent sequences. One way is to use subscript notation, in which each term is written with the variable a followed by a subscript to indicate the place of the term:

- a_1 (pronounced "a-sub-one") represents the first term.
- a_2 (pronounced "a-sub-two") represents the second term.
- a_3 (pronounced "a-sub-three") represents the third term, and so on.
- a_n (pronounced "a-sub-n") represents the nth term, where n is a variable representing any term in the sequence.
- a_{n-1} (pronounced "a-sub-n-minus-one") represents the $(n - 1)$st term, or the term before the nth term.

Here are two other important points about terms in sequences:

- Sometimes, the first term is labeled a_0 (pronounced "a-sub-zero"), so the second term would be a_1, and so on. To reinforce the notion that the subscript represents the place of the term, we recommend labeling the first term as a_1.
- Since subscripts show the place of the term in a sequence, we never use fractions or decimals as subscripts, so you'll never see $a_{\frac{3}{2}}$ or $a_{1.5}$.

We represent the sequence 3, 7, 11, 15, 19, 23, … with two equations (Figure 11.1):

$a_1 = 3$	Start at 3.
$a_n = a_{n-1} + 4\ (n \geq 2)$	To find the next term, add 4 to the previous term. We write the inequality $n \geq 2$ to indicate that we use this equation when $n = 2, 3, 4$, and so on.

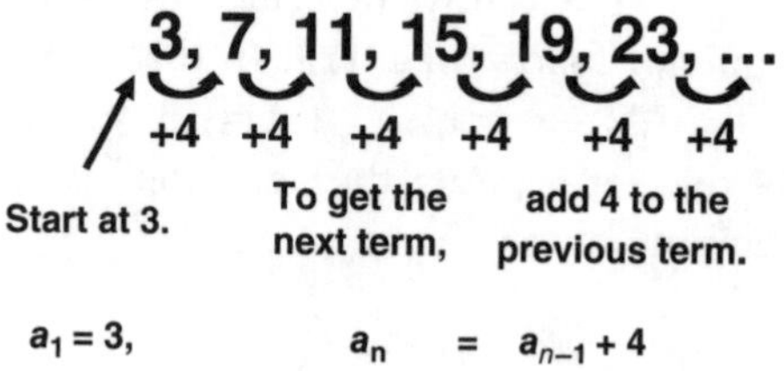

3, 7, 11, 15, 19, 23, …

+4 +4 +4 +4 +4 +4

Start at 3. **To get the next term,** **add 4 to the previous term.**

$a_1 = 3, \qquad a_n = a_{n-1} + 4$

Figure 11.1 Recursive formula for a sequence.

This type of formula is called a **recursive** formula, which contain at least two equations:

- Equations that define the first term or terms
- Equations that define other terms using previous terms (unless otherwise noted, we use only counting numbers for the subscripts)

Because we use subscripts to label each term, we call this way of writing sequences **subscript notation**.

Example 11.1 Write the first four terms of the sequence defined by the equations:

$$a_1 = 7$$

$$a_n = 2a_{n-1} - 1\ (n \geq 2)$$

Solution: Translate the formula into words.

$a_1 = 7$	Start at 7.
$a_n = 2a_{n-1} - 1\ (n \geq 2)$	To find the next term, multiply the previous term by 2 and subtract 1.

Starting with $a_1 = 7$, we use the formula to find the next terms one at a time:

$$a_2 = 2a_{2-1} - 1 = 2a_1 - 1 = 2(7) - 1 = 13$$

$$a_3 = 2a_{3-1} - 1 = 2a_2 - 1 = 2(13) - 1 = 25$$

$$a_4 = 2a_{4-1} - 1 = 2a_3 - 1 = 2(25) - 1 = 49$$

The first four terms are 7, 13, 25, and 49.

Example 11.2 Write the first five terms of the sequence defined by the equations:

$$\mathbf{a_1 = 1}$$

$$\mathbf{a_2 = 1}$$

$$\mathbf{a_n = a_{n-1} + a_{n-2}\ (n \geq 3)}$$

Solution: Translate the formula into words.

$a_1 = 1$	Start with 1.
$a_2 = 1$	The second term is 1.
$a_n = a_{n-1} + a_{n-2}$ $(n \geq 3)$	To find the next term, add the previous two terms.

Starting with $a_1 = 1$ and $a_2 = 1$, we use the formula to find the next terms one at a time:

$$a_3 = a_{3-1} + a_{3-2} = a_2 + a_1 = 1 + 1 = 2$$

$$a_4 = a_{4-1} + a_{4-2} = a_3 + a_2 = 2 + 1 = 3$$

$$a_5 = a_{5-1} + a_{5-2} = a_4 + a_3 = 3 + 2 = 5$$

The first five terms are 1, 1, 2, 3, and 5.

Did You Know?

Indian writers such as Pingala (3rd or 2nd century BCE) and Virahanka (6th or 7th century) wrote about the sequence 1, 1, 2, 3, 5, … while analyzing rhythm and meter in poetry. In Europe, this sequence first appeared in a book on arithmetic by the Italian mathematician Leonardo of Pisa, later known as Fibonacci. Often called the Fibonacci sequence, these numbers appear often in nature, including flower petals, pineapple spirals, and nautilus shells.

To write a recursive formula for a sequence, we need to find a pattern. Two common examples of sequence patterns are adding or multiplying the same number to each term, as shown in Examples 11.3, 11.4, and 11.5:

Example 11.3 Use subscript notation to write a recursive formula for the sequence 2, 11, 20, 29, ...

Solution: We determine the pattern in words and then write it using symbols. We see that $2 + 9 = 11$, $11 + 9 = 20$, and $20 + 9 = 29$, so the pattern is:

$a_1 = 2$	Start at 2.
$a_n = a_{n-1} + 9$	Add 9 to the previous term.

The formula is $a_1 = 2$, $a_n = a_{n-1} + 9$ $(n \geq 1)$.

We also represent sequences using function notation (Table 11.1). Compare the first few terms of the sequence 3, 6, 9, 12, …, written in both subscript notation and function notation:

Table 11.1 Comparing subscript notation and function notation.

Term Number	Subscript Notation	Function Notation
1	$a_1 = 3$	$f(1) = 3$
2	$a_2 = 6$	$f(2) = 6$
3	$a_3 = 9$	$f(3) = 9$
4	$a_4 = 12$	$f(4) = 12$
…	…	…
n	$a_n = a_{n-1} + 3$	$f(n) = f(n - 1) + 3$

Recursive formulas with subscripts of n and $n - 1$ can also be written using subscripts $n + 1$ and n. For example, the following formulas all represent the sequence 3, 6, 9, 12, …

- $a_1 = 3$, $a_n = a_{n-1} + 3$
- $a_1 = 3$, $a_{n+1} = a_n + 3$
- $f(1) = 3, f(n) = f(n - 1) + 3$
- $f(1) = 3, f(n + 1) = f(n) + 3$

Example 11.4 Write the first four terms of the sequence defined by the function:

$$\mathbf{f(1) = 5}$$

$$\mathbf{f(x) = f(x - 1) - 3}$$

Solution: Translate the formula into words.

$f(1) = 5$	The first term is 5.
$f(x) = f(x - 1) - 3$	To find the next term, subtract 3 from the previous term.

Starting with $f(1) = 5$, we use the formula to find the next terms one at a time:

$$f(2) = f(2-1) - 3 = f(1) - 3 = 5 - 3 = 2$$
$$f(3) = f(3-1) - 3 = f(2) - 3 = 2 - 3 = -1$$
$$f(4) = f(4-1) - 3 = f(3) - 3 = -1 - 3 = -4$$

The first four terms are 5, 2, −1, and −4.

Example 11.5 Use function notation to write a recursive formula for the sequence 8, −4, 2, −1, ...

Solution: We determine the pattern in words and then write it using symbols. Since $\frac{8}{-2} = -4$, $\frac{-4}{-2} = 2$, and $\frac{-2}{-2} = 1$, the pattern is:

$f(1) = 8$ — Start at 8.

$f(n) = \dfrac{f(n-1)}{-2}$ — To find the next term, divide the previous term by −2.

The formula is $f(1) = 8, f(n) = \frac{f(n-1)}{-2}$, which we write as $f(1) = 8, f(n) = -\frac{f(n-1)}{2}$.

Watch Out!

- When listing the first few terms or writing the formula of a sequence, make sure you include the first term.
- Don't confuse numbers inside subscripts with numbers outside of subscripts. For example, $a_n - 1$ (which is 1 less than the *n*th term) is different from a_{n-1} (which is the term before the *n*th term). Make sure you write subscripts carefully!
- Don't mix subscript notation and function notation—use one or the other. For example, don't write $f(n) = 2a_{n-1}$.

Exercises

Write the pronunciation of the equations for each sequence.

1. $a_1 = 4, a_n = 3a_{n-1}\ (n \geq 2)$
2. $a_1 = 0, a_n = 4 + a_{n-1}\ (n \geq 2)$
3. $a_1 = 3, a_n = a_{n-1}^2\ (n \geq 2)$
4. $a_1 = 6, a_n = 5a_{n-1} - 2\ (n \geq 2)$

Write the first five terms of each sequence.

5. $a_1 = 8, a_n = a_{n-1} - 6\ (n \geq 2)$
6. $a_1 = 5, a_n = a_{n-1} + 10\ (n \geq 2)$

7. $a_1 = 96,\ a_n = \frac{1}{2}a_{n-1}\ (n \geq 2)$

8. $a_1 = 100,\ a_{n+1} = \frac{1}{5}a_n\ (n \geq 2)$

9. $a_1 = 1,\ a_n = na_{n-1}\ (n \geq 2)$

10. $a_1 = -1,\ a_2 = 3,\ a_{n+1} = a_n + 2a_{n-1}\ (n \geq 2)$

11. $f(1) = 8, f(n) = f(n-1) + 4\ (n \geq 2)$

12. $f(1) = -5, f(x) = 4f(x-1) - 1\ (x \geq 2)$

13. $f(1) = 2, f(n) = 1 + f(n-1)\ (n \geq 2)$

14. $f(1) = -16, f(a+1) = -f(a) - 6\ (a \geq 2)$

15. $f(1) = 20, f(n+1) = \frac{1}{4}f(n) + 16\ (n \geq 2)$

16. $f(1) = \frac{1}{2}, f(2) = \frac{1}{4}, f(x) = f(x-1)\,f(x-2)\ (x \geq 3)$

Write a recursive formula for each sequence using subscript notation.

17. 7, 9, 11, 13, …

18. −3, −4, −5, −6, …

19. 10, 4, −2, −8, …

20. −9, −18, −36, −72, …

21. $6, 2, \frac{2}{3}, \frac{2}{9}, \ldots$

22. 64, −32, 16, −8, …

Write a recursive formula for each sequence using function notation.

23. 18, 20, 22, 24, …

24. 13, 12, 11, 10, …

25. 72, 68, 64, 60, …

26. −20, −15, −10, −5, …

27. 5, −15, 45, −225, …

28. $\frac{1}{5}, -\frac{2}{5}, \frac{4}{5}, -\frac{8}{5} \ldots$

Questions to Think About

29. Brianna wrote the following recursive formula for the sequence 5, 9, 13, 17, … : $f(1) = 5,\ a_n = a_{n-1} + 4\ (n \geq 2)$. Explain the error in her formula.

30. Amari wrote the following recursive formula for the sequence 9, −27, 81, −243, … : $a_n = -3a_{n-1}\ (n \geq 2)$. Explain the error in his formula.

31. To write the second term of the sequence $a_1 = 4,\ a_n = a_{n-1} - 3\ (n \geq 2)$, Makayla did the following: $a_2 = a_{4-1} - 3 = 4 - 1 - 3 = 0$. Explain the error in her reasoning.

11.2 Writing Explicit Formulas for Arithmetic and Geometric Sequences

Recursive formulas have one major limitation—they can't be used to find a term in a sequence quickly. For example, using the recursive formula $a_1 = 2$, $a_n = a_{n-1} + 3$ $(n \geq 2)$ to find the 50th term means that we would have to find $a_2, a_3, a_4, \ldots$, and a_{49} first! Clearly, we need a faster way to find the nth term. If we write out the recursive calculations for each term in the sequence, an interesting pattern emerges. We generate the sequence by starting at 2 and adding 3 to the previous term. For the second term, we add 3 once; for the third term, we add 3 twice; for the fourth term, we add 3 three times; and so on, so for the nth term, we add 3 $(n - 1)$ times. Table 11.2 summarizes this pattern:

Table 11.2 Explicit formula for an arithmetic sequence.

Term	Calculation	Result
1	2	2
2	$2 + 3$	5
3	$2 + 3 + 3 = 2 + 2(3)$	8
4	$2 + 3 + 3 + 3 = 2 + 3(3)$	11
5	$2 + 3 + 3 + 3 + 3 = 2 + 4(3)$	14
...	...	...
n	$2 + 3 + 3 + 3 + \ldots + 3 = 2 + (n - 1)(3)$	$2 + (n - 1)3 = 3n - 1$

A sequence like this, in which the same number is added to each term, is an **arithmetic** (pronounced a-rith-ME-tic, not a-RITH-me-tic) **sequence.** In this case, the sequence 2, 5, 8, 11, ... has the formula $a_n = 2 + (n - 1)3$, which we can simplify to $a_n = 3n - 1$. Unlike the recursive formula $a_1 = 2$, $a_n = a_{n-1} + 3$, this formula allows us to find the nth term of the sequence without having to find previous terms. We call this an **explicit formula**.

In general, the explicit formula for the nth term a_n of an arithmetic sequence is

SUBSCRIPT NOTATION: $a_n = a_1 + (n - 1)d$

FUNCTION NOTATION: $f(n) = f(1) + (n - 1)d$

where $a_1 = f(1)$ is the first term and d is the constant difference between consecutive terms ($d = a_2 - a_1 = a_3 - a_2 = \ldots = a_n - a_{n-1}$).

Some sequences are not arithmetic. For example, the sequence 6, 12, 24, 48, ... has no constant difference ($12 - 6 = 6$, but $24 - 12 = 12$). However, each term is multiplied by the same number to get the next term. We call such a sequence a

geometric sequence. To find an explicit formula, we examine the calculations required to write terms (Table 11.3):

Table 11.3 Explicit formula for a geometric sequence.

Term	Calculation	Result
1	6	6
2	6(2)	12
3	$6(2)(2) = 6(2)^2$	24
4	$6(2)(2)(2) = 6(2)^3$	48
5	$6(2)(2)(2)(2) = 6(2)^4$	96
…	…	…
n	$6(2)(2)(2)\ldots(2) = 6(2)^{n-1}$	$6(2)^{n-1}$

We see that the nth term is obtained by multiplying the first term by the same number $n - 1$ times. In this case, the explicit formula for the sequence 6, 12, 24, 48, … is $a_n = 6(2)^{n-1}$.

In general, the explicit formula for the *n*th term a_n of a geometric sequence can be written as follows:

SUBSCRIPT NOTATION: $a_n = a_1(r)^{n-1}$

FUNCTION NOTATION: $f(n) = f(1)(r)^{n-1}$

where $a_1 = f(1)$ is the first term and r is the constant ratio between consecutive terms $\left(r = \frac{a_2}{a_1} = \frac{a_3}{a_2} = \ldots = \frac{a_n}{a_{n-1}}\right)$**.**

Example 11.6 An arithmetic sequence is defined by the formula $f(n) = 4 + 7n$.

(a) Find $f(1)$.

(b) Find the common difference.

Solution:

(a) To find the first term, substitute $n = 1$ into the formula: $f(1) = 4 + 7(1) = 11$.

(b) The common difference is the coefficient of $7n$, which is 7. We confirm this by finding the next few terms of the sequence (or by computing $d = f(n + 1) - f(n)$):

$$f(2) = 4 + 7(2) = 18$$

$$f(3) = 4 + 7(3) = 25$$

The common difference is $25 - 18 = 18 - 11 = 7$.

Example 11.7 A sequence is defined by the formula $a_n = 4(3)^{n-1}$. Find a_1, a_2, and a_3.

Solution: Substitute $n = 1$, $n = 2$, and $n = 3$ into the formula.

$$a_1 = 4(3)^{1-1} = 4(3)^0 = 4$$

$$a_2 = 4(3)^{2-1} = 4(3)^1 = 12$$

$$a_3 = 4(3)^{3-1} = 4(3)^2 = 36$$

To write an explicit formula for a sequence, test to see if the sequence is arithmetic, geometric, or neither by doing the following:

- If there is a constant difference between consecutive terms, then the sequence is arithmetic. Use the formula $a_n = a_1 + (n - 1)d$.
- If there is a constant ratio between consecutive terms, then the sequence is geometric. Use the formula $a_n = a_1(r)^{n-1}$.
- If there is neither a constant difference nor a constant ratio, then the sequence is neither arithmetic nor geometric. Look for another pattern.

Example 11.8 Given the sequence 18, 13, 8, 3, ...

(a) Write an explicit formula using function notation for the sequence.

(b) Use the formula to find the thirtieth term in the sequence.

Solution:

(a) Determine if there is a constant difference (by subtracting consecutive terms) or constant ratio (by dividing consecutive terms):
$13 - 18 = 8 - 13 = 3 - 8 = -5$, so $d = -5$.
(Since the terms are decreasing, then the constant difference is negative.)
The first term is 18, so $f(1) = 18$.
The explicit formula is $f(n) = 18 + (n - 1)(-5)$.
This can be written as $f(n) = 23 - 5n$, or $f(n) = -5n + 23$.

(b) The thirtieth term is $f(30) = 18 + (30 - 1)(-5) = -127$.

Watch Out!

If the constant difference is negative, put parentheses around it in the formula to clearly indicate multiplication. In Example 11.8, $a_1 = 18 + (n - 1)(-5)$ shows that -5 is multiplied by $n - 1$, but $a_1 = 18 + (n - 1) - 5$ indicates that 5 is being subtracted from $n - 1$.

Example 11.9 Given the sequence −48, 24, −12, 6, ...

(a) Write an explicit formula using subscript notation for the sequence.

(b) Use the formula to find the eighth term in the sequence.

Solution:

(a) Determine if there is a constant difference (by subtracting consecutive terms) or constant ratio (by dividing consecutive terms):

$24 - (-48) = 72$, but $-12 - 24 = -36$, so there is no constant difference.

$\frac{24}{-48} = \frac{-12}{24} = \frac{6}{-12} = -\frac{1}{2}$, so $r = -\frac{1}{2}$.

The first term is 24, so $a_1 = 24$.

The explicit formula is $a_n = 24\left(-\frac{1}{2}\right)^{n-1}$.

(b) The eighth term is $a_8 = 24\left(-\frac{1}{2}\right)^{8-1} = 24\left(-\frac{1}{2}\right)^{7} = -\frac{3}{16}$.

Example 11.10 If the first term of an arithmetic sequence is 8 and the seventh term is 62, find the eighth term.

Solution: Since the sequence is arithmetic, we know that the difference between terms is constant. We also know that $a_1 = 8$ and $a_7 = 62$. We need to find a_8. We can represent this with a diagram in which the unknown terms are blank spaces whose values we need to determine.

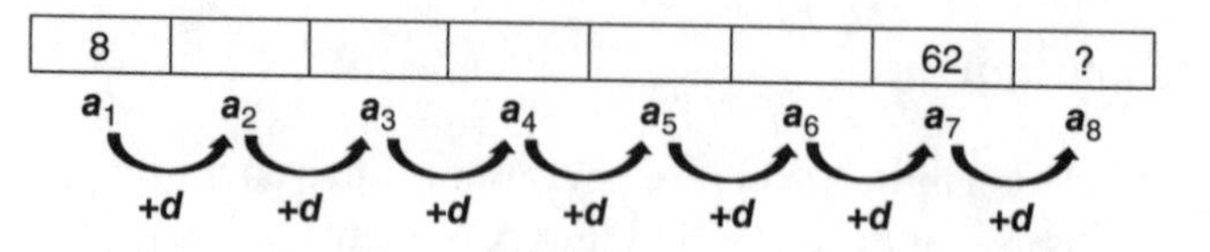

We see from the diagram that $a_1 + d + d + d + d + d + d = 62$, or $a_1 + 6d = 62$. (We could also reason that between a_1 and a_7, we add the constant difference $7 - 1 = 6$ times.) Since $a_1 = 8$, then $8 + 6d = 62$, or $6d = 54$, or $d = 9$. Then $a_8 = 62 + d = 62 + 9 = 71$.

To check, see if a sequence that starts at 8 and a common difference of 9 ends at 62: 8, 17, 26, 35, 44, 53, 62.

Example 11.11 A geometric sequence of positive numbers has a second term of 100 and a fourth term of 25. What is the seventh term of the sequence?

Solution: Since the sequence is geometric, we know that the ratio between terms is constant. We also know that $a_2 = 100$ and $a_4 = 25$. We need to find a_7. We can represent this with a diagram in which the unknown terms are blank spaces whose values we need to determine.

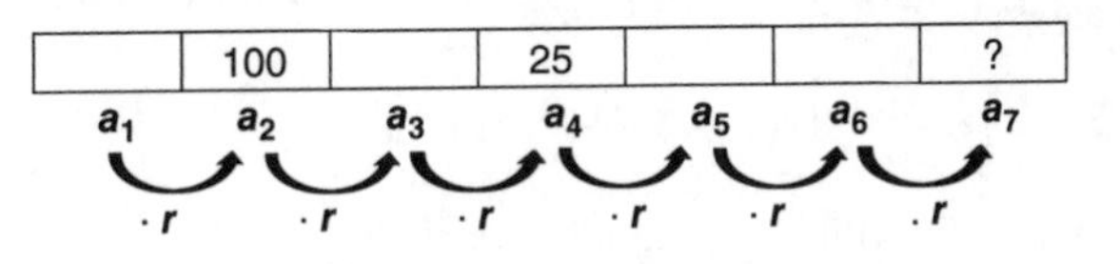

From the diagram, we see that $100(r)(r) = 25$, or $100r^2 = 25$. Solving for r, we get $r^2 = \frac{25}{100}$, or $r = \sqrt{\frac{25}{100}} = \frac{\sqrt{25}}{\sqrt{100}} = \frac{5}{10} = \frac{1}{2}$. (We reject the negative value of r since that would result in negative terms, but the problem stated that all terms are positive.) Then $a_7 = 25(r)(r)(r) = 25\left(\frac{1}{2}\right)\left(\frac{1}{2}\right)\left(\frac{1}{2}\right) = \frac{25}{8}$.

To check, see if a sequence whose second term is 100 and common ratio of $\frac{1}{2}$ has a seventh term of $\frac{25}{8}$: 200, 100, 50, 25, $\frac{25}{2}$, $\frac{25}{4}$, $\frac{25}{8}$.

Table 11.4 summarizes what we've discussed about explicit formulas for arithmetic and geometric sequences and how they relate to the recursive formulas that we discussed in Section 11.1.

Table 11.4 Comparing recursive and explicit formulas.

Characteristic	Recursive Formula	Explicit Formula
Number of Equations	2 or more equations	1 equation
Expression of *n*th Term	Expresses *n*th term in terms of previous terms	Expresses *n*th term in terms of *n*
Arithmetic Sequence: Subscript Notation	a_1 = first term $a_n = a_{n-1} + d$	$a_n = a_1 + (n - 1)d$
Arithmetic Sequence: Function Notation	$f(a_1)$ = first term $f(n) = f(n - 1) + d$	$f(n) = f(1) + (n - 1)d$
Geometric Sequence: Subscript Notation	a_1 = first term $a_n = r \cdot a_{n-1}$	$a_n = a_1(r^{n-1})$
Geometric Sequence: Function Notation	$f(a_1)$ = first term $f(n) = r \cdot f(n - 1)$	$f(n) = f(1)(r^{n-1})$

Exercises

1. If $a_n = 16 + (n - 1)3$, find a_{15}.
2. If $a_n = 7(2)^{n-1}$, find a_7.
3. If $a_n = 4 + (n - 1)(-2)$, find a_{30}.
4. If $a_n = -6(2)^{n-1}$, find a_5.
5. If $a_n = 2 - 7n$, find a_{22}.
6. If $a_n = 243\left(\frac{1}{3}\right)^{n-1}$, find a_6.

Write an explicit formula for each sequence.

7. 12, 18, 24, 30, …
8. 3, 20, 37, 54, …
9. 7, 21, 63, 189, …
10. 6, −12, 24, −48, …
11. 8, 28, 48, 68, …
12. 5, −20, 80, −320, …
13. 26, 23, 20, 17, …
14. 600, 150, 37.5, 9.375, …
15. 625, 125, 25, 5, …
16. 57, 51, 45, 39, …
17. −2, −25, −48, −71, …
18. −686, 98, −14, 2, …
19. In an arithmetic sequence, the third term is 22 and the fifth term is 56. Find the second term.
20. In a geometric sequence of positive numbers, the second term is 18 and the third term is 108. Find the fifth term.
21. In an arithmetic sequence, the fourth term is −15 and the ninth term is −45. Find the eleventh term.
22. In a geometric sequence of negative numbers, the fourth term is −56 and the sixth term is −224. Find the fifth term.
23. In an arithmetic sequence, the fifth term is −30 and the tenth term is −65. Find the eighth term.
24. In a geometric sequence of positive numbers, the second term is 192 and the fourth term is 12. Find the fifth term.

Questions to Think About

25. To describe a sequence whose first term is 5 and common difference is −3, Joseph writes the formula $a_n = 5 + (n - 1) - 3$. Explain the error in his formula.
26. Leslie is thinking of a geometric sequence whose second term is 8 and fourth term is 32. Explain why the third term could be one of two possible numbers.
27. Explain why the sequence 4, 4, 4, 4, … is both arithmetic and geometric.

11.3 Modeling with Sequences

We use sequences to model real-world behavior. Let's begin by comparing the graphs of the sequence $a_n = 2n + 1$ with the equation $y = 2x + 1$ and the sequence $a_n = 3(2)^{n-1}$ with $y = 3(2)^x$ (Figure 11.2).

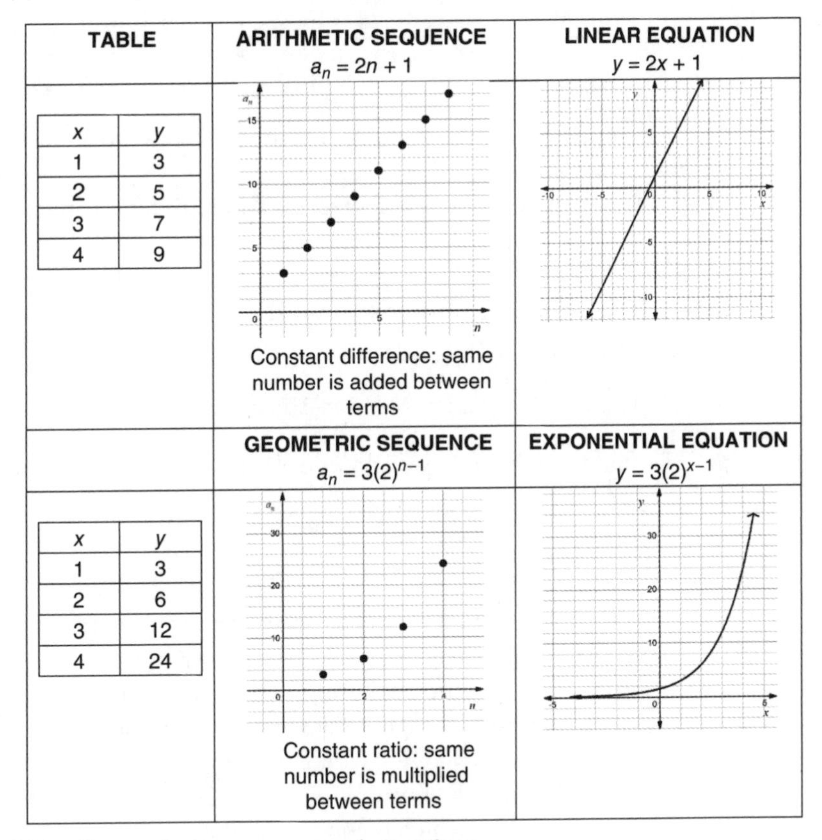

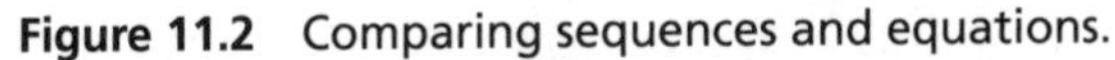
Figure 11.2 Comparing sequences and equations.

Notice the following:

- Arithmetic sequences follow a linear pattern. The same number is added between terms.
- Geometric sequences follow an exponential pattern. The same number is multiplied between terms.
- The graph of a sequence is a set of points, not a smooth curve.
- The graph of a sequence does not extend to the left of the y-axis since the domain does not contain negative integers.

We use sequences to model behavior whose inputs are positive integers, such as students or books.

Example 11.12 A pattern of dots is shown in the accompanying diagram.

Term 1 Term 2 Term 3 Term 4

Assuming this pattern continues, write a recursive formula to represent the number of dots in the nth term.

Solution: Write out the first few terms and look for a pattern.

$n = 1$: The first term has 1 dot.

$n = 2$: The second term has $1 + 2 = 3$ dots.

$n = 3$: The third term has $3 + 3 = 6$ dots.

$n = 4$: The fourth term has $6 + 4 = 10$ dots.

To find the nth term, we add n to the previous term, so the recursive formula is:

$$a_1 = 1$$

$$a_n = n + a_{n-1}$$

When modeling events that take place over time, we're often given information about what happens at the beginning, where the variable representing time equals 0. In these situations, we use inputs that are whole numbers, so the first term is a_0, not a_1. The second term is $a_1 = a_0 + d$ (add d to the first term), the third is $a_2 = a_0 + 2d$ (add d to the second term), the fourth is $a_3 = a_1 + 3d$, and so on. Our explicit formula then becomes $a_n = a_0 + nd$, which is equivalent to $a_n = a_1 + (n - 1)d$. Similarly, we can use function notation to write the explicit formula $f(n) = f(0) + nd$.

Example 11.13 To help control student costs, a public university announces a predictable tuition increase policy, in which tuition will increase by \$750 annually starting next year. The university's annual tuition was \$6,000 when it announced the policy.

(a) Write an explicit function T to represent the university's tuition, in dollars, n years after the policy was announced.

(b) Use the function from part a to determine the university's tuition 15 years after the university announces the policy.

Solution:

(a) The explicit function has the formula $T(n) = T(0) + n \cdot d$, (n is the number of years after the announcement). The university's tuition at the start of the policy is \$6,000, so $T(0) = 6{,}000$. Since the tuition increases by \$750 annually, then $d = 750$. The formula is $T(n) = 6{,}000 + n \cdot 750$, which can also be written as $T(n) = 750n + 5{,}250$.

(b) $T(15) = 750(15) + 6{,}000 = \$17{,}250$.

Exercises

1. A pattern is formed by adding successive rows of 4 squares to form a rectangle, as shown in the accompanying diagram.

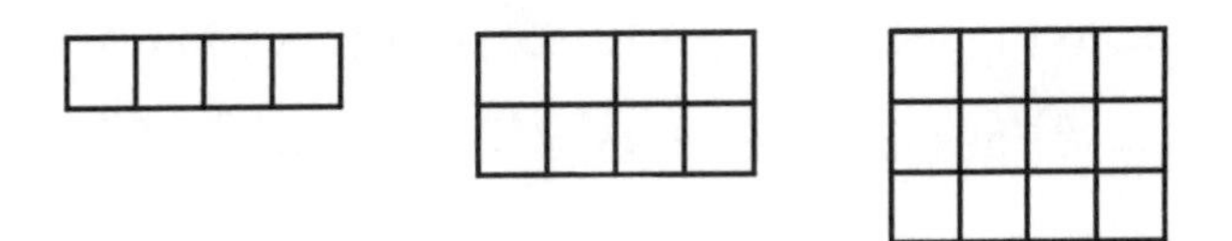

Write a recursive function f that models the number of squares in a rectangle with n rows.

2. For a ride to the airport, a taxicab company charges \$2 for the first mile and \$1.50 for each additional mile. Write an explicit formula a_n that models the cost, in dollars, of a ride that is n miles long.

3. The seats in a large auditorium increase at a constant rate. If the third row of the auditorium has 22 seats and the sixth row has 31 seats, determine the number of seats in the twentieth row.

4. A library gets a grant that allows it to acquire a fixed number of books each month. After 4 months, the library obtained a total of 33 books from the grant. After 7 months, the library obtained a total of 51 books from the grant. At that rate, how many books will the library have obtained from the grant after two years?

5. A restaurant uses square tables that can seat one person on each side. When the restaurant puts two or more square tables together, three people can sit on the tables on each end and two people can sit on tables in between the ends. The accompanying diagram shows the people (represented by circles) that can sit together when 1, 2, and 3 square tables are used.

(a) Write a recursive function f that represents the number of people that can sit when n square tables are put together.

(b) How many people can sit when 6 square tables are put together? Explain.

6. Amelia starts a fitness training program in which she runs 15 miles in the first week and an additional 2 miles per week every week after that.

(a) Write an explicit function f that represents the number of miles that she runs per week n weeks after starting her program.

(b) How many miles will she be running in the sixteenth week of her training program?

CHAPTER 11 TEST

1. In the sequence 5, −15, 45, −135, …, what is the sixth term?

(A) 405 (B) 1,215 (C) −1,215 (D) −3,645

2. Which sequence has a first term of 5 and a common difference of 4?

(A) $a_n = 5(4)^{n-1}$ (C) $a_n = 4 + 5n$

(B) $a_n = 5 + 4(n - 1)$ (D) $a_n = 5 - 4n$

3. If $f(1) = 8$ and $f(n) = 2f(n - 1) + 3$ $(n \geq 2)$, what is $f(3)$?

(A) 7 (B) 9 (C) 19 (D) 41

4. If $a_n = 18\left(\frac{1}{3}\right)^{n-1}$, what is a_4?

(A) 18 (B) 24 (C) $\frac{2}{9}$ (D) $\frac{2}{3}$

5. Which sequences are arithmetic?

I. 7, 12, 17, 22, … II. 22, 17, 12, 7, … III. 7, 9, 13, 19, …

(A) I and II (B) I and III (C) II and III (D) I, II, and III

6. Which sequence is geometric?

   (A) $a_1 = 6,\ a_n = 4a_{n-1}\ (n \geq 2)$

   (B) $a_1 = 6,\ a_n = 4a_{n-1} + 2\ (n \geq 2)$

   (C) $a_1 = 6,\ a_n = 4 + a_{n-1}\ (n \geq 2)$

   (D) $a_1 = 6,\ a_n = na_{n-1}\ (n \geq 2)$

7. Which formula represents the sequence 16, 13, 10, 7, ... ?

   (A) $a_n = 16 + (n - 1)(-3)$

   (B) $a_n = 16 + (n - 1) - 3$

   (C) $a_n = 3n + 13$

   (D) $a_n = 3n - 19$

8. Which formulas could be used to represent the sequence 48, 24, 12, 6, ... ?

   I. $a_n = 48(0.5)^{n-1}$
   II. $a_1 = 6,\ a_n = 2a_{n-1}\ (n \geq 2)$
   III. $a_1 = 48,\ a_n = 0.5a_{n-1}\ (n \geq 2)$

   (A) I and II (B) I and III (C) II and III (D) I, II, and III

9. Which statement about the annual change in a city's population could be modeled by a geometric sequence?

   (A) Every year, the city's population increases by 1,000.

   (B) Every year, 1,000 more people live in the city than lived there the previous year.

   (C) Every year, the city's population increases by 1%

   (D) Every year, the increase in the city's population increases by 1,000.

10. Write a recursive formula using subscript notation for the sequence 11, 16, 21, 26, ...

11. Write the pronunciation of the formula $a_1 = 8,\ a_n = 3a_{n-1}\ (n \geq 2)$.

12. On the coordinate plane, graph the first four terms of the sequence defined by the function
    $f(1) = 7$
    $f(n) = f(n - 1) - 2\ (n \geq 2)$

13. Find the first four terms of the sequence defined by the formula
    $f(1) = 12$
    $f(n) = 2f(n - 1) - n\ (n \geq 2)$

14. In an arithmetic sequence, the fourth term is -2 and the sixth term is -8. Find the fifteenth term of the sequence.

15. To manage crowd control at a busy store, a security guard allows four people to enter every minute. At 2:00 p.m., 77 people are waiting on line to enter. After 2:00 p.m., the security guard prohibits anyone else from joining the line.
    (a) Write an explicitly defined function to model $P(x)$, the number of people waiting on line to enter the store x minutes after 2:00 p.m.
    (b) How many people will be on line after the security guard lets four people in at 2:14 p.m.?

CHAPTER 11 SOLUTIONS

11.1. 1. "*a*-sub-1 equals 4, *a*-sub-*n* equals 3 times *a*-sub-*n*-minus-1"
2. "*a*-sub-1 equals 0, *a*-sub-*n* equals 4 plus *a*-sub-*n*-minus-1"
3. "*a*-sub-1 equals 3, *a*-sub-*n* equals *a*-sub-*n*-minus-1 squared"
4. "*a*-sub-1 equals 6, *a*-sub-*n* equals 5 times *a*-sub-*n*-minus-1 minus 2"
5. 8, 2, −4, −10, −16
6. 5, 15, 25, 35, 45
7. 96, 48, 24, 12, 6
8. 100, 20, 4, $\frac{4}{5}$, $\frac{4}{25}$
9. 1, 2, 6, 24, 120
10. −1, 3, 1, 7, 9
11. 8, 12, 16, 20, 24
12. −5, −21, −85, −341, −1,365
13. 2, 3, 4, 5, 6
14. −16, 10, −16, 10, −16
15. 20, 21, $\frac{85}{4}$, $\frac{341}{16}$, $\frac{1{,}365}{64}$
16. $\frac{1}{2}$, $\frac{1}{4}$, $\frac{1}{8}$, $\frac{1}{32}$, $\frac{1}{256}$
17. $a_1 = 7,\ a_n = a_{n-1} + 2\ (n \ge 2)$
18. $a_1 = -3,\ a_n = a_{n-1} - 1\ (n \ge 2)$
19. $a_1 = 10,\ a_n = a_{n-1} - 6\ (n \ge 2)$
20. $a_1 = -9,\ a_n = 2a_{n-1}\ (n \ge 2)$
21. $a_1 = 6,\ a_n = \frac{a_{n-1}}{3}\ (n \ge 2)$
22. $a_1 = 64,\ a_n = -\frac{a_{n-1}}{2}\ (n \ge 2)$
23. $f(1) = 18,\ f(n) = f(n-1) + 2\ (n \ge 2)$
24. $f(1) = 13,\ f(n) = f(n-1) - 1\ (n \ge 2)$
25. $f(1) = 72,\ f(n) = f(n-1) - 4\ (n \ge 2)$
26. $f(1) = -20,\ f(n) = f(n-1) + 5\ (n \ge 2)$
27. $f(1) = 5,\ f(n) = -3f(n-1)\ (n \ge 2)$

28. $f(1) = \frac{1}{5}$, $f(n) = -2f(n-1)$ $(n \geq 2)$

29. Brianna wrote both subscript notation and function notation in her formula. She should have used one or the other, but not both.

30. Amari did not write the formula for the first term: $a_1 = 9$.

31. Makayla confused numbers in subscripts with numbers outside of subscripts: $a_{n-1} - 3$ means "subtract 3 from the previous term," not "subtract 1 and 3 from the previous term."

11.2. 1. 58 2. 448 3. −54 4. −96 5. −152 6. 1

7. $a_n = 12 + (n-1)6$ or $a_n = 6n + 6$

8. $a_n = 3 + (n-1)17$ or $a_n = 17n - 14$

9. $a_n = 7(3)^{n-1}$

10. $a_n = 6(-2)^{n-1}$

11. $a_n = 8 + (n-1)20$ or $a_n = 20n - 12$

12. $a_n = 5(-4)^{n-1}$

13. $a_n = 26 + (n-1)(-3)$ or $a_n = -3n + 29$

14. $a_n = 600\left(\frac{1}{4}\right)^{n-1}$

15. $a_n = 625\left(\frac{1}{5}\right)^{n-1}$

16. $a_n = 57 + (n-1)(-6)$ or $a_n = -6n + 63$

17. $a_n = -2 + (n-1)(-23)$ or $a_n = -23n + 21$

18. $a_n = -686\left(-\frac{1}{7}\right)^{n-1}$

19. 5

20. 3,888

21. −57

22. −112

23. −51

24. 3

25. Joseph should have written parentheses around the common difference of −3 in order to multiply it by $n - 1$. His formula subtracts 3 from $n - 1$.

26. If $a_2 = 8$ and $a_4 = 32$, then $8(r)(r) = 32$, or $r^2 = 4$. This equation has two possible solutions, +2 or −2, so the third term could be 16 or −16.

27. The sequence is arithmetic because it has a common difference of 0, so $a_n = 4 + (n-1)0$, or $a_n = 4$. It is also geometric since $a_n = 4(1)^{n-1}$, or $a_n = 4$.

11.3. 1. $f(1) = 4$, $f(n) = 4 + f(n - 1)$

2. $a_n = 2 + 1.5(n - 1)$

3. 73

4. 153

5. a. $f(1) = 4$, $f(n) = 2 + f(n - 1)$; b. 14

6. a. $f(n) = 15 + (n - 1)2$ or $f(n) = 2n + 13$; b. 45

CHAPTER 11 TEST SOLUTIONS

1. (C)
2. (B)
3. (D)
4. (D)
5. (A)
6. (A)
7. (A)
8. (B)
9. (C)
10. $a_1 = 11$, $a_n = 5 + a_{n-1}$ $(n \geq 2)$
11. "*a*-sub-1 equals 8, *a*-sub-*n* equals 3 times *a*-sub-*n*-minus-1 for *n* greater than or equal to 2"
12.

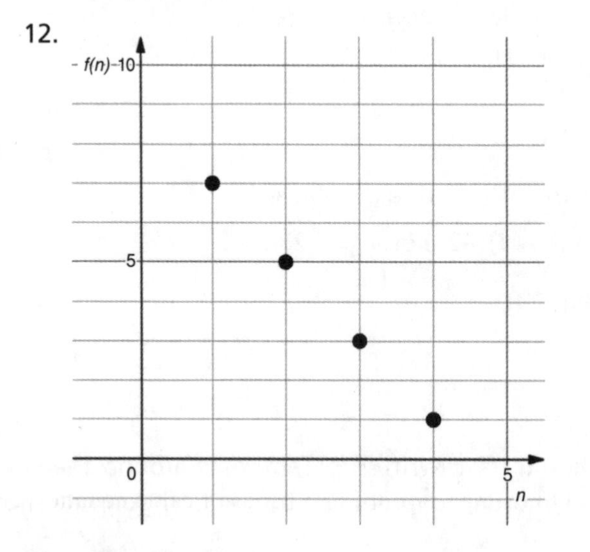

13. 12, 22, 41, 78
14. −35
15. a. $P(x) = 77 - 4x$, b. 21

12 SUMMARY OF FUNCTIONS

In this chapter, we summarize and extend the properties of functions we discussed in previous chapters.

12.1 Cubic, Square Root, and Cube Root Functions

We talked about polynomial functions with degree 1 (in Chapter 7) and degree 2 (in Chapter 9). We can extend these ideas by looking at **cubic functions**, which are polynomial functions with degree 3.

Figure 12.1 shows the graph and equation of the cubic equation $y = x^3 - 2x^2 - 5x + 6$.

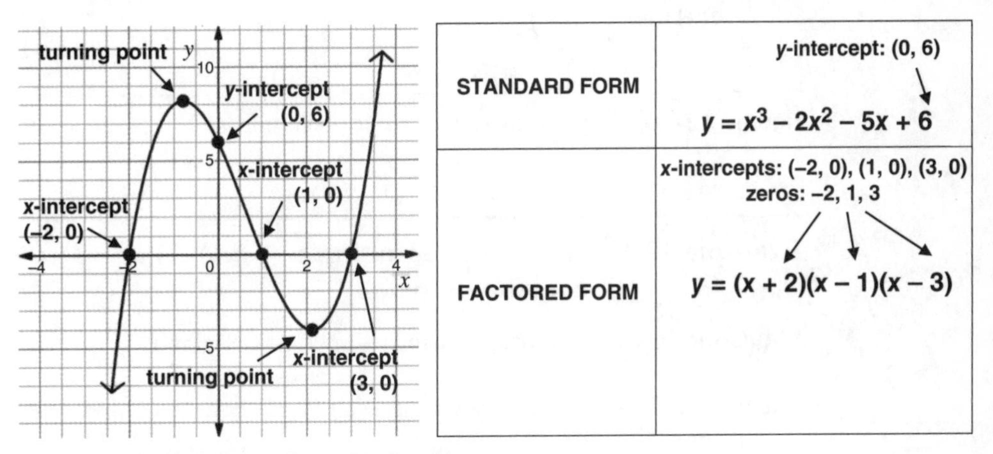

Figure 12.1 Vocabulary for cubic functions.

Note the following characteristics of the graphs of cubic equations:

- They do *not* consist of two parabolas.
- They have no absolute maximum or minimum (the domain and range of cubic functions are the set of real numbers).
- They have one end that points up and another end that points down. (In advanced algebra, you'll learn about a more precise way to describe the behavior of the graphs at each end.)

- For a cubic equation $y = ax^3 + bx^2 + cx + d$, if the leading coefficient $a > 0$, then the graph points down on the left side and up on the right. If $a < 0$, the graph points up on the left and down on the right (Figure 12.2).
- They do *not* have an axis of symmetry. Instead, they have symmetry around a point (informally, this means that if you rotate the graph 180° around this point, it will look the same).
- They usually have two turning points (recall from Section 9.9 that at a turning point, the function changes from increasing to decreasing or vice versa).

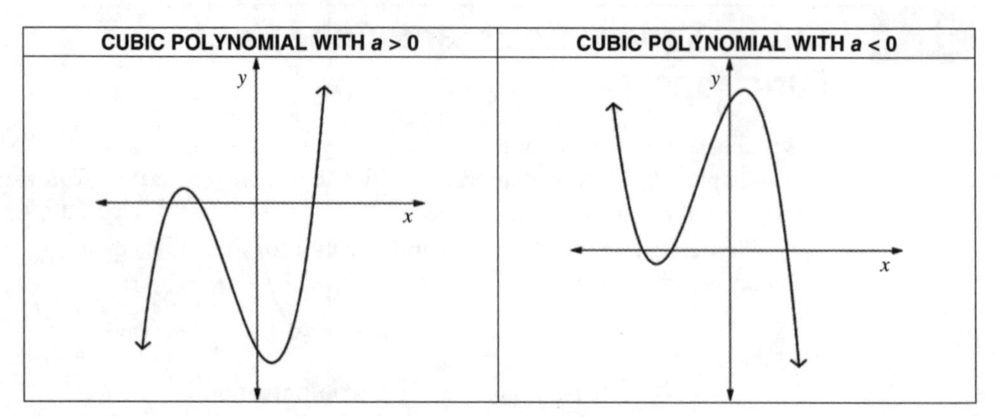

Figure 12.2 Leading coefficients of cubic polynomials.

Example 12.1 Graph the equation $y = (x + 2)(x + 3)(x - 2)$ on the coordinate plane.

Solution: Find the important characteristics on the graph.

- Identify the x-intercepts by setting each factor equal to 0 and solving the resulting equation.
  - $x + 2 = 0 \rightarrow$ zero at $x = -2 \rightarrow$ x-intercept at $(-2, 0)$
  - $x + 3 = 0 \rightarrow$ zero at $x = -3 \rightarrow$ x-intercept at $(-3, 0)$
  - $x - 2 = 0 \rightarrow$ zero at $x = 2 \rightarrow$ x-intercept at $(2, 0)$
- The y-intercept is the point whose x-coordinate is 0 and y-coordinate is $(0 + 2)(0 + 3)(0 - 2) = -12$, or $(0, -12)$.
- Use the trace function on your device to identify the approximate coordinates of the turning points: $(-2.5, 1.1)$ and $(0.5, 13.1)$.

- Since $a > 0$, the graph points down on the left side and up on the right.

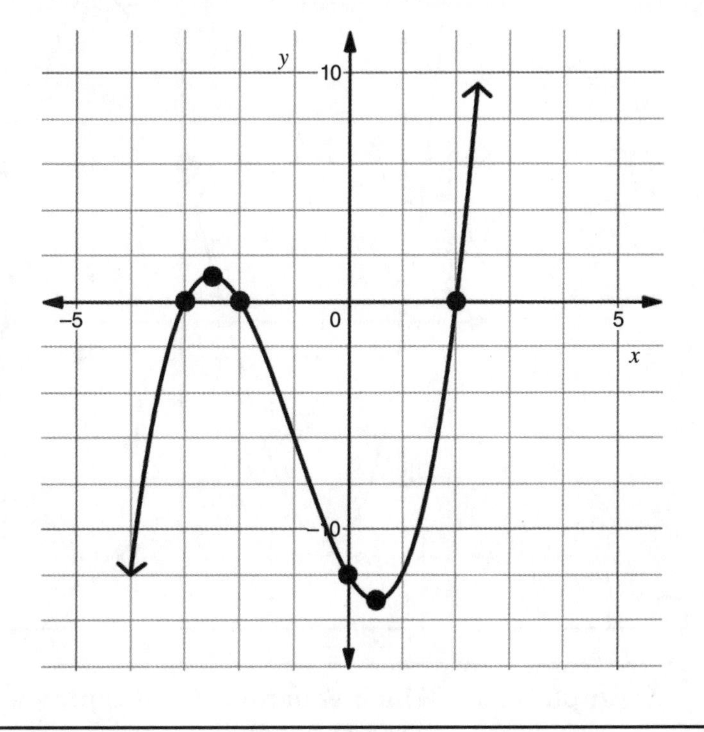

Example 12.2 Graph the equation $f(x) = -(x + 1)(x + 5)(x - 4)$ on the coordinate plane.

Solution: Find the important characteristics on the graph.

- Identify the x-intercepts from the factors in the equation: $(-1, 0)$, $(-5, 0)$, and $(4, 0)$.
- The y-intercept is the point whose x-coordinate is 0 and y-coordinate is $-(0 + 1)(0 + 5)(0 - 4) = 20$, or $(0, 20)$.
- Use the trace function on your device to identify the approximate coordinates of the turning points as $(-3.3, -28.6)$ and $(1.9, 42.0)$.
- Since $a < 0$, the graph points up on the left side and down on the right.

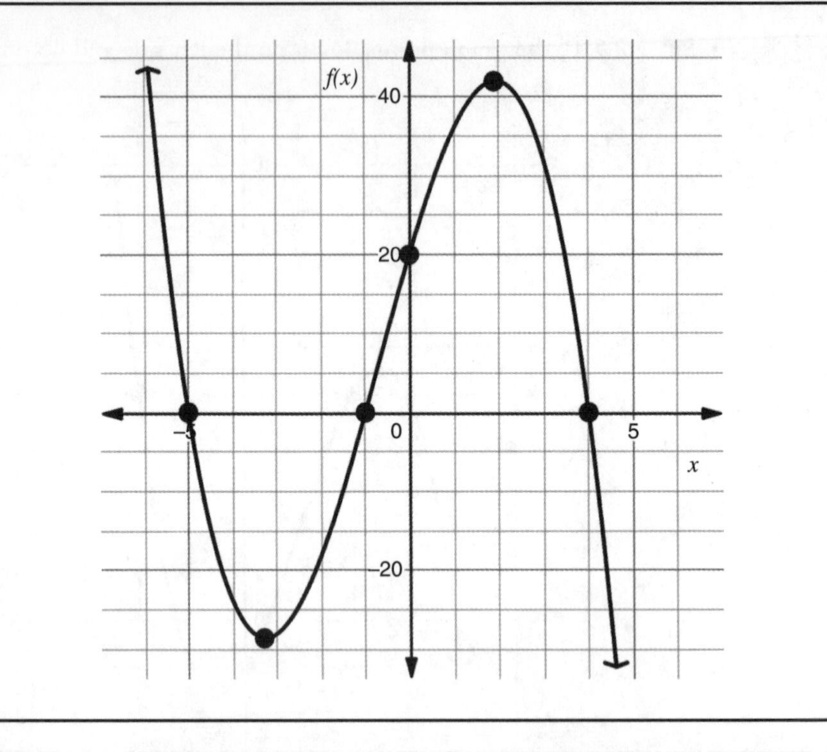

Example 12.3 **Which equation could represent the equation graphed here?**

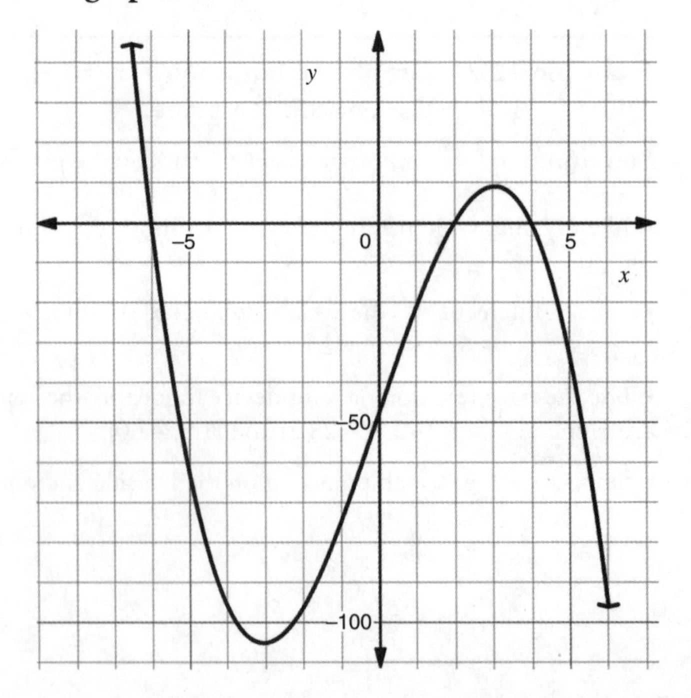

(A) $y = -(x - 6)(x + 2)(x + 4)$ **(C)** $y = -(x + 6)(x - 2)(x - 4)$

(B) $y = (x - 6)(x + 2)(x + 4)$ **(D)** $y = (x + 6)(x - 2)(x - 4)$

Solution: Identify the important characteristics of the graph:

- The x-intercepts are $(-6, 0)$, $(2, 0)$, and $(4, 0)$, so the equation has factors $x + 6$, $x - 2$, and $x - 4$.
- Since the graph points up on the left and down on the right, then $a < 0$. We don't need to find the exact value of a since $a = 1$ or $a = -1$ in all choices.

The only equation that has factors $x + 6$, $x - 2$, and $x - 4$ and has $a < 0$ is choice (C).

If a zero of a function repeats (appears twice), then the function has a double zero at that point. This means that the zero's corresponding factor is raised to the second power. Here, the graph doesn't pass through the x-axis. Instead, the graph just touches the x-axis and then changes direction from increasing to decreasing or vice versa.

Example 12.4 Write the equation of the function f graphed here.

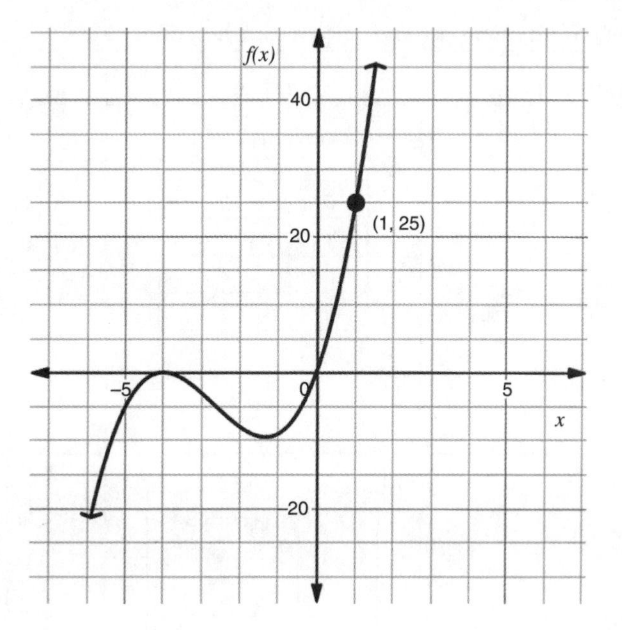

Solution: Identify the important characteristics of the graph:

- The function is cubic since it has two turning points.
- The x-intercepts are $(-4, 0)$ and $(0, 0)$, so the equation has factors $x + 4$ and $x + 0$ (or simply x).

- The graph just touches the x-axis at $(-4, 0)$ and then changes direction, which means that the corresponding factor $x + 4$ is squared. The equation is $f(x) = ax(x + 4)^2$.
- Since the graph passes through $(1, 25)$, then we know $f(1) = 25$. Evaluate $f(1)$ to find a: $f(1) = a(1)(1 + 4)^2 = 25$, so $25a = 25$, so $a = 1$.

The equation is $f(x) = x(x + 4)^2$.

When graphing cubic equations, determine its important characteristics:

- x- and y-intercepts
- approximate location of the turning points (use your device's trace function)
- sign of the leading coefficient

Another type of function relates to inverse operations, which when applied to a number, results in the original number. We can extend the idea of inverse operations to functions. We start by looking at the characteristics of the square root function. Compare the tables and graphs in Table 12.1. What do you notice about the numbers in the tables of values? What do you notice about the graphs?

Table 12.1 Comparing quadratic and square root functions.

<table>
<tr><th>FUNCTION TYPE</th><th>TABLE</th><th>GRAPH</th></tr>
<tr><td>Quadratic</td><td>

x	$f(x) = x^2$
0	0
1	1
2	4
3	9

</td><td> 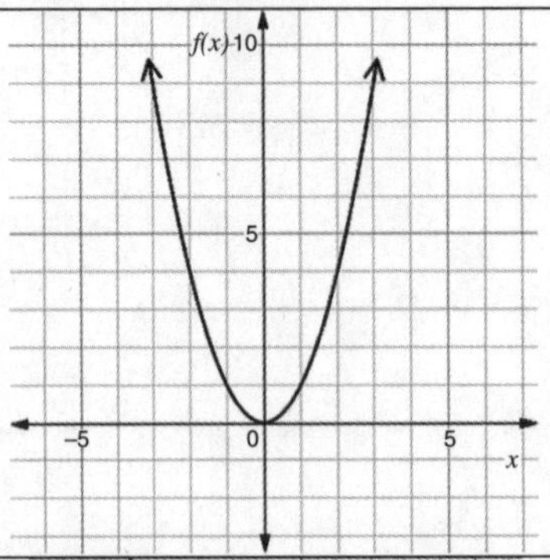f(x) 10, 5, −5, 0, 5, x</td></tr>
<tr><td>Square Root</td><td>

x	$g(x) = \sqrt{x}$
0	0
1	1
4	2
9	3

</td><td> 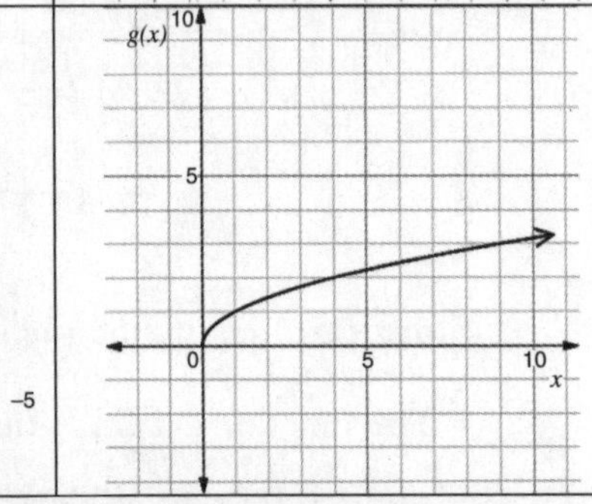g(x) 10, 5, 0, 5, 10, x, −5</td></tr>
</table>

We find points on the graph of $f(x) = \sqrt{x}$ by switching the x- and y-coordinates of points on $f(x) = x^2$. For example, (3, 9) is on the graph of $y = x^2$ since $3^2 = 9$, and (9, 3) is on the graph of $y = \sqrt{x}$ since $3 = \sqrt{9}$. If we square 3 to get 9 and then take the square root of 9, we get 3, the number we started with.

More formally, we say that the **inverse function of a function *f*** (which we can call simply the **inverse of *f***) is a function that undoes the operation of f. In this case, $g(x) = \sqrt{x}$ is the inverse of $f(x) = x^2$ because taking the square root of a nonnegative number undoes squaring it. Note that this undoing doesn't work for negative numbers. If we square -4, we get $(-4)^2 = 16$, but $\sqrt{(-4)^2} = \sqrt{16} = 4$, not -4. Thus, we restrict the domain of f to nonnegative numbers in order for f to have an inverse. (In an advanced algebra course, you'll learn more about the restrictions that we need to impose on functions in order for their inverse functions to exist.)

Here are some other important notes about the square root function:

- Its domain is the set of all nonnegative real numbers.
- Its range is the set of all nonnegative real numbers.
- Informally, we can think of the graph of the square root function as half a parabola, turned on its side. Although it may appear to level off, it does not have an asymptote.

The inverse of a cubic function is the cube root function, shown in Table 12.2:

Table 12.2 Comparing cubic and cube root functions.

FUNCTION TYPE	TABLE		GRAPH
Cubic	x	$f(x) = x^3$	
	–3	–27	
	–2	–8	
	–1	–1	
	0	0	
	1	1	
	2	8	
	3	27	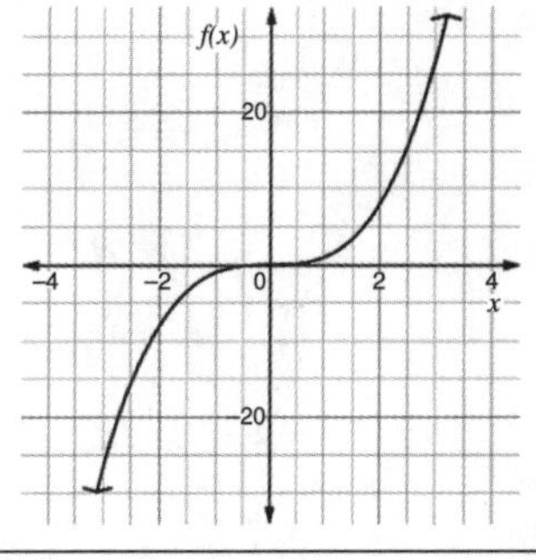
Cube Root	x	$f(x) = \sqrt[3]{x}$	
	–27	–3	
	–8	–2	
	–1	–1	
	0	0	
	1	1	
	8	2	
	27	3	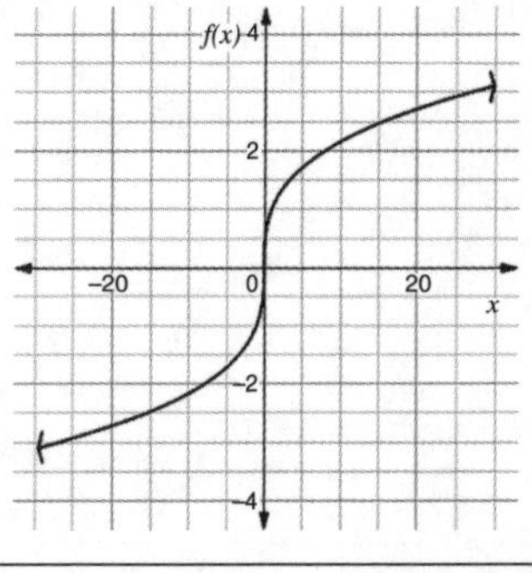

Here are some other important notes about the cube root function:

- Its domain is the set of all real numbers.
- Its range is the set of all real numbers.
- Both the cubic function $f(x) = x^3$ and the cube root function $f(x) = \sqrt[3]{x}$ are always increasing, but they curve in a different way. Informally, we say that $f(x) = x^3$ "flattens horizontally" over the x-axis, while $f(x) = \sqrt[3]{x}$ "flattens vertically" over the y-axis.

Exercises

Graph each equation on the coordinate plane.

1. $y = (x - 1)(x - 2)(x - 4)$
2. $f(x) = (x + 2)(x - 3)(x + 5)$
3. $f(x) = -x(x + 4)(x - 4)$
4. $y = x^2(x - 3)$
5. $f(x) = -(x + 1)^2(x - 6)$
6. $f(x) = -(x + 5)(x - 4)^2$

Write the equation shown in each graph.

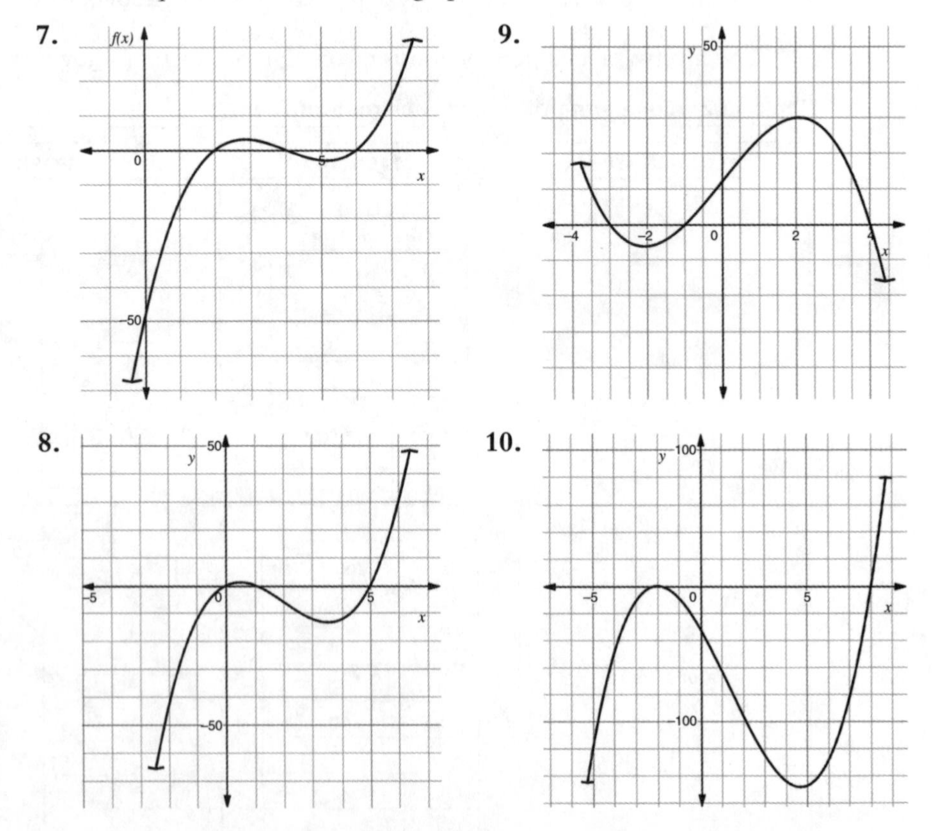

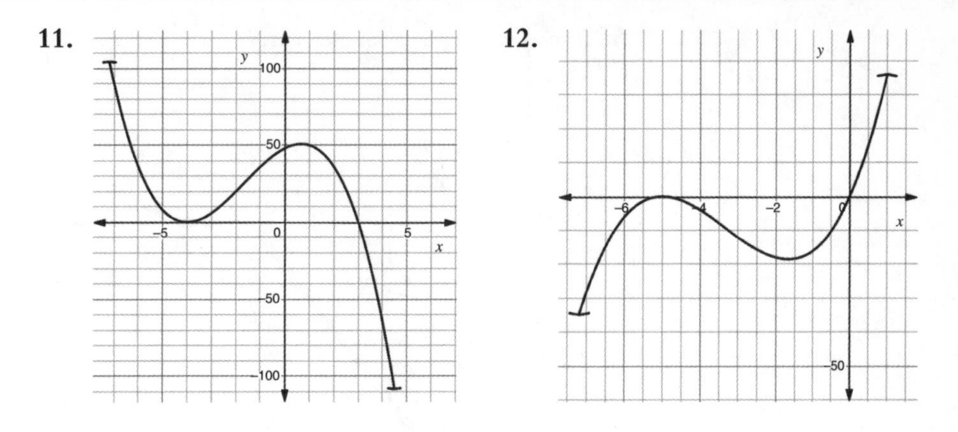

Questions to Think About

13. What similarities do the graphs of first-degree and third-degree functions have?

14. One end of the graph of a cubic function points up and the other end points down. Use this fact to explain informally why a cubic function must have at least one real zero.

15. State two differences between the graphs of quadratic and cubic functions.

16. State the definition of the inverse function of a function.

17. Explain why we need to restrict the domain of $f(x) = x^2$ to nonnegative numbers in order for f to have an inverse. (HINT: What happens to points with negative x-values if the x- and y-values are switched?)

12.2 Piecewise Functions

So far, we have used one equation to model a given situation. Sometimes, we need more than one equation.

Consider the following example: While on vacation, Max drives 200 miles at the same speed for 4 hours, stops for 1 hour to eat lunch, then rides his bicycle 20 miles at the same speed for 1 hour. If we graph the distance $f(x)$ that he travels as a function of time (x), we notice something unusual (Figure 12.3):

We can't use one function to model the distance that Max travels since he travels at different rates. Instead, we need a function with three distinct pieces: one for the first four hours ($0 < x \leq 4$), one for the hour in which he ate lunch ($4 < x \leq 5$), and another for the last hour (from $5 < x \leq 6$). Such a function is called a **piecewise function** – a function that consists of two or more functions defined over different intervals. In this case, the piecewise function f that models Max's distance is:

$$f(x) = \begin{cases} 50x,\ 0 < x \leq 4 \\ 200,\ 4 < x \leq 5 \\ 20x + 100,\ 5 < x \leq 6 \end{cases}$$

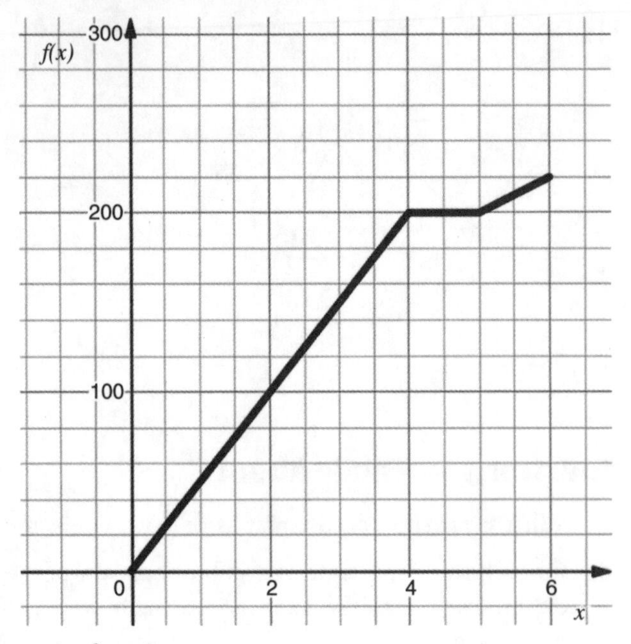

Figure 12.3 Piecewise function.

Note the following important points about this equation:

- The intervals that define each piece of a piecewise function must be specified and can't overlap.
- We use a single left-facing curly bracket to write a piecewise function. We don't pronounce the bracket. We read the function above as "f of x equals $50x$ for 0 is less than x is less than or equal to 4, 200 for 4 is less than x is less than or equal to 5, and $20x$ plus 100 for 5 is less than x is less than or equal to 6."
- Don't draw arrows on the end of the graph unless the interval extends infinitely in that direction.
- If necessary, at the end of each piece of the interval, draw a closed circle if its endpoint is included or an open circle if the endpoint is excluded (we discussed the use of open or closed circles in Section 5.2).

How to Graph a Piecewise Functions

1. On the same coordinate plane, graph the function corresponding to each piece.
2. Select the appropriate part of each graph corresponding to its defined interval.
3. Mark the endpoints appropriately with open circles, closed circles, or arrows.
4. Erase or hide the other parts of each function.

To evaluate a piecewise function, we use only the expression or the part of the graph that corresponds to the input value.

Example 12.5 The piecewise function f is defined as follows:

$$f(x) = \begin{cases} -x, & x < 0 \\ x, & x \geq 0 \end{cases}$$

(a) Graph f on the coordinate plane.

(b) Evaluate $f(-1)$.

Solution:

(a)

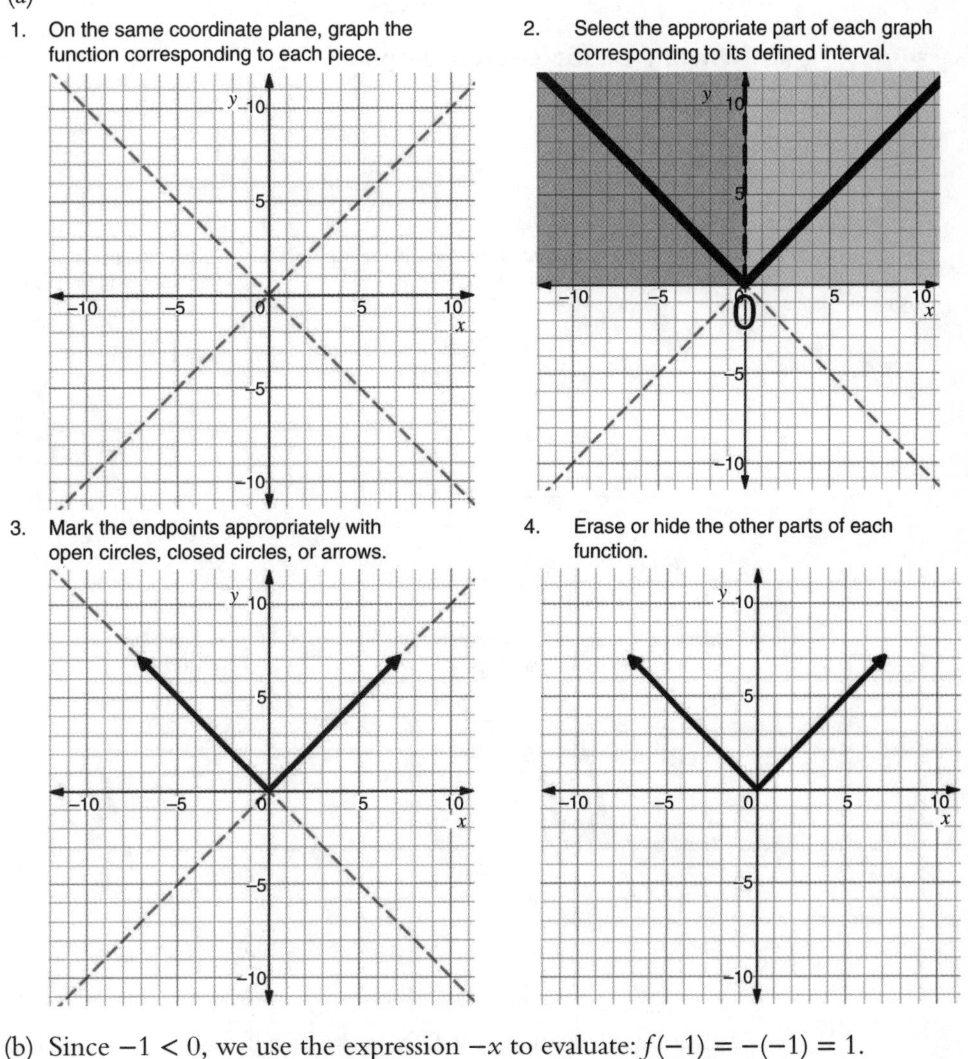

(b) Since $-1 < 0$, we use the expression $-x$ to evaluate: $f(-1) = -(-1) = 1$.

Example 12.5 is the graph of an important piecewise function, the absolute value function. (We defined absolute value in Section 1.1.) We write the equation of this function as $f(x) = |x|$.

When graphing piecewise functions that are discontinuous (not continuous, meaning that the function has gaps), pay particular attention to correctly graphing the endpoints of each piece.

Example 12.6 The piecewise function f is defined as follows:

$$f(x) = \begin{cases} 3x + 7, -3 \le x < -1 \\ x^2, -1 \le x < 2 \\ -x + 6,\ 2 \le x < 4 \\ -2,\ 4 \le x \le 6 \end{cases}$$

(a) Graph f on the coordinate plane.

(b) Evaluate $f(2)$.

Solution:

(a)

1. On the same coordinate plane, graph the function corresponding to each piece.

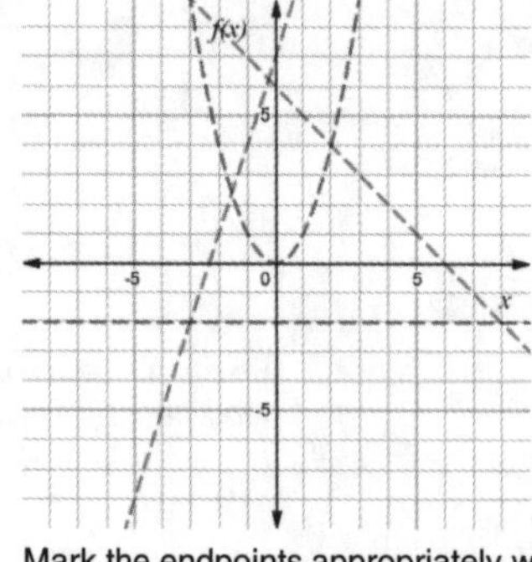

2. Select the appropriate part of each graph corresponding to its defined interval.

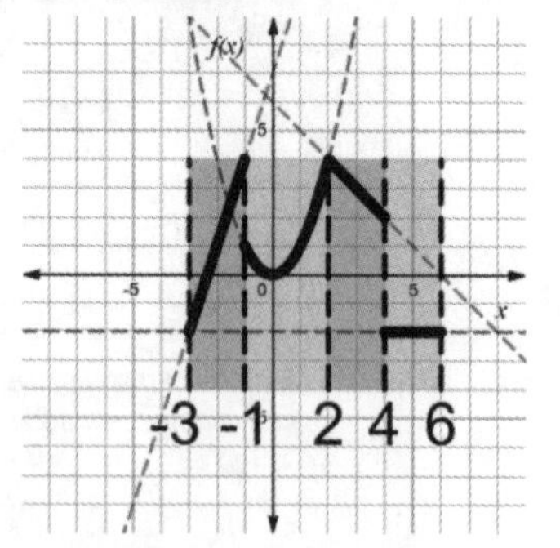

3. Mark the endpoints appropriately with open circles, closed circles, or arrows.

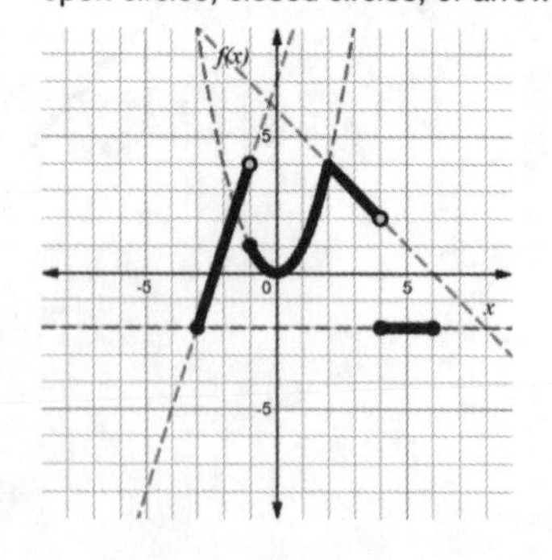

4. Erase or hide the other parts of each function.

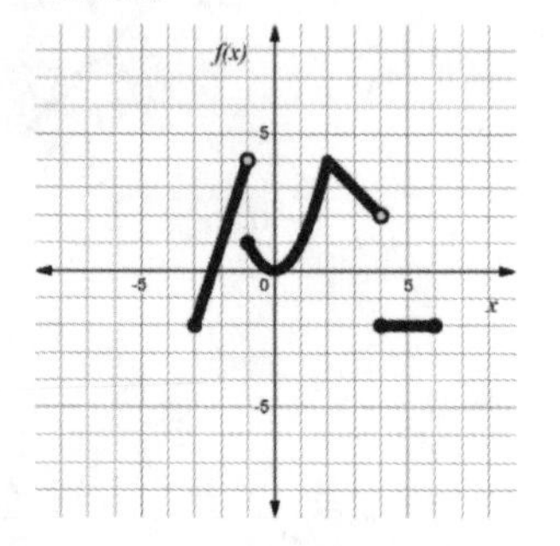

(b) Use the expression defined for $x = 2$, which is $-x + 6$, to evaluate: $f(2) = -(2) + 6 = 4$. We ignore the other expressions in the definition of the piecewise function.

Another type of discontinuous piecewise function is the **step function**, which consists of horizontal pieces that look like steps on a staircase:

Example 12.7 The graph of f is shown below.

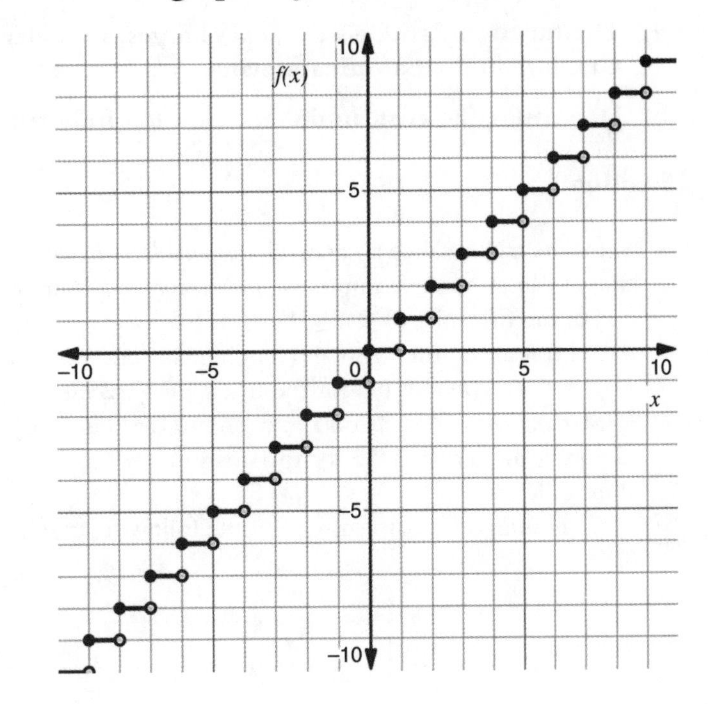

(a) Evaluate $f(1)$.

(b) Evaluate $f(1.5)$.

(c) Evaluate $f(2)$.

Solution:

(a) The graph has an open circle at (1, 0), which means that the point is not on the graph. The graph has a closed circle at (1, 1), which means that the point is on the graph, so $f(1) = 1$.

(b) We see from the graph that over the interval $1 \leq x < 2, f(x) = 1$, so $f(1.5) = 1$.

(c) We see from that the graph has a closed circle at (2, 2), so $f(2) = 2$.

Step functions are often used to model the behavior of taxicab or postage prices, which remain constant over an interval before jumping to another value.

Example 12.8 **A taxi company charges \$3 for the first mile or part of a mile and \$2 for each additional mile or part of a mile after that.**

(a) The function f models the taxi charges, in dollars, for a trip of x miles. Graph f on the coordinate plane.

(b) Determine the cost, in dollars, of a 6.5-mile trip.

Solution:

(a) Since the trip distance and taxi charges can't be negative, the domain and range are the set of nonnegative numbers. The graph of f is limited to Quadrant I.

For the first mile $(0 < x \leq 1)$, $f(x) = 3$.

Starting after the first and ending at (but not including) the second mile $(1 < x \leq 2)$, $f(x) = 5$ (\$3 for the first mile + \$2 for any part of the second mile).

Starting after the second and ending at (but not including) the third mile $(2 < x \leq 3)$, $f(x) = 7$ (\$3 for the first mile + \$2 for the second mile + \$2 for the third mile).

Continuing this pattern, we get the following graph:

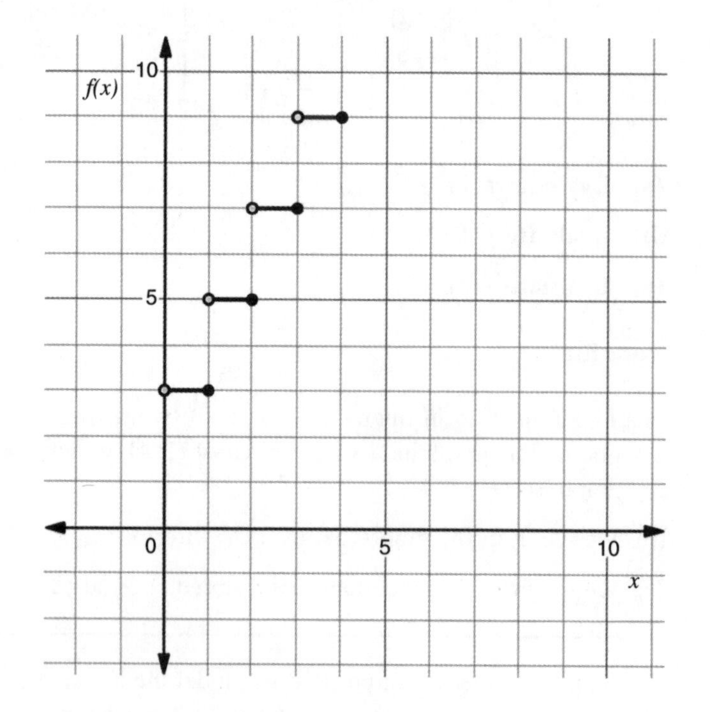

(b) If we write out the intervals in a table, we see that the cost of the trip increases $2 for each interval.

x	$0 < x \le 1$	$1 < x \le 2$	$2 < x \le 3$	$3 < x \le 4$	$4 < x \le 5$	$5 < x \le 6$	$6 < x \le 7$
f(x)	3	5	7	9	11	13	15

The cost of a 6.5-mile trip is $15.

Exercises

1. If $f(x) = \begin{cases} 5x, & x < 2 \\ x, & x \ge 2 \end{cases}$, evaluate $f(-1)$.

2. If $f(x) = \begin{cases} x^2 + x, & 2 < x < 5 \\ -2x + 4, & x \ge 5 \end{cases}$, evaluate $f(6)$.

3. If $f(x) = \begin{cases} 4^x, & -4 < x < 1 \\ x^2 - 3x - 2, & 2 < x < 3 \\ x, & x \ge 5 \end{cases}$, evaluate $f(7)$.

4. Using the accompanying graph of f, evaluate $f(0)$.

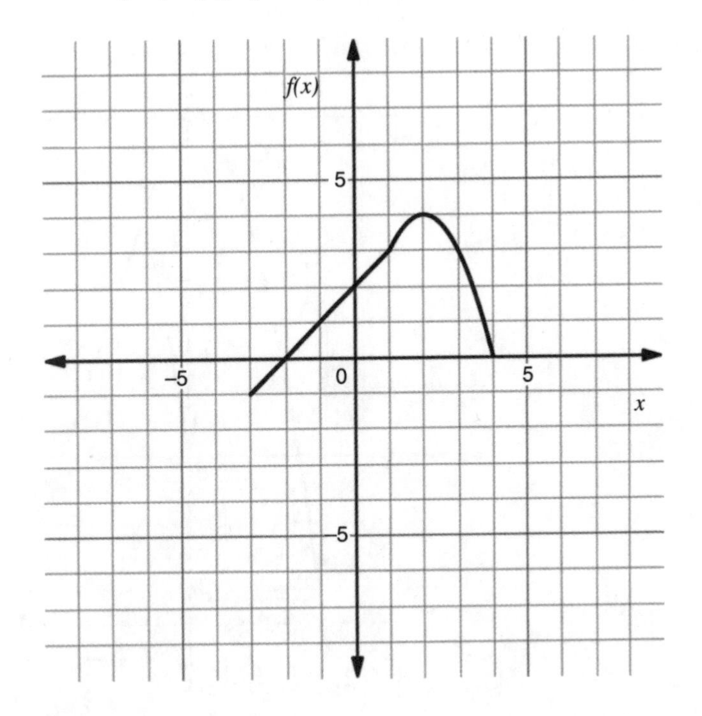

5. Using the accompanying graph of f, evaluate $f(5)$.

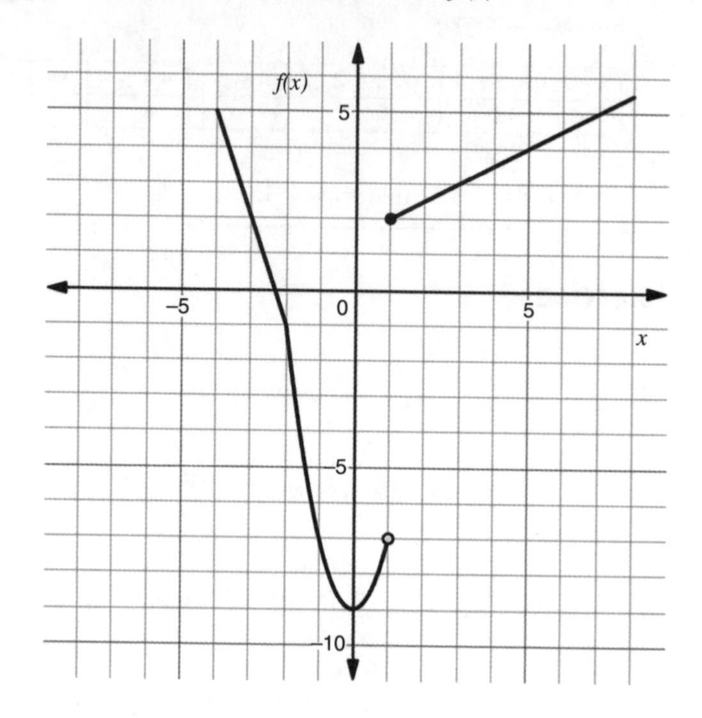

6. Using the accompanying graph of f, evaluate $f(-2)$.

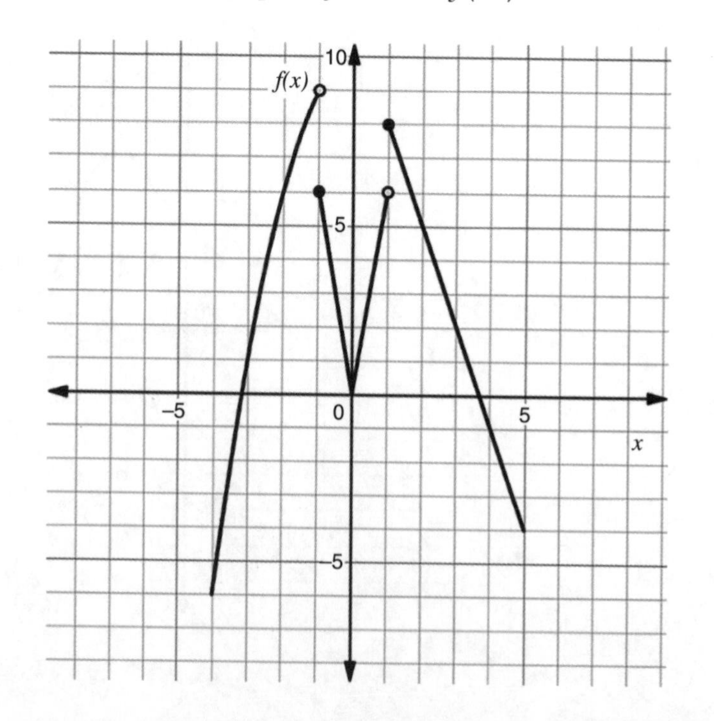

7. Using the accompanying graph of f, evaluate $f(-3)$.

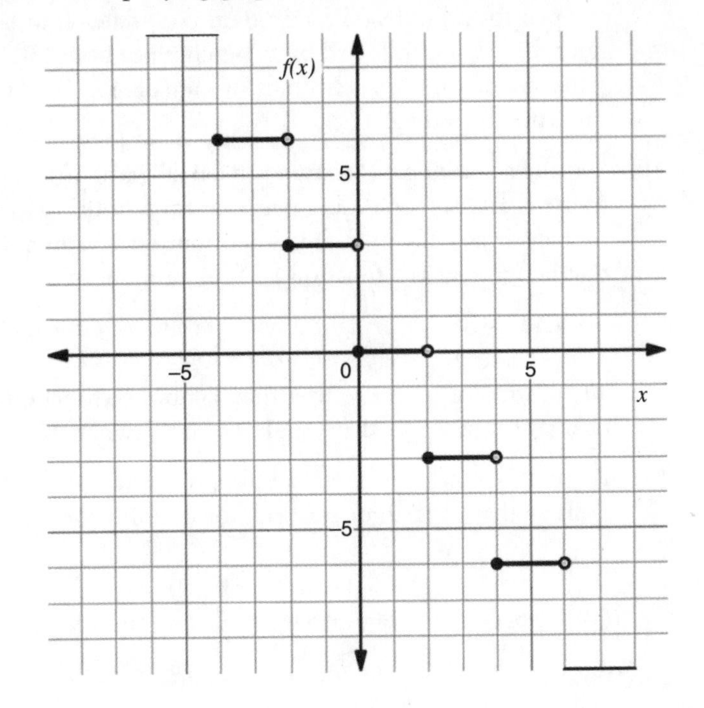

8. Using the accompanying graph of f, evaluate $f(3)$.

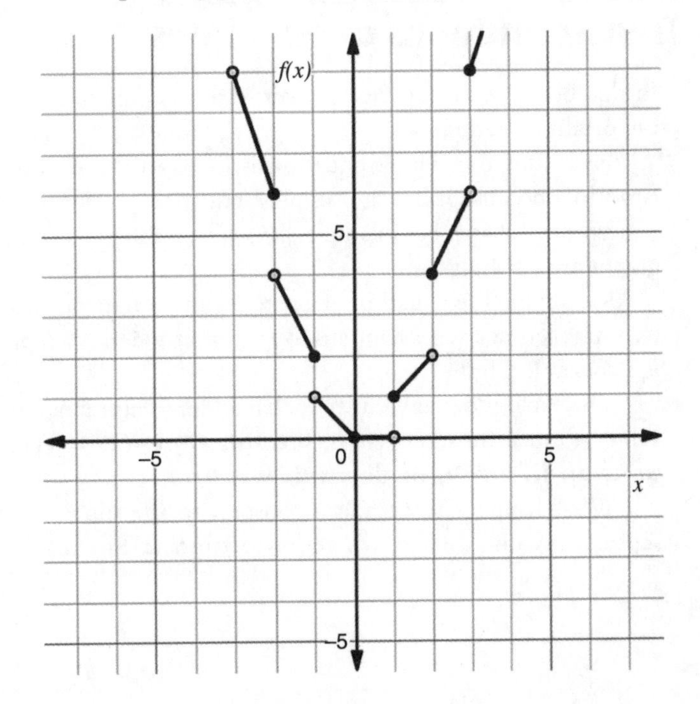

9. A recent storm began with 0.2 inches of rain per hour for 2 hours. The rain then stopped for 2 hours. The rain then continued at a steady rate of 0.5 inches per hour for 1 hour, followed by 0.1 inches per hour for 1 hour. Graph the function f that models the total amount of rainfall as a function of the number of hours since the storm began (x).

10. Starting at 12:00 p.m., Shawn walked at a rate of 3 miles per hour for 1 hour and then ran at a rate of 8 miles per hour for 1 hour. After resting for 1 hour, he rode his bicycle at a rate of 10 miles per hour for half an hour. Graph the total distance that he traveled as a function f of the number of hours (x) after 12:00 p.m.

11. To send a letter overnight anywhere in the United States, a shipping company charges \$10 if it weighs up to 1 ounce and \$1 for every additional 1 ounce or any fraction of an ounce after that. Graph the function f that models the shipping charges for packages that weigh x ounces, where $0 < x < 5$.

12. A taxicab company uses the following function to determine the fares (in dollars) that it charges for trips of x miles that leave from the local airport: $f(x) = \begin{cases} 15, & 0 < x \leq 1 \\ 15 + 2x, & x > 1 \end{cases}$
    (a) Determine the fare of a 0.5-mile trip from the airport.
    (b) Determine the fare of a 15-mile trip from the airport.

12.3 Transformations of Functions

In this book, we've discussed seven types of functions. Table 12.3 summarizes their important characteristics:

Each function whose equation is listed in the second column is called a **parent function** because it has the simplest equation of all functions whose graphs have the same general shape. For example, $f(x) = x^2$ is the parent function of all functions whose graphs are parabolas.

Modifying the equation of a parent function in various ways changes its graph. For example, let's see what happens to $f(x) = x^2$ when we replace x^2 with $x^2 + k$, where k is a constant (Figure 12.4).

This table and graph illustrate that adding a constant k to a function shifts the output values k units. In other words: **the function $g(x) = f(x) + k$ shifts f vertically k units up (if $k > 0$) or down (if $k < 0$).**

What happens if we add a constant to the inputs? Let's see what happens if we replace x^2 with $(x + h)^2$, where h is a constant (Figure 12.5):

Table 12.3 Parent functions.

TYPE	EQUATION	GRAPH	IMPORTANT FEATURES
Linear (First-Degree)	$f(x) = x$	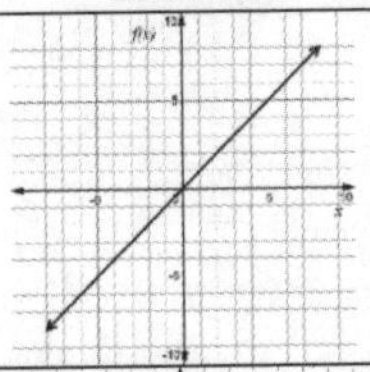	• Domain: $(-\infty, \infty)$ • Range: $(-\infty, \infty)$ • Point symmetry over (0, 0) • Always increasing
Quadratic (Second-Degree)	$f(x) = x^2$	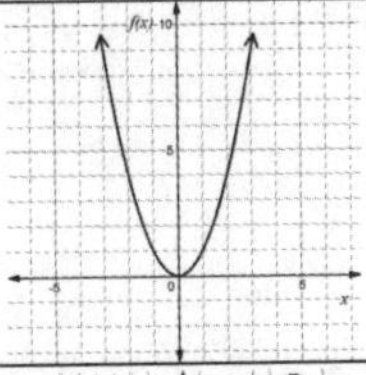	• Domain: $(-\infty, \infty)$ • Range: $[0, \infty)$ • Line symmetry over y-axis • Decreasing when $x < 0$, increasing when $x > 0$
Cubic (Third-Degree)	$f(x) = x^3$	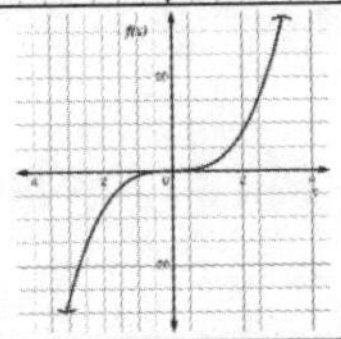	• Domain: $(-\infty, \infty)$ • Range: $(-\infty, \infty)$ • Point symmetry over (0, 0) • Always increasing • "Flattens horizontally" over x-axis
Exponential	$f(x) = b^x$	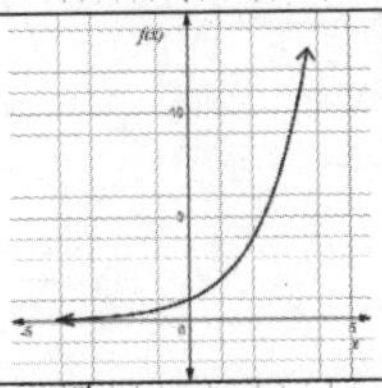	• Domain: $(-\infty, \infty)$ • Range: $(0, \infty)$ • Asymptote at $y = 0$ • Always increasing
Square Root	$f(x) = \sqrt{x}$	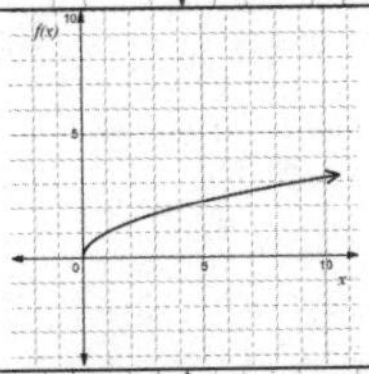	• Domain: $[0, \infty)$ • Range: $[0, \infty)$ • Always increasing
Cube Root	$f(x) = \sqrt[3]{x}$	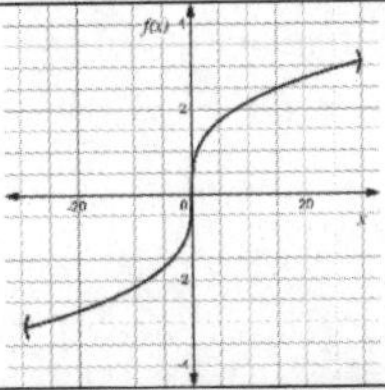	• Domain: $(-\infty, \infty)$ • Range: $(-\infty, \infty)$ • Point symmetry over (0, 0) • Always increasing • "Flattens vertically" over y-axis
Absolute Value	$f(x) = \lvert x \rvert$	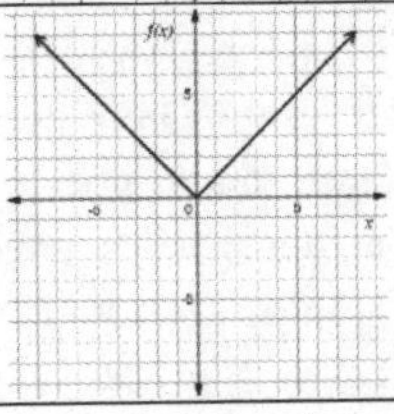	• Domain: $(-\infty, \infty)$ • Range: $[0, \infty)$ • Line symmetry over y-axis • Decreasing when $x < 0$, increasing when $x > 0$

x	$f(x) = x^2$	$g(x) = x^2 - 4$	$h(x) = x^2 + 4$
-3	9	5	13
-2	4	0	8
-1	1	-3	5
0	0	-4	4
1	1	-3	5
2	4	0	8
3	9	5	13

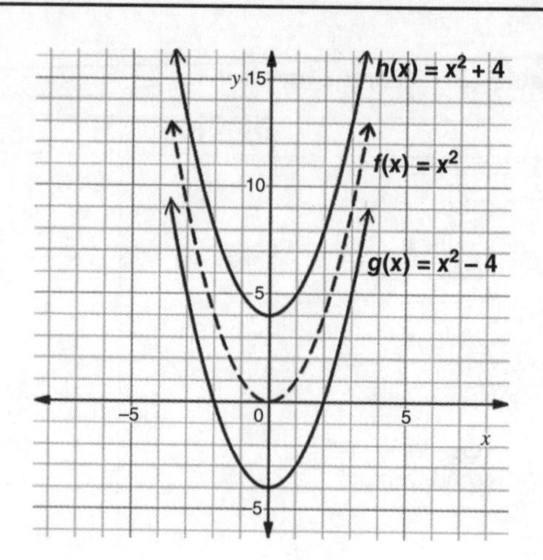

Figure 12.4 Vertical shift.

x	$f(x) = x^2$	$g(x) = (x - 2)^2$	$h(x) = (x + 2)^2$
-3	9	25	1
-2	4	16	0
-1	1	9	1
0	0	4	4
1	1	1	9
2	4	0	16
3	9	1	25

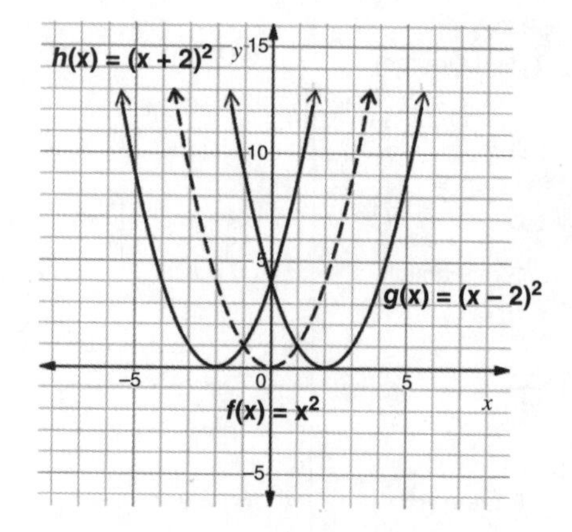

Figure 12.5 Horizontal shift.

The examples here illustrate the following rule: **the function $g(x) = f(x + h)$ shifts f horizontally h units left (if $h > 0$) or right (if $h < 0$).**

Watch Out!

You might expect that adding 2 to the input would shift the graph 2 units to the *right*, but it actually shifts 2 units to the *left*. Why? Think about it this way: in our example above, $f(0) = 0$. When we add 2 to the input, the x-value that gives us the same output of 0 (the value that makes $h(x) = (x + 2)^2 = 0$) is $x = -2$, which is 2 units left of $x = 0$. If we do this for every point in h, then adding 2 to the input shifts the graph 2 units to the left.

If we multiply the inputs by the same number, the graph changes in a different way, as shown in Figure 12.6:

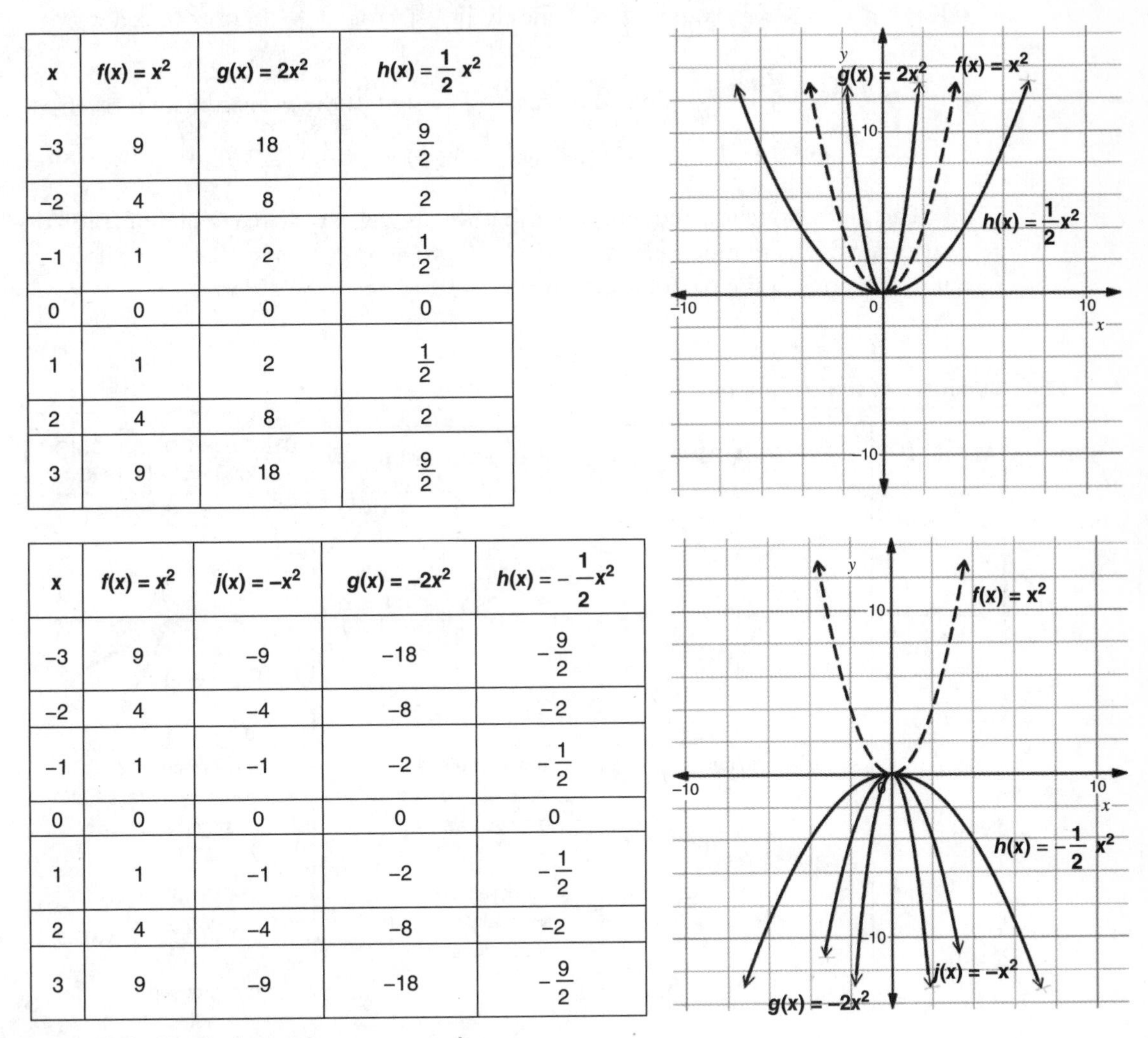

x	$f(x) = x^2$	$g(x) = 2x^2$	$h(x) = \frac{1}{2}x^2$
−3	9	18	$\frac{9}{2}$
−2	4	8	2
−1	1	2	$\frac{1}{2}$
0	0	0	0
1	1	2	$\frac{1}{2}$
2	4	8	2
3	9	18	$\frac{9}{2}$

x	$f(x) = x^2$	$j(x) = -x^2$	$g(x) = -2x^2$	$h(x) = -\frac{1}{2}x^2$
−3	9	−9	−18	$-\frac{9}{2}$
−2	4	−4	−8	−2
−1	1	−1	−2	$-\frac{1}{2}$
0	0	0	0	0
1	1	−1	−2	$-\frac{1}{2}$
2	4	−4	−8	−2
3	9	−9	−18	$-\frac{9}{2}$

Figure 12.6 Vertical stretch or compression.

These examples show that the function $g(x) = af(x)$ changes the graph of f as follows:

- If $a > 1$, f is stretched vertically by a factor of a. (Informally, it gets "narrower.")
- If $a = 1$, f is unchanged.

- If $0 < a < 1$, f is compressed vertically by a factor of a. (Informally, it gets "wider" or it "flattens.")
- If $-1 < a < 0$, f is compressed vertically by a factor of $|a|$ and reflected over the x-axis.
- If $a = -1$, f is reflected about the x-axis. (The sign of the y-coordinate is reversed.)
- If $a < -1$, f is stretched vertically by a factor of $|a|$ and reflected over the x-axis.

The shifts, stretches, and reflections that we discuss here are examples of **transformations**, which are functions that map one set to another. Table 12.4 summarizes the transformations of the parent function $f(x) = |x|$ (shown with dotted lines):

Table 12.4 Summary of transformation.

HORIZONTAL SHIFT	VERTICAL SHIFT	STRETCH/COMPRESSION	STRETCH/COMPRESSION AND REFLECTION
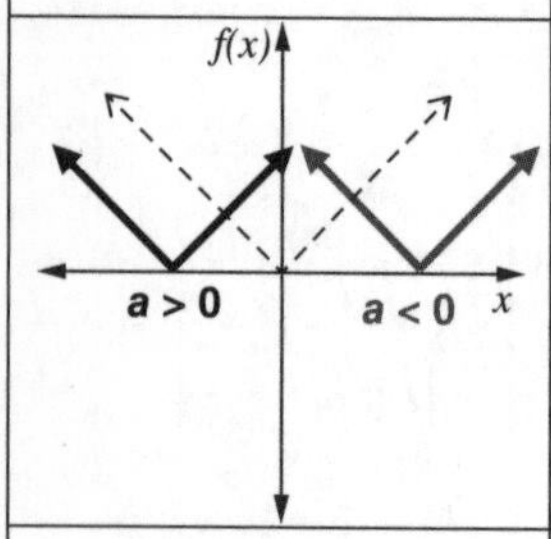	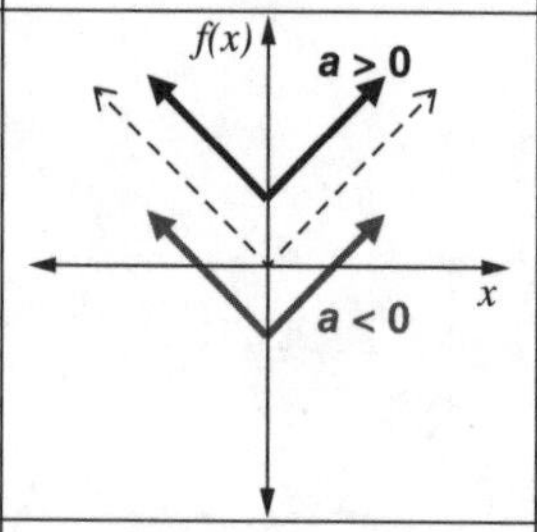	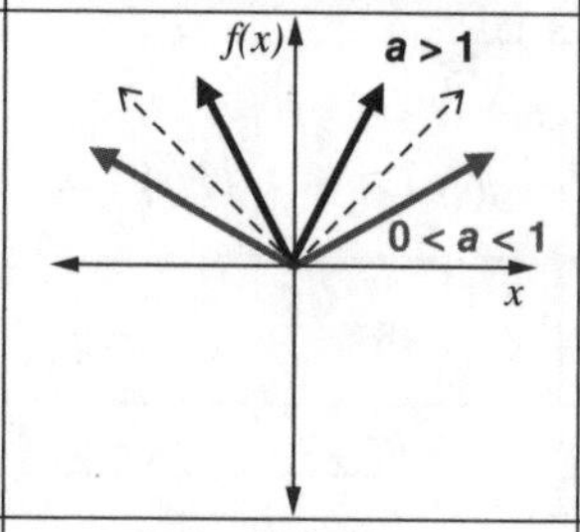	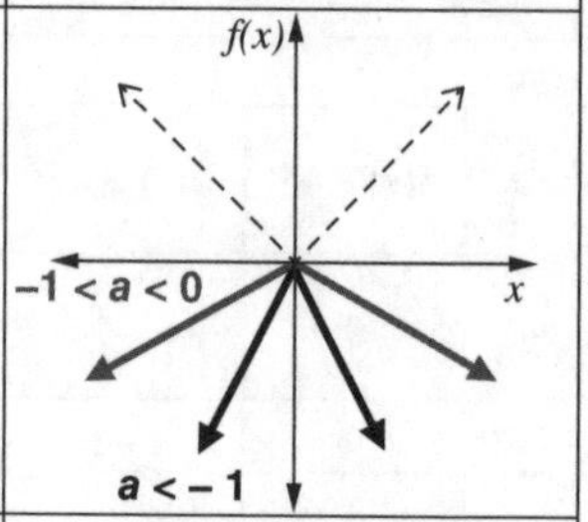
• $a > 0$: shifted a units left • $a < 0$: shifted a units right	• $a > 0$: shifted a units up • $a < 0$: shifted a units down	• $a > 1$, stretched vertically by factor of a ("narrows") • $0 < a < 1$: compressed vertically by factor of a ("flattens")	• $a < -1$: stretched vertically by factor of a and reflected over x-axis • $a = -1$: reflected over x-axis • $-1 < a < 0$: compressed vertically by factor of a and reflected over x-axis

Example 12.9 The graph of f is shown here.

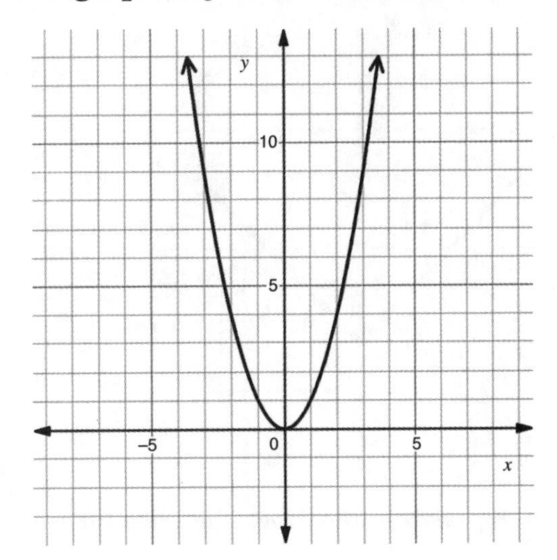

(a) Describe in words the transformation that maps $f(x)$ to $f(x + 2)$.

(b) Graph $y = f(x + 2)$.

Solution:

(a) The graph of $y = f(x + 2)$ is the graph of f shifted 2 units to the left.

(b) To graph $y = f(x + 2)$, we identify points on f and shift them 2 units left. For example, (0, 0) becomes (−2, 0), (2, 4) becomes (0, 4), (3, 9) becomes (1, 9), and so on. The accompanying graph shows $y = f(x + 2)$ as a solid curve and $y = f(x)$ as a dotted curve:

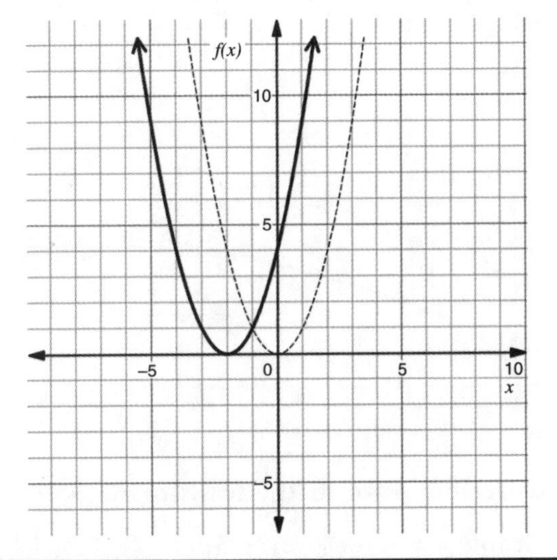

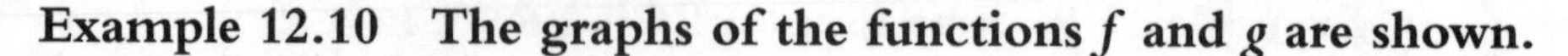

Example 12.10 The graphs of the functions f and g are shown.

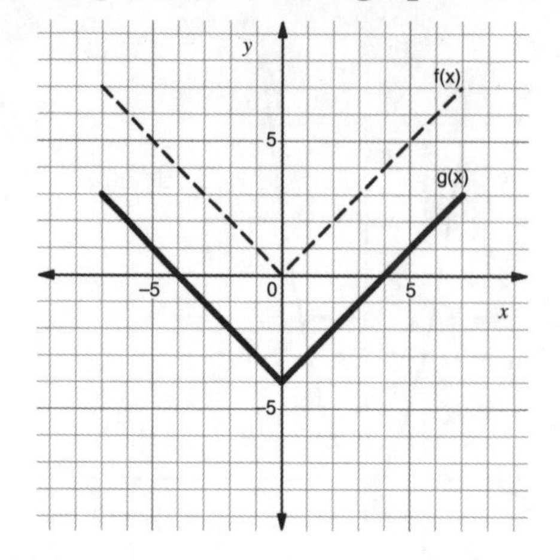

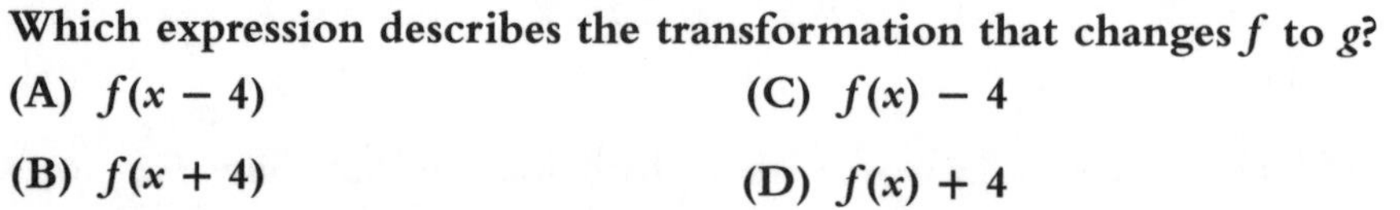

Which expression describes the transformation that changes f to g?

(A) $f(x - 4)$ | **(C)** $f(x) - 4$

(B) $f(x + 4)$ | **(D)** $f(x) + 4$

Solution: The graph of g is the graph of f shifted 4 units down. To shift each point 4 units down, we subtract 4 from each output value, which we create by $f(x) - 4$, or choice (C).

Several transformations can be applied to a parent function to form a new function, as shown in Examples 12.11 and 12.12:

Example 12.11 The graph of f is shown here:

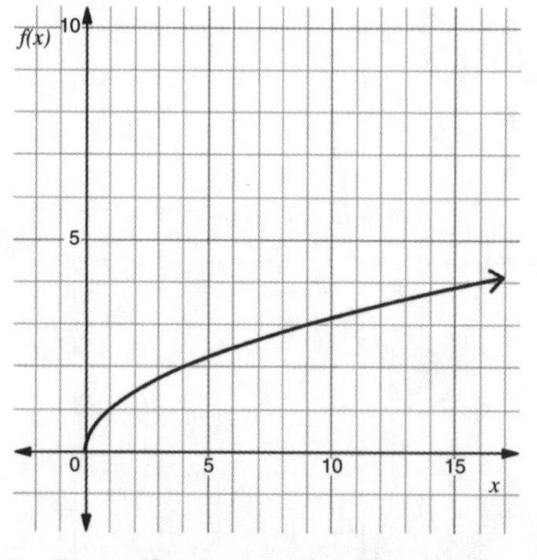

(a) Describe in words the transformations that map $f(x)$ to $-f(x - 2) + 3$.

(b) Graph $y = -f(x - 2) + 3$.

Solution:

(a) $f(x - 2)$ shifts the graph of f to the right 2 units, $-f(x - 2)$ reflects the graph of $f(x - 2)$ about the x-axis, and $-f(x - 2) + 3$ shifts the graph of $-f(x - 2)$ up 3 units. Thus, $-f(x - 2) + 3$ reflects f about the x-axis and shifts f to the right 2 units and up 3 units.

(b) Select points on the graph of f and do the following:

- Add 2 to the x-coordinate, so (1, 1) becomes (3, 1).
- Reflect over the x-axis (this changes the sign of the y-coordinate), so (3, 1) becomes (3, −1).
- Add 3 to the y-coordinate, so (3, −1) becomes (3, 2).

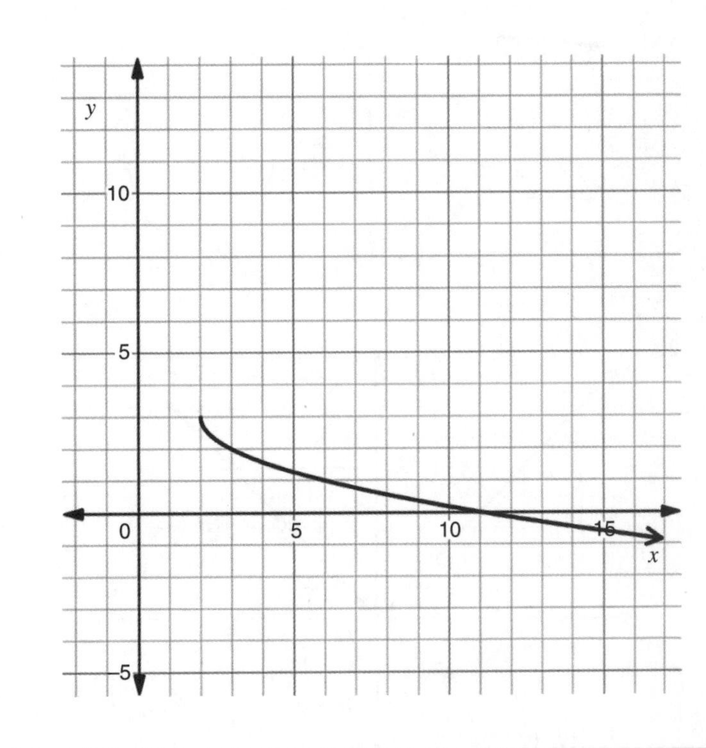

Example 12.12 Describe in words the transformations that map

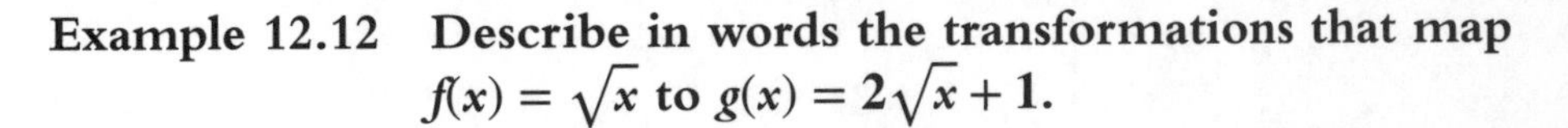

$f(x) = \sqrt{x}$ to $g(x) = 2\sqrt{x} + 1$.

Solution: In the equation for g, the coefficient 2 in front of $\sqrt{x}$ vertically stretches the graph of f by a factor of 2, while the +1 shifts the graph up 1 unit.

Exercises

In each example, the graph of f is shown. Graph the given transformation of f.

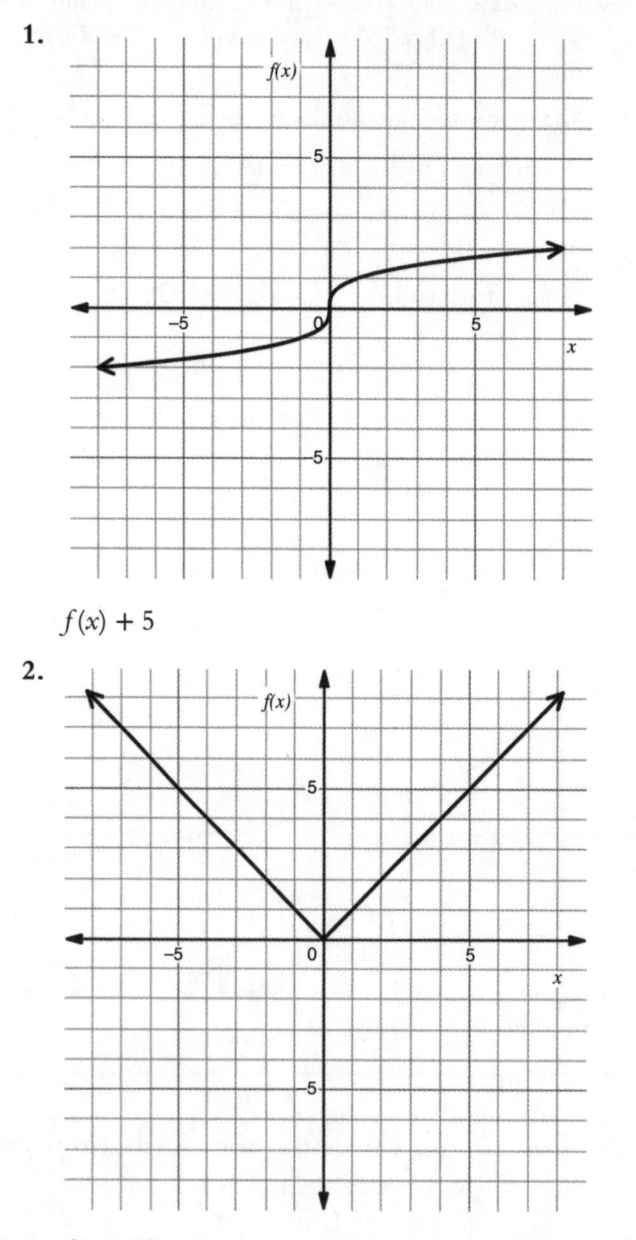

1. $f(x) + 5$

2. $f(x + 3)$

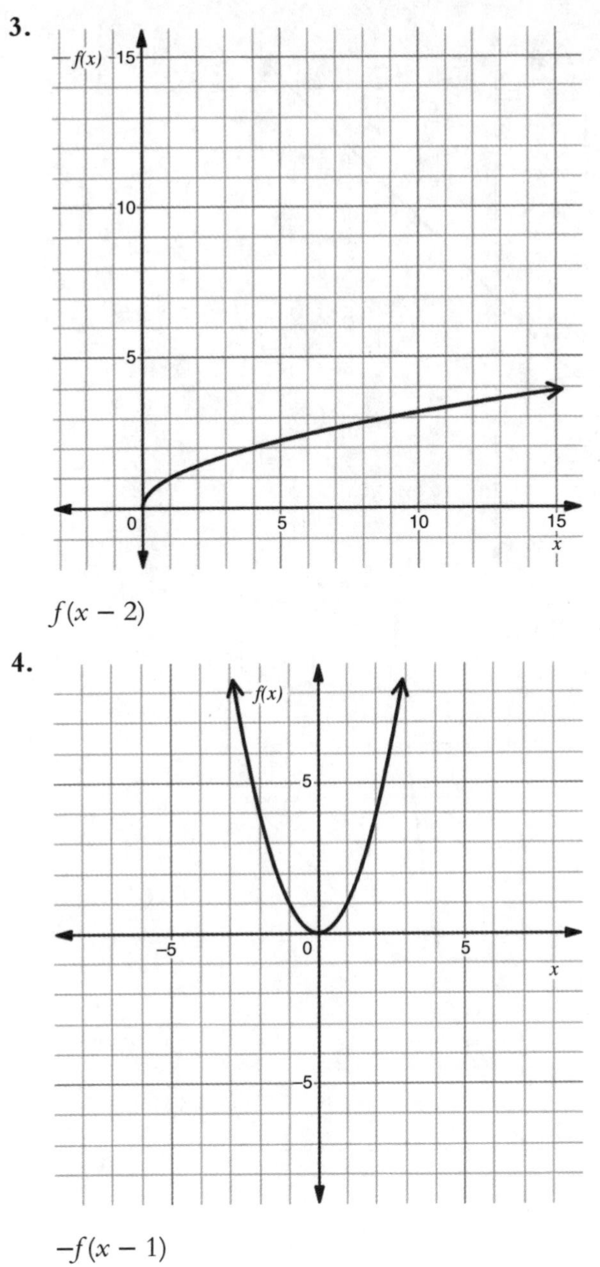

3. $f(x-2)$

4. $-f(x-1)$

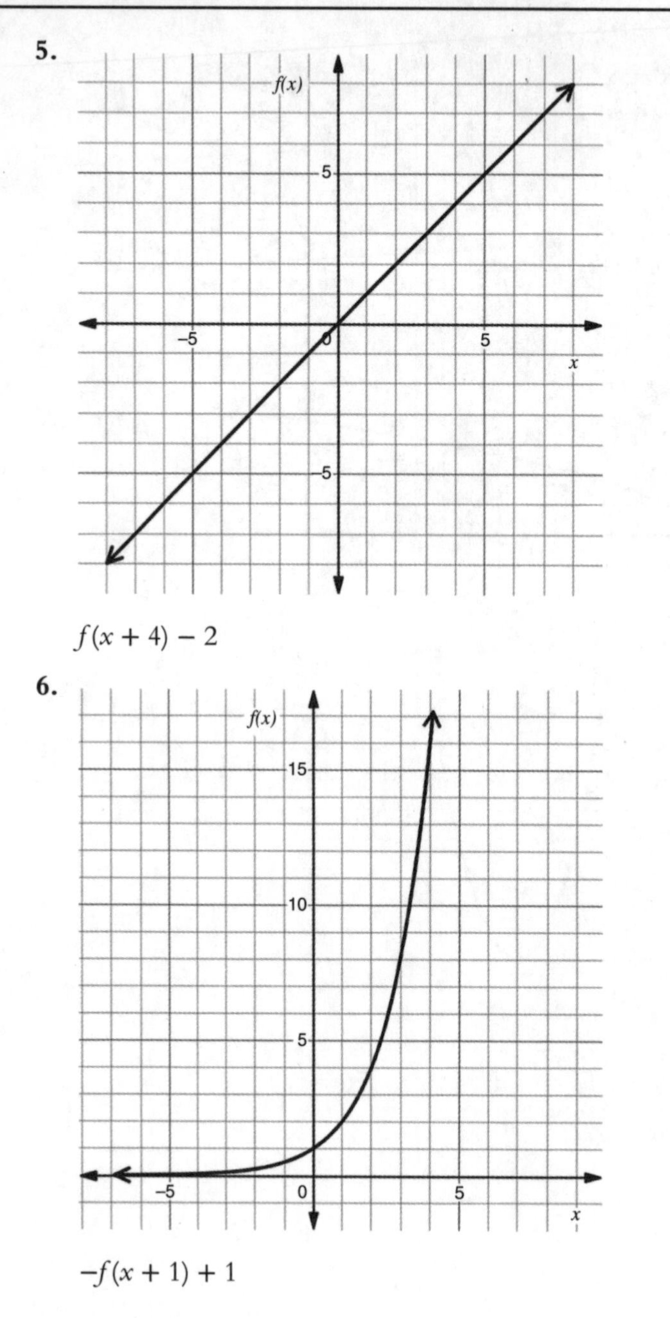

5. $f(x + 4) - 2$

6. $-f(x + 1) + 1$

Describe in words the transformations that map f to g.

7. $f(x) = x$, $g(x) = -x$
8. $f(x) = \sqrt[3]{x}$, $g(x) = \sqrt[3]{x-2}$
9. $f(x) = 2^x$, $g(x) = 3(2^x)$
10. $f(x) = x^2$, $g(x) = -(x-4)^3$
11. $f(x) = |x|$, $g(x) = -|x-6|$
12. $f(x) = x^3$, $g(x) = (x+2)^3 - 1$

Questions to Think About

13. Tina shifts the graph of the function $f(x) = |x|$ 2 units up and then 1 unit to the right. Ratika shifts the graph of f 1 unit to the right and then 2 units up. Will they get the same graph? Explain.
14. Nirupika reflects the graph of the function $f(x) = |x|$ over the x-axis and then shifts the graph 2 units up. Matthew shifts the graph of f 2 units up and then reflects it over the x-axis. Will they get the same graph? Explain.

12.4 Average Rate of Change of Functions

In Chapter 7, we said that slope is the rate of change for a linear function. Recall that the slope (symbolized by m) of the line passing through the points (x_1, y_1) and (x_2, y_2) is calculated using the formula $m = \frac{\text{vertical change}}{\text{horizontal change}} = \frac{y_2 - y_1}{x_2 - x_1}$. We interpret the slope by saying: as x increases by 1 unit, y increases by m units (or decreases by $|m|$ units, if $m < 0$).

For linear functions, the rate of change is constant. However, the rate of change for nonlinear functions is *not* constant (Figure 12.7).

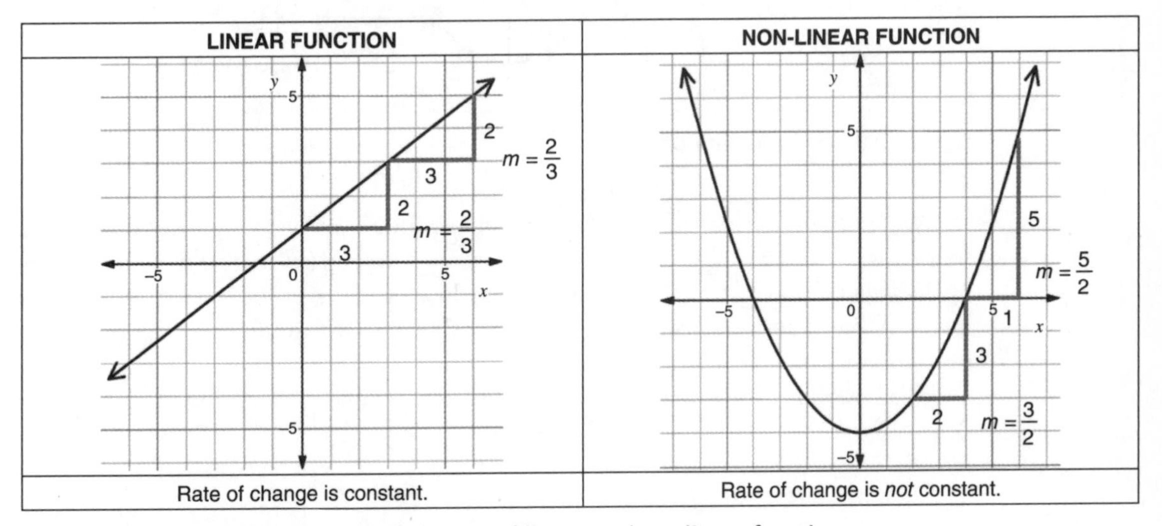

Figure 12.7 Comparing the rate of change of linear and nonlinear functions.

We approximate the rate of change for a nonlinear function over an interval by finding the slope of the line between those two points. We call this slope the **average rate of change**, defined over a specified interval $a \leq x \leq b$ as $\frac{f(b)-f(a)}{b-a}$.

The average rate of change between two points on the graph of a function is the slope of the line that passes through them. For linear functions, the average rate of change equals the slope of the line. Although the average rate of change theoretically applies to linear and nonlinear functions, in practice we only use the term to apply to nonlinear functions. For linear functions, we use the term "rate of change."

The average rate of change approximates how quickly a function changes as x increases (Figure 12.8).

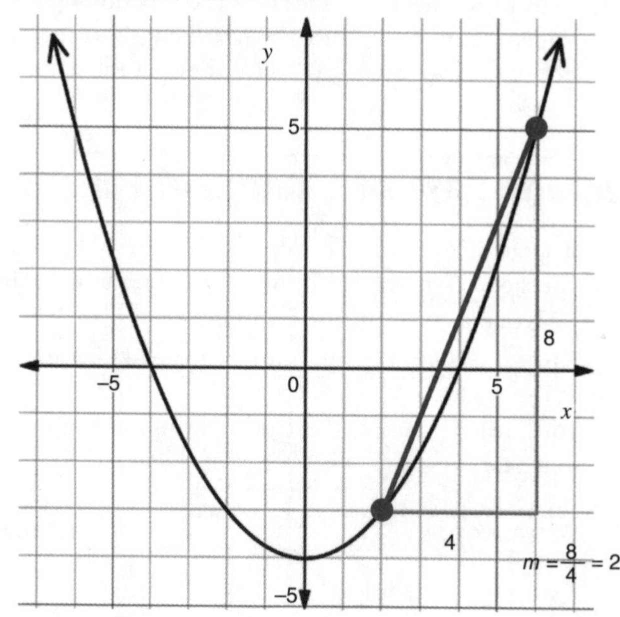

Figure 12.8 Average rate of change.

We interpret the rate of change for nonlinear functions by saying: over the interval from $x =$ ___ to $x =$ ___, as x increases by ___ units, $f(x)$ increases (or decreases) by an average of ___ units. Adding a phrase like "an average of" and specifying the interval clearly communicates that the rate of change for nonlinear functions varies over different intervals.

We determine the average rate of change from a table, a graph, or an equation.

Example 12.13 Find and interpret the average rate of change from $x = 7$ to $x = 11$ for the function f defined in the following table:

x	3	5	7	9	11
$f(x)$	16	48	96	160	240

Solution: Here, $a = 7$ and $b = 11$. From the table, we see that $f(7) = 96$ and $f(11) = 240$. Substitute those values into the formula for the average rate of change: $\frac{f(b)-f(a)}{b-a} = \frac{240-96}{11-7} = 36$. The interpretation is: over the interval from $x = 7$ to $x = 11$, for every increase of 1 unit, $f(x)$ increases by an average of 36 units.

Example 12.14 The accompanying graph shows the number of social media followers (in thousands) for a popular singer over a nine-month period.

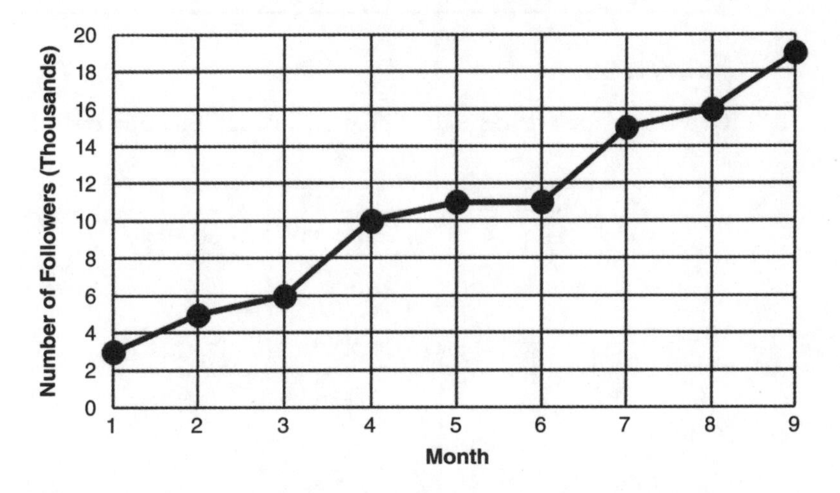

(a) Determine the average rate of change between the fourth and eighth months. Include appropriate units.

(b) Interpret the average rate of change from part *a* in context.

Solution:

(a) In the fourth month, the singer had 10,000 followers. In the eighth month, the singer had 16,000 followers. The average rate of change is $\frac{16{,}000-10{,}000}{8-4} = 1{,}500$ followers per month.

(b) Between the fourth to eighth months, the singer's number of social media followers increased by an average of 1,500 followers per month.

Example 12.15 The height (in feet) of a ball t seconds after being thrown can be modeled by the equation $h(t) = -16t^2 + 50t + 10$.

(a) Determine the average rate of change in the ball's height between the first and third seconds after being thrown. Include appropriate units.

(b) Interpret the average rate of change from part a in context.

Solution:

(a) Substitute $t = 1$ into the equation to find $h(1)$: $h(1) = -16(1)^2 + 50(1) + 10 = 44$. Substitute $t = 3$ into the equation to find $h(3)$: $h(3) = -16(3)^2 + 50(3) + 10 = 16$. The average rate of change is $\frac{h(3)-h(1)}{3-1} = \frac{16-44}{2} = -14$ feet per second.

(b) For every second that the ball travels between the first and third seconds after being thrown, its height decreases by an average of 14 feet per second.

Exercises

Determine the average rate of change for each function over the given interval.

1.

x	−2	0	2	4	6
f(x)	1	4	1	−11	−30

$-2 \leq x \leq 4$

2.

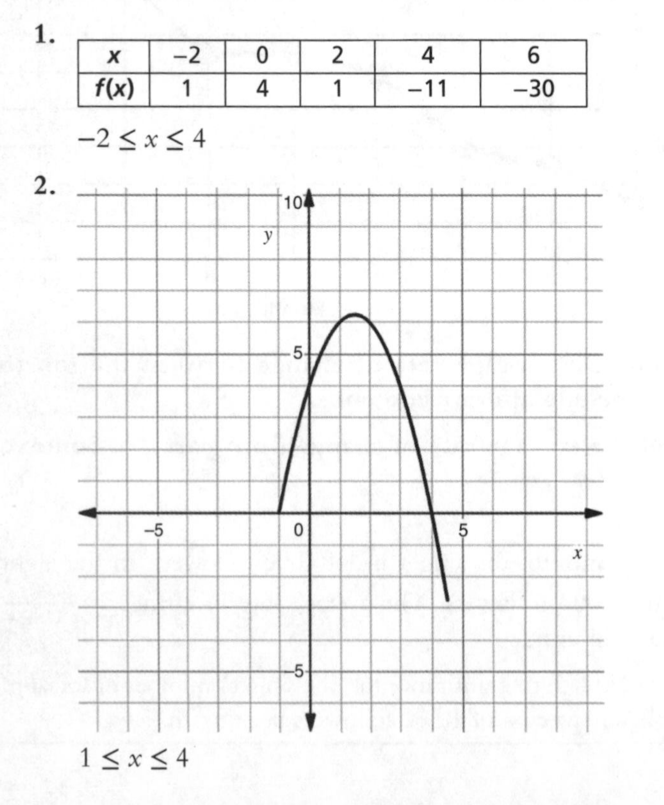

$1 \leq x \leq 4$

3. $f(x) = 10(2^x),\ 1 \leq x \leq 5$

4. The accompanying table shows the temperature, in degrees Fahrenheit, recorded at a weather station starting at 12:00 a.m.

Hours after 12 a.m.	0	3	6	9	12	15	18
Temperature (°F)	75	74	74	82	90	98	90

Find the average rate of change in hourly temperature from 3:00 a.m. to 3:00 p.m. Include appropriate units.

5. The accompanying table shows the percentage of registered voters in a large state who support its governor over an eight-month period.

Month	1	2	3	4	5	6	7	8
Percentage of voters who support the governor	65	64	60	56	55	54	53	55

Find the average rate of change in the monthly percentage of support from the first to the fifth month. Include appropriate units.

6. The accompanying table summarizes the total distance that a family traveled during a recent vacation.

Hours	0	1	2	4	5	6	7	10
Distance traveled (miles)	0	12	55	110	132	170	220	375

What was the average rate of change in the distance traveled between the second and seventh hours traveled? Include appropriate units.

7. The accompanying graph models the monthly income, in millions of dollars, of a large corporation over an eight-month period.

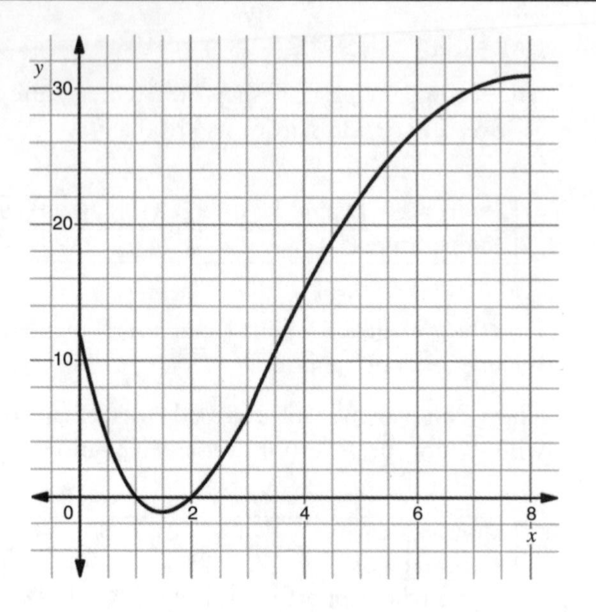

(a) Determine the average rate of change in monthly income between the second and the seventh month. Include appropriate units.

(b) Interpret the average rate of change from part *a* in context.

8. The accompanying graph models the population (in thousands) of a large town x years after 2000.

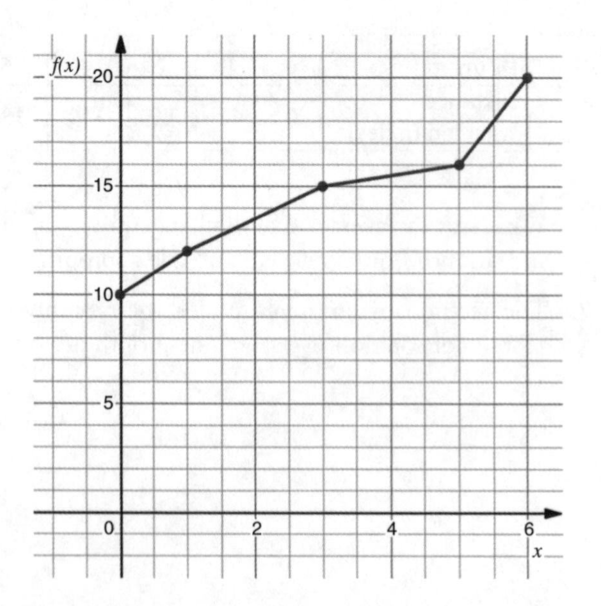

(a) Determine the average rate of change in population from 2010 to 2020. Include appropriate units.

(b) Interpret the average rate of change from part *a* in context.

9. The balance (in dollars) in an investment account x months after being opened can be modeled by the function $f(x) = 500(1.02)^x$.

(a) To the nearest dollar, determine the average rate of change in the account balance between 4 and 9 months after being opened. Include appropriate units.

(b) Interpret the average rate of change from part *a* in context.

10. The height of a projectile, in feet, x seconds after launch can be modeled by the function $h(x) = -16x^2 + 96x$.

(a) Determine the average rate of change in the projectile's height from $x = 2$ to $x = 5$. Include appropriate units.

(b) Interpret the average rate of change from part *a* in context.

12.5 Comparing Functions

As we've seen throughout this book, we represent functions with equations, graphs, tables, or verbal descriptions. We often compare the properties of functions with different representations. If necessary, review the characteristics of linear, quadratic, and exponential functions in Sections 6.3, 7.1, 9.9, and 10.1.

Example 12.16 The functions f and g are shown in the accompanying graph and table.

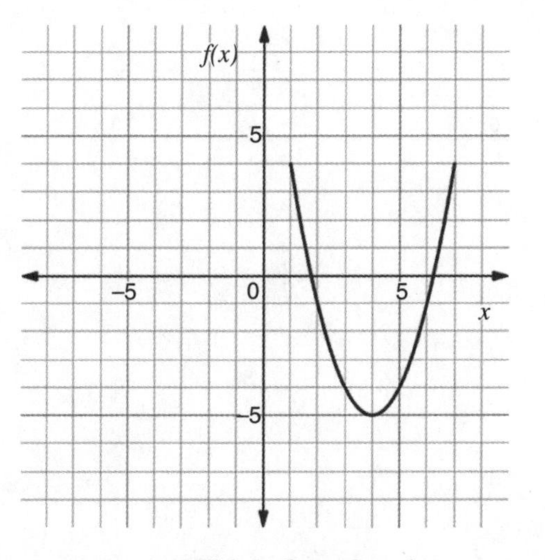

x	–1	0	1	2	3	4
g(x)	–3	–1	1	3	5	7

Which function has a greater minimum?

Solution: Recall from Section 6.3 that a minimum is the y-value of the point where the function changes from decreasing to increasing (in other words, it's the lowest y-value). From the graph, we see that the minimum of f is -5 since the lowest point of the function is $(4, -5)$. From the table, we see that the minimum of g is -3. Since $-3 > -5$, then $g(x)$ has a greater minimum.

Example 12.17 The function f is defined by the equation $f(x) = 2x^2 + 7x + 11$. The function g is a geometric sequence with an initial value of 3 (when $x = 1$) and a common ratio of 2. Which function has a greater average rate of change over the interval $1 \le x \le 3$?

Solution: $f(1) = 2(1)^2 + 7(1) + 11 = 20$ and $f(3) = 2(3)^2 + 7(3) + 11 = 50$. The average rate of change for $f(x)$ over the interval $1 \le x \le 3$ is $\frac{f(3) - f(1)}{3 - 1} = \frac{50 - 20}{2} = 15$.

We represent g explicitly using the formula $g(x) = 3(2)^x$. $g(1) = 3(2)^1 = 6$ and $g(3) = 3(2)^3 = 24$. The average rate of change for g over the interval $1 \le x \le 3$ is $\frac{g(3) - g(1)}{3 - 1} = \frac{24 - 6}{2} = 9$.

Since $15 > 9$, then f has a greater average rate of change over the interval.

Exercises

1. If $f(x) = 2^x - 8$ and g is defined by the following table:

x	0	1	2	3	4	5
g(x)	4	2	0	–2	–4	–6

Which function has an x-intercept with a greater x-coordinate?

2. The functions f and g are defined as follows:

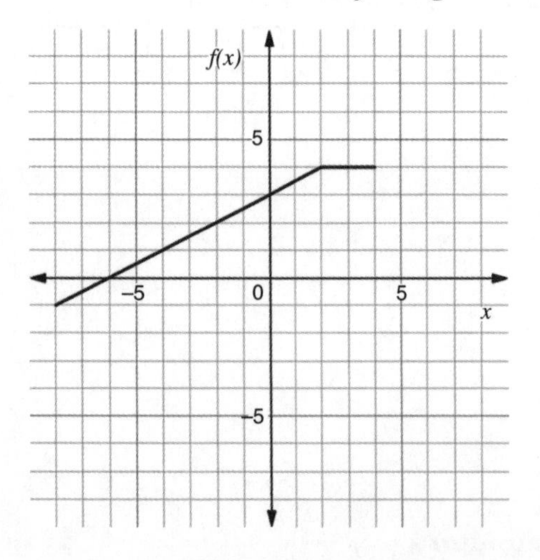

x	–5	–2	0	2	4	5
g(x)	0	3	5	3	1	0

Which function has a greater maximum?

3. The function f is defined by the equation $f(x) = 2(3)^x - 4$, and g is defined by the following table:

x	−4	−3	−2	−1	0	1
g(x)	−3	0	1	2	3	4

Which function has a y-intercept with a smaller y-coordinate?

4. The linear function f has a slope of 2. The function g has the graph shown here:

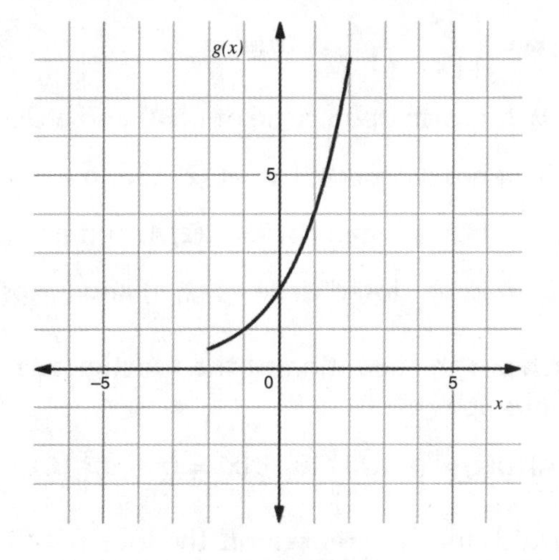

Which function has a greater average rate of change over the interval $-1 \leq x \leq 2$?

5. The function f is defined by the equation $f(x) = 3x^2 - 1$. The function g is shown in the accompanying graph.

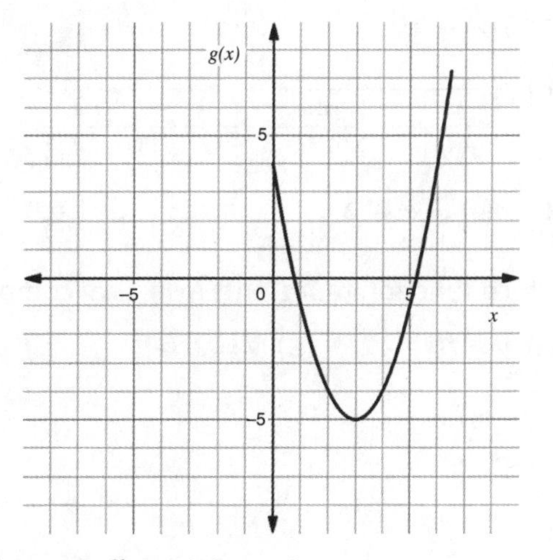

Which equation has a smaller minimum?

6. The graph of a linear function f passes through the points (4, 3) and (6, 17). The function g is defined by the equation $g(x) = |6x| + 5$. Which function has a greater average rate of change over the interval $5 \leq x \leq 7$?

CHAPTER 12 TEST

1. Which statement about the graph of the piecewise function $f(x) = \begin{cases} x^2, x < 2 \\ |x|, x \geq 2 \end{cases}$ is true?

   (A) It has an open circle at (2, 4) and a closed circle at (2, 2).

   (B) It has a closed circle at (2, 4) and an open circle at (2, 2).

   (C) It has an open circle at (2, 4) and an open circle at (2, 2).

   (D) It has a closed circle at (2, 4) and a closed circle at (2, 2).

2. What is the equation of the function that results after $f(x) = x^3$ is shifted 4 units up?

   (A) $g(x) = (x + 4)^3$ (B) $g(x) = (x - 4)^3$ (C) $g(x) = x^3 - 4$ (D) $g(x) = x^3 + 4$

3. Which function represents the inverse of the function $f(x) = x^2$?

   (A) $g(x) = -x^2$ (B) $g(x) = \frac{1}{2}x^2$ (C) $g(x) = \sqrt{x}$ (D) $g(x) = -\sqrt{x}$

4. Which function has the greatest minimum value?

   (A) $f(x) = 3^x$

   (B)

x	−3	−2	−1	0	1	2
g(x)	5	4	3	2	1	2

   (C) $h(x) = x^2 + 2$ (D) $k(x) = -x^3$

5. Which function represents a vertical compression of $y = f(x)$ by a factor of 2?

   (A) $y = f(x) - 2$ (B) $y = 2f(x)$ (C) $y = f(x) + 2$ (D) $y = \frac{f(x)}{2}$

6. If $f(x)$ has a minimum of 2, what is the minimum of $f(x) + 3$?

   (A) 5 (B) -1 (C) 1 (D) 6

7. For the piecewise function $f(x) = \begin{cases} x+5, x<3 \\ \sqrt{x}, x \geq 3 \end{cases}$, what is $f(4)$?

   (A) 2 (B) 9 (C) $\{2, 9\}$ (D) $(2, 9)$

8. Over what interval does the function $f(x) = 2x^2$ have the greatest rate of change?

   (A) $3 \leq x \leq 6$ (B) $4 \leq x \leq 8$ (C) $9 \leq x \leq 11$ (D) $11 \leq x \leq 12$

9. Which function has zeros at $x = a$, $x = b$, and $x = c$?

   I. $f(x) = (x + a)(x + b)(x + c)$

   II. $f(x) = (x - a)(x - b)(x - c)$

   III. $f(x) = -(x - a)(x - b)(x - c)$

   (A) I, only (C) II and III, only

   (B) I and II, only (D) I, II, and III

10. Write the equation of the function g that reflects $f(x) = |x|$ about the x-axis.
11. Describe in words the transformations that map $f(x) = x^2$ to $g(x) = (x - 3)^2 + 5$.
12. A company's shipping fees are based on the weight of a letter. The company charges \$1 for the first ounce or fraction of an ounce and \$0.25 for each additional ounce or fraction of an ounce. Determine the shipping cost of a letter that weighs 5.5 ounces.
13. Graph the piecewise function $f(x) = \begin{cases} 2x, -3 < x < 0 \\ 4, 0 \leq x < 2 \\ -2x, 2 \leq x < 3 \end{cases}$ on the coordinate plane.
14. Graph the equation $f(x) = \sqrt[3]{x}$ on the coordinate plane.
15. The height (in feet) of a ball x seconds after being thrown is modeled by the function $h(x) = -16x^2 + 112x + 30$.

    (a) Determine the average rate of change in the height from the first to the fourth seconds after being thrown. Include appropriate units.

    (b) Interpret the average rate of change in context.

CHAPTER 12 SOLUTIONS

12.1. 1. 4. 2. 5. 3. 6.

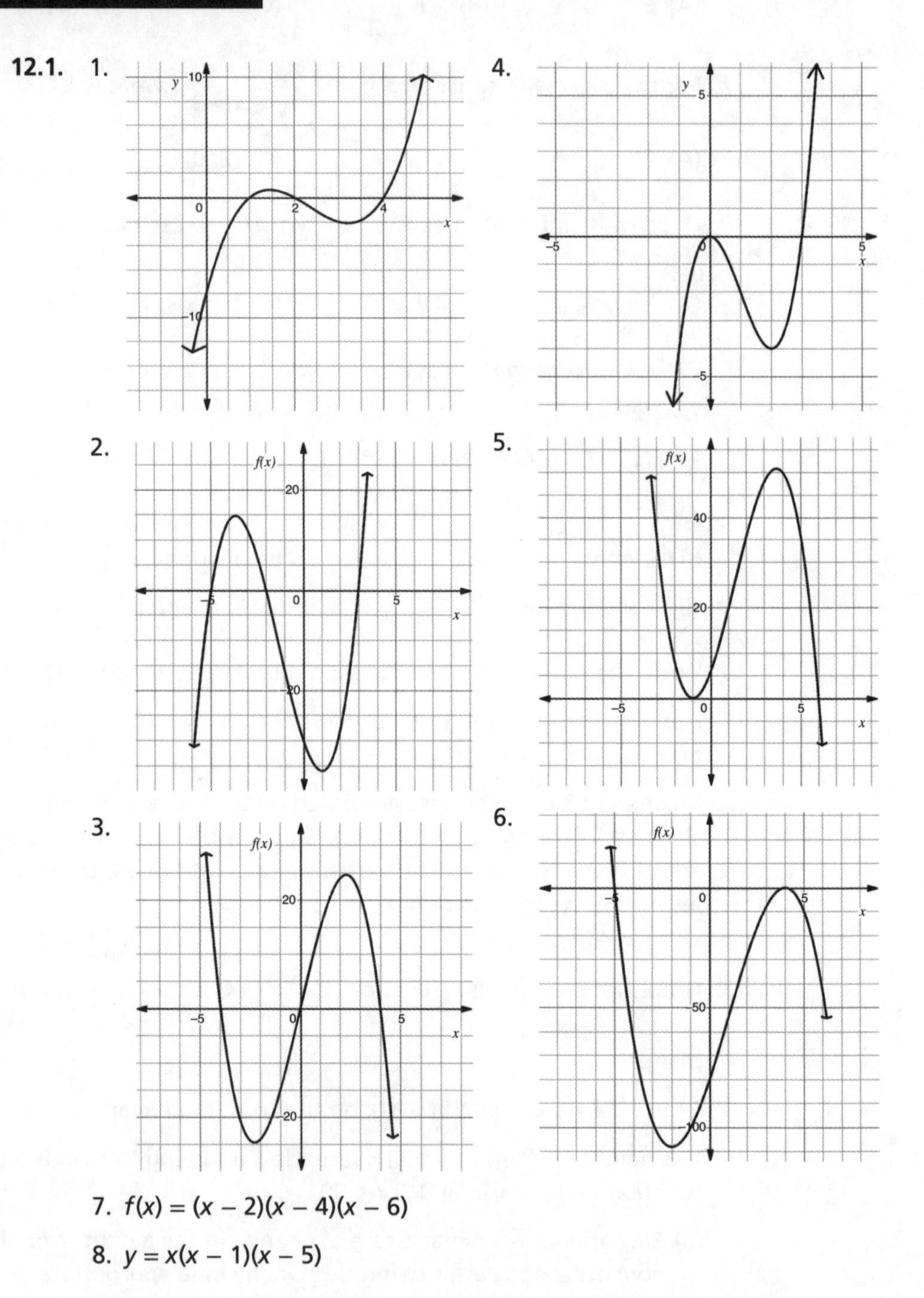

7. $f(x) = (x - 2)(x - 4)(x - 6)$
8. $y = x(x - 1)(x - 5)$

9. $y = -(x + 3)(x + 1)(x - 4)$
10. $y = (x + 2)^2(x - 8)$
11. $y = -(x - 3)(x + 4)^2$
12. $y = x(x + 5)^2$
13. Both first-degree and third-degree functions have symmetry around a point. They also have one end that points up and another that points down.
14. Since one part of the graph is above the x-axis and another is below, the graph must cross the x-axis at some point. This point is the location of a real zero.
15. Quadratic functions have an absolute minimum or maximum, but cubic functions don't. Quadratic functions have symmetry about a line, but cubic functions have symmetry about a point.
16. The inverse of a function f is the function that undoes the operation of f.
17. The inverse of f undoes the operation of f, but only for nonnegative numbers. Squaring a negative number and then taking the square root of the square results in a positive number, not the negative number we started with. For example, if we square −4, we get $(-4)^2 = 16$, but $\sqrt{(-4)^2} = \sqrt{16} = 4$, not −4.

12.2.

1. −5
2. −8
3. 7
4. 2
5. 4
6. 6
7. 6
8. 9
9. 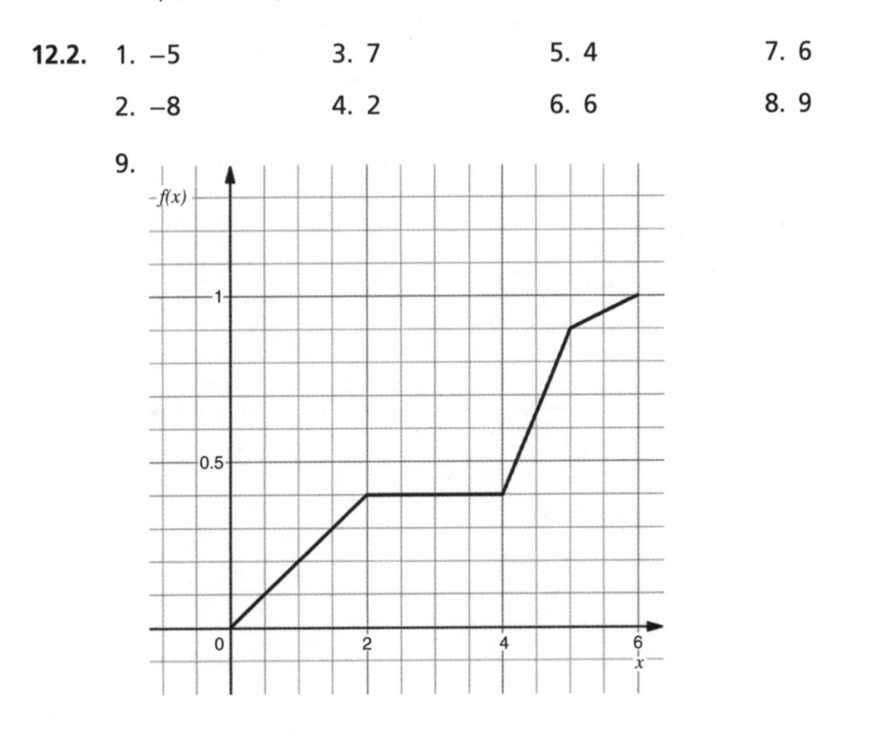

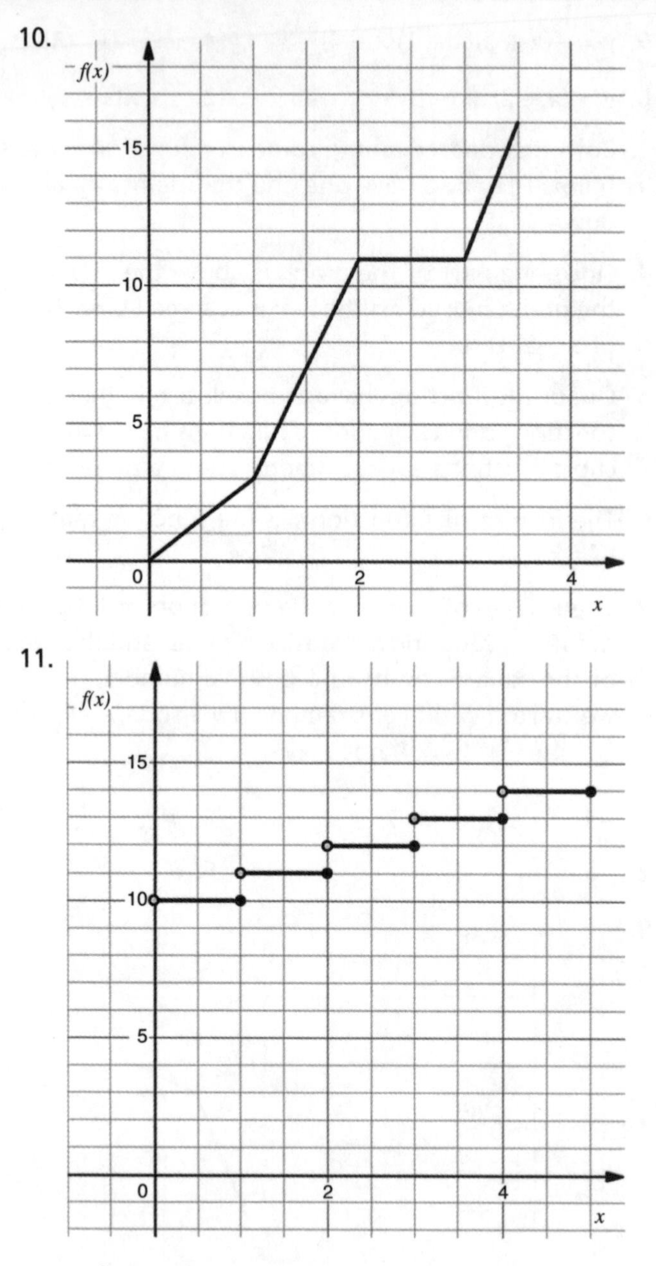

12. a. $15, b. $45

12.3.

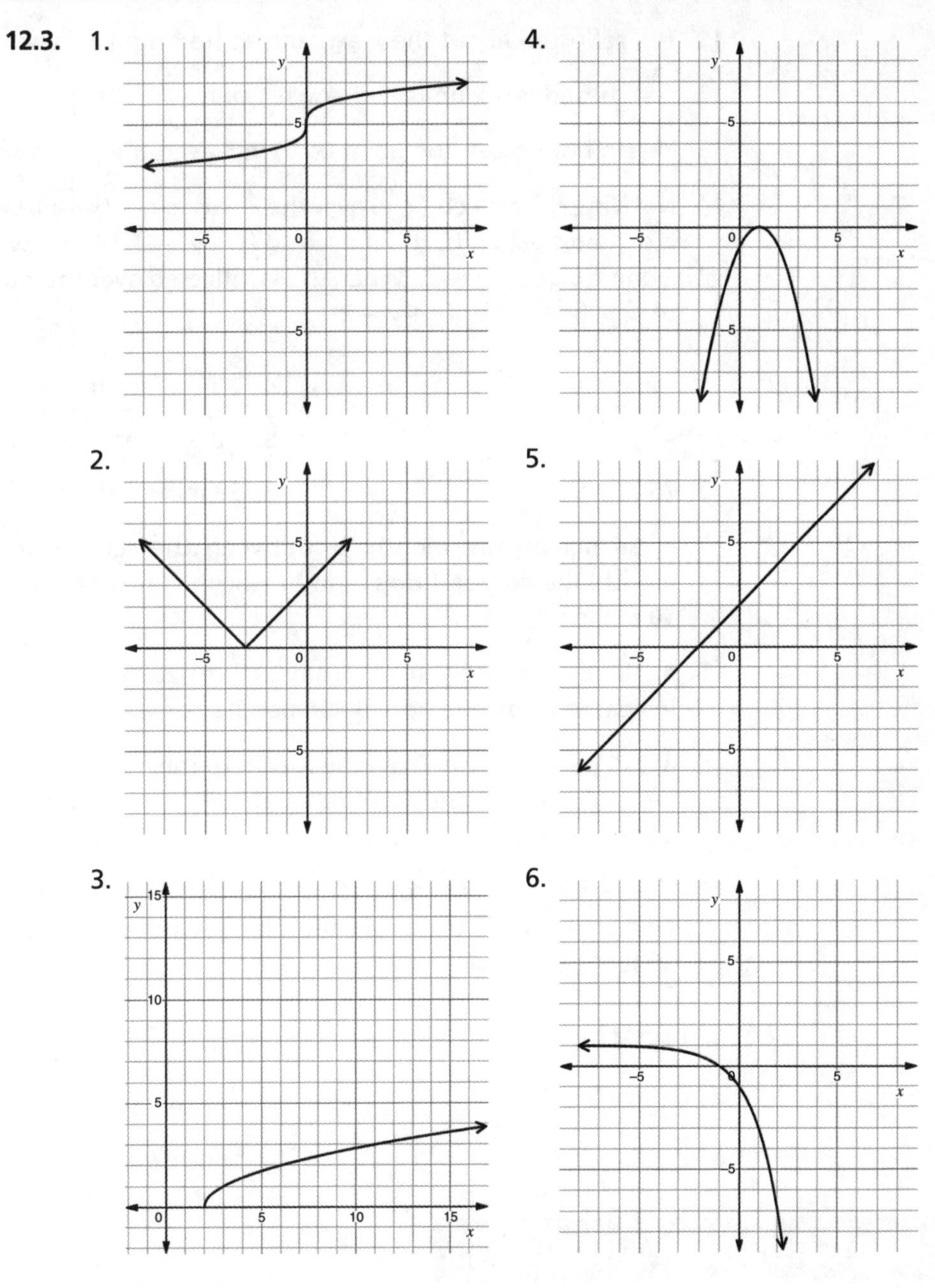

7. $f(x)$ reflected about the x-axis
8. $f(x)$ shifted right 2 units
9. $f(x)$ compressed vertically by a factor of 3
10. $f(x)$ reflected about the x-axis and shifted right 4 units

11. $f(x)$ reflected about the x-axis and shifted right 6 units

12. $f(x)$ shifted left 2 units and down 1 unit

13. Yes. In both cases, the point (x, y) moves to the point $(x - 1, y + 2)$.

14. No. Nirupika reflects (x, y) over the x-axis to get $(x, -y)$. When this point is shifted up 2 units, it is moved to $(x, -y + 2)$. Matthew shifts (x, y) up 2 units to get $(x, y + 2)$. When this is reflected over the x-axis, it is moved to $(x, -(y + 2))$, or $(x, -y - 2)$.

12.4.

1. -2
2. -2
3. 75
4. 2°F. per hour
5. -2.5% per month
6. 33 miles per hour
7. a. \$6 million per month. b. Between the second and the seventh month, the corporation's monthly income increased by an average of \$6 million.
8. a. 400 people per year. b. From 2010 to 2020, the town's population increased by an average of 400 people per year.
9. a. \$11 per month. b. From the fourth to the ninth month after reopening, the investment account's balance increased by an average of \$11 per month.
10. a. -16 feet per second. b. From the second to the fifth second after being launched, the projectile's height decreases by an average of 16 feet per second.

12.5.

1. f since $3 > 2$
2. g since $5 > 4$
3. f since $-2 < 3$
4. g since $\frac{7}{3} > 2$
5. g since $-1 < -5$
6. f since $7 > 6$

CHAPTER 12 TEST SOLUTIONS

1. (A)
2. (D)
3. (C)
4. (C)
5. (D)
6. (A)
7. (A)
8. (D)
9. (C)
10. $g(x) = -|x|$
11. f is shifted to the right 3 units and up 5 units.
12. $\$1 + \$0.25(5) = \$2.25$.

13.

14.

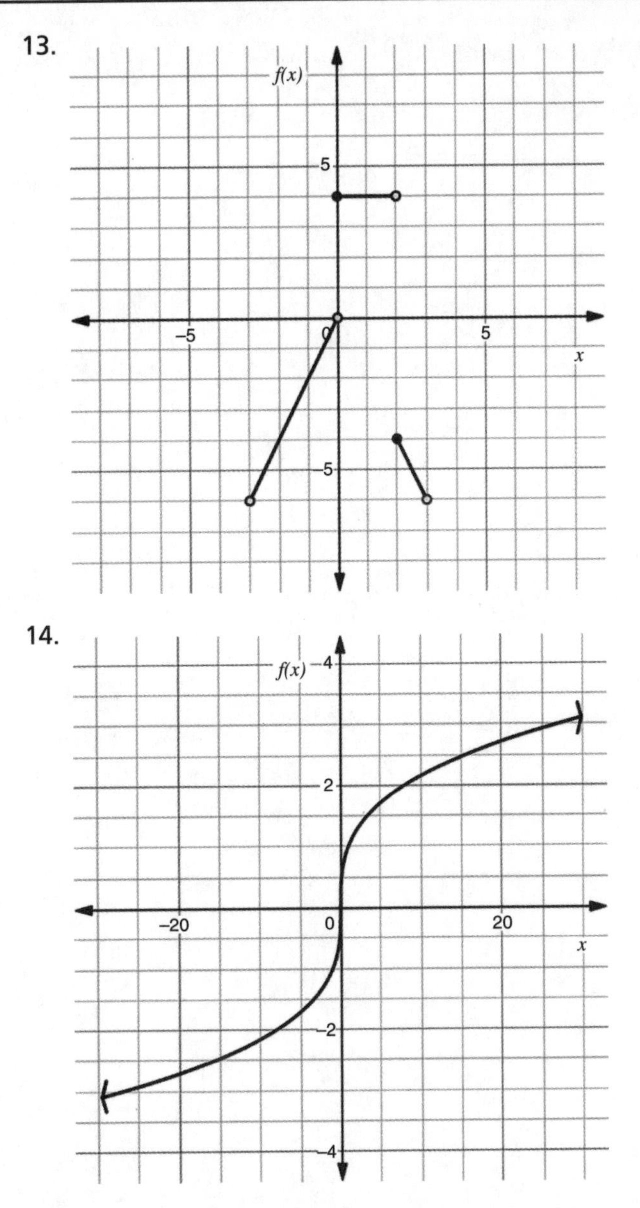

15. a. 32 feet per second. b. Between the first and fourth seconds after being thrown, the ball's height increased by an average of 32 feet per second.

13 STATISTICS

So far, we've discussed how functions can model real-world behavior. To do so, we need **data**—information that comes from observations or measurements. In this chapter, we deal with **statistics**, the science of collecting and analyzing data. We use statistics to tell a story about a **population**, which is a group of individuals, items, or events that we're interested in. Since getting information from everyone in a population is usually unrealistic, we often get information from a **sample**, which is a part of the population.

13.1 Two-Way Tables

Let's say we wanted to know more about the people who live in a small town. We might be interested in their ages, ethnicities, political affiliations, or other traits. Each characteristic that we measure on an individual is called a **variable**. Like a variable in algebra, a variable in statistics can take different values.

We start with data in which we gather *one* characteristic from each individual—what we call **univariate** (meaning one-variable) **data**. For example, we may be interested in the ethnicity of people, the type of computer used at home, or the color of a vehicle. Such variables are called **qualitative** or **categorical variables** because we assign a label (such as "Asian" or "under 18") that represents a quality or category of each individual.

If we looked at census data on the marital status of people in the town, we would see something like this:

Never Married, Never Married, Never Married, Married, Married, Married, Married, Married, Divorced, Divorced, Widowed, …

To get a better understanding of what values of the variable we can expect, we look at its **distribution**, a function whose inputs are the possible values of the variable and outputs are their **frequencies** (number of times that the values occur). In this case, the inputs are the individuals' marital status and the outputs are tallies showing how often each marital status occurs:

Never Married: ||| Married: ~~||||~~ Divorced: || Widowed: |

We write this more compactly with a **frequency table**, which shows each possible value and its corresponding frequency (Table 13.1):

Table 13.1 Frequency table.

Marital status	Frequency
Never married	3
Married	5
Divorced	2
Widowed	1

A **bar graph** enables us to represent this data visually. Each category is listed separately on the horizontal axis, and we draw rectangles whose heights represent the frequency of each category (Figure 13.1):

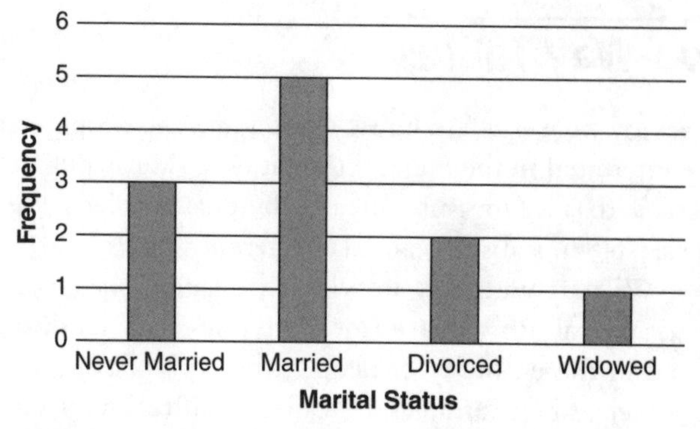

Figure 13.1 Bar graph.

Here are two important points about bar graphs:

- Changing the order in which the categories appear doesn't affect the data. However, when possible, we try to put categories in a logical order, such as "Really disliked," "Disliked," "Liked," and "Really liked."
- Put spaces between the rectangles to separate the values being compared.

We compare two variables to understand the relationship between them. In these situations, we record two characteristics from each individual (we call this information **bivariate** [meaning two-variable] **data**). For example, if we want to see if people's marital status can be used to predict whether or not they own or rent their home, our data would look something like this:

Never Married, Rent; Never Married, Rent; Never Married, Own; Married, Own; Married, Own; Married, Own; Married, Own; Married, Rent; Divorced, Rent, Divorced, Rent; Widowed, Rent; ...

Note that each label "Rent" or "Own" is linked to a marital status, so tallying each category separately loses the connection between the variables:

Never Married: 3, Married: 5, Divorced: 2, Widowed: 1

Rent: 6, Own: 5

Instead, we summarize our data with a special chart called a **two-way table** (Figure 13.2).

How to Make a Two-Way Table

1. Determine which variable is listed in the rows and which is listed in the columns. If we believe that one variable affects the other, we put the first variable in the columns (we call it the **column variable**) and the second in the rows (we call it the **row variable**). In this case, since we're interested in whether marital status affects home ownership, we list marital status in the columns and home ownership in the rows.
2. List the possible values for each variable as the row and column headers. Label the rows and columns with a title for each variable.
3. Fill in the frequencies for each cell inside the table. Since these are frequencies that occur *jointly* with two variables, we call them the **joint frequencies**.
4. Calculate the totals for each row and column and list them in the last row and last column. Since these totals are located on the *margins* of the table, we call them the **marginal frequencies**.
5. In the bottom right, put in the total number of individuals in the data set. This number should be the sum of the row totals or the sum of the column totals.

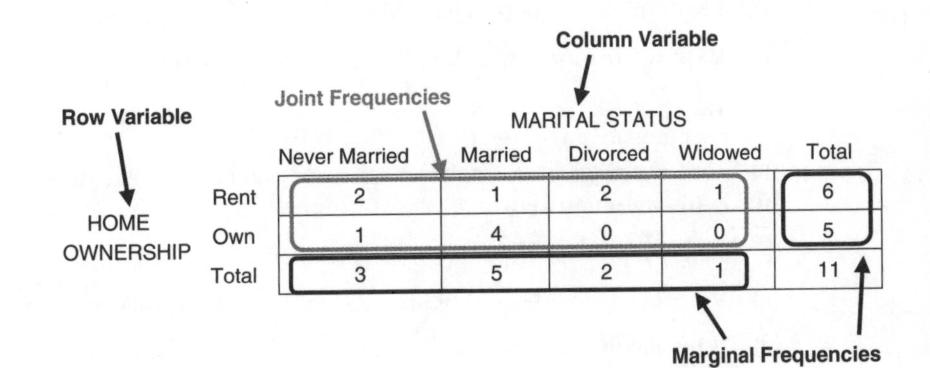

Figure 13.2 Two-way table.

We interpret each joint frequency in the two-way table by reading its corresponding row and column header (Figure 13.3):

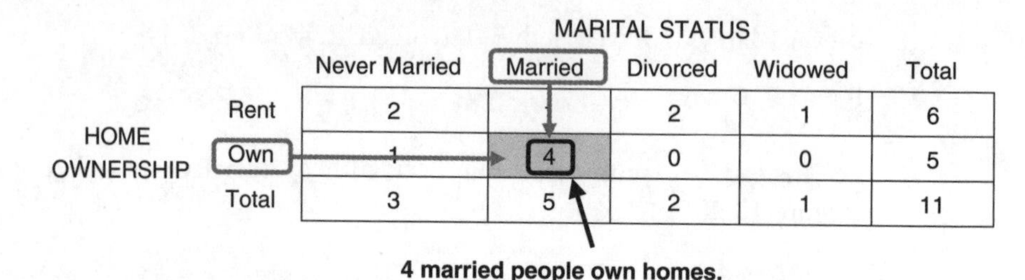

HOME OWNERSHIP	MARITAL STATUS: Never Married	Married	Divorced	Widowed	Total
Rent	2	1	2	1	6
Own	1	4	0	0	5
Total	3	5	2	1	11

Figure 13.3 Joint frequency in a two-way table.

Example 13.1 Researchers want to determine if young people are more likely than older people to use social media as their main source of news. The following two-way table shows the age groups and primary source of news for respondents.

MAIN SOURCE OF NEWS	AGE GROUP: < 18 yrs.	18–39 yrs.	40–64 yrs.	> 64 yrs.	Total
Radio/TV	4	8	14	10	36
Social media	32	16	7	3	y
Print	x	4	15	5	26
Total	38	28	36	18	120

(a) Find the value of x.

(b) Find the value of y.

(c) Explain in context what the frequency 14 represents.

(d) Explain in context what the frequency 28 represents.

Solution:

(a) Since the total in the column where x is located is 38, then $4 + 32 + x = 38$, so $36 + x = 38$, or $x = 2$.

(b) Since y represents the total in its row, then $y = 32 + 16 + 7 + 3 = 58$.

(c) 14 people aged 40–64 rely on radio or TV as their main source of news.

(d) 28 people in the survey were between 18 and 39 years old.

Example 13.2 **In a recent survey, researchers wanted to see if children's anxiety about math affects their grade. They asked a random sample of children in math classes to estimate their level of math anxiety and recorded their final math report card grades. The following two-way table summarizes the results.**

		"MATH ANXIETY" LEVEL		
		High	Medium	Low
MATH GRADE	A–B	3	22	23
	C–D	17	31	21
	F	20	8	5

(a) How many students were in the sample of children?

(b) What percentage of students in the survey failed their math class?

(c) What proportion of students who scored an A or B had high math anxiety?

(d) What proportion of students who had high math anxiety scored an A or B?

Solution:

(a) Since this question asks about the percentages of students in various categories, we determine the marginal frequencies by calculating the row and column totals:

		"MATH ANXIETY" LEVEL			
		High	Medium	Low	Total
MATH GRADE	A–B	3	22	23	48
	C–D	17	31	21	69
	F	20	8	5	33
	Total	40	61	49	150

We see from the bottom right cell in the table that 150 students were in the sample.

(b) The total in the "F" row is 33, meaning that 33 students failed math. Since there were 150 students in the sample, then $\frac{33}{150} = 0.22$, or 22%, of students in the survey failed math.

(c) The denominator is 48, the number of students who scored an A or B (we get this from the row total for "A–B." The numerator is 3, the number of students in the "A–B" row (in other words, from this 48) that had high math anxiety. The proportion is $\frac{3}{48} = 0.0625$.

(d) The denominator is 40, the number of students who had high math anxiety (we get this from the column total for "high"). The numerator is 3, the number of students in the "A–B" row (in other words, from this 40) that had high math anxiety. The proportion is $\frac{3}{40} = 0.075$.

Exercises

1. In April 1912, the ocean liner *Titanic* sank in the Atlantic Ocean after hitting an iceberg. Most of the passengers on board died. The following two-way table summarizes the ticket class and survival status of passengers.

		TICKET CLASS			
		First	Second	Third	Total
SURVIVAL STATUS	Survived	202	118	178	498
	Died	123	167	x	818
	Total	325	y	706	

(a) Find the value of x.

(b) Find the value of y.

(c) How many passengers were on the *Titanic*?

(d) Explain in context what the 167 in the table represents.

2. A national polling company investigated whether people's income level affected their expectations of retirement. They asked respondents to state what their current annual income level is and whether they expected to live comfortably when they retire. The poll results are summarized in the two-way table.

		INCOME LEVEL			
		< \$40,000	\$40,000-\$100,000	> \$100,000	Total
EXPECT TO LIVE COMFORTABLY IN RETIREMENT	Yes	42	x	26	188
	No	80	100	2	182
	Not Sure	5	4	1	10
	Total	127	224	29	

(a) Find the value of x.

(b) How many people were surveyed?

(c) Explain in context what the 42 in the table represents.

(d) How many respondents whose annual incomes were over $100,000 expect to live comfortably?

3. Researchers surveyed 530 Americans to determine if their religious beliefs could be predicted by their ages. The survey results are summarized in the following table.

		AGE GROUP (YEARS)				
		18–29	**30–49**	**50–64**	**≥ 65**	**Total**
RELIGION	Christian	45	120	140	70	375
	Other Religion	6	8	4	3	21
	Unaffiliated	36	50	36	12	134
	Total	87	178	180	85	530

(a) To the nearest tenth of a percent, what percentage of people surveyed are Christian?

(b) To the nearest tenth of a percent, what percentage of people aged 18–29 did not affiliate with any religious group?

(c) Which is greater: the percentage of people aged 18–29 who are Christians or the percentage of Christians who are aged 18–29?

4. To analyze the relationship between the type of neighborhood that people live in and the transportation they use to get to work, researchers survey a random sample of residents who work in a large state. The survey results are summarized in the accompanying two-way table. (Residents who work from home or don't work were excluded.)

		NEIGHBORHOOD TYPE		
		Urban	**Suburban**	**Rural**
Method of commuting to work	Bus/Train	30	22	6
	Car	1	95	64
	Walk	8	3	1

(a) To the nearest hundredth, determine the proportion of residents surveyed who live in a suburban neighborhood.

(b) To the nearest whole number, determine the percentage of suburban residents who commute to work by car.

13.2 Dotplots and Histograms

So far, we've worked with qualitative variables. In the real world, we don't just work with labels. We also work with numbers, such as heights, lengths, or dollars. We call such variables **quantitative variables**, which consist of numerical values acquired through counting or measuring.

As we did with qualitative data, we can create a diagram with quantitative data. The most basic display that we can create is a **dotplot**, in which each value in the data set is represented by a dot or other mark.

How to Make a Dotplot

1. Make an axis (represented by a horizontal line) and write a descriptive name of the variable below it.
2. Label the axis with appropriate values and a scale, ranging from the lowest to highest values.
3. For each value in the data set, put a dot or similar mark above the number on the axis that corresponds to the value.

Example 13.3 The following data show the number of children in 20 families in the town of Summerville: 5, 2, 1, 2, 1, 3, 0, 3, 2, 2, 3, 6, 2, 5, 6, 10, 6, 7, 0, 1.

(a) Create a dotplot of the data.

(b) What is the most common number of children in a family?

Solution:

(a)

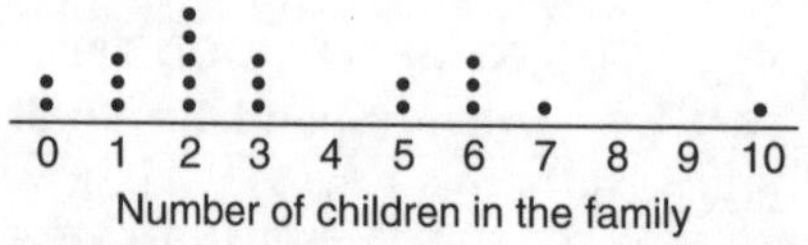

0 1 2 3 4 5 6 7 8 9 10

Number of children in the family

(b) The most common number of children is the value with the highest frequency. In this example, that value is 2.

Example 13.3 illustrates three important points about dotplots:

- Don't omit values on the horizontal axis that have a frequency of 0 (such as 4, 8, and 9 in Example 13.3). Doing so would distort the distribution.

- Don't combine values on the horizontal axis. Labeling the horizontal axis with groups like "0–1," "2–3," and so on would be incorrect.
- Don't draw or label the vertical axis. Instead, make sure that the descriptive name on the horizontal axis starts with a phrase such as "Number of…," which clearly indicates that frequencies are shown.

Some data sets have too many values or values that are spread too far apart to be shown easily on a dotplot. In these situations, we make a **histogram**, a display in which rectangles represent frequencies.

How to Make a Histogram

1. Group the data into intervals (sometimes called **classes**) of equal width with no overlap or gaps between them.
2. Determine the frequency of values in each interval. We can create a frequency table to summarize the counts.
3. Label the horizontal axis with an appropriate scale (ranging from the lowest to highest interval) and descriptive name. Include units, such as miles or thousands of dollars.
4. Label the vertical axis with an appropriate scale (ranging from the lowest to highest frequency). Write a descriptive name, such as "Frequency" or "Percentage."
5. Draw rectangles for each interval. The width of each rectangle corresponds to the width of the interval, while the height corresponds to its frequency.
6. Write the frequency of each interval above its corresponding rectangle.
7. Write a descriptive title for the histogram.

Example 13.4 **The following data show the amount of college loan debt, in thousands of dollars, of a sample of 30 students:**

40.5, 41.1, 42.2, 34.9, 34.6, 38.1, 45.2, 46.1, 37.1, 38.9, 40.2, 38.9, 39.8, 40.4, 34.4, 37.2, 36.3, 37.4, 36.7, 37.5, 44.3, 42.4, 42.4, 50.1, 36.2, 34.1, 33.6, 31.1, 32.2, 42.2

(a) Create a histogram of the data.

(b) What percentage of students in the sample have a college loan debt of between \$40,000 and \$45,000?

Solution:

(a) We start by dividing the data into classes of equal width. Although we can theoretically pick any width, we want an interval that tells us the number of students whose college debt is between 40 and 45 (representing \$40,000 and \$45,000). We see from the data that the minimum is 31.1 and the maximum is 50.1, so if we select a class width of 5 and a starting value of 35, we will include all of our data and easily find the frequency of the interval we want (from 40 to less than 45). We then create a frequency table with the intervals:

Interval	Frequency
30 to < 35	7
35 to < 40	11
40 to < 45	9
45 to < 50	2
50 to < 55	1

From the frequency table, we construct a histogram:

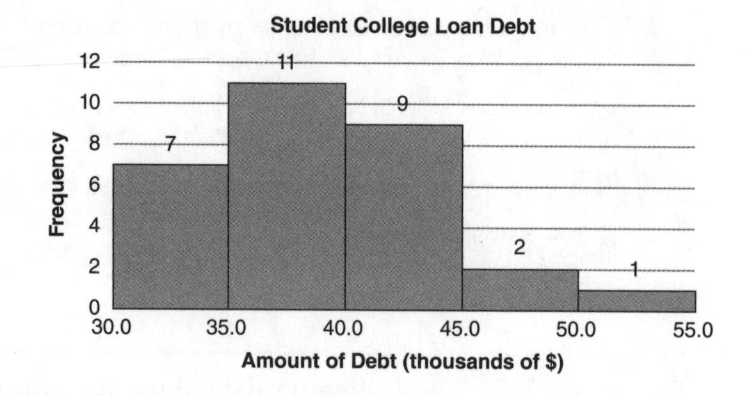

(b) From the frequency table or histogram, we see that 9 out of 30 students have a college loan debt that is between \$40,000 and \$45,000, so the percentage is $\frac{9}{30} =$ 0.3, or 30%.

Note the following about histograms:

- To avoid confusion about the boundaries of each interval, we put labels on the left side of each rectangle to indicate the beginning of each interval.

- We can think of a histogram as a dotplot in which several "columns" of dots are "compressed" into rectangles. This enables histograms to fit more data into a smaller space. However, it also means that unlike dotplots, histograms don't display individual data values. For example, the histogram in Example 13.4 tells us that 2 students had college loan debt between $45,000 and $50,000, but we don't know the exact amount of their debt without looking at the individual values.
- Most technology will allow you to create a frequency table and histogram from data that you enter. See your technology's manual for details.

Watch Out!

Determining whether to make a bar graph, dotplot, or histogram can sometimes be confusing.

- BAR GRAPH OR HISTOGRAM? Bar graphs display qualitative data with categories on the horizontal axis. To show that each category is distinct, we put gaps between the bars. In contrast, histograms display quantitative data on the horizontal axis. To avoid confusion, we don't put gaps between bars in histograms unless there is a gap in the data. Compare the bar graph and histogram in Figure 13.4:
- DOTPLOT OR HISTOGRAM? Dotplots display data sets that have relatively few values and are relatively close together. If a data set has many values or values that are far apart, use a histogram, not a dotplot.

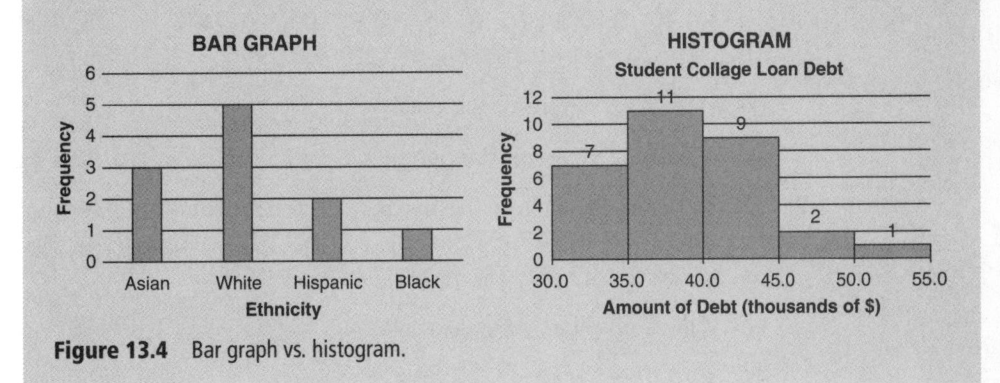

Figure 13.4 Bar graph vs. histogram.

Technology Tip

Use a graphing calculator or spreadsheet to enter data and make a frequency table or histogram. You'll probably need to adjust the interval width, which on many devices is called the bin width.

Exercises

Create the indicated display for each data set.

1. (dotplot) Number of cars owned by urban households: 4, 3, 2, 0, 3, 0, 0, 1, 0, 0, 1, 0, 0, 1, 1, 0, 1, 0, r, 2, 0, 1, 0, 0, 2, 1, 2, 0, 3

2. (histogram) Minutes that students spent doing homework last night: 0, 41, 15, 82, 25, 30, 94, 125, 27, 32, 65, 70, 75, 82, 105, 85, 90, 107, 71, 100

3. (dotplot) Gallons of water used per person per day: 80, 89, 86, 90, 86, 90, 88, 80, 83, 90, 88, 85, 88, 88, 89

4. (histogram) Pounds of food wasted per person per day: 0, 0.1, 0.3, 1.5, 1.7, 2.3, 0.6, 0.3, 1, 1.1, 1.3, 0.9, 1.3, 1.6, 2.5, 3.6, 2.3, 2.2, 1.3, 1.7, 1.6, 2.5, 1.8, 2.2, 2.5, 2.1

5. The following data show the number of books checked out by a group of people at a town public library:

   5, 1, 6, 1, 1, 2, 0, 1, 2, 1, 2, 2, 3, 2, 2, 3, 3, 6, 2, 2

   (a) Make a dotplot for the data.

   (b) What is the most frequent number of books that were checked out?

6. Researchers asked 30 people to estimate how much time (in minutes) they spend commuting to work every day. Here are the results:

   35, 40, 160, 60, 90, 0, 45, 35, 0, 100, 120, 175, 45, 75, 65, 75, 70, 85, 95, 100, 110, 20, 0, 140, 160, 80, 65, 110, 110, 95

   (a) Make a histogram of the data.

   (b) What percentage of the people have a commute of less than 1 hour?

7. The following data show the number of students in 32 high school classes:

   36, 25, 27, 35, 27, 28, 34, 31, 32, 27, 28, 32, 33, 34, 34, 34, 34, 35, 35, 35, 33, 36, 26, 26, 37, 34, 33, 35, 31, 34, 34, 33

   (a) Make a dotplot of the data.

   (b) According to the school district's contract, a class is considered oversized if it has more than 34 students. What percentage of classes in the data are oversized?

8. Here are data on the amount (in dollars) that 11 families spent on groceries last month:

   625, 674, 428, 635, 532, 480, 520, 620, 710, 660, 810

   (a) Make a histogram of the data.

   (b) How many of the families spent more than $750 last month?

Questions to Think About

9. The frequency table in the accompanying diagram shows the height, in inches, of 14 children. Explain the errors in the table.

Heights (inches)	Frequency
0–50	1
50–52	4
52–54	2
54–56	5
58–60	2

10. Giselle used a spreadsheet to graph the lengths, in centimeters, of several insects: 0.8, 1.1, 1.5, 2.3, 5.1, 5.2, 5.3. She obtains the following histogram:

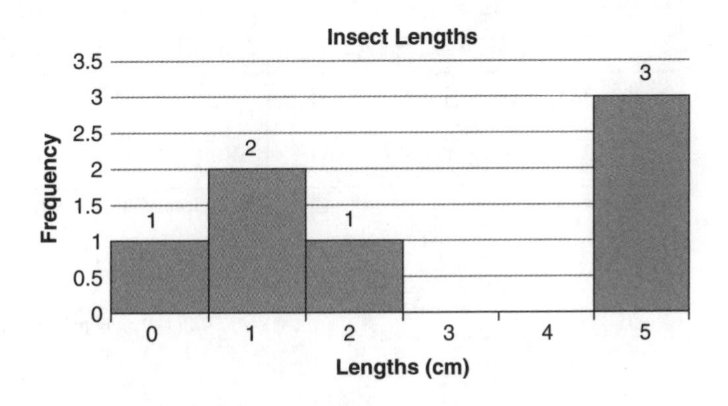

Explain the error in her histogram.

13.3 Shape, Center, and Spread

To understand data, we don't just create displays of it. We also summarize the overall patterns that exist in its distribution. For quantitative data, we describe the following:

- **Shape:** How are the values grouped? Include a description of **outliers** (values that are extremely different from others) and other unusual features.
- **Center:** What is a typical value?
- **Spread:** How far are values from the center?

To describe the shape of a distribution, we look for an overall pattern, including large peaks, clusters, or gaps in the data. Table 13.2 summarizes the measures of shape for a distribution:

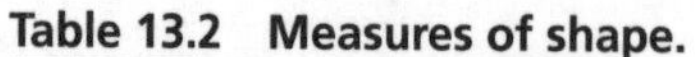

Table 13.2 Measures of shape.

Term	Display	Characteristics
Uniform Distribution		• Frequencies are approximately equal
Symmetric Distribution		• Left and right sides are approximate mirror images of each other
Skewed Left Distribution		• Lower values have lower frequencies • Higher values have higher frequencies • "Tail" of data extends to the left
Skewed Right Distribution		• Lower values have higher frequencies • Higher values have lower frequencies • "Tail" of data extends to the right
Cluster		• Group of values separated from others by a gap
Gap		• "Space" in data where values have frequency of 0
Peak		• Value that is the highest value in a cluster
Outlier		• Value that is very far from other values

Example 13.5 The following dotplot summarizes the retirement ages, in years, of a group of people. Describe the shape of the distribution.

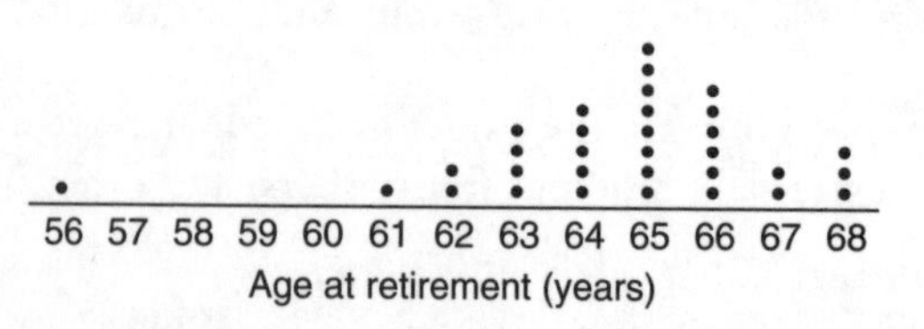

Solution: The data is generally skewed left, clustered from 61 to 68 with a peak at 65. There is an unusual value at 56.

Example 13.6 The following histogram summarizes the prices of items at a discount store. Describe the shape of the distribution.

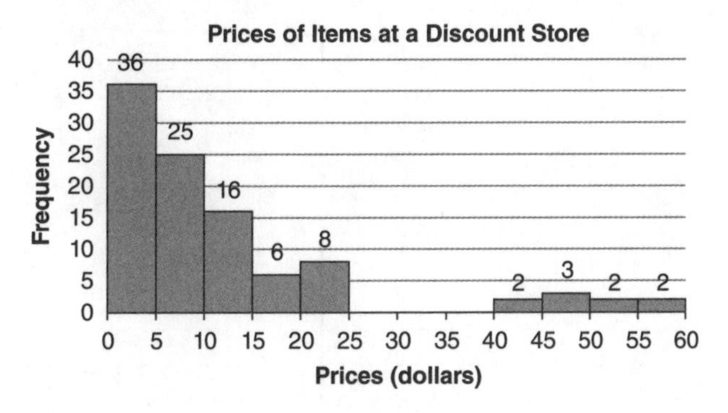

Solution: The data is generally skewed right with a peak price of less than \$5. There is a gap from \$25 to \$40 and a roughly uniform cluster of prices from \$40 to \$60.

We use several methods to describe the center of a distribution. One measure is the **mode**, which is the value(s) with the highest frequency. If several values have frequencies that are noticeably higher than the others, then we call each of those values a mode. The mode is useful when it represents most of the values, such as the data set 70, 70, 70, 70, 34, 35. However, the mode doesn't take any other value into account, so it provides only limited information. For example, the data set 20, 20, 91, 92, 94, 96, 98, 99 has a mode of 20, but this does not describe the other six values in the 90s.

A common measure of center that takes other values into account is the **mean**, or **average**. To find the mean of a set of values $x_1, x_2, \ldots, x_n$, we add the values and divide by the number of values. We use the symbol μ (the Greek letter mu) to represent the mean of a population and $\overline{x}$, pronounced "x-bar," to represent the mean of a sample. The formula for the mean of n values is $\overline{x} = \frac{x_1+x_2+\ldots+x_n}{n}$.

Did You Know?

Methods for estimation and prediction from observations date back thousands of years. Between 500 and 300 BCE, Babylonian astronomers developed a system for predicting the motions of the sun, moon, and planets. Greek (1st century CE) and Arabic (800–1000 CE) astronomers later developed methods for estimating the error from their measurements. The English word *average* (from the Arabic word *awar*, meaning "defect") originally referred to damage that occurred during a sea voyage. By the 18th century, the word average was used in Europe to refer to an equal distribution among interested parties. This helps explain our modern method for calculating the mean. Imagine that you have piles of 1, 4, 5, and 2 books. To distribute them evenly among 4 people, put all the books into one pile ($1 + 4 + 5 + 2 = 12$) and divide the total into 4 even piles ($\frac{12}{4} = 3$) (Figure 13.5):

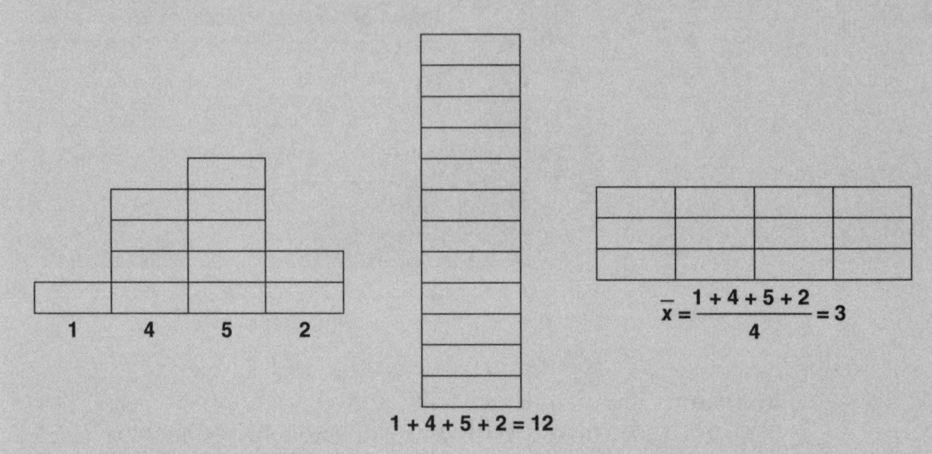

Figure 13.5 The meaning of the mean.

We find the mean of a set of values by dividing the values into groups of the same size. The mean is the number of values in each group.

Although the mean is a common way to summarize a data set, it has its limitations, as shown by Example 13.7:

Example 13.7 The annual income levels of 5 town residents are as follows: \$2,000,000; \$34,000; \$23,000; \$42,000; and \$38,000.

(a) Find the average annual income of all 5 residents.

(b) Exclude the millionaire and find the average annual income of the other residents.

Solution:

(a) The average annual income is $\frac{2{,}000{,}000+34{,}000+23{,}000+42{,}000+38{,}000}{5} = \$427{,}400$.

(b) The average annual income without the millionaire is $\frac{34{,}000+23{,}000+42{,}000+38{,}000}{4} =$ \$34,250.

Example 13.7 shows that a value that is extremely different from other values pulls the mean in its direction. In this case, the average with the millionaire's income is \$427,400, which is higher than the incomes of the other four residents. When a data set has an outlier, we can calculate the average without the outlier. Another option is to use another measure called the **median**, which is calculated as follows:

1. Arrange the values from least to greatest.
2. If the number of values is odd, then the median is the middle value.
3. If the number of values is even (so two values are in the middle), the median is the average of the middle two values.

We can think of the median as the middle number.

Example 13.8 The weights, in pounds, of six people are as follows: 30, 116, 114, 112, 30, 120.

(a) Find the mean, median, and mode of the data.

(b) Which measure of center best summarizes the data? Explain.

Solution:

(a) Mean: The mean is $\frac{30+116+114+112+30+120}{6} = 87$.

Median: Arrange the values in order: 30, 30, 112, 114, 116, 120. The number of values is even, and the two middle values are 112 and 114. Their average, $\frac{112+114}{2} = 113$, is the median.

Mode: The mode is 30 since this value appears more often than other values.

(b) The median is the best measure of center since it is closer than the mean or mode to most of the values.

Example 13.8 shows that outliers affect the median less than the mean. We say that the median is more **resistant** to outliers than the mean.

We can use displays to determine measures of center:

Example 13.9 Determine the mean, median, and mode of the data shown in the following dotplot:

0 1 2 3 4 5 6

Values

Solution: By counting the dots, we determine that the data has 12 values: 0, 0, 0, 1, 1, 1, 1, 3, 4, 5, 5 ,6.

- Mean: $\frac{0+0+0+1+1+1+1+3+4+5+5+6}{12} = 2.25$.
- Median: For a data set with 12 values, the median is the average of the 6th value (1) and the 7th value (1), which is 1.
- Mode: The value with the highest frequency is 1.

Table 13.3 summarizes the measures of center that we discussed.

Table 13.3 Measures of center.

Measure	Calculation	Notes
Mean of n values	If the values are $x_1, x_2, \ldots, x_n$, then $\bar{x} = \frac{x_1 + x_2 + \ldots + x_n}{n}$.	• Best when data has no outliers
Median of n values	1. Order values from least to greatest. 2. Determine the middle value (if n is odd) or the mean of the middle 2 values (if n is even).	• Middle number • Best when data has outliers
Mode	Value with the highest frequency.	• Best when one or more values occur much more frequently than others

To describe the spread of a distribution, we can use the **range**, defined to be the difference between the maximum and minimum values. The range gives us a quick estimate of the spread and provides information about the ends of the distribution. However, it tells us nothing about the rest of the data.

Let's look at the following example: Min has test scores of 80, 80, 90, and 90. Juan has test scores of 80, 84, 86, and 90. Based on their test scores, are they the same kind of student? They both have a mean score of 85 and a range of $90 - 80 = 10$. However, Min's scores vary more than Juan's. We need a more precise method of measuring the spread.

We do this by finding the **standard deviation**, which measures the typical distance of values from the mean. The larger the standard deviation, the greater the spread of the data. We combine the standard deviation with the mean to describe spread. For example, a data set with a mean of 40 and a standard deviation of 2 can be interpreted as follows: values typically differ from the mean of 40 by 2.

To calculate the standard deviation, do the following:

1. For each value, find its **deviation** (the difference between the value and the mean of the data, or $x - \overline{x}$ in symbols).
2. Square each deviation. (We square the deviations because we're only concerned with how far the values are from the mean, not whether they're positive or negative.)
3. Find the average of the squared deviations.
4. Find the square root of the average. (We take the square root of the average to convert the answer from squared units to our original units.)

Watch Out!

The calculation for the standard deviation differs slightly depending on whether our data comes from a population or a sample. We determine this based on the context of the problem. If we're using data to generalize for a larger group, we use the sample standard deviation (represented by the symbol s or s_x) and divide by $n - 1$ (where n is the sample size). If we have data from every person or event of interest, we use the population standard deviation (represented by the Greek letter σ, read as "sigma") and divide by n (the population size). When using technology, make sure you select the correct measure: s for a sample or σ for a population. Since we rarely get data from an entire population, we usually use the sample standard deviation.

We calculate the population standard deviation for Min's and Juan's scores as follows:

- Min's standard deviation: $\sqrt{\frac{(80-85)^2+(80-85)^2+(90-85)^2+(90-85)^2}{4}} = 5.$
- Juan's standard deviation: $\sqrt{\frac{(80-85)^2+(84-85)^2+(86-85)^2+(90-85)^2}{4}} \approx 3.6.$

Example 13.10 The high temperatures (in degrees Fahrenheit) in St. Louis, Missouri, for 8 days in July were: 84, 90, 91, 93, 97, 84, 91, 89. The high temperatures in St. Louis for 8 days in August were: 88, 82, 75, 79, 90, 82, 90, 97.

(a) Use technology to calculate the mean and sample standard deviation for each data set to the nearest tenth.

(b) Interpret each standard deviation in context.

(c) Which month had less variation in high temperatures? Explain.

Solution:

(a) Enter each data set into a separate list on your device. Use appropriate statistical functions to calculate the mean $\overline{x}$ and sample standard deviation s_x. For July, $\overline{x} \approx 89.9$ and $s_x \approx 4.4$. For August, $\overline{x} \approx 85.4$ and $s_x \approx 7.1$.

(b) The high temperatures in July typically varied from the mean of 89.9° by 4.4°. The high temperatures in August typically varied from the mean of 85.4° by 7.1°.

(c) July had less variation in high temperatures since its standard deviation is lower than August's.

Example 13.11 The following table summarizes the monthly rainfall (in inches) in three cities:

City	$\overline{x}$ (inches)	s_x (inches)
A	3.31	1.21
B	5.14	2.73
C	4.67	4.32

(a) Which city has the greatest variation in monthly rainfall? Explain.

(b) Which city typically gets the most rainfall? Explain.

Solution:

(a) City C has the greatest variation since it has the highest standard deviation (4.32 inches) of all 3 cities.

(b) City B gets the most rainfall since it has the highest mean (5.14 inches).

Another way to express spread is based on **percentiles**, which describe the position of a value relative to other values. The *p*th percentile is the value just above *p*% of the data. Percentiles divide data into 100 groups with approximately 1% of the values in each group. For example, if the 90th percentile of home prices in a county is $300,000, then a house that costs $300,000 is more expensive than 90% of homes in the county and cheaper than 10% of homes. Note that the number in the percentile says nothing about the actual value—the 90th percentile of home prices isn't $90!

We use percentiles to describe how an *individual* value relates to others in a data set. To describe the spread of an *entire* data set, we use the following five numbers (called a **five-number summary**), listed in this order:

1. Minimum
2. 1st quartile (Q_1, pronounced "Q-one"): 25th percentile

3. 2nd quartile (Q2, pronounced "Q-two"): 50th percentile
4. 3rd quartile (Q3, pronounced "Q-three"): 75th percentile
5. Maximum

Here are some important notes about five-number summaries:

- Q_1 is the middle of the first half of the data, and Q_3 is the middle of the second half of the data.
- The five-number summary shows the lower end (minimum), upper end (maximum), and middle (median). It also shows the location of the middle half of the data. This interval between Q_3 and Q_1 (in other words, $Q_3 - Q_1$) is known as the **interquartile range** (or IQR).
- We use the five-number summary to determine if an extreme value is an outlier. Any value that is less than $Q_1 - 1.5 \cdot \text{IQR}$ or greater than $Q_3 + 1.5 \cdot \text{IQR}$ is considered an outlier.

We display a five-number summary by creating a **box-and-whisker plot** (or **boxplot** for short). To create a boxplot, do the following:

1. Calculate a five-number summary for the data.
2. Draw a number line that ranges from the minimum to the maximum.
3. Determine if any extreme values are outliers. Indicate outliers with special marks (such as dots or asterisks) above their locations.
4. Draw five small vertical lines above the locations of the non-outlier minimum, Q_1, Q_2, Q_3, and non-outlier maximum.
5. Draw horizontal lines connecting the tops and bottoms of the vertical lines at Q_1, Q_2, and Q_3 to make a box (this shows the IQR).
6. Draw horizontal lines connecting Q_1 to the non-outlier minimum and Q_3 to the non-outlier maximum (this shows the whiskers).

Example 13.12 Construct a boxplot for the following data, which shows the ages (in years) of 15 teachers at a junior high school: 25, 28, 29, 26, 39, 27, 35, 36, 28, 33, 40, 34, 30, 32, 34.

Solution: Arrange the values in numerical order from least to greatest and construct a five-number summary:

- The minimum is 25.
- For 15 values, the median is the 8th value, which is 32.

- Q_1 (the median of the first half of the data) = 28.
- Q_3 (the median of the second half) = 35.
- The maximum is 40.

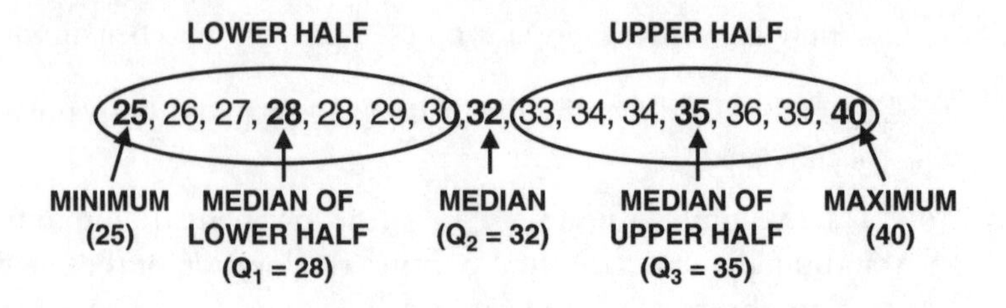

Next, we check for outliers on the left and right sides of the data:

- $IQR = Q_3 - Q_1 = 35 - 28 = 7$
- (Left) Values less than $Q_1 - 1.5 \cdot IQR = 28 - 1.5(7) = 17.5$: none
- (Right) Values greater than $Q_3 + 1.5 \cdot IQR = 35 + 1.5(7) = 45.5$: none

Finally, we construct a boxplot:

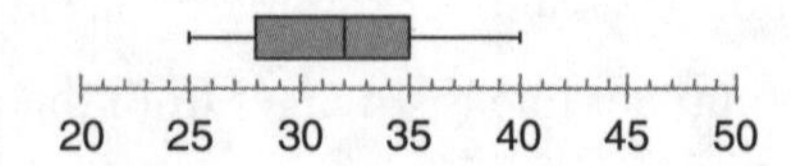

Example 13.13 Latesha uses a spreadsheet to create a boxplot for a data set listing the amount of mercury, measured in parts per million (ppm), contained in a sample of 15 fish. The data set, five-number summary, and boxplot from her spreadsheet are listed below:

0, 0.01, 0.02, 0.05, 0.08, 0.09, 0.11, 0.14, 0.15, 0.18, 0.24, 0.35, 0.37, 0.49, 1.12

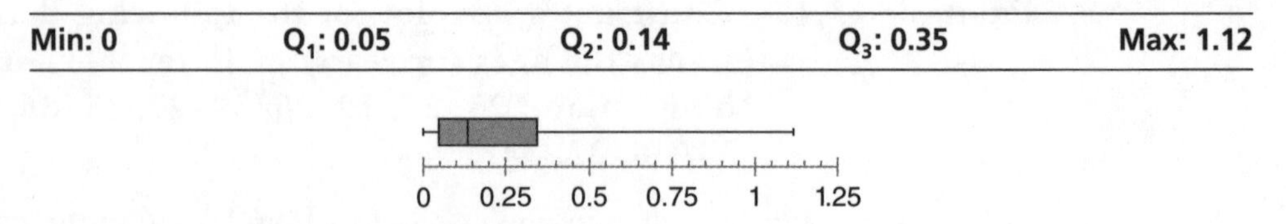

Min: 0	Q_1: 0.05	Q_2: 0.14	Q_3: 0.35	Max: 1.12

(a) Determine if there are any outliers in the data.

(b) Explain the error in the spreadsheet's boxplot.

(c) Construct a corrected boxplot.

Solution:

(a) Check for outliers on the left and right sides:

- $IQR = Q_3 - Q_1 = 0.35 - 0.05 = 0.30$
- (Left) Values less than $Q_1 - 1.5 \cdot IQR = 0.05 - 1.5(0.30) = -0.4$: none
- (Right) Values greater than $Q_3 + 1.5 \cdot IQR = 0.35 + 1.5(0.30) = 0.8$: 1.12 is an outlier.

(b) The spreadsheet's boxplot doesn't show that 1.12 is an outlier. The right whisker should end at the value before 1.12, which is 0.49.

(c) We draw a circle at 1.12 to represent the outlier and end the right whisker at 0.49.

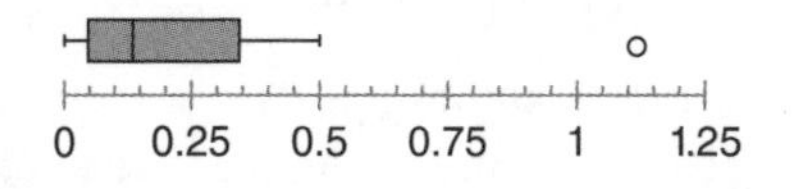

To compare distributions, we create boxplots for them and put them side by side (either horizontally or vertically), as shown in the following example:

Example 13.14 The following boxplots show the number of points scored by the winning team in a large sample of college basketball games (the data are divided into two groups, labeled Group A and Group B):

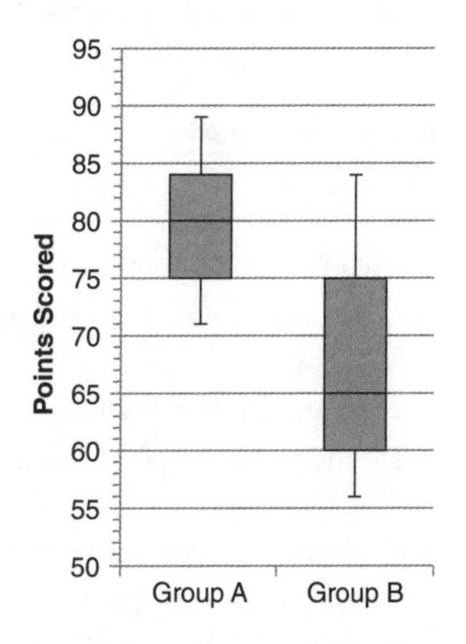

(a) Determine the five-number summary for each group.

(b) Which group has a greater IQR?

(c) What, if anything, can be concluded about the shape of the distribution in Group A? Explain.

(d) What, if anything, can be concluded about the shape of the distribution in Group B? Explain.

Solution:

(a) For Group A: minimum = 71, $Q_1 = 75$, median = 80, $Q_3 = 84$, and maximum = 89. For Group B: minimum = 56, $Q_1 = 60$, median = 65, $Q_3 = 75$, and maximum = 84.

(b) The IQR for Group A is $Q_3 - Q_1 = 84 - 75 = 9$. The IQR for Group B is $75 - 60 = 15$. Group B has a greater IQR.

(c) Although the boxplot appears symmetric, we can't conclude that the distribution is symmetric. We only know that each quartile (each 25% of the data) is spaced out over intervals of approximately equal width. Each quartile could contain different peaks or clusters.

(d) The lower 50% of the data has a range of 9, while the upper 50% of the data has a range of 19. Since the upper half is more spread out, we can conclude that the data is skewed right.

Example 13.14 illustrates that boxplots can show skewness in data, but they can't be used to determine if a data set is symmetric.

Table 13.4 summarizes the measures of spread of a distribution.

Table 13.4 Measures of spread.

Measure	Calculation	Notes
Range	Maximum – minimum	• Always nonnegative
Standard deviation	**1.** For each value, find its deviation $(x - \bar{x})$. **2.** Square each deviation. **3.** Find the average of the squared deviations. **4.** Find the square root of the average.	• Measures the typical distance of values from the mean
Percentile	The *p*th percentile is the value just above *p*% of the data.	• Measures the position of a value relative to other values

Table 13.4 (*Continued*)

Measure	Calculation	Notes
Five-number summary	1. Minimum 2. 1st quartile (Q_1): 25th percentile 3. 2nd quartile (Q_2): 50th percentile 4. 3rd quartile (Q_3): 75th percentile 5. Maximum	• Shows distribution of data divided approximately into quarters • Interquartile range (IQR) = $Q_3 - Q_1$ • Outliers: values less than $Q_1 - 1.5 \cdot \text{IQR}$ or greater than $Q_3 + 1.5 \cdot \text{IQR}$

Technology Tip

Calculating measures of center and spread by hand can get complicated. Most devices enable you to enter the values into a list and use a statistical function such as mean(), stdev(), or range() to determine these measures directly. These devices can also generate boxplots, but they may not determine if a value is an outlier.

Exercises

1. Describe the shape, center (by finding the median), and spread (by finding the range) of the distribution shown in the dotplot below:

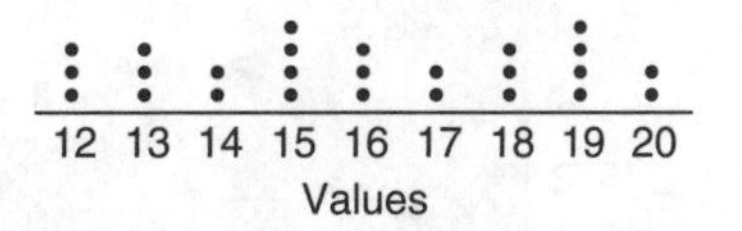

2. The accompanying histogram shows the distribution of a set of 36 values.

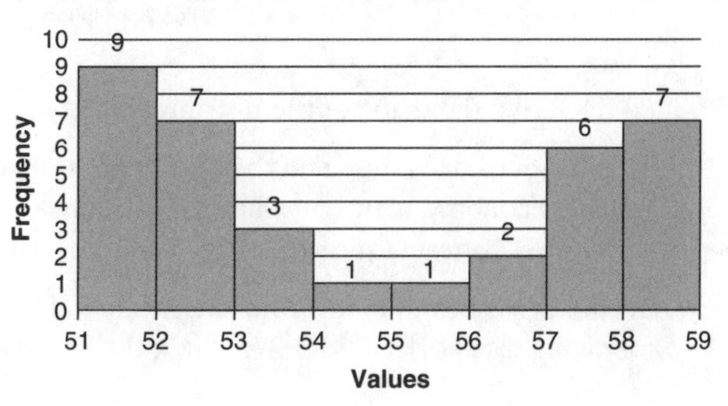

(a) Describe the shape of the distribution.

(b) Find the interval that contains the median.

(c) Find the interval that contains the mode.

3. The accompanying dotplot shows the amount of time, in minutes, that students spent taking a quiz.

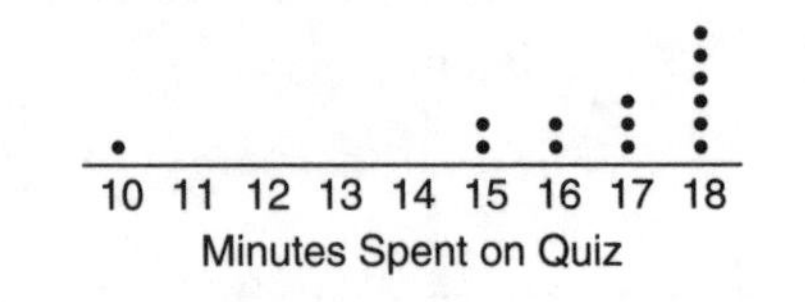

(a) Describe the shape of the distribution.

(b) Find the mean and median of the data.

(c) Which measure of center is more affected by the unusual value at 10 minutes—the mean or the median? Explain.

4. The fuel economy of 30 cars, measured in miles per gallon (mpg), is shown in the accompanying histogram.

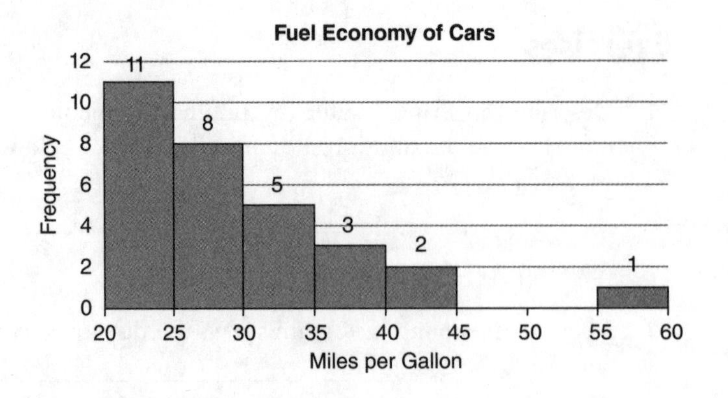

(a) Describe the shape of the distribution.

(b) One of the cars shown in the histogram is a hybrid vehicle, which has a higher fuel economy than conventional cars since it runs on both gasoline and an electric battery. Estimate its fuel economy. Explain your reasoning.

5. To the nearest hundredth, find the mean and sample standard deviation for the following data: 101.68, 103.27, 98.67, 99.34, 104.12, 105.02, 92.06.

6. To the nearest tenth, find the mean and standard deviation of the sample data shown in the accompanying dotplot.

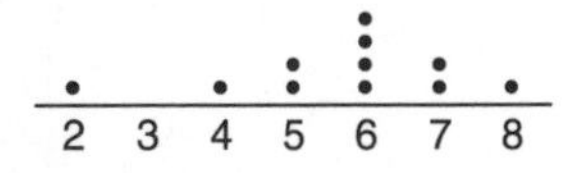

7. The following data show the time, in minutes, that a sample of 12 patients waited in hospital emergency rooms before receiving treatment: 75, 25, 90, 120, 110, 95, 85, 105, 75, 97, 88, 102.
   (a) To the nearest tenth of a minute, calculate the mean and standard deviation of the data.
   (b) Interpret the standard deviation in context.

8. The battery life, in hours, of a sample of mobile phones sold by two different companies is listed below:

   Company A: 12.4, 14.5, 13.8, 13.6, 12.9, 13.2, 14.1

   Company B: 11.8, 14.2, 12.5, 16.1, 16.2, 11.9, 11.1

   (a) To the nearest tenth of an hour, calculate the mean and standard deviation for each data set.
   (b) Which company's phones have more variability in their battery life? Explain.

9. Create a five-number summary and boxplot for the following data: 2.4, 2.8, 2.9, 3.0, 4.4, 4.7, 5.1, 5.1, 5.2, 6.8, 7.2, 7.5, 7.9.

10. Create a five-number summary and boxplot for the following data: 120, 123, 124, 122, 125, 128, 122, 149, 128, 127, 129, 130, 119, 130, 131. Make sure that the outlier is clearly indicated in the boxplot.

11. Researchers conducted a study in which they measured the amount of time (in milliseconds) that people took to press a button after seeing an object. The accompanying boxplot summarizes the data:

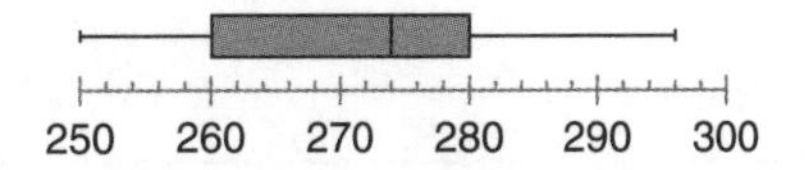

   (a) Determine the five-number summary for the data.
   (b) Determine the interquartile range for the data. Show your work.

12. The following boxplots show the distribution of scores for Ms. Lee's and Mr. Kaur's algebra classes, who took the same test on the same day.

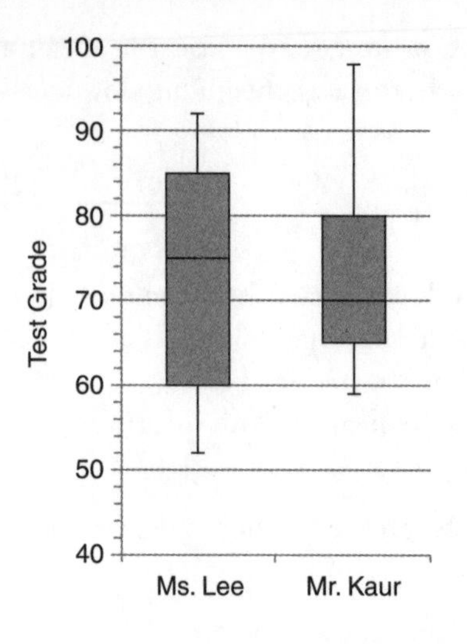

(a) Which class had the higher median?

(b) Which class had a greater interquartile range?

(c) Ms. Lee believes that her students did better on the test. What evidence can she use to support her claim?

(d) Mr. Kaur believes that his students did better on the test. What evidence can he use to support his claim?

Questions to Think About

13. Explain why a boxplot cannot be used to determine if a data set is symmetric.

14. What information about a data set can be easily determined from a dotplot but not from a boxplot?

15. What information about a data set can be easily determined from a boxplot but not from a dotplot?

13.4 Scatterplots and Regression

Just as we compare two qualitative variables, we often compare two quantitative variables. We may want to see if students' math grades affect science grades or if the number of open checkout lines tells us anything about the wait time for shoppers. Instead of creating a two-way table, as we did with qualitative data, we plot each ordered pair of measurements on the coordinate plane so we can graph and analyze the data.

For example, imagine that a teacher wanted to see if the amount of studying is related to their test grades. The teacher asked 10 students to record the number of hours that they spent studying for a test. The teacher then recorded each student's grades on that test. Table 13.5 shows the results:

Table 13.5 Hours studied vs. test grade.

Hours spent studying	0	1	2	4	5	5	6	6	7	8
Test grade	30	40	52	48	65	69	68	90	92	94

When we plot the data on the coordinate plane, we get Figure 13.6:

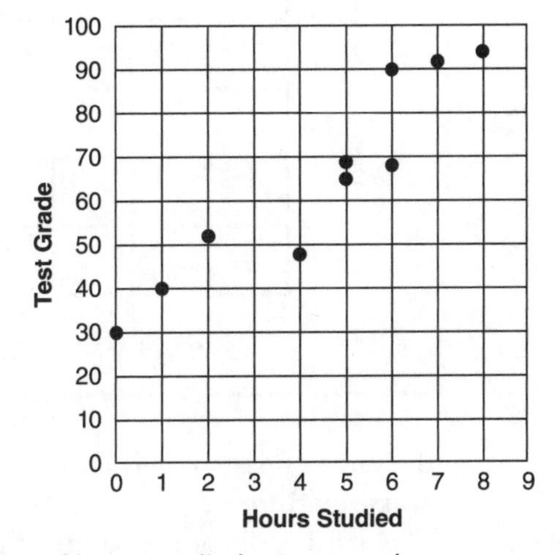

Figure 13.6 Scatterplot of hours studied vs. test grade.

Figure 13.6 is an example of a **scatterplot**, which is a graph of ordered pairs representing two quantitative variables. We put the **response variable** on the vertical axis since it measures the effect of changes in the **explanatory variable**, which is plotted on the horizontal axis. In this case, we want to see whether changes in the number of hours studied (explanatory variable) can be used to predict test grades (response variable), so we plot the hours studied on the horizontal axis and test grade on the vertical axis. Each point represents two measurements on the same individual. If all of the points have nonnegative coordinates, we restrict the scatterplot to Quadrant I.

Example 13.15 Create a scatterplot for the following data, which was gathered to determine the effect of a car's weight on its fuel efficiency.

Weight (thousands of pounds)	3.4	3.9	3.6	2.9	4.4	2.8	5.1	3.3	3	2.8
Fuel efficiency (miles per gallon)	28	25	24	28	21	27	19	30.5	31	32

Solution: First, we determine which variable is the explanatory variable and which is the response variable. Since we're trying to determine the effect of car weight on fuel efficiency, car weight is the explanatory variable (which goes on the horizontal axis) and fuel efficiency is the response variable (which goes on the vertical axis).

Next, we label each axis, write an appropriate scale, and plot each point.

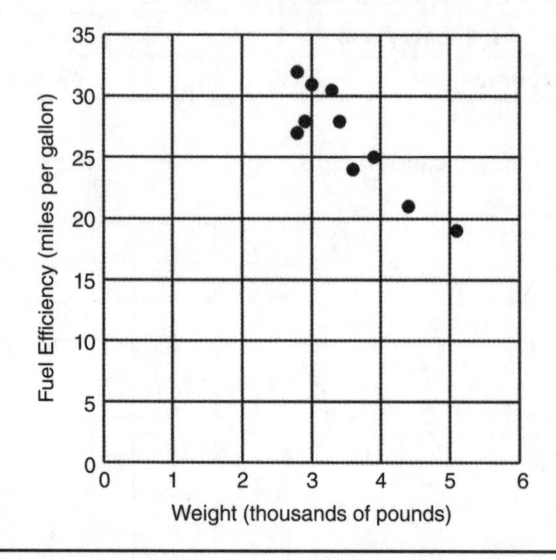

If knowing the value of one variable enables us to predict the value of another, we say that there is an **association** between the two variables. We summarize an association for quantitative data by describing the **form** of the scatterplot, which is the type of function that generally matches the pattern. The most common patterns that you'll see are linear, quadratic, and exponential, as seen in Table 13.6.

Table 13.6 Form of a bivariate relationship.

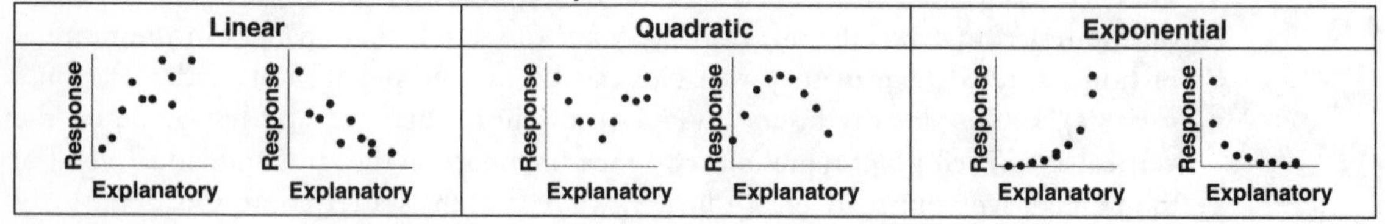

Note the major characteristics of each form:

- Linear: rate of change is approximately constant (generally increasing or generally decreasing)
- Quadratic: decreases and then increases, or increases and then decreases

- Exponential: generally increasing or generally decreasing, but at an increasing rate (for an increasing exponential function) or at a decreasing rate (for a decreasing exponential function)

If an association is linear, we use two additional characteristics to describe it:

- **Direction:** In what direction does the pattern move from left to right on the scatterplot?
- **Strength:** How closely does the pattern match a linear form?

To describe the direction of a linear relationship, we use the terminology in Table 13.7.

Table 13.7 Direction of a bivariate relationship.

Positive Association	Negative Association	No Association
Response / Explanatory	Response / Explanatory	Response / Explanatory
As the explanatory variable increases, the response variable generally increases.	As the explanatory variable increases, the response variable generally decreases.	As the explanatory variable increases, the response variable is generally not affected.

Here are some important points about the direction of a relationship:

- When describing the direction of an association, include a word like "generally," which indicates that the points do not fall exactly on a line. Don't say something like "As the explanatory variable increases, the response variable increases."
- A scatterplot showing a *positive* association has points that slope upward from left to right, like a line with *positive* slope. A scatterplot showing a *negative* association has points that slope downward from left to right, like a line with *negative* slope.

To describe the strength of a linear relationship, we use a quantitative measure called the **correlation coefficient**, which we indicate with the variable r (Figure 13.7). It has the following characteristics:

- It summarizes both the strength and direction of data.
- It is used only for bivariate quantitative data that have a *linear* association and no outliers. (To determine if calculating correlation is appropriate, look at a scatterplot of the data.)
  - If $r < 0$, then the data has a negative association.
  - If $r > 0$, then the data has a positive association.

- It ranges from −1 to 1.
- We use correlation to characterize an association as strong, moderate, or weak.
  - The closer r is to −1 or 1, the stronger the association.
  - The closer r is to 0, the weaker the association.
  - The exact ranges of strong, moderate, and weak associations vary by field. In this book, we define a moderate association as one in which $0.3 < r < 0.7$ or $-0.7 < r < -0.3$ and a strong association as one in which $0.7 \leq r \leq 1$ or $-1 \leq r \leq -0.7$.
- The closer the points are to a straight line, the closer r is to −1 or 1.
- Since the formula for calculating correlation is complicated, we recommend using technology to calculate r.

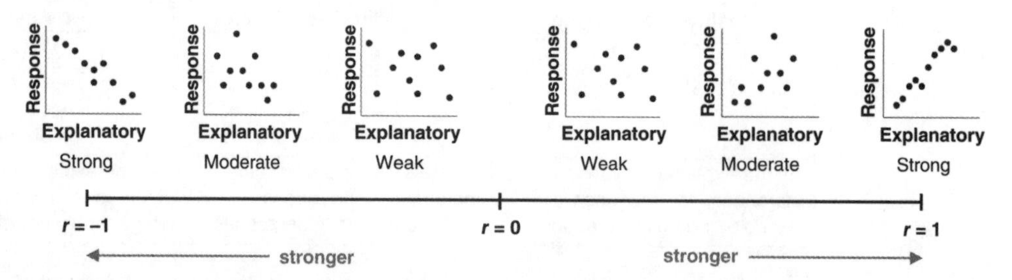

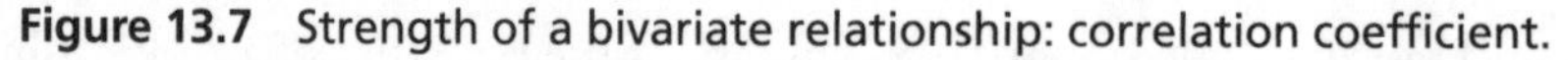

Figure 13.7 Strength of a bivariate relationship: correlation coefficient.

Watch Out!

The words association and correlation have different meanings. *Association* is a general term that refers to a relationship (which may not be linear) between two variables (which may not be quantitative). *Correlation* refers to a specific numerical calculation that measures the strength and direction of a *linear* relationship between two *quantitative* variables. You should only use the word correlation when referring to the number represented by r. Don't say, "Strong correlation"—say, "Strong association" instead.

Example 13.16 The following table represents bivariate quantitative data.

x	32	39	51	60	65	72	78	81
y	11	11	15	14	20	25	20	20

(a) Create a scatterplot of the data.

(b) If appropriate, calculate the correlation coefficient to the nearest hundredth and use it to interpret the strength and direction of the association.

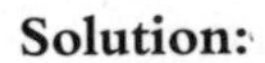

Solution:

(a)

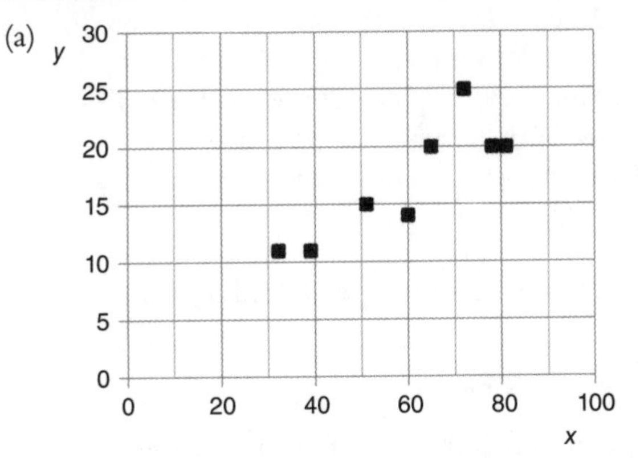

(b) We see from the scatterplot that the form is generally linear with no outliers. Thus, calculating correlation is appropriate. Using technology, we see that $r \approx 0.86$, which indicates a strong positive association between the two variables.

Example 13.17 The following table represents bivariate quantitative data.

x	13	13	17	23	25	30	38	40	48	53
y	3	1	5	6	8	8	9	8	4	2

(a) Create a scatterplot of the data.

(b) If appropriate, calculate the correlation coefficient to the nearest hundredth, and use it to interpret the strength and direction of the association.

Solution:

(a)

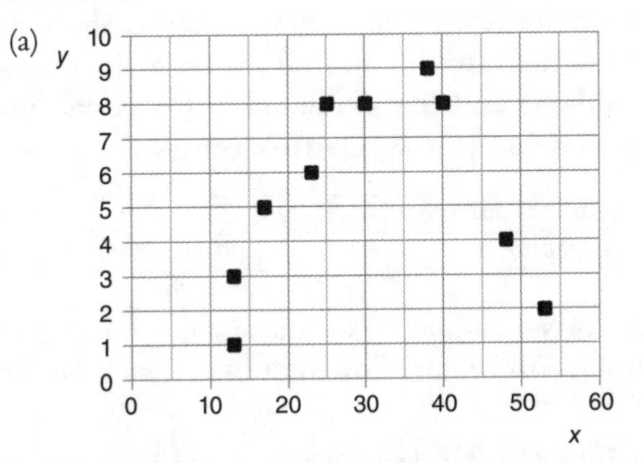

(b) We see from the scatterplot that the form is not linear. Thus, calculating correlation is not appropriate.

So far, we have used scatterplots and correlation to understand the relationship between two variables. When we want to use one variable to predict the other, we need a more precise model—a line that minimizes the distances between data points and the line. For data with a linear association, we call this the **line of best fit, least-squares regression line**, or **regression line.**

The equation of the regression line (sometimes called the **linear regression equation**) has the format $\hat{y} = a + bx$, where

- $\hat{y}$ (pronounced "y-hat") represents the **predicted y-value** (the output value predicted by the model).
- a represents the y-intercept, which is the predicted y-value when $x = 0$.
- b represents the slope, which is the rate of change of the model. We interpret the slope as follows: for every increase of 1 unit in x, the model predicts a change of b units in y.

Watch Out!

The regression equation $\hat{y} = a + bx$ is similar to $y = mx + b$, the equation of a line that we discussed in Chapter 7. Unfortunately, statisticians (and many devices!) use different symbols—they use b to represent the slope, not the y-intercept, and write the y-intercept *before* the slope. Remember that in a linear equation, the coefficient of x is the slope, no matter what term appears first or what variables are used! Also, you should use the symbol $\hat{y}$, not y, for the response variable to clearly indicate that it represents a *predicted* value.

We use technology to find the regression equations, as shown in Examples 13.18, 13.19, and 13.20.

Example 13.18 To study the effect of poverty on rural life expectancy, researchers gathered the following data, which show the annual income (in thousands of dollars) and life expectancy (in years) of 8 individuals in Williams County.

Annual Income (thousands of dollars)	25	20	30	100	60	45	70	80
Life Expectancy (years)	64	62	70	78	72	70	73	75

(a) Write a linear regression equation for this data, rounding all values to the nearest thousandth. Make sure you state what each variable represents in context.

(b) Explain in context what the slope of the regression equation represents.

(c) Explain in context what the y-intercept of the regression equation represents.

(d) Use the regression equation to predict, to the nearest year, the life expectancy of a person whose annual income is \$90,000.

(e) Can the regression equation be used to predict the life expectancy of a person whose annual income is \$500,000? Explain.

Solution:

(a) First, we create a scatterplot of the data to see the overall pattern:

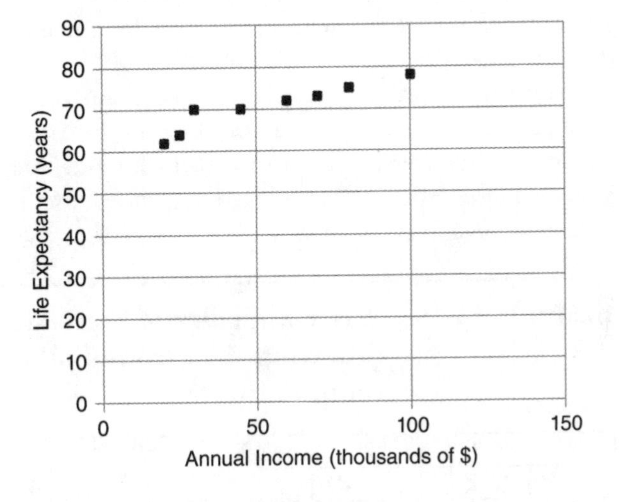

Since the scatterplot shows that data is generally linear with no outliers, we can proceed. We know that life expectancy is the response variable $\hat{y}$ because we are trying to predict it based on annual income, which is the explanatory variable x. When we use technology, we get the following output:

$$\hat{y} = a + bx$$

$a = 61.08496732$ $b = 0.175163399$

Rounding all values to the nearest thousandth gives us the linear regression equation $\hat{y} = 61.085 + 0.175x$, where x = annual income and $\hat{y}$ = life expectancy.

(b) The slope of 0.175 means that an increase of \$1,000 in annual income in Williams County leads to a predicted increase of 0.175 years in life expectancy. (The units of x are thousands of dollars, not dollars, so we use these units when explaining the slope in context.)

(c) The y-intercept of 61.085 means that the life expectancy of a person in Williams County with zero household income is 61.085 years.

(d) Substitute $x = 90$ (not $x = 90{,}000$ since x is in thousands of dollars) into the linear regression equation to find the predicted life expectancy: $\hat{y} = 61.085 + 0.175(90) \approx 77$ years.

(e) No. If we substitute $x = 500$ into the linear regression equation to find the predicted life expectancy, we get $\hat{y} = 61.085 + 0.175(500) \approx 149$ years, which is not reasonable.

Part *e* of Example 13.18 is an example of **extrapolation**, in which a regression line is used to predict values outside the interval of the explanatory variable's given values. When we extrapolate, we assume that the data continues its trend beyond its current range. Since this often yields results that don't make sense, extrapolation is generally discouraged.

Sometimes, you may be asked to determine the type of regression model (such as linear, quadratic, or exponential) that best fits the data. In these cases, create a scatterplot and look at the overall pattern to see which model is appropriate. If you need to write a quadratic or exponential regression equation, use technology to get the coefficients for the regression equation.

Example 13.19 Write a quadratic regression equation for the following data (round all coefficients to the nearest thousandth).

x	30	70	40	60	40	80	90	50
y	2	5	3	6	4	3	2	5

Solution: First, we create a scatterplot of the data to confirm that a quadratic regression model best fits the data:

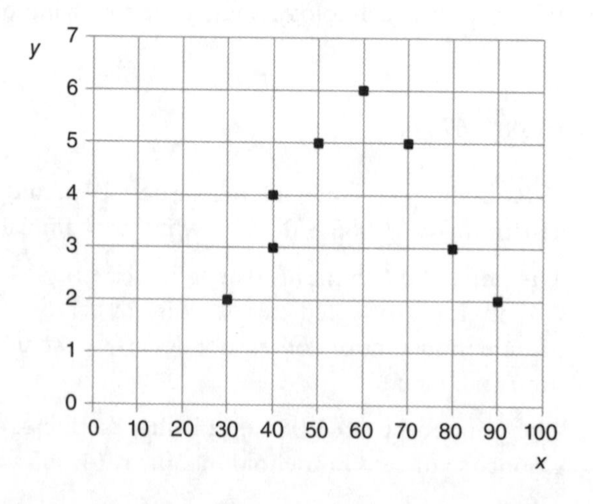

Since the data is decreasing and then increasing, which fits the general pattern of a quadratic model, we can continue. When we use technology, we get the following output:

$$\hat{y} = ax^2 + bx + c$$

$a = -0.004047619$ $b = 0.4841269841$ $c = -9.111111111$

Rounding coefficients to the nearest thousandth gives us $\hat{y} = -0.004x^2 + 0.484x - 9.111$.

Example 13.20 Write an exponential regression equation for the following data (round all coefficients to the nearest hundredth).

x	20	25	10	40	30	45	50	46	35
y	12	14	6	52	28	63	96	78	34

Solution: First, we create a scatterplot of the data to confirm that an exponential regression model best fits the data:

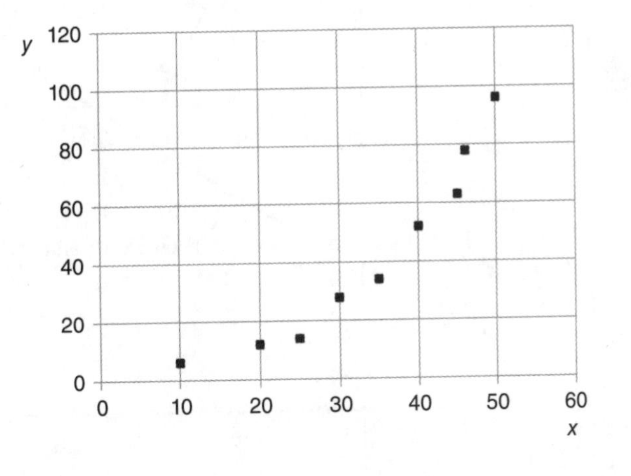

When we use technology, we get the following output:

$$\hat{y} = a(b)^x$$

$a = 2.899510316$ $b = 1.073003029$

Rounding coefficients to the nearest hundredth gives us $\hat{y} = 2.90(1.07)^x$.

If a data set has a strong linear association, then the value of r will be close to -1 or 1. However, a value of r that is close to -1 or 1 may not be linear. For example, the values in Table 13.8 follow an exponential pattern:

Table 13.8 Table with exponential growth.

x	1	2	3	4	5	6
y	2	4	8	16	32	64

However, if we use technology to calculate linear regression, we find that $r \approx 0.9058$, which indicates a strong linear association! Looking at the scatterplot helps, but some non-linear patterns may appear linear. To determine how well a linear regression model fits the data, we look at the distances between each measured y-value and its predicted y-value. We call these distances (calculated as *actual value* − *predicted value*) **residuals**. In symbols, we represent the residual for a point (x, y) as $y - \widehat{y}$ (Figure 13.8).

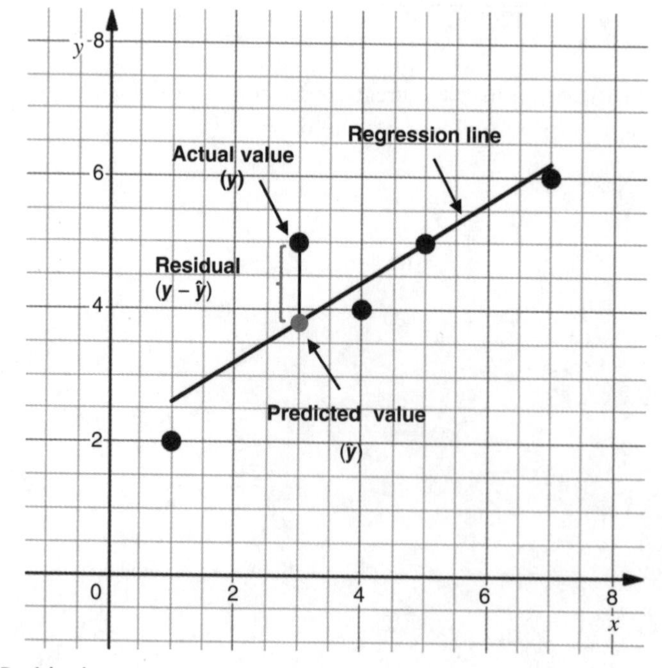

Figure 13.8 Residuals.

This definition gives us another way to understand regression: *the best regression line minimizes the sum of the squares of the residuals for the data points.* (We square the residuals because we're only concerned with how far the points are from the regression line, not whether they're positive or negative.) We say informally that the ideal regression line passes through the "middle" of the points on the scatterplot.

We put all the residuals together on a **residual plot**, a scatterplot whose x-values come from the explanatory variable and y-values are the residuals. Residual plots highlight differences between individual values and the regression line. Points with a positive residual are above the line, while points with a negative residual are below the line. The x-axis on the residual plot corresponds to the regression line of the scatterplot.

When we compare the scatterplots and residual plots of linear associations with those of nonlinear associations, we notice the following in Table 13.9:

Table 13.9 Scatterplots and residual plots.

FORM	SCATTERPLOT	RESIDUAL PLOT
Linear	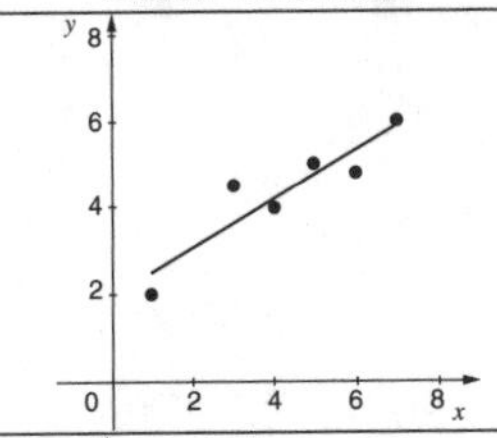	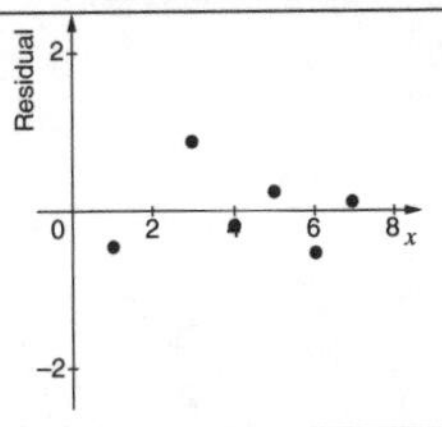
Nonlinear (Quadratic)	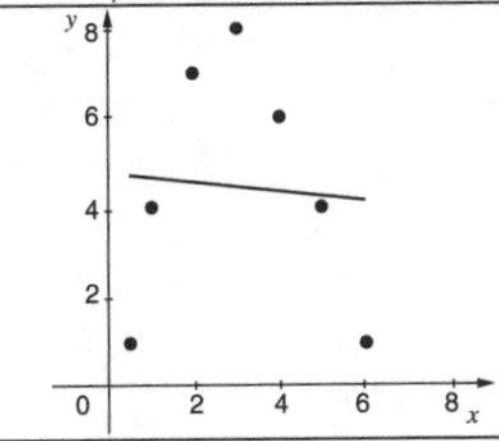	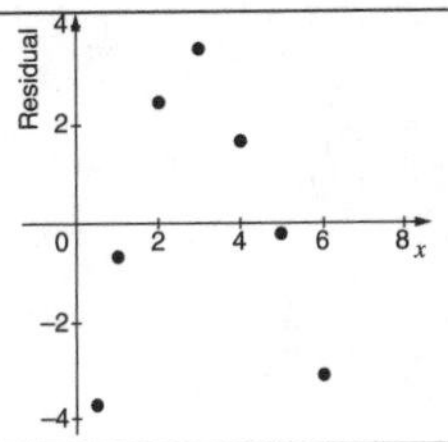
Nonlinear (Exponential)	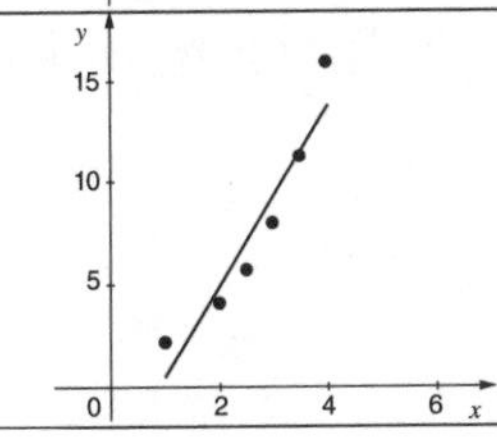	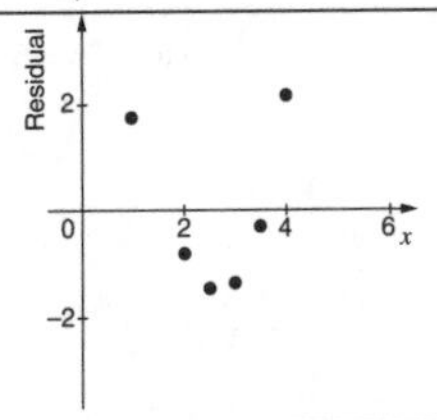

- If a residual plot has no clear curved pattern, then the association is linear.
- If a residual plot has a clear curved pattern, then the association is *not* linear.

Example 13.21 Create a residual plot for the following data and use it to determine if a linear model is appropriate.

x	5	2	4	3	1	6
y	6	5	7	7	3	8

Solution: A scatterplot of the data indicates that the form is generally linear with no outliers.

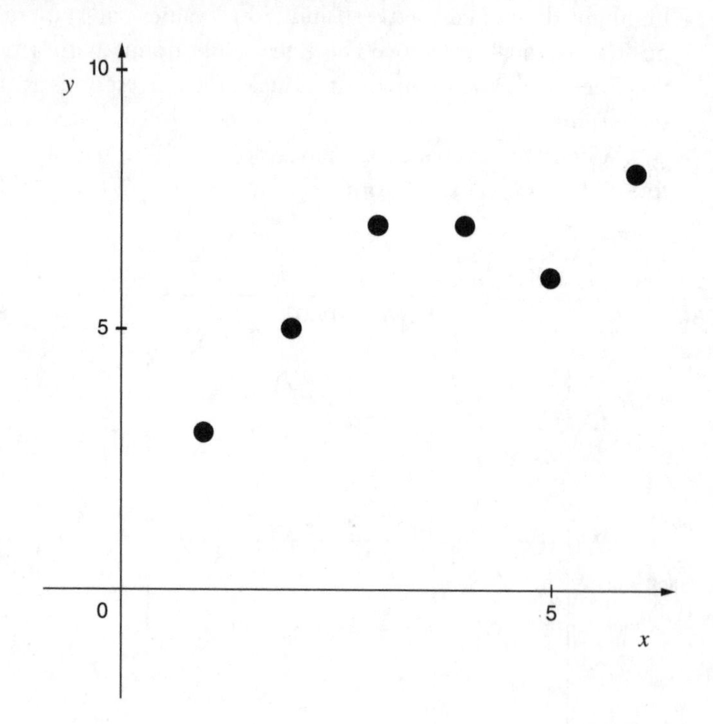

When we use technology to calculate the regression equation, we get the following output:

$$\hat{y} = a + bx$$

$a = 3.2$ $b = 0.8$

This tells us that the regression equation is $\hat{y} = 3.2 + 0.8x$.

To calculate the residuals, we substitute each x-value into the regression equation and find its corresponding predicted value. We then find the differences of the actual and predicted values:

x	5	2	4	3	1	6
y	6	5	7	7	3	8
$\hat{y} = 3.2 + 0.8x$	7.2	4.8	6.4	5.6	4	8
Residual $(y - \hat{y})$	−1.2	0.2	0.6	1.4	−1	0

Finally, we create a residual plot. We plot the x-values on the horizontal axis and the residuals on the vertical axis:

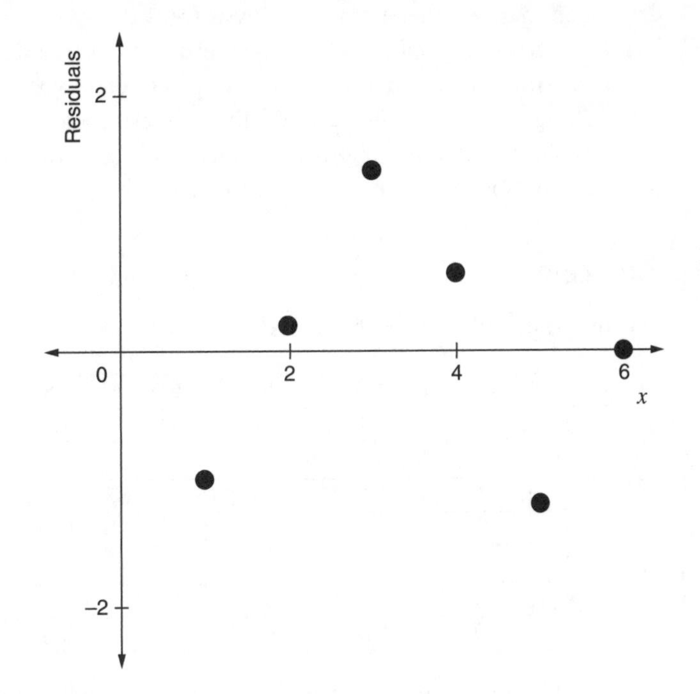

Since the points do not follow a clear curved pattern, we may conclude that a linear model is appropriate.

Technology Tip

Use technology to calculate the residuals and create a residual plot directly from the data.

We conclude this section with one final note about association: the techniques described here enable us to show that a strong association between two variables exists. But do these variables have a **causal relationship**? (In other words, does the explanatory variable cause the change in the response variable?) Not necessarily. Another variable may be causing the changes in one or both variables or the change may simply be a coincidence. The relationship between association and causation has long been controversial. For example, in the 20th century, researchers began publishing studies noting a strong association between smoking and cancer. For decades, tobacco companies denied that smoking actually *caused* cancer, claiming that more evidence was needed and that other factors could be contributing. (In fact, Ronald A. Fisher, who developed many of the statistical analysis tools that we use today, questioned whether health experts could validly state such a claim. He was also a heavy smoker.)

The bottom line is that the relationship between two variables is often complicated. Establishing an association between them doesn't address why this association exists or what should be done about it. Understanding both the power and limitations of statistical tools is important for using them wisely.

Exercises

Create a scatterplot for each set of data.

1.

x	4	4.5	5	5.5	6.5	6	7	7.5
y	75	71	64	78	73	65	48	59

2.

x	10	30	20	15	35	25	40	42	45
y	6	15	20	8	25	15	20	30	45

3.

x	50	250	100	350	150	200	250	300
y	30	15	40	10	20	30	25	25

If appropriate, calculate the correlation coefficient to the nearest thousandth for each set of data. Use r to interpret the strength and association of the association.

4.

x	3	7	3	8	1	6	4	10	9
y	20	30	20	45	30	40	50	55	50

5.

x	0.1	0.9	0.3	1.2	0.4	1.4	0.5	1.5	0.7
y	40	16	25	30	40	40	30	20	50

6.

x	50	85	30	45	95	62	85	100	70
y	4.5	2	6	8	1	4	4	2	6

7. Researchers examined the relationship between the poverty rate and childhood asthma in 9 neighborhoods in a large state. The accompanying table summarizes some of the data that they gathered.

Poverty Rate (%)	12	35	50	30	45	20	15	5	22
Childhood Asthma Cases (per 10,000)	200	400	580	165	420	410	184	50	220

(a) Write a linear regression equation for the data. State what your variables represent and round all values to the nearest thousandth.

(b) Explain in context what the slope of the regression equation represents.

(c) To the nearest whole number, predict the number of childhood asthma cases per 10,000 in a neighborhood with a 25% poverty rate.

8. The age and value of a random sample of a popular model of cars sold from a major dealer is summarized in the following table:

Age of car (years)	5	2	8	6	5	7	6	9	3
Value of car (thousands of dollars)	20	40	15	25	32	24	37	10	30

(a) Write a linear regression equation for the data. State what your variables represent and round all values to the nearest hundredth.

(b) Explain in context what the slope of the regression equation represents.

(c) Explain in context what the y-intercept of the regression equation represents.

(d) To the nearest thousand dollars, predict the value of a car that is four years old.

9. Write a quadratic regression equation for the following data (round all values to the nearest thousandth).

x	2	3	4	5	6	7	8
y	8	3	1	1	3	6	12

10. Write a quadratic regression equation for the following data (round all values to the nearest thousandth).

x	10	30	40	60	70	90	110	130
y	23	18	15	11	10	13	18	22

11. Write an exponential regression equation for the following data (round all values to the nearest thousandth).

x	3	3.5	2	2.5	4.5	5	3	1
y	54	94	18	32	285	475	50	6

12. Write an exponential regression equation for the following data (round all values to the nearest hundredth).

x	80	20	60	10	40	30	15	45	5
y	2	60	15	200	20	20	110	20	350

13. For the following set of data:

x	1	2	3	4	5	6
y	30	26	20	16	7	3

(a) Write a linear regression equation for the data.

(b) Calculate the residuals for the data.

(c) Create a residual plot for the data.

14. The accompanying table represents a bivariate data set:

x	1	2	3	5	6	4
y	2	4	12	11	7	4.5

The equation of the regression line is $\hat{y} = 2.9 + 1.1x$.

(a) Determine the predicted value when $x = 3$.

(b) Calculate the residual when $x = 3$.

(c) Is the point (3, 12) above or below the regression line? Explain.

15. Researchers survey 100 mothers, asking them how many children they have and how old they were when they had their first child. A residual plot of the data (with age as the explanatory variable) shows no curved pattern. The researchers found a strong negative linear association between the number of children and the age of mothers when they had their first child, with $r = -0.897$. Can the researchers reasonably conclude that having children at a younger age causes mothers to have more children? Explain.

Questions to Think About

16. If a bivariate quantitative data set has a generally linear association with $r = 1$, what is true about the points in the scatterplot?

17. Explain why the regression line can't be above all the points on a scatterplot.

18. State a real-life example of two quantitative variables that have a strong association but do not have a causal relationship.

CHAPTER 13 TEST

1. The five-number summary for a data set is as follows:
minimum = 6, $Q_1 = 8$, $Q_2 = 16$, $Q_3 = 25$, maximum = 33.
What is the interquartile range for the data?

(A) 2 (B) 17 (C) 27 (D) 33

2. A teacher computes the linear regression equation $\hat{y} = 15 + 0.85x$ to predict students' final exam score based on their first test score (x). What is the predicted final exam for a student who scores an 80 on the first test?

   (A) 76.5 (B) 80 (C) 83 (D) 85

3. For the data set 3, 3, 4, 5, 6, which statement is true?

   (A) mean $<$ median
   (B) mean $>$ median
   (C) mean $<$ mode
   (D) mean $=$ mode

4. Which variable is considered qualitative?

   (A) number of books owned by 100 households
   (B) heights of 100 adults
   (C) number of meals eaten per day by 100 children
   (D) genders of 100 voters

5. If a bivariate quantitative data set with a linear association and no outliers has $r = -0.88$, which statement *must* be true?

   (A) The distribution is skewed left.
   (B) The association is negative.
   (C) The distribution is skewed right.
   (D) The association is positive.

6. Which statement about the data sets shown in the accompanying diagram *must* be true?

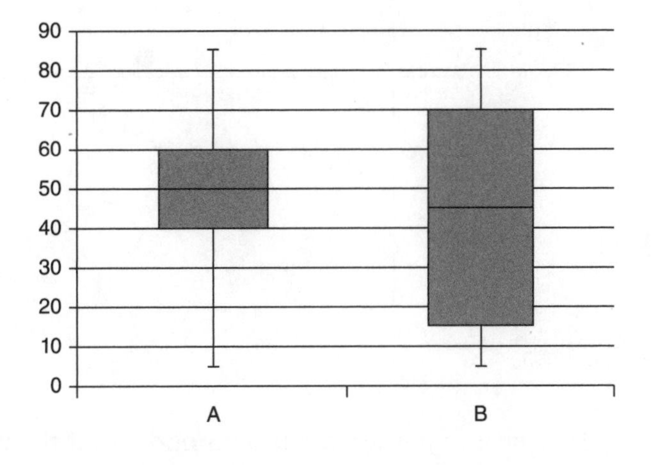

   (A) Set B has a greater mean.
   (B) Set B has a lower median.
   (C) Set B has a greater range.
   (D) Set B is symmetric.

7. Which measures can be determined from a dotplot?

   I. Mean   II. Median   III. Mode

   (A) I, only
   (B) I and II, only
   (C) II and III, only
   (D) I, II, and III

8. Which measures can be determined from a boxplot?

   (A) mean and median
   (B) mean and mode
   (C) median and range
   (D) standard deviation and range

9. If an outlier is removed from a data set, which measure is *least* likely to be affected?

   (A) median
   (B) range
   (C) mean
   (D) standard deviation

10. A bivariate quantitative data set has the following residual plot:

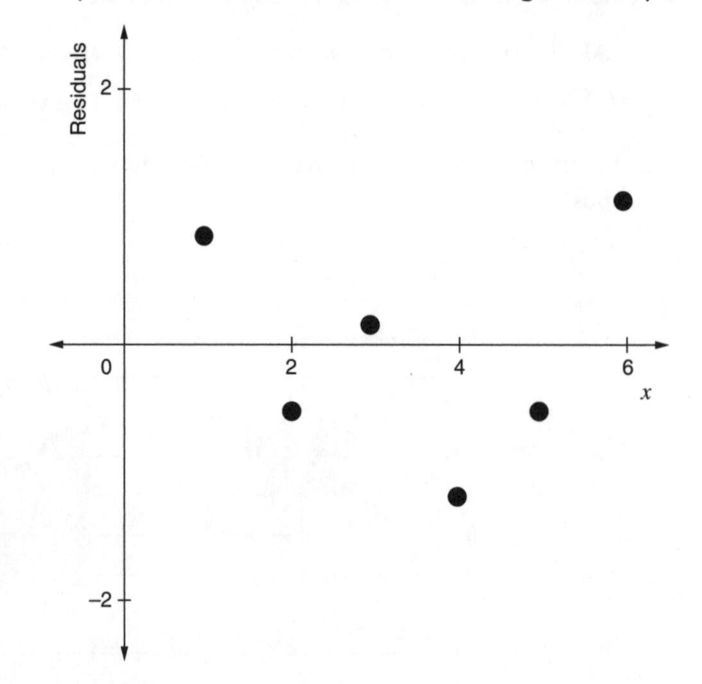

    Is a linear regression model appropriate for the data? Explain.

11. In a survey of registered voters in a large city, a polling company found that 52% of respondents supported the Democratic candidate in the upcoming mayoral election, 40% supported the Republican candidate, and 8% supported other candidates. Explain why this distribution is *not* right-skewed.

12. The number of overtime hours that a sample of 8 police officers worked per year are as follows: 200, 198, 190, 130, 150, 140, 110, 250.

    (a) Find the mean number of hours.

    (b) To the nearest whole number, find the standard deviation.

    (c) Interpret the standard deviation in context.

13. The political affiliations of voters who live in two counties are summarized in the accompanying two-way table.

		COUNTY		
		Greene	Williams	Total
	Liberal	44	17	61
POLITICAL	Moderate	19	9	28
BELIEFS	Conservative	10	51	61
	Total	73	77	150

    (a) Explain in context what the "9" in the table represents.

    (b) To the nearest tenth of a percent, what percentage of voters in Greene County are liberal?

    (c) To the nearest tenth of a percent, what percentage of voters surveyed were conservative?

14. The following histogram shows the heights, in inches, of a random sample of adults:

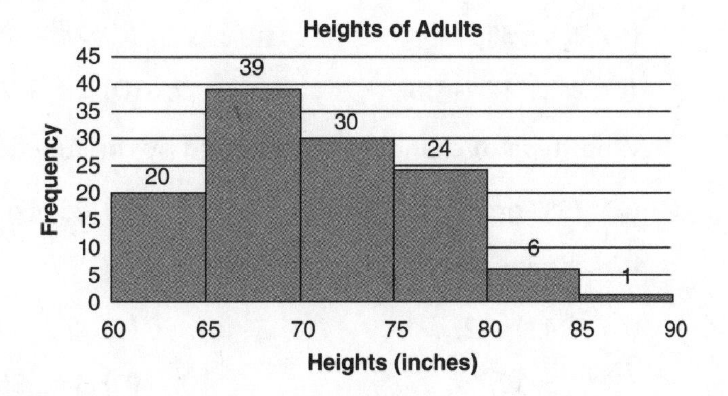

    (a) How many adults are in the sample?

    (b) What percentage of adults are between 70 and 80 inches tall?

    (c) What interval contains the median? Explain your answer.

15. The following data shows the heights and number of points scored by 8 basketball players.

Height (x)	78	78	78	81	70	80	73	81
Points scored (y)	214	205	220	232	175	250	198	250

(a) Create a scatterplot of the data.

(b) Write a linear regression equation for the data. Round all values to the nearest hundredth.

(c) Use your equation from part *b* to predict, to the nearest whole number, how many points a player who is 75 inches tall would score.

REVIEW TEST 2: CHAPTERS 8–13

1. Which expression is equivalent to $(5y^3)^2$?

(A) $25y^6$ (B) $5y^6$ (C) $5y^5$ (D) $5y^5$

2. What are the solutions to the equation $(m + 4)(m - 3) = 0$?

(A) −4 and 3 (B) 4 and −3 (C) −4 and −3 (D) 4 and 3

3. Which expression is equivalent to $2(r + 6)^2$?

(A) $2r^2 + 36$
(B) $2r^2 + 12r + 36$
(C) $2r^2 + 24r + 72$
(D) $2r^2 + 72$

4. What type of change is represented by the function $f(x) = 72(0.75)^x$?

(A) 75% growth (B) 25% decay (C) 25% growth (D) 54% decay

5. Which sequence is geometric?

(A) 7, 4, 1, −2, ...
(B) 2, 5, 10, 17, ...
(C) 4, 9, 16, 25, ...
(D) 320, 80, 20, 5, ...

6. Which function shifts the graph of $y = f(x)$ 3 units to the left and 2 units down?

(A) $f(x - 3) + 2$ (B) $f(x - 3) - 2$ (C) $f(x + 3) - 2$ (D) $f(x + 3) + 2$

7. $f(x) = x^2 + 2x - 7$, and the graph of g is shown below:

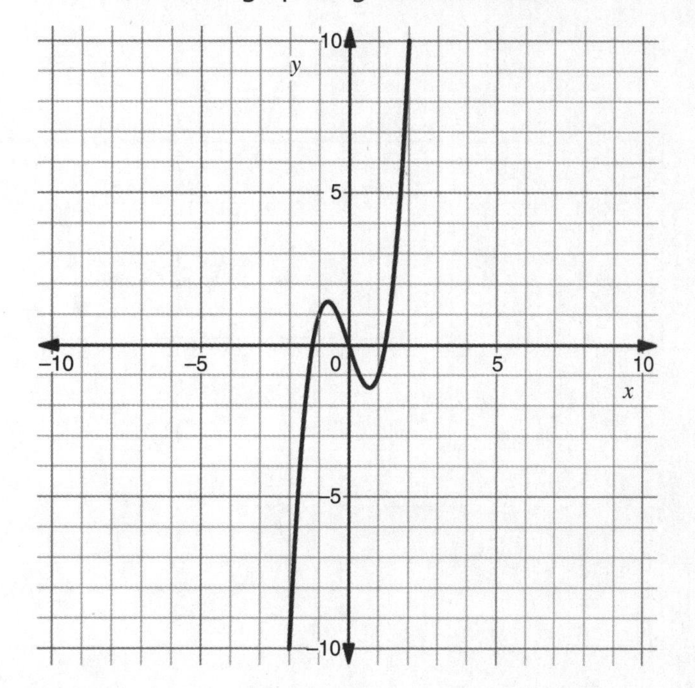

Which statement about the average rate of change of f and g over the interval (–2, 1) is correct?

(A) f has a greater average rate of change.

(B) g has a greater average rate of change.

(C) f and g have the same average rate of change.

(D) The average rate of change cannot be calculated over the interval.

8. The amount of annual rainfall, in inches, for 8 cities in Florida are as follows: 52.2, 66.5, 70.4, 61.9, 60.0, 46.0, 62.3, 51.7. What is the approximate mean, in inches, of the data?

(A) 7.70 (B) 8.24 (C) 58.88 (D) 67.29

9. A bivariate quantitative data set with no outliers has a correlation coefficient of $r = 0.967$. Which statement about the slope of the regression line for the data is *most* likely to be true?

(A) The slope is positive.

(B) The slope is negative.

(C) The slope is 0.

(D) The slope cannot be determined.

10. Using the accompanying graph of f, evaluate $4f(2) + 1$.

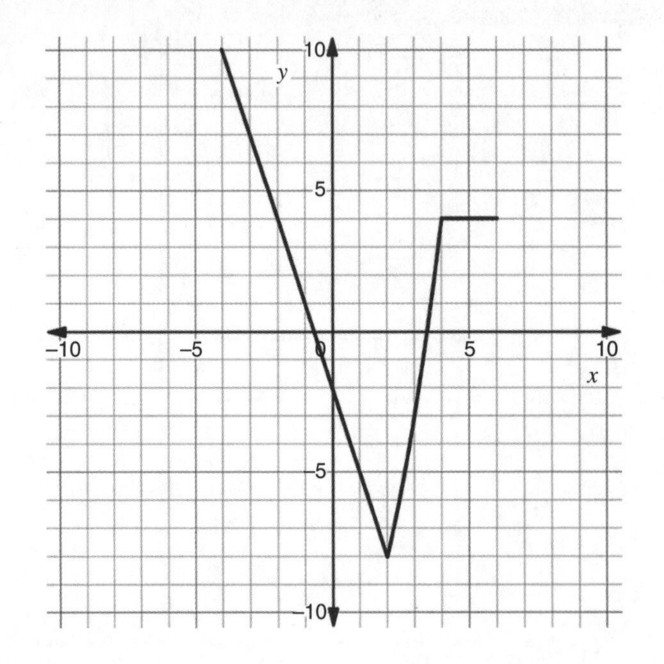

11. Write the pronunciation of the equation $f(n) = 3(2)^{n-1}$
12. Solve $2d^2 + 3d - 10 = 0$ for d.
13. Express $3m(m + 1)^2 - 2(m + 4)$ in standard form.
14. The height h, in feet, of a ball x seconds after being kicked can be modeled by the function $h(x) = -16x^2 + 45x + 2$.

    (a) Calculate $h(0)$ and interpret it in context.

    (b) To the nearest tenth of a foot, determine the maximum height that the ball reaches.

15. The ages of the women's Olympic gymnastics champions at the time they won the gold medal is listed below:

    30, 21, 25, 22, 26, 20, 14, 18, 16, 19, 15, 17, 16, 16, 18, 16, 19, 18

    (a) Determine the five-number summary for the data.

    (b) A journalist looking at the data argues that 30 is an outlier. Use your answer from part *a* to support this claim.

CHAPTER 13 SOLUTIONS

13.1. 1. a. 528; b. 285; c. 1,316; d. 167 second-class passengers died.

2. a. 120; b. 380; c. 42 people had incomes less than \$40,000 and expect to live comfortably in retirement; d. 26.

3. a. $\frac{375}{530} \approx 0.7075$, or 70.8%. b. $\frac{36}{87} \approx 0.41379$, or 41.4%. c. The percentage of people aged 18-29 who are Christians is $\frac{45}{87} \approx 0.517$, or 51.7%, which is greater than the percentage of Christians who are aged 18-29 ($\frac{45}{375} =$ 0.12, or 12%).

4. a. $\frac{120}{230} \approx 0.52$, b. $\frac{95}{120} \approx 0.7917 \approx 79\%$.

13.2. 1.

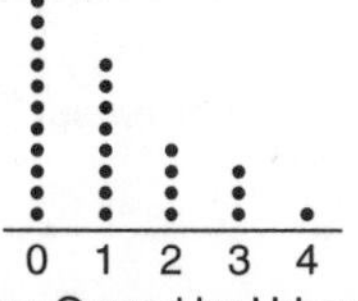

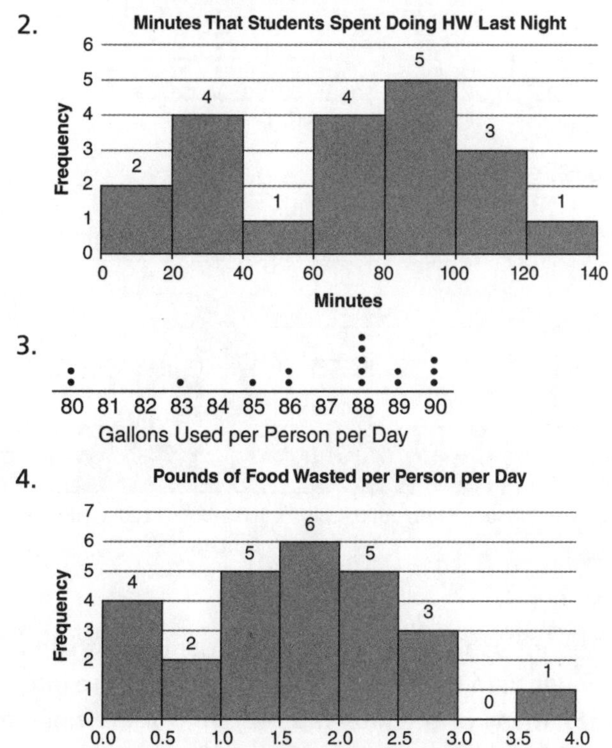

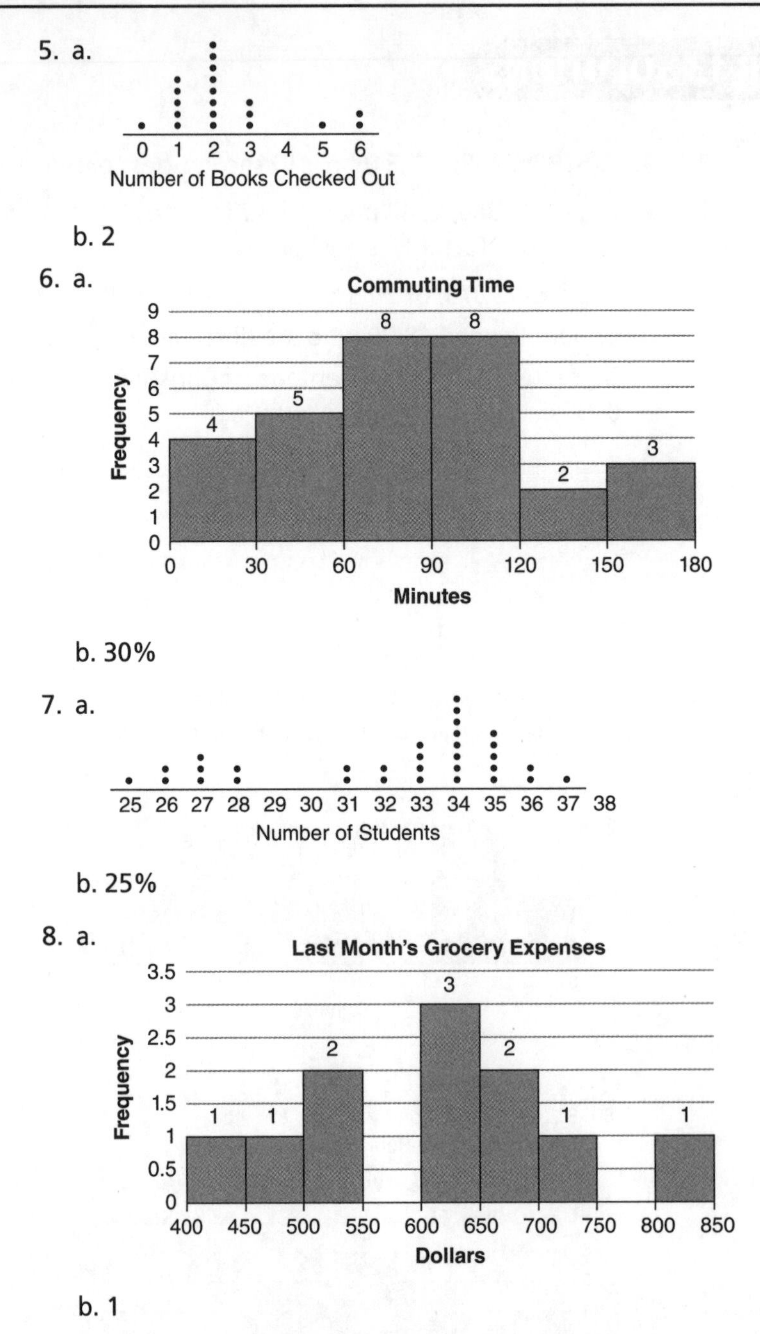

5. a.

   b. 2

6. a.

   b. 30%

7. a.

   b. 25%

8. a.

   b. 1

9. The intervals don't all have the same width (the first interval 0–50 is wider than the others). There is a gap in the intervals (between 56 and 58). Many of the intervals overlap (for example, the first interval ends at 50 and the second begins at 50).

10. The horizontal axis is incorrectly labeled, giving the impression that 1 insect has a length of 0 centimeters, 2 have lengths of 1 centimeter, and so on. Instead, the corners of each rectangle should be labeled 0, 1, 2, and so on to indicate that each rectangle represents the interval between those values.

13.3. 1. Shape: approximately uniform, center: median = 16, spread: range = 8

2. a. roughly symmetric, b. 53–54, c. 51–52

3. a. Skewed left with a cluster from 15 to 18 minutes and an unusual value at 10 minutes, b. Mean = 16.5, median = 17, c. The unusual value at 10 minutes lowers the mean (which would be 17 without the outlier) but does not affect the median.

4. a. Skewed right with an unusual value between 55 and 60 mph, b. 55–60 mph since a hybrid vehicle has a fuel economy that is higher than the others.

5. $\bar{x} = 100.59$, $s_x = 4.44$

6. $\bar{x} = 5.6$, $s_x = 1.6$

7. a. $\bar{x} = 88.9$, $s_x = 24.2$; b. The wait time for hospital patients typically varied by 24.2 minutes from the mean of 88.9 minutes.

8. a. Company A: $\bar{x} = 13.5$, $s_x = 0.7$; Company B: $\bar{x} = 13.4$, $s_x = 2.1$. b. Company B has more variability since its standard deviation is greater.

9. Minimum = 2.4, $Q_1 = 2.95$, $Q_2 = 5.1$, $Q_3 = 7$, maximum = 7.9,

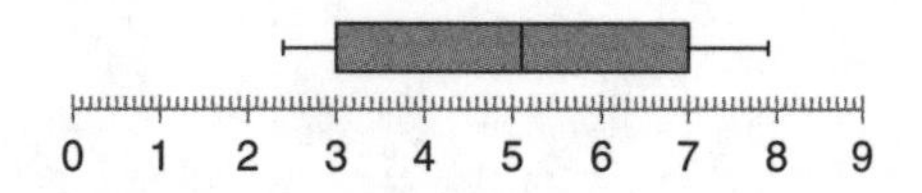

10. Minimum = 119, $Q_1 = 122$, $Q_2 = 127$, $Q_3 = 130$, maximum = 149. IQR = $Q_3 - Q_1 = 130 - 122 = 8$. Since outliers are greater than $Q_3 + 1.5 \cdot \text{IQR} = 130 + 12 = 142$, then 149 is an outlier.

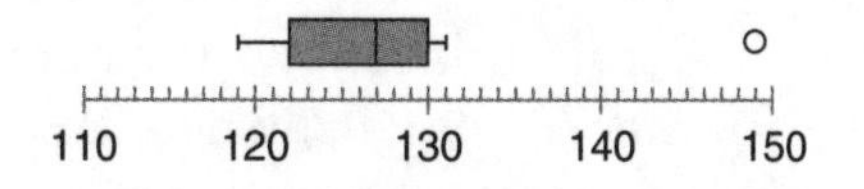

11. a. Minimum = 256, $Q_1 = 260$, $Q_2 = 274$, $Q_3 = 280$, maximum = 296; b. The IQR = $Q_3 - Q_1 = 280 - 260 = 20$.

12. a. The median of Ms. Lee's class (75) is higher than the median of Mr. Kaur's class (70). b. Ms. Lee's class has a greater interquartile range (25) than Mr. Kaur's class (15). c. Ms. Lee's has a higher median. d. Mr. Kaur's class has a higher maximum and minimum, and the bottom 25% of his class did better than the bottom 25% of Ms. Lee's.
13. Although the boxplot appears symmetric, we can't conclude that the distribution is symmetric. We only know that each quartile (each 25% of the data) is spaced out over intervals of approximately equal width. Each quartile could contain different peaks or clusters.
14. Unlike boxplots, dotplots show individual values, clusters, and gaps.
15. Unlike dotplots, boxplots provide a quick overview of the distribution. They clearly show the median and outliers.

13.4. 1. 2. 3.

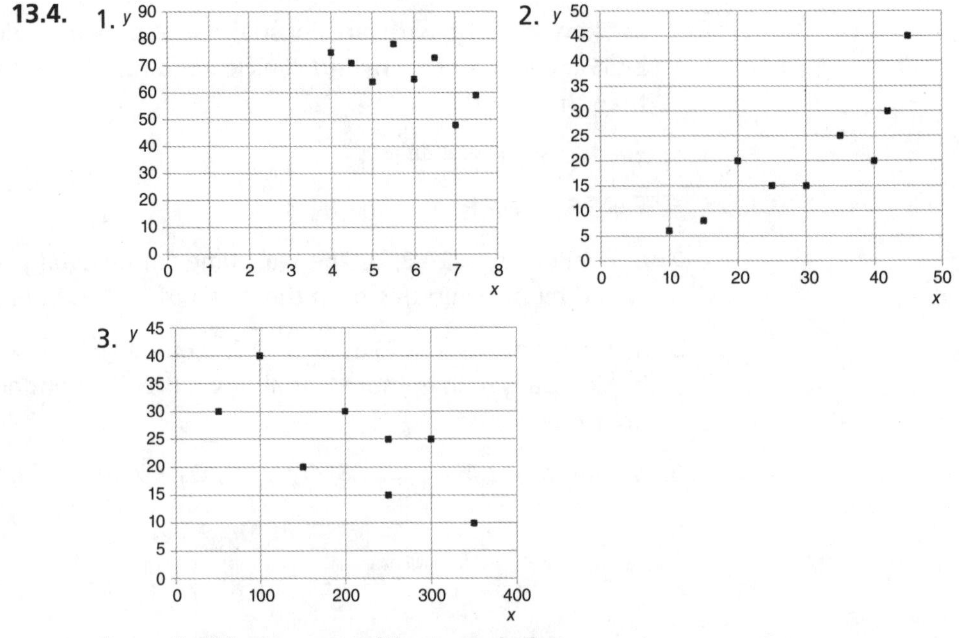

4. $r \approx 0.714$, strong positive association
5. $r \approx -0.281$, weak negative association
6. $r \approx -0.809$, strong negative association
7. a. A scatterplot shows the data is generally linear with no outliers. $\hat{y} = 53.176 + 9.190x$, where x = poverty rate (%) and $\hat{y}$ = number of childhood asthma cases per 10,000. b. For every increase of 1% in the poverty rate, the number of childhood asthma cases is predicted to increase by about 9.190 per 10,000. c. $\hat{y} = 53.176 + 9.190(25) \approx 283$.

8. a. $\hat{y} = 45.77 - 3.51x$, where x = age of car (in years) and $\hat{y}$ = value of car (in thousands of dollars). b. For every additional year that the car ages, the value of the car decreases by about \$3,510. c. The value of the car when it is sold is about \$45,770. d. $\hat{y} = 45.77 - 3.51(4) = 31.73$, or about \$32,000.
9. $\hat{y} = x^2 - 9.286x + 22.286$
10. $\hat{y} = 0.003x^2 - 0.463x + 27.808$
11. $\hat{y} = 2.002(2.992)^x$
12. $\hat{y} = 282.53(0.94)^x$
13. a. $\hat{y} = 36.6 - 5.6x$.

b.

x	1	2	3	4	5	6
y	30	26	20	16	7	3
Residual	−1	0.6	0.2	1.8	−1.6	0

c.

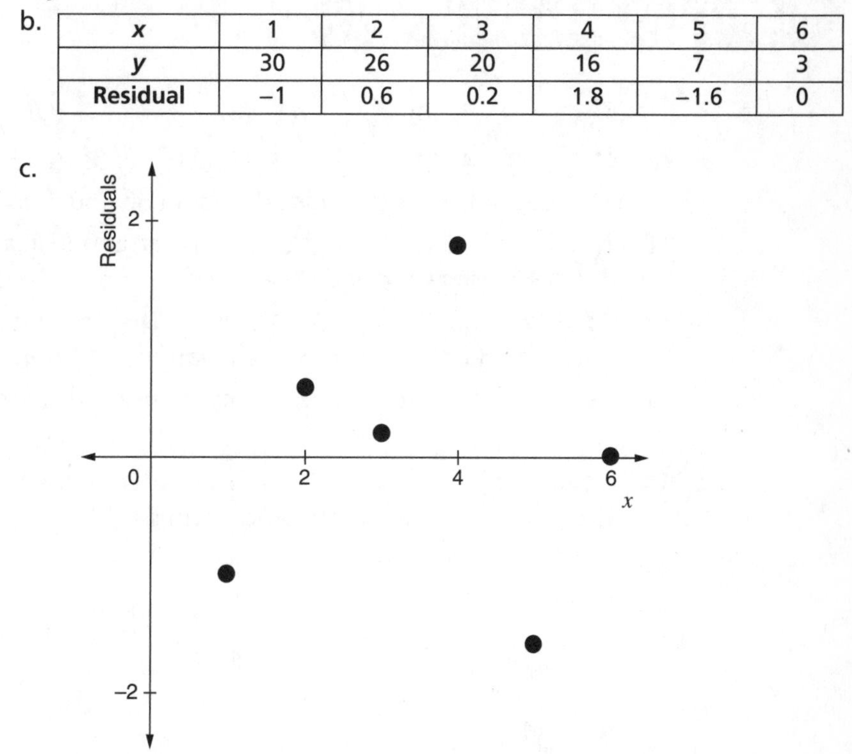

14. a. When $x = 3$, $\hat{y} = 6.2$. b. The residual is $12 - 6.2 = 5.8$. c. The point is above the regression line since the residual is positive.
15. Not necessarily. Although there is a strong association between the age of mothers when they have their first child and the number of children they have, other factors (such as the mothers' religion, careers, or beliefs about the importance of large families) may affect both the number of children they have and when they have children.

16. The points lie on a straight line.

17. A regression line that is above all the points on a scatterplot wouldn't minimize the sum of the squares of the residuals. It would be farther from some points than a line that passes through the middle of the points.

18. Examples include the number of flu cases and number of sunburn cases (both are affected by the weather), or the height of children and number of words in their vocabulary (both are affected by their age).

CHAPTER 13 TEST SOLUTIONS

1. (B)
2. (C)
3. (B)
4. (D)
5. (B)
6. (B)
7. (D)
8. (C)
9. (A)
10. Yes because the residual plot shows no obvious curved pattern.
11. The data is qualitative, so they can't be arranged in a natural order. Measures of shape (such as skewness) can only be applied to quantitative data.
12. a. $\overline{x} = 171$ hours. b. $s_x = 46$ hours. c. The amount of overtime that police officers worked varied from the mean of 171 hours by 46 hours.
13. a. 9 surveyed voters live in Williams County and are moderates. b. $\frac{44}{73} \approx 60.3\%$. c. $\frac{61}{150} \approx 40.7\%$.
14. a. 120, b. $\frac{30+24}{120} = 45\%$, c. 70–75 since it contains the 60th and 61st heights when they are listed in numerical order.
15. a.

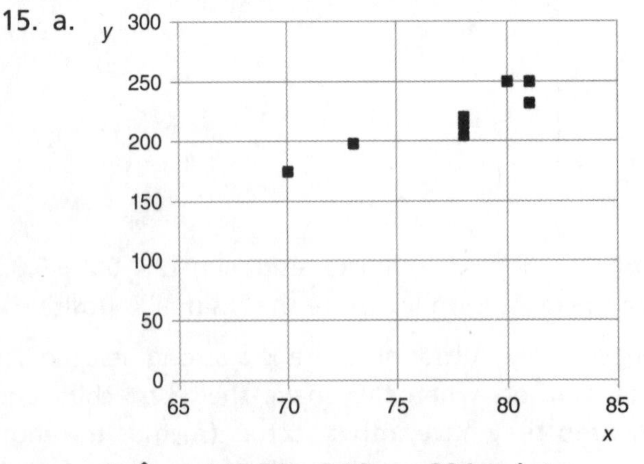

b. $\hat{y} = -245.35 + 5.99x$. c. 204 points.

REVIEW TEST 2 SOLUTIONS

1. (A)
2. (A)
3. (C)
4. (B)
5. (D)
6. (C)
7. (A)
8. (C)
9. (A)
10. −31
11. "*f* of *n* equals 3 times 2 to the *n* minus 1 power"
12. $d = \frac{-3 \pm \sqrt{89}}{4}$
13. $3m^3 + 6m^2 + m - 8$
14. a. 2; the ball was 2 feet high when it was kicked. b. 33.6 feet
15. a. Minimum = 14, $Q_1 = 16$, $Q_2 = 18$, $Q_3 = 21$, maximum = 30; b. IQR = 21 − 16 = 5, so outliers are greater than 21 + 1.5(5) = 28.5. Since 30 > 28.5, 30 is considered an outlier.

FORMULAS

Linear Equations

Slope: $m = \frac{y_2 - y_1}{x_2 - x_1}$

Equation of a Line:

- Point-slope Form: $y - y_1 = m(x - x_1)$
- Slope-intercept Form: $y = mx + b$

Exponents

$$x^a(x^b) = x^{a+b}$$

$$(x^a)^b = x^{ab}$$

$$(x^a y^b)^c = x^{ac} y^{bc}$$

$$\frac{x^a}{x^b} = x^{a-b} \; (x \neq 0)$$

$$\left(\frac{x}{y}\right)^a = \frac{x^a}{y^a} \; (y \neq 0)$$

$$x^0 = 1 \; (x \neq 0)$$

$$x^{-a} = \frac{1}{x^a} \; (x \neq 0)$$

Radicals

$$a\sqrt{x} + b\sqrt{x} = (a + b)\sqrt{x}$$

$$a\sqrt{x} - b\sqrt{x} = (a - b)\sqrt{x}$$

$$a\sqrt{x}(b\sqrt{y}) = ab\sqrt{xy}$$

$$\frac{a\sqrt{x}}{b\sqrt{y}} = \frac{a}{b}\sqrt{\frac{x}{y}} \; (b \neq 0, y \neq 0)$$

Quadratic Equations

Standard Form: $y = ax^2 + bx + c$

Factored Form: $y = a(x - x_1)(x - x_2)$

Vertex Form: $y = a(x - h)^2 + k$

Quadratic Formula: $x = \dfrac{-b \pm \sqrt{b^2 - 4ac}}{2a}$

Axis of Symmetry: $x = -\dfrac{b}{2a}$

Exponential Equations

Exponential Change: $y = a(1 + r)^x$

Sequences

Arithmetic: $a_n = a_1 + (n - 1)d$

Geometric: $a_n = a_1(r)^{n-1}$

Functions

Average Rate of Change: $\dfrac{f(b) - f(a)}{b - a}$

Statistics

Mean: $\overline{x} = \dfrac{x_1 + x_2 + \ldots + x_n}{n}$

Deviation: $x - \overline{x}$

Population Standard Deviation: $\sigma = \sqrt{\dfrac{(x_1 - \mu)^2 + \ldots + (x_n - \mu)^2}{n}}$

Sample Standard Deviation: $s = \sqrt{\dfrac{(x_1 - \overline{x})^2 + \ldots + (x_n - \overline{x})^2}{n - 1}}$

Interquartile Range (IQR): $Q_3 - Q_1$
Outliers:

- Less than $Q_1 - 1.5 \cdot \text{IQR}$
- Greater than $Q_3 + 1.5 \cdot \text{IQR}$

Line of Best Fit: $\hat{y} = a + bx$
Residual: $y - \hat{y}$

GLOSSARY OF MATHEMATICAL SYMBOLS

Basic Operations

Symbol	Pronunciation	Meaning
$3 + 5$	"three plus five"	addition
$+3$	"positive three"	positive
$3 - 5$	"three minus five"	subtraction
-5	"negative five"	negative
3 ± 5	"three plus or minus five"	plus or minus
± 3	"positive or negative 3"	positive or negative
$\lvert 3 \rvert$	"the absolute value of three"	absolute value
$3(5)$	"three times five"	multiplication
$(3)(5)$		
$3 \cdot 5$		
3×5		
3^5	"three to the fifth power" "three to the fifth"	exponent
$\frac{3}{5}$	"three divided by five" "three over five" "three-fifths"	division

Ratios and Proportions

Symbol	Pronunciation	Meaning
3:5	"three to five"	ratio
3:5::6:10	"three is to five as six is to ten"	proportion
15%	"fifteen percent"	percentage

Polynomial and Radical Expressions

Symbol	Pronunciation	Meaning
$3x$	"three x" "three times x"	multiplication with variables
x^2	"x to the second power" "x to the second" "x squared"	exponent
x^3	"x to the third power" "x to the third" "x cubed"	exponent
$\sqrt{x}$	"the square root of x"	square root
$\sqrt[3]{x}$	"the cube root of x"	cube root

Equations and Inequalities

Symbol	Pronunciation	Meaning
$3(5) = 15$	"three times five equals fifteen" "three times five is equal to fifteen"	equality
$15.1 \approx 15$	"fifteen point one is approximately equal to fifteen"	approximate equality
$3(5) \neq 14$	"three times five is not equal to fourteen"	inequality
$3 < 5$	"three is less than five"	less than
$3 \leq 5$	"three is less than or equal to five"	less than or equal to
$5 > 3$	"five is greater than three"	greater than
$5 \geq 3$	"five is greater than or equal to three"	greater than or equal to
∞	"infinity"	infinity
$(3, 5)$	"the open interval from three to five"	interval notation
$(3, 5]$	"the interval from three to five, including five"	interval notation
$[3, 5]$	"the closed interval from three to five"	interval notation
$\{3, 5\}$	"the set three, five"	set

Symbol	Pronunciation	Meaning
$\{x \mid x < 10\}$	"the set of all *x*s such that *x* is less than ten"	set-builder notation
$A \cap B$	"*A* intersection *B*"	intersection of sets
$A \cup B$	"*A* union *B*"	union of sets
$\mathbf{Z}^+$ $\mathbb{Z}^+$ $\mathbf{N}$ $\mathbb{N}$	"the set of positive integers" "the set of natural numbers"	set-builder notation
$\mathbf{W}$ $\mathbb{W}$	"the set of whole numbers"	set-builder notation
$\mathbf{Z}$ $\mathbb{Z}$	"the set of integers"	set-builder notation
$\mathbf{Q}$ $\mathbb{Q}$	"the set of rational numbers"	set-builder notation
$\mathbf{R}$ $\mathbb{R}$	"the set of real numbers"	set-builder notation

Functions and Graphs

Symbol	Pronunciation	Meaning
(3, 5)	"the point three, five"	ordered pair on the coordinate plane
$f(x)$	"*f* of *x*"	function
x_1	"*x*-sub-one"	subscript (used to differentiate between values of *x*)
a_n	"*a*-sub-*n*"	Subscript: *n*th term in a sequence
a_{n-1}	"*a*-sub-*n*-minus-one"	Subscript: $(n - 1)$st term in a sequence
$f(x) = \begin{cases} 5x, & x < 2 \\ x, & x \geq 2 \end{cases}$	"*f* of *x* equals five *x* for *x* less than two, and *x* for *x* greater than or equal to two"	piecewise function

Statistics

Symbol	Pronunciation	Meaning
μ	"mu"	population mean
$\overline{x}$	"*x*-bar"	sample mean
σ_x	"sigma" "population standard deviation of *x*"	population standard deviation
s_x	"*s*-sub-*x*" "sample standard deviation of *x*"	sample standard deviation
Q_1	"*Q*-one"	first quartile
Q_2	"*Q*-two"	second quartile (median)
Q_3	"*Q*-three"	third quartile
$\hat{y}$	"*y*-hat"	predicted value of *y*

GLOSSARY OF MATHEMATICAL TERMS

The section where each term is introduced in the book is listed in parentheses.

absolute maximum (6.3): a point that has a y-coordinate that is greater than or equal to the y-coordinates of all other points on the graph (or the highest point on the graph)

absolute minimum (6.3): a point that has a y-coordinate that is less than or equal to the y-coordinates of all other points on the graph (or the lowest point on the graph)

absolute value (1.1): the distance of a number from 0 on a number line

arithmetic sequence (11.2): a sequence that has a common difference between consecutive terms

association (13.4): a relationship between two qualitative or quantitative variables

asymptote (10.1): a line or curve that a graph approaches

axis (6.2): a horizontal or vertical number line on the coordinate plane used as a reference to indicate position

axis of symmetry of a parabola (9.9): a line that divides a parabola into two identical parts

binomial (8.1): an expression consisting of 2 monomials linked by addition or subtraction

bivariate data (13.1): data consisting of two characteristics from each individual

boxplot (box-and-whisker plot) (13.3): a graphical representation of a five-number summary

Cartesian plane (6.2): the coordinate plane

coefficient (3.5): the number that is being multiplied by variables in a term

compound inequality (5.5): a combination of two or more inequalities

consecutive integers (3.5): integers that follow each other in order (such as 3, 4, 5, ...)

consecutive even integers (3.5): even integers that follow each other in order (such as 2, 4, 6, 8, ...)

consecutive odd integers (3.5): odd integers that follow each other in order (such as 1, 3, 5, 7, ...)

constant (3.1): a fixed value that does not change

coordinate plane (6.2): the entire two-dimensional surface defined by the axes

coordinates (6.2): individual numbers in an ordered pair that are used to indicate position on the coordinate plane
correlation coefficient (13.4): the measure (represented by r, where $-1 \leq r \leq 1$) that summarizes both the strength and direction of the linear association of the two variables in bivariate quantitative data
counting numbers (natural numbers) (1.3): the set of numbers 1, 2, 3, 4, 5, …
cubic equation (12.1): an equation whose degree is 3
data (13.1): information that comes from observations or measurements
degree of a polynomial (8.1): the largest of all degrees of terms in a polynomial
degree of a term (8.1): the sum of the exponents of the variables in a term
dependent variable (6.1): the variable that represents the output values of a relation
domain (5.3): the set of all possible values that can be used to replace the variable, or the set of input values in a relation
dotplot (13.2): a graphical representation of quantitative data in which each value in the data set is represented by a dot or other mark
equation (1.1): a mathematical statement containing an equal sign
equivalent equations (3.1): equations that have the same solutions as the original
equivalent functions (10.2): functions whose expressions are equivalent
evaluate (1.1): to perform mathematical operations in order to get a single answer
explanatory variable (13.3): a variable used to predict the response variable
explicit formula (11.2): a formula for a sequence in which the nth term is defined in terms of n
exponential decay (10.2): a situation that can be modeled by the function $f(x) = a(1 + r)^x$, where $r < 0$
exponential function (10.1): a function in which the input variable is in the exponent
exponential growth (10.2): a situation that can be modeled by the function $f(x) = a(1 + r)^x$, where $r > 0$
expression (1.1): a mathematical statement with quantities connected by operations
extrapolation (13.4): a situation in which a regression line is used to predict values outside the interval of the domain
factor (1.1): (noun) one or more quantities that multiply together to form a product; (verb) to rewrite a quantity as a product of other quantities
five-number summary (13.3): a summary of the spread of a data set, consisting of the minimum, first quartile (Q_1), second quartile (Q_2), third quartile (Q_3), and maximum
function (6.1): a relation in which every input gets exactly one output, or a relation in which every element of the domain is mapped to exactly one element of the codomain
geometric sequence (11.2): a sequence that has a common ratio between consecutive terms
graph (6.2): a representation of ordered pairs on a coordinate plane

histogram (13.2): a graphical representation of quantitative data in which frequencies are represented by rectangles

identity (3.6): an equation that is true for any allowed value of the variable that can be substituted

independent variable (6.1): a variable that represents the input values of a relation

inequality (5.1): a statement that says that one expression is greater than or less than another expression

integers (1.3): the set of whole numbers and their opposites $(\ldots, -3, -2, -1, 0, 1, 2, 3, \ldots)$

interquartile range (13.3): the range of the middle half of the data $(Q_3 - Q_1)$

interval notation (5.2): a shorthand way of describing all numbers between two boundary points, such as $[5, 8)$

inverse of a function (12.2): a function that undoes the operation of f

inverse operations (1.1): operations that when applied to a number result in the original number (addition and subtraction are inverse operations since $4 + 3 - 3 = 4$)

irrational numbers (9.6): numbers that cannot be expressed as an integer divided by a nonzero integer

leading coefficient (3.5): the coefficient of the term in which the variable is raised to the highest power

line of best fit: see *regression line*

like terms (3.5): terms that have the same variables raised to the same powers

literal equation (3.1): an equation with two or more variables

mean (13.3): in statistics, the measure of center for quantitative data consisting of the sum of the values divided by the number of values

median (13.3): the measure of center for quantitative data consisting of the middle value when there are an odd number of values or the mean of the two central values when there are an even number of values after the values have been arranged in order from least to greatest

mode (13.3): the most common value(s) in a data set

monomial (8.1): a quantity with one term; a number, a variable or variables, or the product of both

operation (1.1): a mathematical process that is performed on two or more quantities to get a result

opposites (1.1): numbers that are the same distance from 0 on a number line but have different signs

ordered pair (6.2): two quantities written so that their order has meaning

parabola (9.9): the curve of a quadratic equation on the coordinate plane

parent function (12.3): the function that has the simplest equation of all functions of that type

percentile (13.3): 1 of 100 equal groups into which a data set has been divided when organized from least to greatest

piecewise function (12.2): a function composed of two or more functions defined over different intervals
polynomial (8.1): an expression consisting of 1 or more monomials linked by addition or subtraction
population (13.1): all of the individuals of interest
principal square root of a number (9.6): a positive quantity that when multiplied by itself produces the original number (the principal square root of 9 is 3 since $3^2 = 9$)
proportion (4.3): a statement that two ratios are equal
quadrants (6.2): 1 of 4 sections of the coordinate plane formed by the axes
quadratic equation (9.1): equation whose degree is 2
qualitative data (13.1): data that represents qualities or categories of individuals
quantitative data (13.2): data consisting of numerical values acquired through counting or measuring
quartile (13.3): 1 of 4 equal groups into which a data set has been divided when organized from least to greatest
radical expression (9.6): an expression containing a radical symbol $(\sqrt{\ })$
radicand (9.6): the expression underneath the radical symbol in a radical expression
rational numbers (1.3): numbers that can be expressed as an integer divided by a nonzero integer
range (6.1): the set of output values for a function
rate (3.8): the frequency with which an event occurs
ratio (3.1): a quantity that indicates how many times one number contains another
real numbers (9.6): the union of the set of rational numbers and the set of irrational numbers
recursive formula (11.2): a formula for a sequence in which the nth term is defined in terms of previous terms
regression line (least-squares regression line) (13.4): the line that minimizes the sum of the squares of the vertical distances between itself and data points and is used to predict outputs for values in the domain
relation (6.1): a relationship that connects elements of one set (called the domain) to elements of another set (called the codomain), or a set of ordered pairs
replacement set (5.2): the set of all possible values that can be used to replace the variable
residual (13.4): the difference between an actual value y and a predicted value $\hat{y}$ (residual $= y - \hat{y}$)
residual plot (13.4): the scatterplot whose x-values come from the explanatory variable and y-values are the residuals for all data values
response variable (13.3): the variable used to state the outcome of a study
root (3.1): the value of the variable that makes an equation true
sample (13.1): a part of a population

scatterplot (13.4): a graph of ordered pairs representing two quantitative variables
sequence (11.1): a list of ordered numbers, or a function whose domain is the set of whole numbers or counting numbers
set-builder notation (5.2): a shorthand way of describing a set of numbers by stating their properties (such as $\{x \mid x < 10\}$)
slope (7.1): the ratio of the amount of vertical change (or change in y, often called *rise*) to the amount of horizontal change (or change in x, often called *run*)
solution set (5.2): the set of numbers from the domain that make a mathematical statement true
solution set to a system of equations (7.5): the set of all points that make both equations true
square root of a number (9.6): a quantity that when multiplied by itself produces the original number (the square root of 9 is 3 or -3 since $3^2 = 9$ and $(-3)^2 = 9$)
standard deviation (13.3): a measure of the typical distance of values from the mean
standard form of a polynomial (8.1): a polynomial in which terms decrease in degree from left to right and all like terms are combined
system of equations (7.5): a collection of two or more equations that have the same variables
term of a polynomial (3.5): a number, a variable or variables, or the product of both
term of a sequence (11.1): a number in a sequence
transformation (12.3): a function that maps one set to another
trinomial (8.1): an expression consisting of 3 monomials linked by addition or subtraction
turning point (9.9): a point where the graph of a function changes from decreasing to increasing or vice versa
univariate data (13.1): data consisting of one characteristic from each individual
variable (1.3): a letter or other symbol that represents a quantity or quantities that can change in value
variable (13.1): a characteristic that is measured on individuals
vertex (9.9): the point on a parabola where the graph changes from decreasing to increasing
vertical line test (6.2): a method of drawing vertical lines from each input value to determine if the graph of a relation is a function
whole numbers (1.3): the set of counting numbers and 0
x-intercept (6.3): a point where a graph intersects the x-axis
y-intercept (6.3): a point where a graph intersects the y-axis
zero of a function (6.3): an input value that makes the output value of a function equal to 0, or the x-coordinate of an x-intercept of a function
zero-product property (9.5): if a product equals 0, then at least one of its factors equals 0

ABOUT THE AUTHORS

Bobson Wong and **Larisa Bukalov** have been math teachers in New York City public high schools since 2005 and 1998, respectively. They have taught all levels of high school math from pre-algebra to calculus and statistics, coached their school's math team, and supervised students in a math research program. Their extensive writing and speaking experience is based on techniques developed while teaching a combined 37,000 lessons to a diverse group of over 7,000 students over the years. They are the co-authors of *The Math Teacher's Toolbox: Hundreds of Practical Ideas to Support Your Students* (Jossey-Bass, 2020) as well as articles for such publications as the National Council of Teachers of Mathematics' (NCTM) *Mathematics Teacher* journal, *Education Week*, and Edutopia. Bobson and Larisa have extensive experience providing professional development to pre-service and in-service teachers. They have mentored student teachers, presented at local and national math conferences, and led professional development sessions for New York City teachers. Their work and influence in math education helped them win the Math for America Master Teacher Fellowship multiple times.

Bobson is a recipient of the New York State Master Teacher Fellowship and the New York Educator Voice Fellowship (2014–15). He is a member of NCTM's Nominations & Elections Committee and the Advisory Council of the National Museum of Mathematics. He was chosen to serve on state committees to revise the Common Core standards and reexamine teacher evaluations. As an educational specialist for the New York State Education Department, he writes and edits questions for high school math Regents exams. Bobson graduated from the Bronx High School of Science, Princeton University (BA, history), the University of Wisconsin-Madison (MA, history), and St. John's University (MS Ed., adolescent education, mathematics), where he received his teacher training through the New York City Teaching Fellows program.

Larisa has won several awards for excellence in classroom teaching, including the Queens College Mary Fellicetti Memorial Award (2009) for excellence in mentoring and supervising student teachers, and the Queens College Excellence in Mathematics Award (2017) for promoting mathematics teaching as a profession. A fourth-generation math teacher, she simultaneously earned degrees from a specialized math high school in Ukraine and a distance learning high school at Moscow State University. After emigrating to the United States, she learned English while earning both her bachelor's degree in math and her master's degree in math education from Queens College, City University of New York. She also teaches math education courses at Touro College.

Both Bobson and Larisa live in New York City with their families.

Index